LEGACY, PATHOGENIC AND EMERGING CONTAMINANTS IN THE ENVIRONMENT

Sustainable Water Developments
Resources, Management, Treatment, Efficiency and Reuse

Series Editor

Jochen Bundschuh
University of Southern Queensland (USQ), Toowoomba, Australia Royal Institute of Technology (KTH), Stockholm, Sweden

ISSN: 2373-7506

Volume 7

Legacy, Pathogenic and Emerging Contaminants in the Environment

Editors

Manish Kumar

University of Petroleum and Energy Studies, School of Engineering, Sustainability Cluster, Dehradun, India

Meththika Vithanage

University of Sri Jayewardenepura, Ecosphere Resilience Research Centre, Faculty of Applied Sciences, Sri Lanka

Sanjeeb Mohapatra

National University of Singapore (NUS), Environmental Research Institute, Singapore

Kishor Acharya

Newcastle University, School of Engineering, UK

CRC Press
Taylor & Francis Group
Boca Raton London New York

CRC Press is an imprint of the
Taylor & Francis Group, an **informa** business

Cover image: photo by Manish Kumar

First published 2022
by CRC Press/Balkema
Schipholweg 107C, 2316 XC Leiden, The Netherlands
e-mail: enquiries@taylorandfrancis.com
www.routledge.com – www.taylorandfrancis.com

CRC Press/Balkema is an imprint of the Taylor & Francis Group, an informa business

Library of Congress Cataloging-in-Publication Data
A catalog record has been requested for this book

ISBN: 978-0-367-74363-5 (hbk)
ISBN: 978-0-367-74365-9 (pbk)
ISBN: 978-1-003-15746-5 (ebk)

DOI: 10.1201/9781003157465

Typeset in Times New Roman
by codeMantra

Dedicated to all the loving souls that departed owing to the COVID-19 pandemic

About the book series

Augmentation of freshwater supply and better sanitation are two of the world's most pressing challenges. However, such improvements must be done economically in an environmental and societally sustainable way.

Increasingly, groundwater – the source that is much larger than surface water and which provides a stable supply through all the seasons – is used for freshwater supply, which is exploited from ever-deeper groundwater resources. However, the availability of groundwater in sufficient quantity and good quality is severely impacted by the increased water demand for industrial production, cooling in energy production, public water supply and in particular agricultural use, which at present consumes on a global scale about 70% of the exploited freshwater resources. In addition, climate change may have a positive or negative impact on freshwater availability, but which one is presently unknown. These developments result in a continuously increasing water stress, as has already been observed in several world regions and which has adverse implications for the security of food, water and energy supplies, the severity of which will further increase in the future. This demands case-specific mitigation and adaptation pathways, which require a better assessment and understanding of surface water and groundwater systems and how they interact with a view to improve their protection and their effective and sustainable management.

With the current and anticipated increased future freshwater demand, it is increasingly difficult to sustain freshwater supply security without producing freshwater from contaminated, brackish or saline water and reusing agricultural, industrial and municipal wastewater after adequate treatment, which extends the life cycle of water and is beneficial not only to the environment but also leads to cost reduction. Water treatment, particularly desalination, requires large amounts of energy, making energy-efficient options and use of renewable energies important. The technologies, which can either be sophisticated or be simple, use physical, chemical and biological processes for water and wastewater treatment, to produce freshwater of a desired quality. Both industrial-scale approaches and smaller-scale applications are important but need a different technological approach. In particular, low-tech, cost-effective, but at the same time sustainable water and wastewater treatment systems, such as artificial wetlands or wastewater gardens, are options suitable for many small-scale applications. Technological improvements and finding new approaches to conventional technologies (e.g., those of seawater desalination), and development of innovative processes, approaches and methods to improve water and wastewater treatment and sanitation, are needed. Improving economic, environmental and societal sustainability needs research and development to improve process design, operation, performance, automation and management of water and wastewater systems considering aims, and local conditions.

In all freshwater-consuming sectors, the increasing water scarcity, and correspondingly increasing costs of freshwater, calls for a shift toward more water efficiency and water savings. In the industrial and agricultural sectors, it also includes the development of technologies that reduce contamination of freshwater resources, e.g., through the development of a chemical-free agriculture. In the domestic sector, there are plenty of options for freshwater saving and improving efficiency such as water-efficient toilets, water-free toilets, or on-site recycling for uses such as toilet flushing, which alone could provide an estimated 30% reduction in water use for the average household. As already mentioned, in all water-consuming sectors, the recycling and reuse of the respective wastewater can provide an important freshwater source. However, the rate at which these water-efficient technologies and water-saving applications are developed and adopted depends on the behavior of individual consumers and requires favorable political, policy and financial conditions.

Due to the interdependency of water and energy (water–energy nexus), i.e., water production needs energy (e.g., for groundwater pumping) and energy generation needs water (e.g., for cooling), the management of both commodities should be more coordinated. This requires integrated energy and water planning, i.e., management of both commodities in a well-coordinated form rather than managing water and energy separately as is routine at present. Only such integrated management allows reducing trade-offs between water and energy use.

However, water is not just linked to energy, but must be considered within the whole of the water–energy–food–ecosystem–climate nexus. This requires consideration of what a planned water development requires from the other sectors or how it affects – positively or negatively – the other sectors. Such integrated management of water and the other interlinked resources can implement synergies, reduce trade-offs, optimize resources use and management efficiency, all in all improving security of water, energy and food security and contributing to protection of ecosystems and climate. Corresponding actions, policies and regulations that support such integral approaches, as well as corresponding research, training and teaching, are necessary for their implementation.

The fact that in many developing and transition countries, women are disproportionately disadvantaged by water and sanitation limitation requires special attention to this aspect in these countries. Women (including schoolgirls) often spend several hours a day fetching water. This time could be much better used for attending school or working to improve knowledge and skills as well as to generate income and so to reduce gender inequality and poverty. Absence of in-door sanitary facilities exposes women to potential harassment. Moreover, missing single-sex sanitation facilities in schools and absence of clean water contributes to diseases. This is why women and girls are a critical factor in solving water and sanitation problems in these countries and necessitates that men and women work alongside to address the water and wastewater-related operations for the improvement of economic, social and sustainable freshwater provision and sanitation.

Individual volumes published in the series are spanning the wide spectrum between research, development and practice in the topic of freshwater and related areas such as gender and social aspects as well as policy, regulatory, legal and economic aspects of water. It covers all fields and facets in optimal approaches to the:

- Assessment, protection, development and sustainable management of groundwater and surface water resources, thereby optimizing their use.
- Improvement of human access to water resources in adequate quantity and good quality.
- Meeting of the increasing demand for drinking water, and irrigation water needed for food and energy security.

  Treatment of water and wastewater also including its reuse.
  Implementation of water-efficient technologies and water-saving measures.

A key goal of the series is to include all countries of the globe in jointly addressing the challenges of water security and sanitation. Therefore, we aim to a balanced choice of authors and editors originating from developing and developed countries as well as gender equality. This will help society to provide access to freshwater resources in adequate quantity and good quality, meeting the increasing demand for drinking water, domestic water and irrigation water needed for food security while contributing to social and economically sound development.

This book series aims to become a state-of-the-art resource for a broad group of readers, including professionals, academics and students dealing with groundwater and surface water resources, their assessment, exploitation and management as well as the water and wastewater industry. This comprises especially hydrogeologists, hydrologists, water resources engineers, wastewater engineers, chemical engineers and environmental engineers and scientists.

This book series provides a source of valuable information on surface water but especially on aquifers and groundwater resources in all their facets. Thereby, it covers not only the scientific and technical aspects but also environmental, legal, policy, economic, social and gender aspects of groundwater resources management. Without departing from the larger framework of integrated groundwater resources management, the topics are centered on water, solute and heat transport in aquifers, hydrogeochemical processes in aquifers, contamination, protection, resources assessment and use.

This book series constitutes an information source and facilitator for the transfer of knowledge, for small communities with decentralized water supply and sanitation as well as large industries that employ hundreds or thousands of professionals in countries worldwide, working in the different fields of freshwater production, wastewater treatment and water reuse as well as those concerned with water-efficient technologies and water-saving measures. In contrast to many other industries, suffering from the global economic downturn, water and wastewater industries are rapidly growing sectors providing significant opportunities for investments. This applies especially to those using sustainable water and wastewater technologies, which are increasingly favored. The series is also aimed at communities, manufacturers and consultants as well as a diversity of stakeholders and professionals from governmental and nongovernmental organizations, international funding agencies, public health, policy, regulators and other relevant institutions, and the broader public. It is designed to increase awareness of water resources protection and understanding of sustainable water and wastewater solutions including the promotion of water and wastewater reuse and water savings.

By consolidating international research and technical results, the objective of this book series is to focus on practical solutions in better understanding groundwater and surface water systems, the implementation of sustainable water and wastewater treatment and water reuse and the implementation of water-efficient technologies and water-saving measures. Failing to improve and move forward would have serious social, environmental and economic impacts on a global scale.

This book series includes books authored and edited by world-renowned scientists and engineers and by leading authorities in economics and politics. Women are particularly encouraged to contribute, either as author or as editor.

Jochen Bundschuh
(Series Editor)

Editorial board

Contents

Contributors

S. Aishwarya	School of Chemical Engineering, Vellore Institute of Technology, Vellore, India
Muneesh Kumar Barman	Laboratory for HIV Research, National Centre for Cell Sciences, Pune, India
Unnati Bhalerao	National Institute of Virology, Pune, India
Mahalaxmi U Bhat	National Institute of Virology, Pune, India
Prudhvilal Bhukya	Transplant Immunology and Immunogenetics Lab (HLA), ACTREC, Tata Memorial Centre Navi Mumbai, India and National Institute of Virology, Pune, India
Trinath Biswal	Veer Surendra Sai University of Technology, Burla, India
Muskan Chakraborty	Department of Microbiology, Deccan Education Society's Fergusson College, Pune, India
Samarshi Chakraborty	School of Chemical Engineering, Vellore Institute of Technology, Vellore, India
Kailash Chand	Laboratory for HIV Research, National Centre for Cell Sciences, Pune, India
Shreeja Datta	Department of Biotechnology, Delhi Technological University, Delhi, India
Anshula Dhiman	School of Environmental Sciences, Jawaharlal Nehru University, Delhi, India
M. Divyagnaneswari	Department of Zoology, Gargi College, New Delhi, India
Govindaraj Divyapriya	Department of Civil and Environmental Engineering, Virginia Polytechnic Institute and State University, Blacksburg, Virginia
Kiruthika Eswari V.	Environmental and Water Resources Engineering Division, Department of Civil Engineering, Indian Institute of Technology Madras, Chennai, India
Abhranil Gangopadhayya	National Institute of Virology, Pune, India
Neha Garg	Environment Science and Engineering Department, Indian Institute of Technology, Bombay, India
Suresh Ghotekar	Department of Chemistry, Smt. Devkiba Mohansinhji Chauhan College of Commerce and Science, Silvassa, India
Shivaraju Harikaranahalli Puttaiah	Department of Environmental Sciences, Faculty of Natural Sciences, JSS Academy of Higher Education and Research, Mysore, India

Akshay Jakhete	Environment Science and Engineering Department, Indian Institute of Technology, Bombay, India
Prabagar Jijoe Samuel	Department of Environmental Sciences, Faculty of Natural Sciences, JSS Academy of Higher Education and Research Mysore, India
S. Karthika	School of Chemical Engineering, Vellore Institute of Technology, Vellore, India
Umesh Kumar	IMS Ghaziabad University Courses Campus, Ghaziabad, India
Stuti Kushwaha	School of Environmental Sciences, Jawaharlal Nehru University, Delhi, India
Sai Tejaswi Lavuri	Chalmeda Ananda Rao Institute of Medical Sciences, Bommakal, India
Madhu Balaji C. K.	Plant Cell Biotechnology Department, CSIR- Central Food Technological Research Institute, Mysore, India
Piyush Malaviya	Department of Environmental Science, University of Jammu Jammu, India
Salonee Martins	National Institute of Virology, Pune, India and Transplant Immunology and Immunogenetics Lab (HLA), ACTREC, Tata Memorial Centre, Navi Mumbai, India
Deepa Minakshi	Department of Environmental Science, Graphic Era Deemed to be University, Dehradun, India
Brijesh Kumar Mishra	Indian Institute of Technology (ISM) Dhanbad, Dhanbad, India
Nitish Venkateswarlu Mogili	Department of Biotechnology, National Institute of Technology Andhra Pradesh, Tadepalligudem, India
Vijay Laxmi Mohanta	Indian Institute of Technology (ISM) Dhanbad, Dhanbad, India
Sanjeeb Mohapatra	NUS Environmental Research Institute, National University of Singapore, Singapore, Singapore
Subham Mukherjee	Institute of Geographical Sciences, Physical Geography, Freie Universität Berlin, Berlin, Germany
Aiswarya Cholayampara Murukadasan	Department of Environmental Sciences, Faculty of Natural Sciences, JSS Academy of Higher Education and Research, Mysore, India
Indumathi M. Nambi	Environmental and Water Resources Engineering Division, Department of Civil Engineering, Indian Institute of Technology Madras, Chennai, India
Vinod Kumar Nigam	Department of Bioengineering, Birla Institute of Technology, Mesra, India
Sukdeb Pal	Wastewater Technology Division, CSIR-National Environmental Engineering Research Institute (CSIR-NEERI), Nagpur, India and Academy of Science and Innovative Research (AcSIR), New Delhi, India

Aishwarya Pathare	National Institute of Virology, Pune, India
Amit Prasad	School of Basic Sciences, IIT Mandi Kamand, India
A.L. Ramanathan	School of Environmental Sciences, Jawaharlal Nehru University, Delhi, India
Ramesh Babu P.	School of Chemistry, Trinity College Dublin, Dublin, Ireland and AMBER Centre, CRANN Institute, Trinity College Dublin, Dublin, Ireland and BiOrbic Bioeconomy Centre, University College Dublin, Dublin, Ireland
Anju Rani	Department of Life Sciences, Graphic Era Deemed to be University, Dehradun, India
Arpita Roy	Department of Biotechnology, School of Engineering & Technology, Sharda University, Noida, India
Purva Salvi	National Institute of Virology, Pune, India
Sarat Chandra T.	School of Chemistry, Trinity College Dublin, Dublin, Ireland
Sasikaladevi R.	Environmental and Water Resources Engineering Division, Department of Civil Engineering, Indian Institute of Technology Madras, Chennai, India and Department of Biotechnology, Bhupat and Jyoti Mehta School of Biosciences, Indian Institute of Technology Madras, Chennai, India
Krushna Prasad Shadangi	Veer Surendra Sai University of Technology, Burla, India
Pradeep Kumar Sharma	Department of Environmental Science, Graphic Era Deemed to be University, Dehradun, India
Shubha Rani Sharma	Department of Bioengineering, Birla Institute of Technology, Mesra, India
Ujjawal Sharma	Department of Community Medicine and School of Public Health, Postgraduate Institute of Medical Education and Research, Chandigarh, India
Yashas Shivamurthy Ravindra	Department of Environmental Sciences, Faculty of Natural Sciences, JSS Academy of Higher Education and Research, Mysore, India and Center for Water, Food and Energy, GREENS Trust, Tumkur, India
Pradip Kumar Sikdar	Department of Environment Management, Indian Institute of Social Welfare and Business Management, Kolkata, India
Avinash Singh	School of Basic Sciences, IIT Mandi, Kamand, India
Meenakshi Singh	Transplant Immunology and Immunogenetics Lab (HLA), ACTREC, Tata Memorial Centre, Navi Mumbai, and India National Institute of Virology, Pune, India
Neha Singh	School of Basic Sciences, IIT Mandi, Kamand, India
Swati Singh	Department of Environmental Science, Graphic Era Deemed to be University, Dehradun, India
K. Sivagami	School of Chemical Engineering, Vellore Institute of Technology, Vellore, India

Ramya Srinivasan	Environmental and Water Resources Engineering Division, Department of Civil Engineering, Indian Institute of Technology Madras, Chennai, India
Manjita Srivastava	National Institute of Virology, Pune, India
G. Sruthi	School of Chemical Engineering, Vellore Institute of Technology, Vellore, India
Parasuraman Aiya Subramani	(Formerly) Centre for Fish Immunology, Vels Institute of Science, Technology and Advanced Research, Chennai, India
Saumya Vyas	Department of Microbiology, Kristu Jayanti College, Bengaluru, India
Sneha Yadav	Department of Environmental Sciences, Faculty of Natural Sciences, JSS Academy of Higher Education and Research, Mysore, India

Foreword

I am delighted to note that CRC Press will publish a book project entitled *Legacy, Pathogenic and Emerging Contaminants in the Environment* under the dynamic editorship of Prof. Manish Kumar, Prof. Meththika Vithanage, Dr. Sanjeeb Mohapatra and Dr. Kishor Acharya. The focal theme is aptly chosen as all over the world people are facing a bunch of new challenges with the rise of emerging contaminants every day. Groundwater (and/or surface water) is considered one of the integral parts of the hydrological cycle, where a mass balance of energy flow and matter lies in a state of equilibrium in the ocean, atmosphere and land. This book is aiming to highlight the most recent advances on distribution, occurrence and the fate of geogenic and emerging chemical and microbial contaminants in the surface and subsurface system and techniques to study their transport behavior, which will determine the sustainable environmental assessment in order to comply with the strategic environmental planning and management and associated human health risk. In a nutshell, this book is designed to fill the gap between contamination events, source tracking and aquatic toxicity by considering the seasonal trends and variations through coordination between multidisciplinary, multitemporal, multinational and multispatial research approaches. Since the spatio-temporal variation in the concentration of contaminants is the core focus of this book, all the contributed chapters lie within the aspects and themes of geogenic, anthropogenic and biological contaminants management.

Scanty information is available on the surface water-groundwater contaminant interactions under different environmental conditions across the world, as it is an important driver for predicting risk and management of water cycle through hydrogeochemical events. I am happy to know that around 15 contributed chapters are embedded under the three sections in this book project by the learned experts of their respective fields. This book depicts the distribution, occurrence and fate of emerging contaminants in the first section, whereas the second section on microbial contaminants has highlighted indicator organisms, detection techniques and microbial source tracking for early prediction of risk and management purposes. On the other hand, the final section explores the leading approaches in science and engineering of environmental risk management and human behavior and decision-making based on water security and sustainable governance. Therefore, this book will enlighten a comprehensive and succinct idea about legacy, pathogenic and emerging contaminants and further finding suitable ways for the scientific evaluation of complex multidisciplinary problems.

Dr. Devesh Kumar Avasthi
Professor Emeritus
Dean, Research & Development
UPES Dehradun, India

Editors

Manish Kumar is a Professor and Head of Sustainability Cluster at School of Engineering, University of Petroleum & Energy Studies, Dehradun, India. He earned his Ph.D. in Environmental Engineering from the University of Tokyo, Japan, and as a faculty he has supervised seven awarded Ph.D. thesis and over 28 master dissertations. He is the recipient of recognitions like Expert Panel for UNEP on antimicrobial resistance, Fellow of the Royal Society of Chemistry (FRSC), Best Research Award at 4th International Water Association (IWA) Asia Pacific Water Young Professional Conference, Global Change Research (GCR) Grant from Asia Pacific Network, DST young scientist grant, JSPS Research Fellowship, Centre of Excellence (CoE) Young Researcher Fund, Linnaeus-Palme Grant from SIDA, Sweden, and others. He renders the editorial service to several reputed journals like *Nature Partner Journals* (NPJ) *Clean Water, Science of the Total Environment, ACS ES&T-Engineering, Current Pollution Report* (Springer), and others. He featured among world's top 2% of researcher list prepared by Stanford University. He is an executive committee member of IWA – India Chapter and represents South-Asia in IWA specialist group for METRAL and related substances. Prof. Kumar is active in the fields of hydro-geochemistry, contaminant transport and modeling, metal speciation and toxicity, and wastewater surveillance of COVID-19. He has more than 150 international journal publications to his credit.

Meththika Vithanage is a professor at the Ecosphere Resilience Research Centre, University of Sri Jayewardenepura, Sri Lanka. Her academic background covers environmental remediation of toxic metals, antibiotics, agrochemicals, serpentine soil chemistry and waste biomass conversion. She is a Young Affiliate of the Third World Academy of Sciences, and she became the Chairperson of the Young Scientists Forum in 2017. She was awarded as the Best Young Scientist, 2018, by the Young Scientist Forum of the National Science and Technology Commission, Sri Lanka, and in 2016, she became the Young Scientist in Chemistry by the National Science Foundation of Sri Lanka. She was selected as one of the Early Career Women Scientists by the Organization for Women Scientists in the Developing Countries, Italy. Further, Prof. Meththika Vithanage is considered the topmost scientist in Sri Lanka for Environmental Science based on the scientific output in the International scientific publishing website SCIVal. She has contributed over 125 Science Citation Indexed journal articles, over 30 book chapters and four co-edited books published by Elsevier Inc. She is an associate editor for the *Journal of Groundwater for Sustainable Development*. She served as the guest editor for virtual special issues for international journals, including *Journal of Hazardous Material* (Elsevier), *Soils and Sediments* (Springer) and *Water* (MDPI).

Sanjeeb Mohapatra completed his Ph.D. degree from Indian Institute of Technology Bombay (IIT Bombay), India, and is now working at Environmental Research Institute, National University of Singapore (NUS), Singapore. His research interest broadly covers the monitoring of emerging contaminants (ECs), photo-degradation and enzymatic degradation of ECs, and the role of dissolved organic matter in deciding the fate of such contaminants in the environment. He is a recipient of the Water Advanced Research Innovation (WARI) Fellowship awarded by the Department of Science and Technology (DST), India, University of Nebraska Lincoln, USA; Daugherty Water for Food Global Institute (DWFI), USA; and Indo-U.S. Science and Technology Forum (IUSSTF). He is also a recipient of the Newton-Bhabha Fellowship jointly awarded by DST, India, and British Council, UK and INSPIRE fellowship offered by DST, India. Dr. Mohapatra also won several international awards for his research work and has several publications in reputed journals. He served as the guest editor for international journals, including *Microorganisms* (MDPI) and *Frontiers in Environmental Science.*

Kishor Acharya is a postdoctoral researcher at the School of Engineering, Newcastle University, UK, specialized in public health engineering. His research interest varies from developing and applying portable molecular tools for on-site microbial hazard assessment to design novel techniques/methods for bioremediation to remove organic pollutants. Dr. Acharya is a recipient of the prestigious Rapid Response Grant Award by the Global Challenges Academy to help the global dissemination of research skills and tools and capacity building among researchers of Nepal working in WASH sector. Dr. Acharya was recently endorsed as an exceptional talent by the Royal Society. He has published his research work in leading environmental engineering journals such as *Water Research* and *Environmental Science and Technology*.

Acknowledgments

This book project would not have been possible without the financial contribution from UNICEF-India, Kiran C Patel Centre for Sustainable Development (KPCSD)-IIT Gandhinagar, Asia Pacific Network (APN)-Grant Number CRRP2016-06MY-Kumar under the aegis of which all contributed chapters have been compiled as a comprehensive reference book on legacy, emerging and pathogenic contaminant status and management in the environment of post-COVID-19 Anthropocene. We acknowledge the constant support and co-operation of Indian Institute of Technology (IIT) Gandhinagar, University of Sri Jayewardenepura, Sri Lanka, National University of Singapore, Singapore, and Newcastle University. We would like to express our sincere thanks to the CRC Press/Balkema book series editor on sustainable water development, Prof. Jochen Bundschuh of the University of Southern Queensland, Australia; and gratitude to all the CRC employee, Mr. Janjaap Blom, Debakanya Haldar and others for their persistent and hard efforts in the successful coordination of this book project. We would like to thank Dr. Nirav Raval, the chief operating office of Encore Insoltech, Pvt. Ltd.

The editors would like to express their sincere gratitude to the authors who have submitted their work on time and revised it appropriately at short notice. We would like to express our sincere thanks and gratitude to all the domain experts: Prof. Algappan Ramanathan, Dr. Gayathri Menon, Dr. Nitish Venkateswarlu, Dr. MS Priyanka, Dr. Santanu Mukherjee, Dr. Richa Singh, Dr. Anirudh Gupta, Dr. Arpita Dasgupta, Dr. Vaibhav Srivastava, Dr. Lakshmi Pisharody, Dr. Arbind Patel, Dr. Ansaf V Karim, Mr. Niroj Amatya, Mr. Kalyan Pantha, Dr. Aom Jidapa and Ms. Jiaa Hee Yo, who reviewed various chapters and provided their valuable suggestions to improve this book. We also acknowledge every small help received from our colleagues, students, peers and notable academicians, without whom we would not have been completed our editorial responsibilities on time. Finally, we acknowledge the constant care and support of our families, parents and well-wishers with which this book project came into reality. Although it is impossible to list all the names, we are thankful to our colleagues, students, peers and notable academicians, without whom we would not have been completed our editorial responsibilities in time. Last but not the least, we acknowledge the constant care and support of Prof. Devesh Kumar Avasthi, Dean Research & Development, University of Petroleum and Energy Studies (UPES) Dehradun, India, who throughout have guided us to complete this book and obliged us to have written a foreword for this book.

Preface

It has been an exciting endeavor to contribute as the editors for the book *Legacy, Pathogenic and Emerging Contaminants in the Environment* published by CRC Press, Taylor & Francis Group. At the same time, it was a daunting task for us editors to thoroughly select only the ones we thought were unique and matched the scope of this book. With an increased demand for wastewater reuse, groundwater recharge with treated wastewater has been practiced worldwide. Additionally, rising threat of novel viruses and emerging contaminants is a setback in the modern world. As a result, groundwater quality further deteriorates by microbial and emerging micropollutants from the various anthropogenic origins, including untreated wastewater, seepage of landfill leachate, eutrophied water and runoff from agricultural lands. The fate of such geogenic, microbial and anthropogenic contaminants in subsurface systems, especially in the groundwater, depends on several factors. Physicochemical properties of emerging contaminants such as octanol-water partition coefficient, dissociation constant, water solubility, susceptibility to biodegradation under anaerobic conditions and environmental persistence under diverse geological and pH conditions play an acritical role during subsurface mass flow.

To address those issues, this book selected 15 excellent chapters distributed into three broad sections. This book comprised topics dealing with the subject mentioned above and intended for the students, professionals and researchers working on various legacy, emerging chemical and microbial contaminants. The first section on *Distribution, Occurrence and Fate of Emerging Contaminants* deals with an array of emerging contaminants such as pharmaceutical and personal care products with a particular focus on antibiotics, fate and transport of micro- and nanoplastics in freshwater, solid waste and landfill leachate as a transient source of emerging and legacy contaminants and finally their manifestation and consequences in surface and groundwater. The second section on the Techniques to Study the *Fate, Transport and Removal* majorly targets emerging microbial contaminants such as occurrence and fate of virus, novel next-generation sequencing for their characterization, microbial indicators and source tracking strategies, microbial migration with a special emphasis on the runoff from the aquaculture industries and finally efficacy of natural wetland systems. Finally, the last section on *Environmental and Health Risk of Anthropogenic and Geo-genic Contaminants* depicts the potential concern of emerging contaminants to microalgae, comprehensive understanding of contaminants in the urban environment and their sustainable governance in the post-Covid-19 world, and advanced treatment strategies. Various case studies have been discussed to present a cutting-edge scenario of the problem, perspective and challenges of emerging water quality issues in the present context of climate change and significant human interventions.

This book explains the occurrence, fate and transport of geogenic, microbial and anthropogenic contaminants in the groundwater. Thereby, it covers not only the scientific and technical aspects but also environmental, legal and policy aspects for groundwater resources management. Without departing from the broader framework of integrated groundwater resources management, the topics are centered on sources of contaminants, their partitioning and mass flow behavior in subsurface systems, hydrogeochemical processes in aquifers, and finally, groundwater resources assessment and policy issues. The main aim will be to transfer knowledge to a broader audience. This book seeks its impact from its diverse topic coverage, revealing different contemporary issues, such as groundwater management, emerging and microbial contamination in the groundwater and freshwater.

Section 1

Distribution, occurrence, and fate of emerging contaminants

Chapter 1

Occurrence and fate of emerging contaminants in groundwater

Anshula Dhiman, Stuti Kushwaha, and A.L. Ramanathan
Jawaharlal Nehru University

CONTENTS

1.1 INTRODUCTION

Over the decades, the ever-growing demand for agricultural, livestock, medical, and industrial products resulted in the overconsumption of water resources leading to scarcity and pollution of the water bodies. Approximately 80,000 synthetic chemicals recognized as "emerging contaminants" (since 1990) (Naidu et al., 2016) are discharged into the natural surroundings annually as a result of the wastewater release (Lapworth et al., 2012; Stuart and Lapworth, 2013; Kumar et al., 2021). "Emerging contaminants" (ECs) are a group of synthetic and natural organic chemicals that are currently not incorporated in conventional monitoring plans at the European level and can be candidates for future regulation, depending on the investigation on their (eco)toxicity, potential health consequences, and their existence in the multiple environmental compartments (Rosenfeld and Feng, 2011; Norman, 2012). Over 1,036 compounds have been distinguished as ECs by the NORMAN network and are classified into 30 classes based on their origin and kind (Dey et al., 2019). Prominent categories that are usually identified in aquifers are pharmaceuticals and personal care products (PPCPs), lifestyle compounds, plasticizers (bisphenol A (BPA)), industrial solvents, flame retardants, surfactants, benzothiazoles, polyaromatic hydrocarbon (PAHs), pesticides, food additives, and transformed products of parent organic chemical (Menon et al., 2020;

DOI: 10.1201/9781003157465-2

Mohapatra et al., 2018, 2021; Priyanka and Mohapatra, 2020; Kumar et al., 2019 a,b). They are detected in studies at concentrations varying from µg/L to ng/L in aquatic environments, and the degree of contamination depends on the properties of both contaminants and aquifers, and the pathway through which contaminants reach groundwater. Many of these organic contaminants and their by-products are persistent and bio-accumulative in nature (Stefanakis and Becker, 2015). This chapter discusses the regularly occurring ECs in groundwater from the year 2015 across the globe and identifying the areas with the highest concentrations. Their sources, pathways, fate, and condition of the aquifer that results in contamination are reviewed. Based on this information risk assessment, methods are reported from different studies. Still, many unknown organic compounds are being discharged, and fate and toxicological studies of the identified ones are poorly understood.

1.2 SOURCES AND PATHWAYS

Based on the conventional classification, sources of ECs are described as point or non-point sources. Point sources are characterized as the discernible, confined point that can be spatially isolated, identified like the plume with the highest concentration near the source and decreasing as drifting further away (Thomaidis et al., 2012). Point sources include wastewater discharge from different industries (such as tanneries, paper, and pulp industries, steel plants, food processing industries, industrial-scale animal feeding operations, manufacturing factories, pharmaceutical industrial), municipal wastewater treatment plant and combined sewage-stormwater overflow, healthcare facilities for animals and humans (veterinary hospitals, nursing homes, hospitals), domestic sewage, mining activities, waste disposal sites (such as landfill settings, agricultural waste pools, and industrial confinements), aquacultural activities, livestock farms, buried septic tanks, and artificial recharge from undertreated water (Lapworth et al., 2012; Stefanakis and Becker, 2015; Pal et al., 2014). Non-point sources refer to pollution from indistinct and diffused sources occurring over a larger geographical area and are difficult to demarcate. Examples include stormwater runoff; surface runoff from agrarian land, urban areas, and roads; use of biosolids and manure to the soil; diffuse aeolian deposition; leakage from urban sewer settings; and flash floods eroding landfill sites (Stuart et al., 2012; Lapworth et al., 2012). Diffuse sources are difficult to trace back to the polluter due to lower loading of contaminants compared to point sources and are more prone to attenuation (Lapworth et al., 2012). ECs introduced from different sources are summarized in Table 1.1.

Pathways include both direct and indirect passages of contamination in groundwater. Direct passages include leakages from point sources, permeation of chemicals applied to the soil and landfill sites, artificial recharge, and indirectly from groundwater-surface water (GW-SW) interaction. The route favored will depend upon the physicochemical properties of the contaminant in the environment, such as its solubility in water (Lapworth et al., 2012; Stuart et al., 2012). For example, plasticizers with low molecular weight have high mobility and easily diffuse to the environment (Yan et al., 2010). Organic manure widely used in agricultural fields liberates pharmaceutical residues, which may infiltrate and pollutes groundwater directly or indirectly through surface water contamination during runoff. The movement of contaminants occurs through the soil zone, the unsaturated zone, and the saturated zone

Table 1.1 Different sources of ECs for groundwater

Type of emerging contaminant	*Sources*	*References*
Pharmaceuticals	Discarding unused and expired medicine in drainage systems and at landfill sites releases antibiotics and inflammatory drugs. Hospitals wastewater discharge of iodinated X-ray contrast media and from human excretion via waste water treatment plant (WWTP) effluents. Animal feeding operations (CAFOs) and aquaculture produce contaminants and biosolids applied to cultivated field such as veterinary antibiotics, hormones. Artificial recharge and septic tanks release ibuprofen and paracetamol.	Lapworth et al. (2012), Stuart and Lapworth (2013), Pal et al. (2014), Naidu et al. (2016)
Personal care products (PCPs)	Perfumes, cosmetic soaps, and lotions release musk, and fragrances like galaxolide, tonalide, nitro musk, phthalates, etc. via WWTP effluents. UV filters like benzophenone, antimicrobial agents (triclosan), and preservatives (parabens and derivatives) are released from sunscreens, moisturizers, lipsticks, and haircare products. Mosquito repellants (only <20% is absorbed by the skin) contaminate bathing water and are discharged via WWTP effluents.	
Lifestyle compound and food additives	Caffeine and nicotine from sewage effluent. Food additives like artificial sweeteners (like aspartame saccharine, cyclamate, and sucralose) are discharged directly from food industries, domestic sewage, livestock raising, and excreted unaltered.	
Industrial compounds	Perfluorinated compounds (PFCs) used in chromium plating, aviation hydraulic fluid, photography industry, semiconductor industries, and firefighting foam released via wastewater discharge. Flame retardants, plasticizers, and solvents like polychlorinated alkanes (PCAs) from industries such as plastics and textiles, and surface finishes and coatings present in WWTP effluents, combustion, and the natural breakdown of the plastic. Polychlorinated naphthalenes (PCNs) are used as dielectric fluids, engine oil additives, electroplating masking compounds, wood preservatives, and dye production. Benzothiazoles employed for corrosion inhibition, herbicides, antialgal agents, in paper and pulp industry and azo dyes, chemotherapeutics and fungicides in lumber, and leather production release via WWTP.	Walters and Santillo (2006) Clarke and Smith (2011), Pal et al. (2014)
Quaternary ammonium compounds (QACs)	Household usage of QACs as cationic surfactant is the chief source that enters the WWTP. Also sourced from shampoos, toilet cleansers, soaps, shaving cream, baby wipes, sunscreens, lotions, sanitizing sprays, fabric softeners, disinfectants, and spermicidal jellies.	Clarke and Smith (2011)
Pesticides	Pesticides are introduced in water by agricultural runoff and from industrial manufacturing of pesticides, spraying of insecticides in urban households, and atmospheric deposition.	Stuart et al. (2012) Pal et al. (2014)

in a well-entrenched way. Recharging activities through rainwater and roof runoff can also contaminate groundwater (Bucheli et al., 1998, Patel et al., 2019).

The presence of some compounds, i.e., carbamazepine, sulfamethoxazole, benzophenone, 5-methyl-1H-benzotriazole, N, N-diethyl-meta-toluamide (DEET), and tributyl phosphate, after soil-aquifer passage, suggests that these compounds can reach subsurface under recharge conditions (Mohapatra et al., 2021). Insufficient removal of estrogenic steroids during wastewater treatment can pollute groundwater via AR activities. Urban shallow aquifers are more susceptible to contamination as rapidly expanding mega-cities resulting in the drying up of these aquifers that needed to be artificially recharged.

A study (Carrara et al., 2008) pointed that high concentrations of various pharmaceutical composites in septic tanks may lead to leakage and direct intrusion. Superficial waters carry elevated concentrations, and a wider array of organic contaminants than groundwaters (Lapworth et al., 2012), and are hence a key cause of ECs presence in aquifers as a result of both natural and engineered GW-SW exchange processes (Lapworth et al., 2009; Drewes, 2009). Lewandowski et al. (2011) insinuated that after the intrusion of ECs in the surface watercourse, they may stay in the water body or may be transported by lateral or vertical hydraulic exchange to groundwater body. When the local water table is underneath the surface water level, like in more arid regions, the transfer may happen. The presence of ECs in these aquatic systems are markers of wastewater origins and also determine the SW-GW exchange process. Diffused spillage from sewerage networks can be a subject of attention as ECs avoid natural attenuation mechanisms in the subsurface (Ellis, 2006).

1.3 DISTRIBUTION AND CONCENTRATION IN G.W

Over the years due to the advancement of technology and the invention of sensitive analytical instruments, researchers are able to study the presence of ECs in groundwater. Generally, the concentration and diversity of the contaminants are lesser in the groundwater due to natural attenuation and dilution mechanism (Lapworth et al., 2012) but there are cases where values are significantly present, i.e., more than 100 ng/L. A total of 29 studies (2015 onward) are reviewed to understand the occurrence and distribution of ECs in groundwater. The most common EC is carbamazepine followed by sulfamethoxazole, caffeine, BPA, and perfluorooctane sulfonate (PFOS). This distribution is consistent with the observations by Lapworth et al. (2012) (Table 1.2).

Table 1.2 Comparison of the common ECs concentrations from past studies

Compound	*Before 2012 (Lapworth et al., 2012)*	*2012–2014 (Sui et al., 2015)*	*2015–2020*
Carbamazepine	1.64–99,194	BDL-136	BDL-620
Sulfamethoxazole	5.7–1110	BDL-124.5	BDL-1820
Caffeine	14–9774	BDL-16249	BDL-3100
BPA	14–9774	-	BDL-10^5
Triclosan	6–509	-	ND-83
DEET	4–2251	BDL-3481	BDL-1610

In India also, carbamazepine is more frequent along with PFOS and sucralose. BPA and sulfamethoxazole have higher concentrations than reported in studies before 2015 (Table 1.2). Common ECs in each category are shown in Figure 1.1a and b. The concentrations of 13 compounds are above 397.4 ng/L, which is the upper limit detected by calculating the interquartile range (IQR).

$$\begin{gathered} IQR = Q3 - Q1 \\ \left(Q1 = 4.205, Q3 = 161.5\right) \end{gathered} \tag{1.1}$$

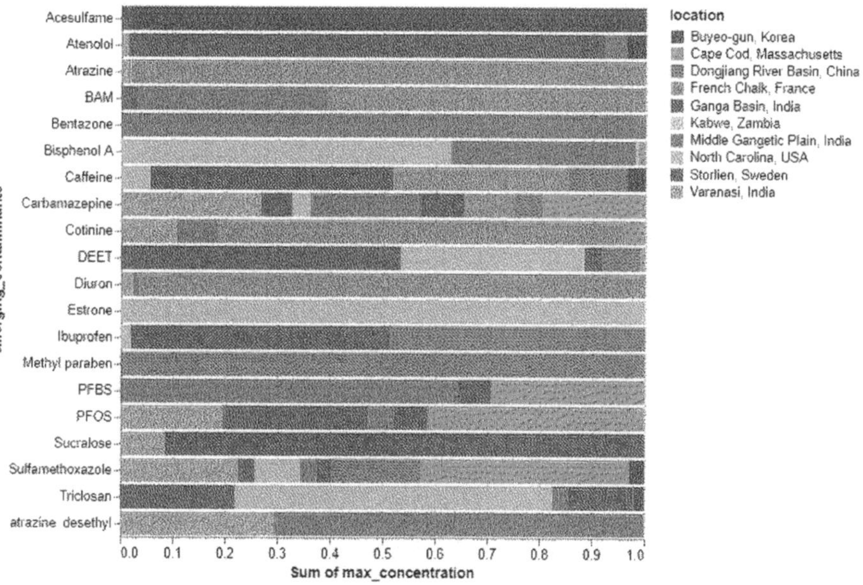

(a)

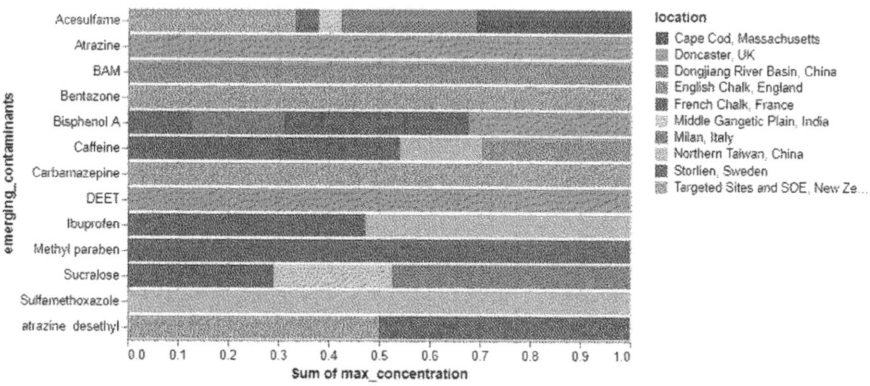

(b)

Figure 1.1 (a) Normalized outlier concentrations of ECs and (b) normalized concentrations of ECs with no outliers.

$$\text{Upper Limit} = Q3 + 1.5 * \text{IQR} \tag{1.2}$$

$$\text{OUTLIER} > \text{Upper Limit} \tag{1.3}$$

A total of 30 concentrations are designated as outliers. Outlier concentrations are more for acesulfame (5) and BPA (4) shown in Figure 1.1a. The maximum concentration of seven ECs is observed below 397.4 ng/L (Figure 1.1b), and collectively, 122 concentrations lie in the IQR. Concentrations and occurrence of different classes of contaminants are discussed below.

1.3.1 Pharmaceuticals, personal care products, and lifestyle compounds (PPCPLs)

Pharmaceuticals can be classified based on chemical compositions, mode of action, and therapeutic application. Some of the classes are antibiotic, antipyretic, analgesic, antihistamine, antiparasitic, beta-blockers, stimulants, antiepileptic, blood thinners, nutritional supplements, hormones, antiseptic, and tranquilizers. The maximum number of compounds (44) studied worldwide belong to this group. Among them, carbamazepine, which is used to treat epilepsy, is the most common drug and EC in groundwaters (Figure 1.1b). Antibiotic sulfamethoxazole is the next frequently found contaminant that appeared in 11 studies. The highest and lowest concentrations of these composites detected are 0.1–620 (Richards et al., 2021; Moreau et al., 2019) and 0.1–1,820 ng/L (Lin et al., 2015), respectively. Both of them along with ibuprofen have some extreme maximum concentrations (Figure 1.1a). Detection frequencies (DFs) of sulfamethoxazole and carbamazepine are reported to be high in many groundwater analyses (Lapworth et al., 2018; Lesser et al., 2018; Gao et al., 2019; Lin et. al., 2015; Sharma et al., 2019; Kuroda and Kobayashi, 2021). Other pharmaceuticals that are investigated in five or more studies are included in classes such as antibiotics (trimethoprim and sulfadimethoxine), nonsteroidal anti-inflammatory drugs (NSAIDs) (diclofenac, naproxen, and ibuprofen), analgesic (acetaminophen), and beta-blockers (atenolol) with their maximum concentration mentioned in Table 1.3. Most of them show their peak concentration in aquifers of Storlien, Sweden (Gao et al., 2019). Pharmaceuticals like hormones, antipruritic, antidepressants, diuretics, antidiabetic, anesthetic, cocaine, antiprotozoal, statins (antilipemic agents), antifungal, bronchodilators, and angiotensin II receptor blocker (ARB) are also discovered from at least two research studies (Lesser et al., 2018; Schaider et al., 2016; Kong et al., 2016; Lesser et al., 2018; Lapworth et al., 2018; Yang et al., 2018; Moreau et al., 2019; Gao et al., 2019; Richards et al., 2021).

Personal care products are consumed every day by humans. There is an increase in the number of varieties of chemicals that are incorporated in cosmetic products. Antimicrobial agent triclosan and mosquito repellent-active agent DEET are the most abundant PCPs (Figure 1.1b) and two of the top six commonly found ECs. Their concentration ranged from not detected (ND)-83 and ND-1610 ng/L, respectively, in groundwater of Mezquital Valley (Lesser et al., 2018). DEET has higher DF (100% and 92.3%) in Kabwe, Zambia, and Mezquital Valley system, Mexico, respectively (Sorensen et al., 2015; Lesser et al., 2018). The prescribed value for the compound by the Minnesota Department of Health is 2*10^5 ng/L, which is much higher than the reported maximum concentration (Table 1.3). Antimicrobial (triclocarban), preservatives (parabens), synthetic musk (galaxolide), UV filter (benzophenone), and phthalates (dibutyl phthalates, diethyl phthalates (DEPs), and dimethyl phthalates) are other PCPs that are identified by the researchers (Sorensen et al., 2015; Lapworth et al., 2015; Kong et al., 2016; Lesser et al., 2018; Yang et al., 2018; Moreau et al., 2019; Gao et al., 2019; Kuroda and Kobayashi, 2021). Among them, DEP (22*10^3 ng/L) is recorded with a very high concentration in English Chalk (Lapworth et al., 2015). Methyl parabens is one of the ECs recognized as an outlier with a maximum concentration of 437.4 ng/L in groundwaters of Storlien, Sweden (Gao et al., 2019).

Lifestyle compounds are associated with today's modern way of life. Caffeine and nicotine are two stimulants that pertain to this category. Caffeine is among the top five regularly encountered EC and shown (Figure 1.1) to have very high concentrations in aquifers at three places Storlien, Sweden (3,100 ng/L); English Chalk, England; and Northern Taiwan (930.7 ng/L) (Gao et al., 2019; Lapworth et al., 2015; Li et al., 2015). Cotinine is the metabolite of nicotine obtained in tobacco. It is published in five studies with the highest concentration (20 ng/L) in French Chalk, France (Lapworth et al., 2015).

In India, EC research is mainly restricted to Ganga River Basin. Lapworth et al. (2018) investigated 13 pharmaceuticals (0.3–120 ng/L) in deep and shallow aquifers in Varanasi and Ramnagar, India. The groundwater of Varanasi is discovered with the maximum values of sulfamethoxazole, carbamazepine, and cotinine (Table 1.3). Elevated concentrations are mostly observed in the middle and lower reaches of the Ganges River, e.g., triclosan, atenolol, DEET, ibuprofen, and sulfamethazine (Sharma et al., 2019; Richards et al., 2021). Caffeine with 100% DF is seen to be contaminating groundwaters across the basin (Sharma et al., 2019). In Punjab also, pharmaceutical pitavastatin (blood cholesterol regulator) is found with concentrations ranging from below ND to 4,800 ng/L (Jindal et al., 2015).

1.3.2 Industrial compounds

A wide range of industrial compounds and their transformed products are detected from subsurface water in many countries. Most often identified compound is BPA, a plasticizer, produced in huge quantities to incorporate into polycarbonate-made plastic and epoxy resins (Thomaidis et al., 2012). Majority values of BPA are outliers (Figure 1.1a) with four studies observing the maximum concentration above 1,000 ng/L in groundwaters of English Chalk, England (10^5 ng/L); Berkshire, UK (39,000 ng/L); French Chalk, France (1,360 ng/L); Doncaster, UK (1,200 ng/L) (Lapworth et al., 2015; Manamsa et al., 2016a,b; White et al., 2016). Storlien, Sweden, and Dalian rural area, China, discovered BPA with 100% DF (Li et al., 2016; Gao et al., 2019). PFCs are other industrial compounds that occurred in more than five surveys, including perfluorononanoic acid (PFNA), perfluorooctanoic acid (PFOA), perfluorohexane sulfonate (PFHxS), perfluorohexanoic acid (PFHxA), perfluorodecanoic acid (PFDA), PFOS, and perfluorobutane sulfonate (PFBS). These are used as surfactants to reduce tension between two liquids. Their highest and recommended values are mentioned in Table 1.3. The maximum concentrations of PFOS and PFOA are greater than the standard limits specified by the USEPA but much lower than the EU limits. An industrial solvent such as chlorodibromomethane ($19*10^3$ ng/L in English Chalk and $2.9*10^3$ ng/L in French Chalk) and di(2-ethylhexyl) phthalate (DEHP) (2,700 ng/L in French Chalk and 1,600 ng/L in Doncaster) are also discovered at very high concentrations (Lapworth et al., 2015; White et al., 2016). The categories of industrial compounds that appeared less commonly in groundwater are PAHs (naphthalene, dimethyl naphthalene, and anthracene), industrial solvents (tetrachloroethylene, isophorone, and PCAs such as chlorobenzene, dichlorobenzene), flame retardants and plasticizer (triphenyl phosphate, tributyl phosphate, and n-butyl benzenesulfonamide), surfactants benzotriazoles, and benzothiazoles (Lapworth et al., 2015; Sorensen et al., 2015; Kong et al., 2016; Wang et al., 2016; Schaider et al., 2016; Lesser et al., 2018; Yang et al., 2018; Gao et al., 2019). Benzothiazoles, 2-methylnaphthalene (a PCN), PFOS, and PFOA are recorded with higher DF (70% and beyond) (Wang et al., 2016; Gao et al., 2019; Kong et al., 2016; Liu et al., 2018). In India, concentrations of BPA (0.7–97 ng/L) and PFCs (0.54–55 ng/L) are comparatively low in the groundwaters of the Ganga River Basin (Lapworth et al., 2018 and Richards et al., 2021). Two more PFCs surfactant (perfluorododecanoic acid (PFDoDA) and perfluorooctane sulfonamide (PFOSA)) and plasticizer triphenyl phosphate are also found in Ganga Basin aquifers (Sharma et al., 2016; Richards et al., 2021).

1.3.3 Pesticides

Pesticides are the synthetic organic compounds that are used in agriculture to prevent, destroy, and mitigate pest, as plant regulators, defoliants, or desiccants (US EPA, 2007a). Currently, the most common types of pesticides being used are herbicides, insecticides, fungicides, and bactericides. Frequently occurring pesticides such as 2,6-dichlorobenzamide (BAM), atrazine, and its metabolite atrazine desethyl and bentazone (a polar pesticide) have concentrations above the upper limit (Figure 1.1a). Besides these three mentioned herbicides, chloridazon-desphenyl exceeds 1,000 ng/L level in groundwaters (Lapworth et al., 2015; White et al., 2016; Moreau et al., 2019). Diuron is another herbicide reported in six studies (Lapworth et al., 2015; Kong et al., 2016; Richards et al., 2021) with the maximum concentration below advised values by European Union (EU) and New Zealand Drinking water standards (Table 1.3). Atrazine, however, shows the concentration beyond EU limits. Eight other herbicides and five fungicides are detected from more than one study (Lapworth et al., 2015; Sorensen et al., 2015; Kong et al., 2016; Lee et al., 2019; Lesser et al., 2018; Yang et al., 2018; Moreau et al., 19; Gao et al., 2019; Richards et al., 2021). Lee et al. (2019) also found an insecticide (carbofuran) with DF 62% in agricultural groundwater, South Korea. In India, pesticides are extensively used for food production as well as indoor insecticides for mosquito, ant, and termite control. Major cities like Varanasi and Patna alongside the Ganga River observe a wide range of pesticides (11 and 27 pesticides, respectively) in the groundwater. Their concentration ranges from 0.1 ng/L to 2,100 ng/L. BAM, diuron, and carbendazim are most abundantly discovered in Patna, and chlorpyrifos and phenoxy acetic in Varanasi (Lapworth et al., 2018; Richards et al., 2021). The reported concentration of atrazine in Varanasi is within BIS prescribed limits (Table 1.3).

Table 1.3 Maximum concentration and detection frequency (DF) of the most common ECs (detected at least in five studies globally and two studies India) from 2015 to 2020

ECs (no. of studies)	*Standard limits*	*Max. Concentration (DF)*	*References (Location)*
PPPCPL			
Antibiotics			
Sulfamethoxazole (11)		1820(61.8)	Lin et al. (2015) (Northern Taiwan, China)
Trimethoprim (5)		17.8(8.8)	Lin et al. (2015) (Northern Taiwan, China)
Sulfadimethoxine (5)		4.3(20)	Lin et al. (2015) (Northern Taiwan, China)
Antiepileptic			
Carbamazepine (15)		620(12.5)	Moreau et al. (2019) (Waikato, New Zealand)
NSAID			
Diclofenac (6)		94 (100)	Gao et al. (2019) (Storlien, Sweden)
Naproxen (5)		200 (50)	Gao et al. (2019) (Storlien, Sweden)
Ibuprofen (6)		1006(58)	Kuroda and Kobayashi (2021) (Tokyo, Japan)
Analgesic			
Acetaminophen (5)		1880(80)	Gao et al. (2019) (Storlien, Sweden)
Beta-blocker			
Atenolol (6)		18(100)	Gao et al. (2019) (Storlien, Sweden)
Antimicrobial			
Triclosan (9)		83(7)	Lesser et al., 2018 (Mezquital Valley, Mexico)
Insect repellent			
DEET (8)	200000[4]	1610(92.3)	Lesser et al. (2018) (Mezquital Valley, Mexico)
Stimulant			
Caffeine (10)		3100(100)	Gao et al. (2019) (Storlien, Sweden)
Cotinine (5)		20(2.2)	Lapworth et al. (2015) (French Chalk, France)
Industrial compounds			
Plasticizer			
BPA (10)		10*5 (2.7)	Lapworth et al. (2015) (English Chalk, France)
Surfactants			
PFBS (8)		50(25)	Richards et al. (2021) (Middle Gangetic Plain, India)
PFDA (7)		14.6(13.5)	Chen et al. (2016) (Rural areas, Eastern China)
PFHxA (7)		357.8 (90)	Lin et al. (2015) (Northern Taiwan, China)
PFOS (10)	70[1], 36000[2]	110(16)	Moreau et al. (2019) (Waikato, New Zealand)
PFHpA (6)		99.7 (38.4)	Chen et al. (2016) (Eastern China)
PFHxS (8)		820	Moreau et al. (2019) (Waikato, New Zealand)
PFNA (7)		22 (42.3)	Chen et al. (2016) (Eastern China)
PFOA (8)	70[1], 36000[2]	475(69.2)	Chen et al. (2016) (Eastern China)

(Continued)

Table 1.3 (Continued) Maximum concentration and detection frequency (DF) of the most common ECs (detected at least in five studies globally and two studies India) from 2015 to 2020

ECs (no. of studies)	*Standard limits*	*Max. Concentration (DF)*	*References (Location)*
Pesticides (herbicide)			
BAM (6)		10000(6.3)	Lapworth et al. (2015) (English Chalk, England)
Diuron (6)	1800[2], 20000[3]	373(7.4)	Kong et al. (2016) (Beijing and Tianjin, China)
Atrazine (8)	200[2], 2000[3]	1800(63.1)	White et al. (2016) (Doncaster, UK)
Atrazine desethyl (6)		470(28.6)	Moreau et al. (2019) (Waikato, New Zealand)
Food additives (AS)			
Acesulfame (7)		5700(25)	Moreau et al. (2019) (Waikato, New Zealand)
Sucralose (5)		2440(100)	Yang et al. (2018) (Dongjiang River Basin, China)
ECs in India			
Sulfamethoxazole (2)		34(77)	Lapworth et al. (2018) (Varanasi)
Carbamazepine (2)		86(27)	Lapworth et al. (2018) (Varanasi)
Triclosan (2)		10(7)	Sharma et al. (2019) (Ganga Basin)
Cotinine		2.8(4)	Lapworth et al. (2018) (Varanasi)
PFBS		50(25)	Richards et al. (2021) (Middle Gangetic Plain)
PFDA		5.7(4)	Lapworth et al. (2018) (Varanasi)
PFHxA		55(6)	Richards et al. (2021) (Middle Gangetic Plain)
PFOS		3.3(12)	Lapworth et al. (2018) (Varanasi)
PFNA		0.54 (4)	Lapworth et al. (2018) (Varanasi)
PFOA		0.76 (92.8)	Sharma et al. (2016) (Ganga Basin)
PFHxS		5.5 (12)	Richards et al. (2021) (Middle Gangetic Plain)
Atrazine	2000[5]	1.9(12)	Lapworth et al. (2018) (Varanasi)
Atenolol		0.8(2)	Richards et al. (2021) (Middle Gangetic Plain)
Diuron		2.0(27)	Lapworth et al. (2018) (Varanasi)
Atrazine desethyl		1.2(2)	Richards et al. (2021) (Middle Gangetic Plain)
Acesulfame		760(22)	Richards et al. (2021) (Middle Gangetic Plain)
Sucralose		1200(41)	Richards et al. (2021) (Middle Gangetic Plain)

1.3.4 Food additives

Food additives include artificial sweeteners (ASs), preservatives, foam stabilizers, and artificial coloring agents to enhance taste, smell, appearance, and shelf life. ASs are reported as the most investigated group of ECs in groundwater from 2015. Acesulfame is the frequently encountered AS, and the maximum values of the contaminant are outliers (Figure 1.1a) (Lee et al., 2019; Yang et al., 2018; Schaider et al., 2016; Gao et al., 2019; Moreau et al., 2019; Richards et al., 2021). The highest concentration of acesulfame of 5,700 ng/L is observed in New Zealand, Waikato Regional Council's State of the Environment (SOE) (Moreau et al., 2019), and in other studies, above 100 ng/L. Sucralose is the second most common AS detected with a concentration range (46.8–2,440 ng/L) in Dongjiang River Basin, China (Yang et al., 2018). Different ASs like saccharine and cyclamate are also regularly found in groundwater but are comparatively less common (Richards et al., 2021; Lee et al., 2019; Moreau et al., 2019; Yang et al., 2018; Sharma et al., 2019). Among all four ASs, cyclamate has the highest concentration (58–24,000 ng/L) in New Zealand (Moreau et al., 2019). Acesulfame, sucralose, and cyclamate are determined with 100% DF in some studies (Gao et al., 2019; Yang et al., 2018). Food additives like benzaldehyde (1.2 ng/L) and triacetin (0.04 ng/L) are discovered in Kabwe groundwater (Sorensen et al., 2015), and methylparaben (437.4, ng/L) in Sweden sourced from septic tanks (Gao et al., 2019). In India, elevated concentration is observed for sucralose (19–1,200 ng/L) and acesulfame (51–760 ng/L) in the middle Gangetic Plain (Richards et al., 2021). Also, sucralose is commonly reported AS in groundwater in the basin (Sharma et al., 2019; Richards et al., 2021; Lapworth et al., 2018). Sharma et.al. (2019) found sucralose with 100% DF with 25 ng/L concentration.

1.4 OCCURRENCE OF ECs WITH RESPECT TO LAND USE, WELL DEPTH, AND LITHOLOGY

ECs are introduced into the environment by various anthropogenic activities, and their concentration indicates specific land use in the recharge area. The concentration and diversity of EC depend upon land-use activities and lithology of the aquifers. Most of the studies conclude that groundwater in or near the urban centers with higher population density and industries are more susceptible to contamination by a wide range of ECs and zones with agricultural activities are the second most contaminated sites (Lopez et al., 2015; Sorensen et al., 2015; Manamsa et al., 2016a; Sharma et al., 2019; Lee et al., 2019; Richards et al., 2021). Some researchers identify sites with intensive cultivation and horticultural practices that are most polluted with pesticides, pharmaceuticals, and hormones (Kong et al., 2016; Moreau et al., 2019). With an exception, Lapworth et al. (2015) observed the highest detections for pesticides and pharmaceuticals in natural areas. In the Gangetic Basin, greater concentrations are observed in middle and lower reaches with few specific ECs like caffeine and AS that are found in Himalayan reaches due to tourist activities (Sharma et al., 2016, 2019; Richards et al., 2021). A range of organic contaminants and their elevated concentrations are more common in shallow aquifers with sandy gravel and sandy in composition due to higher permeability (Li et al., 2015; Kong et al., 2016; Schaider et al., 2016; Liu et al., 2018; Lapworth et al., 2018; Gao et al., 2019; Kuroda and Kobayashi, 2021). Other aquifers like chalk (Lapworth et al., 2015; Manamsa et al., 2016b; Moreau et al., 2019) are also more exposed to contamination due to the easy leaching of contaminants. Lopez et al. (2015) report a greater number of detections in sedimentary formation than alluvial aquifers, and Moreau et al. (2019) in basalt than in sandy aquifer. The vulnerability assessment of deep aquifers in Varanasi (Lapworth et al., 2018) shows an increased concentration of ECs due to the high organic carbon content. However, the argillaceous nature of the shallow aquifer at the Ramnagar side forms a thicker layer, leading to less permeability and contamination. A similar assessment in Zambia (Sorensen et al., 2015) suggests a low risk of contamination as subsurface waters are protected by extensive overlying clay sheets. The lower concentration in the deep aquifer can also be attributed to potential attenuation due to longer hydraulic retention and dilution, while water in shallow aquifer follows the shorter flow paths that allow less time for sorption and biodegradation processes (Gao et al., 2019).

1.5 FATE AND RECEPTORS

The fate of any EC will be governed by both the properties of the contaminant, such as polarity, octanol/water partition coefficient (K_{OW}), and persisting environmental states in the subsurface, which include hydraulic properties of the aquifer, redox potential, residence time, type of aquifer (confined and unconfined), and overlying soil texture (Lapworth et al., 2012, 2018). Several attenuation processes, physical (adsorption onto clay and organic matter), chemical (ion exchange), and biological (microbial transformation) take place when a contaminant migrates from the soil zone to the confined zone (Lapworth et al., 2012; Jurado et al., 2012). Redox processes (aerobic and anaerobic processes) are related to the microbial activity that is also temperature dependent (Díaz-Cruz and Barceló, 2008 and Sui et al., 2015). For example, PPCPs are converted into hazardous metabolites after incomplete degradation in groundwaters or remain unchanged due to less microbial activities when compared to soils. But a temperature change leads to more efficient attenuation of metoprolol, iopromide, and diclofenac during summers (Burke et al., 2014). Though, hydrophobic ECs (high log K_{OW}) tend to bioaccumulate and bind to soil because of higher soil/sediment adsorption coefficients. But hydrophobic pesticides like simazine, atrazine, and diuron due to greater (than 2.8) Groundwater Ubiquity Score (GUS) index (Köck-Schulmeyer et al., 2012) are found in groundwater. BPA is detected with greater frequency in exposed aquifers; on the other hand, carbamazepine values are significantly elevated in the concealed sites where suboxic groundwater conditions are more likely; e.g., aquifers under Varanasi facilitate the persistence of sulfonamides and carbamazepine. This persistence of carbamazepine is because of low degradation and lack of sorption in the water-sediment system and hydrolysis following no adsorption to negatively charged particles in aquifer sediments (Lapworth et al., 2015; Lapworth et al., 2018). During recharge, anoxic conditions along with temperature more than 5°C, many halogenated C1-C2 aliphatics, nitro- and chloro-benzene exhibit sequential dehalogenation; however, at lower temperatures, degradation rates decrease, and by-products are formed (such as dichloro- and bromodichloromethane, and dicholoroanilines). Higher elimination rates of trihalomethanes and phenoxy pesticides under aerobic condition lead to the formation of hazardous chlorophenols. The transition from anoxic to anaerobic conditions occurs during the infiltration process at the bank that can reduce the concentration of sulfamethoxazole by 20%. The half-lives of sulfa antibiotics are noticed to be shorter in silt compared to silty and fine sand aquifers due to differences in organic matter content (Zhou, 2015). DEET is also shown to be persisting in the bedrock and fipronil with no transformation products in aquifers, indicating much lower degradation rates than half-life (Weeks et al., 2012; Sorensen et al., 2013).

One of the groundwater-dependent receptors can be humans for using this source of water for drinking and irrigation purposes, affecting both fauna (also fishes, invertebrates, and microbial community) and flora (Stuart et al., 2012; Stefanakis and Becker, 2015). ECs have a drastic impact on the aquatic ecosystem as compared to humans (Stefanakis and Becker, 2015). There are two main hazardous effects, i.e., toxicity and endocrine disruption. Toxicity can be acute and chronic depending upon the concentration and duration of exposure. Endocrine disruption occurs when a compound indirectly interferes with the normal functioning of the hormonal system. Endocrine disruptors include plasticizers (such as BPA and phthalates), flame retardants, domestic and industrial cleaning agents, steroids, pesticides, and some other PPCPs (Mathew and Kanmani, 2020; Stuart et al., 2012). Pharmaceuticals like antibiotics cause bacterial communities to be more resistant to specific drugs (Stuart et al., 2012; Kumar et al., 2019 a,b). PFCs bioaccumulate in fishes and human kidneys, serum, and liver and potentially affect the developmental and reproductive systems (US EPA, 2014). Endocrine disruptors may result in sexual dimorphism, neuroendocrine disorder, hormonal imbalance, diabetes, and cardiovascular problems in humans. Synthetic steroids cause feminization in fishes and mimic estrogen to non-targets (Frye et al., 2012; Braun and Hauser, 2011; Vandenberg et al., 2013). Caffeine causes endocrine disruption in goldfishes, and carbamazepine results in oxidation stress in trout fishes (Li et al., 2010). Certain antibiotics, gemfibrozil (a blood lipid regulator), and antimicrobial agents like triclosan and triclocarban cause the growth inhibition of algae. Pesticides also bioaccumulate in aquatic fauna and cause necrosis of the gastrointestinal tract, kidney tubules, liver, and lung from oxidative stress (Pereira et al., 2015). Because of the broad class of the ECs, toxicological studies for the majority of the organic contaminants and the synergetic effects of the cocktail of the contaminants in groundwaters are still not investigated.

1.6 REGULATION AND RISK ASSESSMENT OF ECs

Risk assessment is a scientific procedure to evaluate the adverse effects of a known or potentially hazardous compound to human health and environment and also prioritize the compound. It includes four steps: hazard identification, exposure assessment: estimating contaminant release, its pathways, and exposure concentration, toxicity assessment, and risk characterization (US EPA, 1989b). For a vast array of ECs, it will be a big task to produce risk assessment methods and databases. There is a need to focus on the toxic effects of a single compound, and it transformed products as well as joint toxic effects of ECs as they can be synergistic or antagonistic (Di Poi et al., 2018). The procedure of risk assessment includes factors (US EPA): how much of a chemical is present in the groundwater, the amount of exposure of the receptor with the contaminated groundwater, and the inherent toxicity of the chemical.

The two most widely methods applied are (i) risk quotient (RQ) and (ii) hazard quotient (HQ). RQ defines the ratio of a point evaluation of exposure and point estimation of effects, while HQ ratio measures environmental concentration (MEC) and an effect concentration. Risk assessment can be performed for acute (short-term) exposure, chronic (long-term) exposure, or a combination of both. Majolagbe et al. (2016) carried out a risk assessment using the DRASTIC model to study the impact of dumping sites on groundwater quality, concluding that clayey soil hinders the infiltration of contaminants into the groundwater. Chonattu et al. (2016) studied the gaps by developing a water quality index, using physicochemical and biological parameters to study the potential threats caused by landfill leachates on groundwater quality in Njeliyanparamba, India. The European Commission instructs drug manufacturing companies for the environmental risk assessment (ERA) for products used by humans that includes initial screening, determining the fate of the compounds, and possible impacts on the aquatic ecosystem. Ranking and scoring systems (RSS) can be developed for groundwater to learn the potential of a chemical to cause environmental effects (Mitchell et al., 2002). Conceptual models quantify groundwater systems, processes, and their interactions and are developed to different incremental levels of complexity (Lamastra et al., 2016). Environmental tracers provide information on the time scales of the solute transport and maximum concentration of a contaminant at the discharge area, which can support the development of such models. There are many studies for prioritizing compounds but still lack research that focuses on the exposure of subsurface water and quantifying hazards.

In the USA, Environmental Protection Act (US EPA, 2021a) defines the primary standards to control the contaminants in drinking water and made a contaminant candidates list (CCL) for priority compounds. The CCL3 list includes pesticides, pharmaceuticals, chemicals used in commerce, pathogens, and biological toxins. CCL4 list includes 97 chemicals from industrial use, pesticides, disinfected by-products, and pharmaceuticals (US EPA, 2021b). Preliminary regulations are taken for PFOS and PFOAs in the CCL4 list. Most of the European countries implement the rules of the European Water Framework Directive (2000; 2000/60/EC) and the Groundwater Daughter Directive (2006; 2006/118/EC). The new ECs included in European Commission 2012 list are cypermethrin, dicofol, heptachlor and quinoxyfen, hexabromocyclododecane (HBCDD), dioxin and dioxin-like PCBs, and PFOS. There is a group called Endocrine Disrupters Expert Advisory Group of the European Commission to deal with endocrine disrupter chemicals (EDCs). The Indian government has proposed some set of rules for pharmaceuticals (antimicrobial-resistant) and are updated with time. Chennai Declaration has been proposed in 2012 to manage antimicrobial drugs in the country (Ghafur et al., 2013). A new schedule (schedule H1) is added to the Drugs and Cosmetic Rules, 1945, including 46 drugs implemented in 2014 by the Central Drug Standard Control Organization (PIB, 2015; Kumar et al., 2014). Recently, a Nation Action Plan on antimicrobial resistance 2017–2021 has been launched with six strategic priorities to effectively curd antimicrobial resistance in India (MoHFW, 2017).

1.7 CONCLUSION

The main source of contamination for groundwater remains wastewater effluents, which directly or indirectly reach groundwater. From the past studies, it can be concluded that carbamazepine continues as the most frequently occurring organic compound followed by sulfamethoxazole, caffeine, and BPA. The present concentrations are still not proven to be toxic due to the absence of standard values from authorities. Artificial sweeteners are extensively reported after 2015 or are more focused upon recently as previously reviewed studies did not mention much about their distribution. In India, PFCs are widely distributed in the Ganga Basin. Generally, the concentrations of ECs are low in groundwater due to attenuation and dilution of the contaminants but still, toxicological studies are important and have identified the likely environmental effect of some contaminants. In addition to other factors, the mobility and possible contamination of the groundwater depends on the permeability of aquifers. From the previous investigation, the sand gravel aquifer and chalk aquifers are realized to be more prone to contamination. Deep aquifers are vulnerable if not protected by an impervious layer of clay. Concerning land use, specific contaminants are released from different sources, which can be beneficial to trace back to the source. The regulatory authorities choose priority compounds based on their toxic effect, information gap, and lack of monitoring data. An interdisciplinary approach combining science-policy and advocacy can be helpful for the regulation of ECs and introducing threshold values.

REFERENCES

Braun, J.M., Hauser, R., 2011. Bisphenol A and Children's Health. *Curr. Opin. Pediatr.* 23, 233–239. https://doi.org/10.1097/MOP.0b013e3283445675.

Bucheli, T.D., Müller, S.R., Heberle, S., Schwarzenbach, R.P., 1998. Occurrence and behavior of pesticides in rainwater, roof runoff, and artificial stormwater infiltration. *Environ. Sci. Technol.* 32, 3457–3464. https://doi.org/10.1021/es980317n.

Burke, V., Greskowiak, J., Asmuß, T., Bremermann, R., Taute, T., Massmann, G., 2014. Temperature dependent redox zonation and attenuation of wastewater-derived organic micropollutants in the hyporheic zone. *Sci. Total Environ.* 482–483, 53–61. https://doi.org/10.1016/j.scitotenv.2014.02.098.

Carrara, C., Ptacek, C. J., Robertson, W. D., Blowes, D. W., Moncur, M. C., Sverko, E., Backus, S., 2008. Fate of pharmaceutical and trace organic compounds in three septic system plumes, Ontario, Canada. *Environ. Sci. and Technol.* 42, 2805–2811. https://doi.org/10.1021/es070344q.

Chen, S., Jiao, X.C., Gai, N., Li, X.J., Wang, X.C., Lu, G.H., Piao, H.T., Rao, Z., Yang, Y.L., 2016. Perfluorinated compounds in soil, surface water, and groundwater from rural areas in eastern China. *Environ. Pollut.* 211, 124–131. https://doi.org/10.1016/j.envpol.2015.12.024.

Chonattu, J., Prabhakar, K., Pillai, H.P.S., 2016. Geospatial and statistical assessment of groundwater contamination due to landfill leachate—A case study. *J. Water Resour. Prot.* 08, 121–134. https://doi.org/10.4236/jwarp.2016.82010.

Clarke, B.O., Smith, S.R., 2011. Review of "emerging" organic contaminants in biosolids and assessment of international research priorities for the agricultural use of biosolids. *Environ. Int.* 37, 226–247. https://doi.org/10.1016/j.envint.2010.06.004.

Dey, S., Bano, F., Malik, A., 2019. Pharmaceuticals and personal care product (PPCP) contamination-a global discharge inventory. In: *Pharmaceuticals and Personal Care Products: Waste Management and Treatment Technology Emerging Contaminants and Micro Pollutants.* Elsevier Inc. https://doi.org/10.1016/B978-0-12-816189-0.00001-9.

Di Poi, C., Costil, K., Bouchart, V., Halm-Lemeille, M.P., 2018. Toxicity assessment of five emerging pollutants, alone and in binary or ternary mixtures, towards three aquatic organisms. *Environ. Sci. Pollut. Res.* 25, 6122–6134. https://doi.org/10.1007/s11356-017-9306-9.

Díaz-Cruz, M.S., Barceló, D., 2008. Trace organic chemicals contamination in ground water recharge. *Chemosphere* 72, 333–342. https://doi.org/10.1016/j.chemosphere.2008.02.031.

Drewes, J. E., 2009. Ground water replenishment with recycled water - Water quality improvements during managed aquifer recharge. *Ground Water* 47, 502–505. https://doi.org/10.1111/j.1745-6584.2009.00587_5.x.

Ellis, J. B., 2006. Pharmaceutical and personal care products (PPCPs) in urban receiving waters. *Environ. Pollut.* 144, 184–189. https://doi.org/10.1016/j.envpol.2005.12.018.

EPA, 1989b. Risk Assessment Guidance for Superfund. Volume I Human Health Evaluation Manual (Part A) I, 289. https://doi.org/EPA/540/1-89/002.

Frye, C., Bo, E., Calamandrei, G., Calzà, L., Dessì-Fulgheri, F., Fernández, M., Fusani, L., Kah, O., Kajta, M., Le Page, Y., Patisaul, H.B., Venerosi, A., Wojtowicz, A.K., Panzica, G.C., 2012. Endocrine disrupters: A review of some sources, effects, and mechanisms of actions on behaviour and neuroendocrine systems. *J. Neuroendocrinol.* 24, 144–159. https://doi.org/10.1111/j.1365–2826.2011.02229.x.

Gao, Q., Blum, K.M., Gago-Ferrero, P., Wiberg, K., Ahrens, L., Andersson, P.L., 2019. Impact of on-site wastewater infiltration systems on organic contaminants in groundwater and recipient waters. *Sci. Total Environ.* 651, 1670–1679. https://doi.org/10.1016/j.scitotenv.2018.10.016.

Ghafur, A., Mathai, D., Muruganathan, A., Jayalal, J.A., Kant, R., Chaudhary, D., Prabhash, K., Abraham, O.C., Gopalakrishnan, R., Ramasubramanian, V., Shah, S.N., Pardeshi, R., Huilgol, A., Kapil, A., Gill, J., Singh, S., Rissam, H.S., Todi, S., Hegde, B.M., Parikh, P., 2013. The Chennai Declaration: a roadmap to tackle the challenge of antimicrobial resistance. *Indian J. Cancer* 50, 71–73. https://doi.org/10.4103/0019-509X.104065.

Jindal, K., Narayanam, M., Singh, S., 2015. A systematic strategy for the identification and determination of pharmaceuticals in environment using advanced LC-MS tools: Application to ground water samples. *J. Pharm. Biomed. Anal.* 108, 86–96. https://doi.org/10.1016/j.jpba.2015.02.003.

Jurado, A., Vàzquez-Suñé, E., Carrera, J., López de Alda, M., Pujades, E., Barceló, D., 2012. Emerging organic contaminants in groundwater in Spain: A review of sources, recent occurrence and fate in a European context. *Sci. Total Environ.* 440, 82–94. https://doi.org/10.1016/j.scitotenv.2012.08.029.

Köck-Schulmeyer, M., Ginebreda, A., de Alda, M.L., Barceló, D., 2012. Fate and risks of polar pesticides in groundwater samples of Catalonia. *Handb. Environ. Chem.* 20, 375–394. https://doi.org/10.1007/698_2011_132.

Kong, L., Kadokami, K., Duong, H.T., Chau, H.T.C., 2016. Screening of 1300 organic micro-pollutants in groundwater from Beijing and Tianjin, North China. *Chemosphere* 165, 221–230. https://doi.org/10.1016/j.chemosphere.2016.08.084.

Kumar, M., Kuroda, K., A. K., Patel, N., Bhattacharya, P., Joshi, M., Joshi, C.G. 2021. Decay of SARS-CoV-2 RNA along the wastewater treatment outfitted with Upflow Anaerobic Sludge Blanket (UASB) system evaluated through two sample concentration techniques. *Sci. Total Environ.* 754, 142329.

Kumar, M., Ram, B., Honda, R., Poopipattana, C., Canh, V.D., Chaminda, T., Furumai, H., 2019a. Concurrence of antibiotic resistant bacteria (ARB), viruses, pharmaceuticals and personal care products (PPCPs) in ambient waters of Guwahati, India: Urban vulnerability and resilience perspective. *Sci. Total Environ.* 693, 133640. https://doi.org/10.1016/j.scitotenv.2019.133640.

Kumar, M., Snow, D. D., Li, Y., Shea, P. J., 2019b. Perchlorate behavior in the context of black carbon and metal cogeneration following fireworks emission at Oak Lake, Lincoln, Nebraska, USA. *Environ. Pollut.* 253, 930–938.

Kumar, S.G., Adithan, C., Harish, B.N., Sujatha, S., Roy, G., Malini, A., 2014. Antimicrobial resistance in India: A review. *J. Nat. Sci. Biol. Med.* 4, 286–291. https://doi.org/10.4103/0976-9668.116970.

Kuroda, K., Kobayashi, J., 2021. Pharmaceuticals, personal care products, and artificial sweeteners in Asian groundwater: A review. In: *Contaminants in Drinking and Wastewater Sources.* Springer, Singapore. https://doi.org/10.1007/978-981-15-4599-3_1.

Lamastra, L., Balderacchi, M., Trevisan, M., 2016. Inclusion of emerging organic contaminants in groundwater monitoring plans. *MethodsX*, 3, 459–476. https://doi.org/10.1016/j.mex.2016.05.008.

Lapworth, D.J., Baran, N., Stuart, M.E., Manamsa, K., Talbot, J., 2015. Persistent and emerging micro-organic contaminants in Chalk groundwater of England and France. *Environ. Pollut.* 203, 214–225. https://doi.org/10.1016/j.envpol.2015.02.030.

Lapworth, D.J., Baran, N., Stuart, M.E., Ward, R.S., 2012. Emerging organic contaminants in groundwater: A review of sources, fate and occurrence. *Environ. Pollut.* 163, 287–303. https://doi.org/10.1016/j.envpol.2011.12.034.

Lapworth, D.J., Das, P., Shaw, A., Mukherjee, A., Civil, W., Petersen, J.O., Gooddy, D.C., Wakefield, O., Finlayson, A., Krishan, G., Sengupta, P., MacDonald, A.M., 2018. Deep urban groundwater vulnerability in India revealed through the use of emerging organic contaminants and residence time tracers. *Environ. Pollut.* 240, 938–949. https://doi.org/10.1016/j.envpol.2018.04.053.

Lapworth, D.J., Gooddy, D.C., Allen, D., Old, G.H., 2009. Understanding groundwater, surface water, and hyporheic zone biogeochemical process in a Chalk catchment using fluorescence properties of dissolves and colloidal organic matter. *J. Geophys. Res. Biogeosci.* 114, 1–10. https://doi.org/10.1029/2009JG000921.

Lee, H.J., Kim, K.Y., Hamm, S.Y., Kim, M.S., Kim, H.K., Oh, J.E., 2019. Occurrence and distribution of pharmaceutical and personal care products, artificial sweeteners, and pesticides in groundwater from an agricultural area in Korea. *Sci. Total Environ.* 659, 168–176. https://doi.org/10.1016/j.scitotenv.2018.12.258.

Lesser, L.E., Mora, A., Moreau, C., Mahlknecht, J., Hernández-Antonio, A., Ramírez, A.I., Barrios-Piña, H., 2018. Survey of 218 organic contaminants in groundwater derived from the world's largest untreated wastewater irrigation system: Mezquital Valley, Mexico. *Chemosphere* 198, 510–521. https://doi.org/10.1016/j.chemosphere.2018.01.154.

Lewandowski, J., Putschew, A., Schwesig, D., Neumann, C., & Radke, M., 2011. Fate of organic micropollutants in the hyporheic zone of a eutrophic lowland stream: Results of a preliminary field study. *Sci. Total Environ.* 409, 1824–1835. https://doi.org/10.1016/j.scitotenv.2011.01.028.

Li, Z., Xiang, X., Li, M., Ma, Y., Wang, J., Liu, X., 2015. Occurrence and risk assessment of pharmaceuticals and personal care products and endocrine disrupting chemicals in reclaimed water and receiving groundwater in China. *Ecotoxicol. Environ. Saf.* 119, 74–80. https://doi.org/10.1016/j.ecoenv.2015.04.031.

Li, J.J., Hartono, D., Ong, C.N., Bay, B.H., Yung, L.Y.L., 2010. Autophagy and oxidative stress associated with gold nanoparticles. Biomaterials 31, 5996–6003. https://doi.org/10.1016/j.biomaterials.2010.04.014.

Li, X., Shang, X., Luo, T., Du, X., Wang, Y., Xie, Q., Matsuura, N., Chen, J., Kadokami, K., 2016. Screening and health risk of organic micropollutants in rural groundwater of Liaodong Peninsula, China. Environ. Pollut. 218, 739–748. https://doi.org/10.1016/j.envpol.2016.07.070.

Lin, Y.C., Lai, W.W.P., Tung, H. H., Lin, A.Y.C., 2015. Occurrence of pharmaceuticals, hormones, and perfluorinated compounds in groundwater in Taiwan. *Environ. Monit. Assess.* 187. https://doi.org/10.1007/s10661-015-4497-3.

Liu, Y., Ma, L., Yang, Q., Li, G., Zhang, F., 2018. Occurrence and spatial distribution of perfluorinated compounds in groundwater receiving reclaimed water through river bank infiltration. *Chemosphere* 211, 1203–1211. https://doi.org/10.1016/j.chemosphere.2018.08.028.

Lopez, B., Ollivier, P., Togola, A., Baran, N., Ghestem, J.P., 2015. Screening of French groundwater for regulated and emerging contaminants. *Sci. Total Environ.* 518–519, 562–573. https://doi.org/10.1016/j.scitotenv.2015.01.110.

Majolagbe, A. O., Adeyi, A. A., Osibanjo, O., 2016. Vulnerability assessment of groundwater pollution in the vicinity of an active dumpsite (Olusosun), Lagos, Nigeria. *Chem. Int.* 2, 232–241.

Manamsa, K., Crane, E., Stuart, M., Talbot, J., Lapworth, D., Hart, A., 2016a. A national-scale assessment of micro-organic contaminants in groundwater of England and Wales. *Sci. Total Environ.* 568, 712–726. https://doi.org/10.1016/j.scitotenv.2016.03.017.

Manamsa, K., Lapworth, D.J., Stuart, M.E., 2016b. Temporal variability of micro-organic contaminants in lowland chalk catchments: New insights into contaminant sources and hydrological processes. *Sci. Total Environ.* 568, 566–577. https://doi.org/10.1016/j.scitotenv.2016.01.146.

Mathew, R.A., Kanmani, S., 2020. A review on emerging contaminants in indian waters and their treatment technologies. *Nat. Environ. Pollut. Technol.* 19, 549–562. https://doi.org/10.46488/NEPT.2020.V19I02.010.

Menon N.G., Mohapatra S., Padhye L.P., Tatiparti S.S.V., Mukherji S., 2020. Review on Occurrence and toxicity of pharmaceutical contamination in Southeast Asia. Springer Transactions in Civil and Environmental Engineering. 63–91. https:// doi.org/10.1007/978-981-32-9771-5_4.

Mitchell, R.R., Summer, C.L., Blonde, S.A., Bush, D.M., Hurlburt, G.K., Snyder, E.M., Giesy, J.P., 2002. SCRAM: A scoring and ranking system for persistent, bioaccumulative, and toxic substances for the North American Great Lakes-resulting chemical scores and rankings. *Hum. Ecol. Risk Assess.* 8, 537–557. https://doi.org/10.1080/10807030290879817.

Mohapatra, S., Padhye, L.P., Mukherji, S., 2018. Challenges in Detection of Antibiotics in Wastewater Matrix. *Energy, Environ. Sustain.* 3–20. https://doi.org/10.1007/978-981-10-7332-8_1.

Mohapatra, S., Menon, N., Padhye, L.P., Tatiparti, S.S.V., Mukherji, S., 2021. Natural attenuation of pharmaceuticals in the aquatic environment and role of phototransformation. Springer Transactions in Civil and Environmental Engineering. 65–94. https://doi.org/10.1007/978-981-15-4599-3_3.

MoHFW, 2017. National action plan on antimicrobial resistance (NAP-AMR) 2017–2021. https://ncdc.gov.in/WriteReadData/l892s/File645.pdf (Accessed April 2017).

Moreau, M., Hadfield, J., Hughey, J., Sanders, F., Lapworth, D.J., White, D., Civil, W., 2019. A baseline assessment of emerging organic contaminants in New Zealand groundwater. *Sci. Total Environ.* 686, 425–439. https://doi.org/10.1016/j.scitotenv.2019.05.210.

Naidu, R., Arias Espana, V.A., Liu, Y., Jit, J., 2016. Emerging contaminants in the environment: Risk-based analysis for better management. *Chemosphere* 154, 350–357. https://doi.org/10.1016/j.chemosphere.2016.03.068.

NORMAN, 2012. “Emerging substances” Network of reference laboratories, research centres and related organisations for monitoring of emerging environmental substances. https://www.norman-network.net/?q=node/19/ (Accessed Feb, 2016).

Pal, A., He, Y., Jekel, M., Reinhard, M., Gin, K.Y.H., 2014. Emerging contaminants of public health significance as water quality indicator compounds in the urban water cycle. *Environ. Int.* 71, 46–62. https://doi.org/10.1016/j.envint.2014.05.025.

Patel, A. K., Das, N., & Kumar, M., 2019. Multilayer arsenic mobilization and multimetal co-enrichment in the alluvium (Brahmaputra) plains of India: a tale of redox domination along the depth. *Chemosphere*, *224*, 140–150.

Pereira, L.C., de Souza, A.O., Bernardes, M.F.F., Pazin, M., Tasso, M.J., Pereira, P.H., Dorta, D.J., 2015. A perspective on the potential risks of emerging contaminants to human and environmental health. *Environ. Sci. Pollut. Res.* 22, 13800–13823. https://doi.org/10.1007/s11356-015-4896-6.

PIB, 2015. Antibiotic policy. https://pib.gov.in/newsite/PrintRelease.aspx?relid=117474 (Accessed 20 March 2015).

Priyanka M.S., Mohapatra S., 2020. Shifts and trends in analysis of contaminants of emerging concern: Sulfonamides. In: Kumar M., Munoz-Arriola F., Furumai H., Chaminda T. (eds.) *Resilience, Response, and Risk in Water Systems*. Springer Transactions in Civil and Environmental Engineering. Springer, Singapore. https://doi.org/10.1007/978-981-15-4668-6_14.

Richards, L.A., Kumari, R., White, D., Parashar, N., Kumar, A., Ghosh, A., Kumar, S., Chakravorty, B., Lu, C., Civil, W., Lapworth, D.J., Krause, S., Polya, D.A., Gooddy, D.C., 2021. Emerging organic contaminants in groundwater under a rapidly developing city (Patna) in northern India dominated by high concentrations of lifestyle chemicals. *Environ. Pollut.* 268, 115765. https://doi.org/10.1016/j.envpol.2020.115765.

Rosenfeld, P.E., Feng, L.G.H., 2011. Emerging contaminants. Risks Hazard. *Wastes* 215–222. https://doi.org/10.1016/b978-1-4377-7842-7.00016-7.

Schaider, L.A., Ackerman, J.M., Rudel, R.A., 2016. Septic systems as sources of organic wastewater compounds in domestic drinking water wells in a shallow sand and gravel aquifer. *Sci. Total Environ.* 547, 470–481. https://doi.org/10.1016/j.scitotenv.2015.12.081.

Sharma, B.M., Bečanová, J., Scheringer, M., Sharma, A., Bharat, G.K., Whitehead, P.G., Klánová, J., Nizzetto, L., 2019. Health and ecological risk assessment of emerging contaminants (pharmaceuticals, personal care products, and artificial sweeteners) in surface and groundwater (drinking water) in the Ganges River Basin, India. *Sci. Total Environ.* 646, 1459–1467. https://doi.org/10.1016/j.scitotenv.2018.07.235.

Sharma, B.M., Bharat, G.K., Tayal, S., Larssen, T., Bečanová, J., Karásková, P., Whitehead, P.G., Futter, M.N., Butterfield, D., Nizzetto, L., 2016. Perfluoroalkyl substances (PFAS) in river and ground/drinking water of the Ganges River basin: Emissions and implications for human exposure. *Environ. Pollut.* 208, 704–713. https://doi.org/10.1016/j.envpol.2015.10.050.

Sorensen, J.P.R., Lapworth, D.J., Nkhuwa, D.C.W., Stuart, M.E., Gooddy, D.C., Bell, R.A., Chirwa, M., Kabika, J., Liemisa, M., Chibesa, M., Pedley, S., 2015. Emerging contaminants in urban groundwater sources in Africa. *Water Res.* 72, 51–63. https://doi.org/10.1016/j.watres.2014.08.002.

Sorensen, J.P.R., Maurice, L., Edwards, F.K., Lapworth, D.J., Read, D.S., Allen, D., Butcher, A.S., Newbold, L.K., Townsend, B.R., Williams, P.J., 2013. Using boreholes as windows into groundwater ecosystems. *PLoS One* 8. https://doi.org/10.1371/journal.pone.0070264.

Stefanakis, A.I., Becker, J.A., 2015. A review of emerging contaminants in water: Classification, sources, and potential risks. *Impact Water Pollut. Hum. Heal. Environ. Sustain.* 55–80. https://doi.org/10.4018/978-1-4666-9559-7.ch003.

Stuart, M., Lapworth, D., 2013. Smart Sensors for Real-Time Water Quality Monitoring 4, 259–284. https://doi.org/10.1007/978-3-642-37006-9.

Stuart, M., Lapworth, D., Crane, E., Hart, A., 2012. Review of risk from potential emerging contaminants in UK groundwater. *Sci. Total Environ.* 416, 1–21. https://doi.org/10.1016/j.scitotenv.2011.11.072.

Sui, Q., Cao, X., Lu, S., Zhao, W., Qiu, Z., Yu, G., 2015. Occurrence, sources and fate of pharmaceuticals and personal care products in the groundwater: A review. *Emerg. Contam.* 1, 14–24. https://doi.org/10.1016/j.emcon.2015.07.001.

Thomaidis, N.S., Asimakopoulos, A.G., Bletsou, A.A., 2012. Emerging contaminants: A tutorial mini-review. *Glob. Nest J.* 14, 72–79. https://doi.org/10.30955/gnj.000823.

US EPA, 2014. Emerging Contaminants and Federal Facility Contaminants of Concern. Technical fact sheets. http://www.epa.gov/sites/production/files/2014-03/documents/ffrrofactsheet_contaminant_perchlorate_january2014_final.pdf/ (Accessed 5/01/2016).

US EPA, 2021a. Basic information on the CCL and regulatory determination. Drinking water contaminant candidate list (CCL) and regulatory determination. https://www.epa.gov/ccl basic-information-ccl-and-regulatory-determination (Assessed 19/02/2021)

US EPA, 2021b. Regulatory Determination 4. https://www.epa.gov/ccl/regulatory-determination-4 (Assessed 4/02/2021).

Vandenberg, L.N., Hunt, P.A., Myers, J.P., Saal, F.S., 2013. Human exposures to bisphenol A: mismatches between data and assumptions. *Rev. Environ. Health* 28, 37–58. https://doi.org/10.1515/reveh-2012-0034.

Walters, A., Santillo, D., 2006. Uses of Perfluorinated Substances 1–20.
Wang, L., Zhang, J., Sun, H., Zhou, Q., 2016. Widespread occurrence of benzotriazoles and benzothiazoles in tap water: Influencing factors and contribution to human exposure. *Environ. Sci. Technol.* 50, 2709–2717. https://doi.org/10.1021/acs.est.5b06093.
Weeks, J.A., Guiney, P.D., Nikiforovz, A.I., 2011. Assessment of the environmental fate and ecotoxicity of N, N-diethyl-m-toluamide (DEET). *Integr. Environ. Assess. Manag.* 8, 120–134. https://doi.org/10.1002/ieam.1246.
White, D., Lapworth, D.J., Stuart, M.E., Williams, P.J., 2016. Hydrochemical profiles in urban groundwater systems: New insights into contaminant sources and pathways in the subsurface from legacy and emerging contaminants. *Sci. Total Environ.* 562, 962–973. https://doi.org/10.1016/j.scitotenv.2016.04.054.
Yan, S., Subramanian, S.B., Tyagi, R.D., Surampalli, R.Y., Zhang, T.C., 2010. Emerging contaminants of environmental concern: Source, transport, fate, and treatment. *Pract. Period. Hazardous, Toxic, Radioact. Waste Manag.* 14, 2–20. https://doi.org/10.1061/(ASCE)HZ.1944-8376.0000015.
Yang, Y.Y., Zhao, J.L., Liu, Y.S., Liu, W.R., Zhang, Q.Q., Yao, L., Hu, L.X., Zhang, J.N., Jiang, Y.X., Ying, G.G., 2018. Pharmaceuticals and personal care products (PPCPs) and artificial sweeteners (ASs) in surface and ground waters and their application as indication of wastewater contamination. *Sci. Total Environ.* 616–617, 816–823. https://doi.org/10.1016/j.scitotenv.2017.10.241.
Zhou, A., 2015. Research on the migration and transformation mechanism and remediation technique of sulfa antibiotics in phreatic water (inChinese). PhD Thesis, Jilin University, Changchun, China.

Chapter 2

Occurrence, sources, and the fate of antibiotics, antibiotic-resistant bacteria and their resistance genes in groundwater and subsurface environment

Sasikaladevi R and Kiruthika Eswari V
Indian Institute of Technology Madras

Govindaraj Divyapriya
Virginia Polytechnic Institute and State University

Ramya Srinivasan and Indumathi M. Nambi
Indian Institute of Technology Madras

CONTENTS

DOI: 10.1201/9781003157465-3

2.1 INTRODUCTION

There is a widespread use of antibiotics for the treatment of diseases in animals and humans, which has led to the presence of the antibiotic residues (AB) across various spheres of the ecosystem. Besides, antibiotics are heavily used for metaphylaxis, prophylaxis, and treatment of diseases across livestock, poultry, and aquaculture; they are also being used as growth promoters for enhancing feed efficiency and weight gain in animals (Walia et al., 2019). Such rampant use of antibiotics has resulted in the development of resistance among bacteria, leading to the prevalence of antibiotic-resistant bacteria (ARB) and antibiotic resistance genes (ARGs). The bacteria, which have acquired resistance against antibiotics, are called ARB and the associated gene is called ARG. The World Health Organization (WHO) has published a list of global-priority antibiotic-resistant pathogens and has classified them into critical, high, and medium priorities as applicable (Amarasiri et al., 2020; WHO, 2017). Moreover, WHO has categorized this global issue as one among the third most serious threats to human health in this century. By the year 2050, 10 million human deaths are estimated due to ARB-based infections (Amarasiri et al., 2020).

In recent times, the presence of antibiotics has been detected in water, which is of great health concern. This is due to inadequate and/or inappropriate treatment of wastewater containing pharmaceutical compounds. The entry of antimicrobial resistance (AMR) into wastewater treatment system is through several sources, namely, through discharge of hospital waste, discharge of untreated pharmaceutical wastewater, improper disposal of drugs, and unmetabolized antibiotics from human and animal excreta (from household and farmlands), as shown in Figure 2.1.

The indirect sources of antibiotics have increased because of drug misuse, which is more prevalent in developing countries where the regulations for purchase of antibiotics are not in place. Malpractices such as over-the-counter sales of antibiotics and lack of awareness resulting in the consumption of an incomplete course of antibiotics are two of the important contributors for the development of antibiotic resistance in

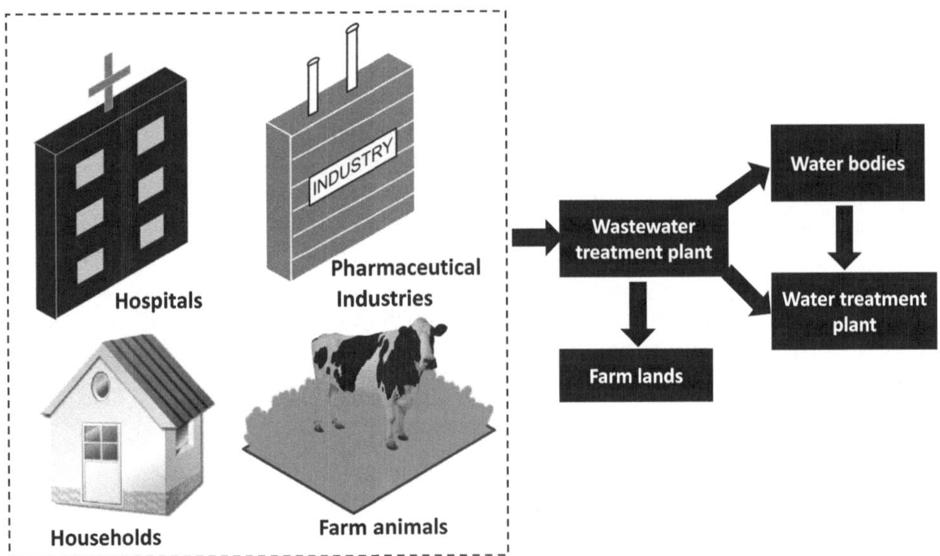

Figure 2.1 Sources of antibiotics and their dissemination routes.

microorganisms. The conventional wastewater treatment plans have predominantly biological treatment processes, which cannot remove antibiotics and associated ARB and ARGs. The treated effluents containing ARB and ARGs end up in either water bodies due to their discharge, in water treatment plants, in groundwater, or in farmlands during reuse. It has recently been established that such improperly treated wastewater (TWW) is an important vehicle for dissemination of microbial resistance to antimicrobial drugs. The discharge or land application of such inadequately and/or inappropriately TWW or sludge leads to the contamination of surface and ground water, respectively, resulting in frequent exposures to humans and animals. This will eventually result in the development of resistance in bacteria and further the antimicrobials becoming ineffective (due to the development of antibiotic resistance), leading to thousands of deaths due to infections (which were previously curable). Since groundwater is one of the major drinking water sources in developing countries, the presence of ARB and ARGs in them is a cause of serious health concern.

Common classes of antimicrobials are currently being used widely in as that of humans. A few examples are indicated in Table 2.1. This would help in understanding the possible direct impacts of improper use or disposal of drugs. Use of similar drugs in humans and animals can further accelerate the process of antibiotic resistance dissemination, since in animals the administration of drugs is more indiscriminate. Moreover, antibiotics are also used as growth promoters, increasing the overall amount of administration.

Dissemination of resistance has caused the development of superbugs having multidrug resistance, which on infecting humans make it challenging to treat. WHO proclaims that with the development of resistance to antimicrobials and hence the reduction of their efficacy, several diseases such as pneumonia, tuberculosis, gonorrhea, and salmonellosis are becoming harder to treat. With such improper usage of drugs, we are moving toward a post-antibiotic era, where even minor infections can cause fatality.

Several aspects about dissemination of resistance remain less understood. Water characteristics such as the presence of antimicrobial drugs, metals, pH, and chemical stressors may influence ARG selection and horizontal gene transfer. The role of mobile genetic elements such as plasmids and transposons in ARG transfer and dissemination is an important aspect to consider. Sources, occurrence, dissemination, and the fate of antibiotics, ARB, and ARGs in groundwater environments are crucial to understand to minimize resistance transfer and cease their spread in the environment. This chapter reviews in detail about (i) occurrence of different types of antibiotics, ARB,

Table 2.1 Mapping of antibiotics used for animals and those prescribed for human consumption

Drugs classification	*Drugs used in livestock/poultry/aqua culture*	*Cross-resistance*
Sulfonamides (class)	**Sulfonamides (Class)**	All sulfonamides
	Aqua culture – sulfamerazine, sulfadimethoxine	
	Swine – sulfamerazine, sulfadimethoxine as antimicrobial growth promoter	
Lincosamide (class)	**Lincosamide (class)**	
	In chickens and swine used as AGP	Erythromycin
	Therapy for swine dysentery	Other macrolides
	Pneumonia	Lincosamide
	Chicken necrotic enteritis	Clindamycin
	Respiratory disease	

ARGs, and the type of drug resistance; (ii) their sources like urban wastewater, activated sludge, animal wastes and manure, landfills, septic and sewer systems; (iii) the fate and transport of AMR in the subsurface by means of sorption, transformation, degradation, and their mechanisms; and (iv) future perspectives in this field.

2.2 OCCURRENCE OF ANTIBIOTICS, ARB, AND ARGs

Tables 2.2–2.4 represent the occurrence of AB, ARB, and ARG in groundwater around the world, respectively, from the recent studies (2015–2020). With the advanced analytic techniques, the trace-level (ng/L) detection of emerging pollutants has been made possible. The occurrence data specific to the country is significant to make an

Table 2.2 Occurrence of antibiotic residue in groundwater

Location	*Antibiotic*	*Class*	*Concentration (ng/L)*	*References*
Lower Saxony, Germany	Trimethoprim		12	Burke et al. (2016)
Baix Fluvià Aquifer, Spain	Ciprofloxacin	Fluoroquinolones	298	Boy-Roura et al. (2018)
	Enrofloxacin		63	
	Norfloxacin		48	
	Danofloxacin		68	
	Ofloxacin		18.2	
	Orbifloxacin		5.2	
	Azithromycin	Macrolide	13	
	Flumequine	Quinolones	8.9	
	Oxolinic acid		2.3	
	Pipemidic acid		29	
	Sulfamethoxazole	Sulfonamide	29	
Jianghan Plain, Central China	Erythromycin	Macrolide	28.9	Yao et al. (2017)
	Chlortetracycline	Tetracycline	24	
	Tetracycline	Fluoroquinolones	25.2	
	Ciprofloxacin		26.3	
	Norfloxacin		96.9	
Cluj-Napoca, Romania	Cefepime	Cephalosporins	917	Szekeres et al. (2018)
	Piperacillin		517	
	Trimethoprim		730	
Beijing and Changzhou, China	Ofloxacin	Fluoroquinolones	36.2	Chen et al. (2016)
	Enrofloxacin		70.9	
	Norfloxacin		96.8	
Netherlands	Sulfamethoxazole	Sulfonamides	18	Kivits et al. (2018)
	Sulfamethazine		12.58	
	Lincomycin	Macrolide	2.5	
Hubei Province, China	Sulfaquinoxaline	Sulfonamides	26.2	Tong et al. (2020)
	Sulfamethoxazole		30.4	
	Ciprofloxacin	Fluoroquinolones	28.2	
	Enrofloxacin		9.9	
	Lomefloxacin		15.6	
	Norfloxacin		31	
	Tetracycline	Tetracycline	170.6	
	Doxycycline		10.8	
	Oxytetracycline		11.1	
	Chlortetracycline		28.9	
	Spiramycin	Macrolide	18.2	

Table 2.3 Occurrence of antibiotic-resistant bacteria in groundwater

Location	*ARB*	*Resistance*	*Type of resistance*	*References*
North West Province, South Africa	*Enterococcus faecalis, Enterococcus faecium*	Vancomycin	MDR	Matlou et al. (2019)
Ireland	*Escherichia coli*	Ampicillin, Ticarcillin, Cefpodoxime, Ciprofloxacin, Norfloxacin, Nitrofurantoin, Trimethoprim	MDR	O'Dwyer et al. (2017)
Adelaide, South Australia	Enterobacteriaceae, Pseudomonadaceae, Bacillaceae, Rhodobacteraceae	-	β-Lactam	Sidhu et al. (2020)
Eastern Ontario, Canada	*Campylobacter* spp.	Tetracycline	-	Frey et al. (2015)
Central Valley of California, USA	*Escherichia coli*	Zithromycin, Chloramphenicol, Trimethoprim, Sulfamethoxazole, Tetracycline	MDR	Li et al. (2015)
	Enterococcus	Tigecycline, Linezolid, Chloramphenicol, Erythromycin, Ciprofloxacin, Tetracycline		

* MDR-Multiple Drug Resistance.

important decision to frame monitoring plan, priority pollutants list, and regulation work in each region.

2.3 SOURCES OF AB, ARB, AND ARG IN GROUNDWATER ENVIRONMENTS

Anthropogenic activities are the major sources of AB pollution in groundwater. ARB and ARG are inherent to microbial community. However, excessive use of AB selects ARB and induces AMR in bacteria and invariably increases the proportion of ARB and ARG in groundwater (Sasikaladevi et al., 2020). Figure 2.2 represents different sources and pathways of AB, ARB, and ARG in groundwater.

2.3.1 Wastewater-based dissemination

AB, ARB, and ARG reach wastewater treatment plants through sewage and waste generated from domestic, medical, and industrial uses. Biological treatment processes are not efficient in removing low concentration AB due to their recalcitrant

Table 2.4 Occurrence of antibiotic resistance genes in groundwater

Location	*ARG*	*Class*	*Gene copy*	*Reference*
Cluj-Napoca, Romania	*aac*C2	Aminoglycoside	-	Szekeres et al. (2018)
	bla_{SHV}	β-Lactam	2.30×10^{-1} copies/16S rRNA	
	*flo*R	Chloramphenicol	-	
	*erm*A, *mef*A	MLSB*	-	
	*sul*I,	Sulfonamide	1.58×10^{-1} copies/16S rRNA	
	*sul*II		-	
	*tet*A,	Tetracycline	-	
	*tet*C,		-	
	*tet*O,		6.79×10^{-2} copies/16S rRNA	
	*tet*W		-	
Estarreja, Portugal	*BlaTEM*	β-Lactams	-	Araújo et al. (2017)
	tetA, *tetB*	Tetracycline	-	
	sul1, *sul2*, *dfrA1*	Sulfonamide	-	
	strA, *strB*, *aadA1*, *aadA2*	Aminoglycoside	-	
Eastern Ontario, Canada	*tetO*	Tetracycline	1.68×10^{1} gene copies /100 mL	Szekeres et al. (2018)
South Germany	*blaVIM-1*	β-Lactams	-	Alexander et al. (2015)
	vanA	Vancomycin	-	
	ampC	Ampicillin	-	
	ermB	Erythromycin	-	
Shenzhen, China	*sul*1, *sul*2	Sulfonamide	1.23×10^{5} to 8.89×10^{6} copies/mL	Wu et al. (2020)
	*aad*A1	Aminoglycoside		
	ereA	MLSB		
	*cml*A	Chloramphenicols		
	*tet*C, *tet*O	Tetracyclines		
California, USA	oxa-1	β-Lactams	1.7×10^{4} and 5.7×10^{4} copies/L	Harb et al. (2019)
Hubei Province, China	*erm*C	Macrolide	-	Tong et al. (2020)
	*tet*Q	Tetracycline	-	
Jhizhon Village, China	*sul*1, *sul*2	Sulfonamide	5.27×10^{7} copies/mL	Gao et al. (2020)
	tetC, *tetG*,	Tetracycline		
	floR,	Florfenicol		
	cmlA	Chloramphenicol		
	aadA	Aminoglycoside		

* MLSB- macrolide-lincosamide-streptogramin B.

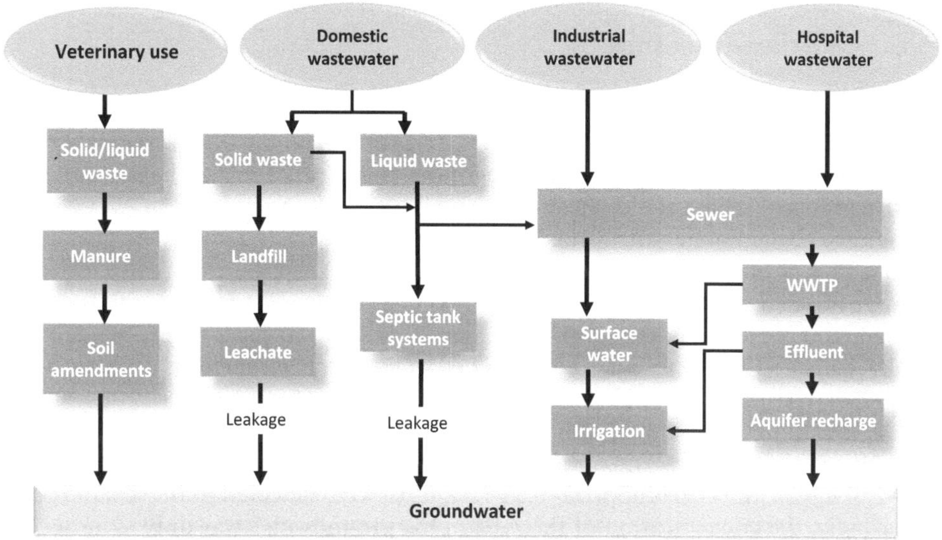

Figure 2.2 Sources and pathways of antibiotics, antibiotic-resistant bacteria, and antibiotic resistance genes in groundwater

nature and the removal efficiency varies with type of AB. To exemplify, Al Qarni et al. (2016) observed the removal of sulfamethoxazole (SMX) and ciprofloxacin between 95% and 99% in hospital wastewater treatment plant. However, only 86% removal of carbamazepine and clarithromycin was reported in conventional activated sludge process (Al Qarni et al., 2016). Class and nature of AB as well as type of biological process is known to influence the removal of AB by the biological system. On the other hand, biological treatment processes are not effective in removing ARB and ARGs.

If the secondary TWW is subjected to disinfection, the TWW might be free of ARB. Chlorination, ultraviolet radiation, and ozonation are the most widely employed disinfection processes. However, re-growth, reversal, and resistance toward disinfectant is a potential problem, which compromises the quality of the treated effluent (Huang et al., 2011; Yuan et al., 2015). Moreover, ARGs are not completely removed and any DNA damage caused could be reversed after disinfection (Yuan et al., 2015). It is important to note that the effluents of wastewater treatment plants are rich in ARGs (Rodriguez-Mozaz et al., 2015). Hence, TWW serves as a reservoir of AB, ARB, and ARGs at varying concentrations depending on the extent of treatment and treatment process. Wastewater reuse has become important due to increasing sustainable wastewater management practices across the globe especially in arid region and other water-scarce region (Fahrenfeld et al., 2013). Therefore, it becomes important to understand the role of TWW in AMR spread in groundwater.

2.3.1.1 Reclaimed wastewater reuse

2.3.1.1.1 IRRIGATION/AGRICULTURE USE

In countries like Israel, TWW comprises 50% of the water used for agriculture and irrigation (Jueschke et al., 2008). Continuous irrigation of land with TWW could result in the accumulation of AB in the soil (Negreanu et al., 2012). AB in the soil can exert selective pressure, which could select ARB and increase the abundance of ARGs in soil microbiome (Negreanu et al., 2012). Moreover, AB in the soil could enter groundwater via a variety of routes, including, but not limited to sorption to soil and hydraulic exchange in the hyporheic zone (Lewandowski et al., 2011; Navon et al., 2011). Avisar et al. (2009) reported antibiotic SMX concentration up to 37 ng/L in the groundwater samples collected from phreatic aquifer. The aquifer was situated in an area where the land was irrigated with TWW for over five decades (Avisar et al., 2009). Kibuye et al. (2019) studied the fate of AB and other pharmaceuticals during spray irrigation of TWW for over a year. Ampicillin (0.20–1.22 μg/L), ofloxacin (1.5–7.1 μg/L), SMX (0.05–27.4 μg/L), and trimethoprim (0.05-1.51 μg/L) were detected in the groundwater. Surprisingly, the concentration of the ABs in the groundwater was only ~2 order less than the concentration in the irrigated effluent (Kibuye et al., 2019).

Since the secondary TWW is normally disinfected, one could assume that the TWW is free of live ARB and other microorganisms. However, Katz and Griffin (2008) demonstrated the contamination of groundwater with bacteria due to land application of TWW. Elevated levels of fecal and enterococci were reported in the groundwater samples (Katz and Griffin, 2008). ARGs in the TWW are not completely eliminated. In a simulated laboratory study, Fahrenfeld et al. (2013) periodically irrigated the soil with TWW and observed a significant increase of *sul*1 and *sul*2 genes in the soil. However, no increase in *tet*(O) or *tet*(W) was observed (Fahrenfeld et al., 2013). These ARGs could be horizontally transferred to soil microbiome resulting in increased ARB abundance in the soil (Negreanu et al., 2012). Elevated levels of fecal and total coliforms in the groundwater have been recorded in the past due to the application of wastewater for irrigation (Gallegos et al., 1999). This confirms that the AB and ARB could reach groundwater by subsurface transport. The level of contamination, on the other hand, may vary due to both spatial and temporal factors (Gallegos et al., 1999).

2.3.1.1.2 AQUIFER/POND RECHARGE

Recharge with TWW mitigates groundwater depletion and is a means to sustainable and integrated water management practices. However, studies on the introduction of AB, ARB, and ARGs to aquifer by TWW recharges are limited. Rozman et al. (2017) studied the transport of pharmaceuticals from recharge ponds to groundwater. It was reported that the concentration of the pharmaceutical compounds gradually decreased along the soil depth. No compounds were detected in the groundwater samples since most of the compounds were trapped in the sediment and unsaturated zone. However, this was not the case with all the pharmaceuticals (Rozman et al., 2017). Böckelmann et al. (2009) observed *tetO* and *ermB* genes reclaimed wastewater used for aquifer recharge (Böckelmann et al., 2009). Harb et al. (2019) reported 10^4–10^6 copies/L of *sul1*, *sul2*, and *oxa-1* conferring resistance to sulfonamides and *β*-lactams in groundwater samples recharged with effluents from advanced wastewater treatment facility (Harb et al., 2019). Aquifer recharge with storm water is also reported to introduce ARB and ARGs in the subsurface. Sidhu et al. (2020) studied the prevalence of ARGs in aquifer recharged with storm water. Several families of pathogenic ARB resistant to multiple drugs were reported in the samples (Sidhu et al., 2020). However, the transport of AB, ARB, and ARG to groundwater during aquifer recharge with TWW is rather poorly understood and not investigated.

2.3.1.2 Contamination in urban river basin

TWW and untreated wastewater is directly or indirectly discharged into surface water bodies. Rodriguez-Mozaz et al. (2015) observed that the samples collected from downstream to discharge point of treated urban and domestic wastewater supposedly had high concentration of AB and ARG than in the upstream to the discharge point (Rodriguez-Mozaz et al., 2015). Harb et al. (2019) pointed out that the groundwater quality is greatly influenced by river in the proximity to groundwater that receives effluents from WWTP. This is predominant in areas where groundwater recharge is due to percolation in the river basin (Harb et al., 2019). Kruć et al. (2019) also observed the infiltration of SMX from Warta River to aquifer. A maximum of 20 ng/L of SMX was detected in a well located at 82 m from the Warta River, while no SMX was detected in a well located at 250 m (Kruć et al., 2019). Bradley et al. (2014) observed a hydrologic connection between surface water and shallow groundwater. The changes in hydrology conditions were driven by effluent discharge in the surface water, which resulted in changes in groundwater level with seasons. Both lateral and horizontal exchanges were recorded between surface water and groundwater. Moreover, the concentration of antibiotic SMX in the discharged effluent, surface water, and groundwater showed correlation. Besides, SMX concentration was greater than 0.02 μg/L in the groundwater sample collected from distance up to 6 m from the riverbank (Bradley et al., 2014).

ARB could be inherently present in groundwater environment. However, infiltration could increase ARB prevalence in groundwater. Moreover, resistance gain in microbial community due to AB exposure in water environment is a well-recognized scenario. This is further confirmed in a 30-day simulated study by Haack et al. (2012). Change in resistance and bacteria community structure in the aquifer due to SMX exposure at doses less than clinical concentration was observed (Haack et al., 2012). Seasonal and temporal variations affect the contamination level in groundwater. Severe rain promotes sewage contamination of groundwater. Gowrisankar et al. (2017) observed an increase in groundwater contamination in the vicinity of Adyar River during flood. Further, characterization of the isolated bacteria revealed the presence of ARB pathogens. Moreover, isolated pathogens showed resistance to ceftriaxone, doxycycline, and nalidixic acid (Gowrisankar et al., 2017). Further, Harb et al. (2019) reported ubiquitous presence of ARGs *sul1*, *sul2*, and *oxa-1* up to 10^6 copies/L in the groundwater samples collected from percolation basin (Harb et al., 2019). These genes could be horizontally shared among microbial community resulting in resistance spread (Böckelmann et al., 2009).

2.3.1.3 *Sewage/drainage systems*

Sewage exfiltration into groundwater is a common phenomenon in urban settlements due to weep holes and defective sewer networks. However, the extent of groundwater contamination depends on spatial, temporal variations, wastewater quality, groundwater hydrology, and soil geology (Reynolds and Barrett, 2003). In developing countries, sewage collection systems are decentralized, incomplete, or damaged (Sui et al., 2015). This often results in leakage and direct disposal into nearby areas contaminating shallow groundwater and surface water (Ly and Chui, 2012). The sewer systems carrying domestic, hospital, and industrial wastewater are a rich source of AB, ARB, and ARGs in groundwater. Reynolds and Barrett (2003) studied the exfiltration of sewage from leaky sewers into shallow aquifers, resulting in microbial contamination of the aquifer. It was concluded that urban aquifers are more prone to microbial contamination due to leaky sewer networks. It was also observed that the level of contamination varies with season and rainwater increases the exfiltration rate (Reynolds and Barrett, 2003). A study by Powell et al. (2003) confirmed the vulnerability of deep aquifer to microbial contamination due to leaky sewers (Powell et al., 2003). Several research groups have confirmed groundwater contamination with pharmaceuticals and chemicals due to sewer leakage. In a specific study conducted in Tokyo, both unconfined (shallow, <30m) and confined (deep, up to 500m) aquifers were evaluated for pharmaceutical and microbial contaminations due to sewer leakage. The study confirmed the vulnerability of unconfined aquifer and springs and confined aquifer to both pharmaceutical and microbial contaminations (Kuroda et al., 2012). However, an exclusive study on groundwater contamination with AB, ARB, and ARG is not available to confirm the same. Nevertheless, available studies are enough to speculate direct groundwater contamination with AMR components due to leaky sewers and drainage systems.

2.3.2 Sludge/solid waste-based dissemination

2.3.2.1 *Sludge from wastewater treatment plant*

Sludge/biosolids generated in WWTP have very diverse bacteria. Sludge generated is rich in organic matter and minerals essential for plant growth. Hence, both processed and unprocessed biosolids are widely used as soil amendments (Rahube et al., 2014). AB, ARB, and ARGs are widely reported in activated sludge/biosolids generated in the WWTP. Gao et al. (2012) reported tetracycline and sulfonamide antibiotics of 1,150.0 and 76.0 µg/kg dry weight of sludge, respectively. Besides, ARB and ARGs were also reported to be 7.08×10^{6}–1.91×10^{8} CFU/100mL and 1.00×10^{8}–1.78×10^{9} gene abundance/100mL, respectively, in the treated sludge. Application of class B solids and untreated sludge could increase the abundance of AMR components in soil. Rahube et al. (2014) studied the impact of application of raw and anaerobically digested sewage sludge on the prevalence of ARB and ARG in soil. Antibiotic-resistant coliform bacteria (ARCB) resistant to nitrofurantoin, tetracycline, chloramphenicol, ceftiofur, cefotaxime, SMX, co-trimoxazole, and trimethoprim were detected less frequently in amended soil. Frequent detection of ARCB resistant to ampicillin, amikacin, gentamicin, cefoxitin, and chlortetracycline was observed in amended soil. ARGs, namely, *inc*W, *int3*, *erm*(C), bla_{OXA-1}, *fos*(A2), and *van*(C2/C3), were exclusively detected in amended soil (Rahube et al., 2014). Soil amended with raw sludge was reported to increase ARG abundance. ARGs such as *intI1*, *aadA*, *strA*, and *tet*(W) were persistent and detectable up to 4 months. On the other hand, ARGs such as *aacC*, *aadE*, *ermB*, *ermF*, *tet*(L), *tet*(M), and *tet*(X) were present abundantly after amendment (Wolters et al., 2019). Digested sludge is rich in ARB and ARG. Hence, soil amendment with digested sludge will alter soil microbial community and ARG prevalence. Long-term application of sewage sludge to soil was reported to impact soil antibiotic resistome. Genes such as *blaCTX-M-02*, *ampC-09*, *aacC*, *emrD*, and *aac(6')-Ib(akaaacA4)-02* were reportedly increased by 55.0-fold, 38.0-fold, 40.7-fold, 99.2-fold, and 64.1-fold, respectively, in amended soil (Xie et al., 2016).

2.3.2.2 Manure and animal waste

Manures are usually solid waste generated by animals commonly applied to the agricultural land to increase the nutrient value of the soil as a soil amendment. The expansion of the large-scale animal breeding industry has increased the use of antibiotics for therapeutic, subtherapeutic, and nontherapeutic purposes in the animal food industry (Sasikaladevi et al., 2020). Fluoroquinolones, sulfonamides, and tetracycline have been widely identified in poultry, swine, and cattle manure. The soil amended with livestock manure is found to have different antibiotics with the concentration ranging widely from 0.1 to 8,400 µg/Kg (Zhang et al., 2016). The efficiency of current treatment methods is not enough to eliminate the dissemination of antibiotic resistance. Anaerobic digestion and aerobic lagoons show comparatively high removal of antibiotics in manure than conventional composting practice. The antibiotic with high solubility leach deep into the soil eventually reaches groundwater. Spielmeyer et al. (2017) conducted a long-term lysimeter experiment to reveal the permanent transfer of the once-applied sulfonamides into leachate (Spielmeyer et al., 2017). The application of manure to cropland around the poultry barn showed the effect on nearby streams with cephalosporin-resistant *E. coli* (Amato et al., 2020). Li et al. (2018) reported the contamination of groundwater near the swine feedlot with AB and ARGs due to the penetration of wastewater from swine feedlots. The most common antibiotics found in the groundwater samples are tetracycline, fluoroquinolones, and sulfonamides with an average concentration of 19.9, 11.8, and 0.3 µg/L, respectively (Li et al., 2018). The ARGs in groundwater near feedlot had *qnr*A, *sulI*, *sulII*, *tetG*, *tetM*, and *tetO* with a concentration of 10^2–10^3 gene copies. Li et al. (2014) explained the chances of contamination of groundwater proximal to the diary operation and reported the prevalent antibiotic resistance in *E. coli* (Li et al., 2014).

2.3.2.3 Septic and on-site wastewater treatment systems

Septic systems and drain fields do not guarantee complete removal and degradation of emerging contaminants such as an AB (Godfrey et al., 2007). Subsurface wastewater treatment systems are one of the direct sources of contamination in the groundwater. Demographic factors greatly influence the level of contamination in the aquifers.

On-site and decentralized septic systems can still contaminate the groundwater with pharmaceutical compounds even though these are usually located downstream to the drinking water wells. On-site effluent from septic systems has been widely reported to impact the water quality of aquifers. Godfrey et al. (2007) assessed the presence of AB in two shallow-unconfined aquifers due to percolation. Despite the fact that the concentrations of carbamazepine and SMX were reduced to 1.2–5 and 15–1,200 times, respectively, measurable quantities of SMX and trimethoprim were still reported in the aquifers. Moreover, SMX and trimethoprim in the aquifer reportedly correlated with local usage and prescription volume (Godfrey et al., 2007).

On-site wastewater treatment, subsurface wastewater treatment, and rapid infiltration basins are preferred over direct sewage discharge into streams (Elliott et al., 2018). Discharge of water from on-site wastewater treatment supposedly contaminates aquifers downstream to the discharge site. Elliott et al. (2018) studied and characterized the occurrence of AB and other micropollutants in groundwater samples downstream to seven on-site treatment systems, five community large subsurface treatment systems, and two municipal rapid infiltration basins in the United States. SMX was detected in all the groundwater samples in the range of 7–965 ng/L (Elliott et al., 2018).

Septic tanks are one of the hotspots of ARB. Further, on-site treatment systems could introduce ARB into aquifer. Scandura and Sobsey (1997) reported the contamination of groundwater with fecal coliforms due to on-site septic tank-soil absorption systems (Scandura and Sobsey, 1997). The presence of fecal coliform in the groundwater could indicate the contamination of aquifers with potential pathogens and antibiotic-resistant coliforms. Similarly, ARGs are abundant in wastewater. Hence, it is safe to assume the presence of ARGs in septic system, which could promote resistance gain and transfer to groundwater.

2.3.2.4 Landfill

The groundwater near landfill is the undervalued reservoir for antibiotic resistance and dissemination happens due to the flow of groundwater (Chen et al., 2017a,b). The common practice for the disposal of municipal solid waste is landfill. Solid waste that contains unused or expired medication, sewage sludge, and illegal biomedical waste that has antibiotics residue ends up in landfills that may get adsorbed and metabolized by microorganisms; at the same time, the anaerobic condition in landfills slows down the degradation so the majority of pollutants get dissolved into leachate due to its acidic pH. The release of antibiotics from landfill depends on the age of the landfill also. The level of antibiotics in the leachate from the aged landfill is considerably low when compared to that of younger landfills, which may be due to the degradation and usage pattern. The pollutants leached from refuse include AB, heavy metals, ARGs, and a variety of chemicals. Wang et al. (2019) highlighted the extensive distribution of antibiotics, heavy metals, and ARGs in the landfill system (Wang et al., 2019). The antibiotic concentration increases ARGs because exposure to antibiotics causes selective pressure for ARGs. Similarly, heavy metals occurrence was found to increase the abundance of *sul* gene (Yu et al., 2016). The average concentration of oxytetracycline, tetracycline, and SMX in the landfill refuse was found to be 100.9 ± 141.81, 63.8 ± 37.7, and 47.9 ± 8.1 μg/kg, respectively, along with ARGs *sul*I and *tet*O with an abundance of 10^2–$10^7$and 10^4–10^5copies/g (Song et al., 2016). The condition prevailing in the landfill also affects the susceptibility of bacteria to antibiotics in the leachate. The percentage of *E.* coli isolates resistant to antibiotics in anaerobic landfill leachate is higher than that of semi-aerobic landfill leachate due to the development of tolerance to antibiotics at low oxygen conditions (Threedeach et al., 2012). Leakage of leachate into subsurface environment poses a serious threat to both ecosystem and public health. Sulfonamides of concentration of 10 μg/L–1 mg/L were found in the groundwater down gradient to the landfill sites (Sui et al., 2015). Chen et al. (2017a,b) reported the detection of 171 unique ARGs with the abundance range 2.5×10^9–1.27×10^{11}copies/L, which confers resistance to multidrug, β-lactams, and tetracycline (Chen et al., 2017a,b). The pollutant concentration significantly reduces with the distance from landfills (Peng et al., 2014). Chung et al. (2018) emphasized the need to include antibiotics in leachate quality monitoring and intervention strategies to treat landfill leachate (Chung et al., 2018).

2.4 FATE AND TRANSPORT OF AMR COMPONENTS IN SUBSURFACE WATER

AB, ARB, and ARG have been reported in groundwater. Fate and transport of antibiotics is well studied, whereas the fate and transport of ARB and ARG is poorly understood. Fate of AB, viz, sorption, leaching, and biodegradation mechanisms and fate of ARB and ARG, viz, degradation, horizontal gene transfer is discussed in detail.

2.4.1 Antibiotics

2.4.1.1 Sorption

Sorption is a significant process that controls the reach of AB to groundwater. The sorption of AB to soil depends on the physicochemical properties of AB, nature and texture of the soil, and hydrochemical properties like soil organic matter (SOM), pH, and ionic strength. Pan and Chu (2016) compared adsorption affinities of tetracycline, norfloxacin, erythromycin, chloramphenicol, and sulfamethazine and reported that tetracycline has strong adsorption behavior and sulfamethazine is highly mobile (Pan and Chu, 2016). Tetracyclines have multiple functional groups, whereas sulfonamides have only aniline and amide (Wang and Wang, 2015). The molecular structure, especially the structural group and the functional group of the AB, determines the predominant adsorption mechanisms, which include cation exchange, surface complexation, cation bridging mechanism, and hydrogen bonding. Sorption capacity is highly determined by the characteristics and composition of soil. Silty clay loam type of soil has higher adsorption capacity than loamy sand (Chen et al., 2016) due to the variation in the percentage of sand, silt, and clay. High sand content is not conducive to AB adsorption. Srinivasan et al. (2014) observed a remarkable adsorption affinity of tylosin to the soil, which is attributed to the presence of 44% of montmorillonite in the clay fraction. The presence of divalent and trivalent ions even at low concentrations influence the adsorption of AB (Srinivasan et al., 2014). In the case of monovalent ion, the adsorption is hindered due to the competition for adsorption sites. Chen et al. (2013) observed an increased mobility of ciprofloxacin in sand media with the presence of Ca and Cu in the aqueous phase (Chen et al., 2013). The presence of Cu^{2+} suppressed the adsorption of sulfamethazine and sulfadiazine on to the soil but enhanced the sorption of SMX due to the formation of Cu^{2+} complex with SMX (Liu et al., 2017). The stability of the AB complex formed depends on the type of metal cation (Chen et al., 2013). Humic acid (HA) contains a functional group with large negative ions such as a carboxyl group, a phenolic hydroxyl group, carbonyl group, an amino group, and a methoxy group, and the AB speciation to anion, zwitterions, or cations is pH dependent. Liu et al. (2017) investigated the effect of HA on the adsorption of sulfonamides and observed the hindrance due to the complex formation with HA at different pH. At acidic pH of 5.13, sulfonamides are cationic and zwitterionic and form a complex with HA, which is responsible for an increased adsorption, and at basic pH of 7.33, sulfonamides are anionic and become repulsive with HA, which ultimately hinders sorption to the soil (Liu et al., 2017). The distribution coefficient k_d is the sorptive exchange of AB between soil and water and describes the extent to which the pollutant moves to groundwater. K_d is correlated with K_{oc} and f_{oc} to understand the partitioning between water and organic carbon.

$$K_d = K_{oc} * f_{oc} \tag{2.1}$$

where K_{oc} is the soil carbon partition coefficient and f_{oc} is the fraction of organic carbon.

The species dissociation in groundwater is controlled pK_a of the AB and pH that exist. SOM is the active compound in soil, which is organic in nature that is composed mainly of carbon. The soil containing higher SOM may lead to a decrease in adsorption (Chen et al., 2017a,b). In contrast, a good affinity of avermectins to the soil with high SOM was reported due to an interaction between uncharged nonpolar compounds and SOM (Rath et al., 2019).

2.4.1.2 Leaching

Hydrochemical conditions of groundwater and hydrogeological condition influence the transport of AB to vadose zone, partially saturated zone, or aquifer. The horizontal movement of AB may affect the uncontaminated sites also. Chen et al. (2011) observed the effect of pH and ionic concentration on SMZ and CIP mobility in saturated porous media (Chen et al., 2011). The rainfall intensity and the frequency of rainfall affect the mobility and the concentration of the AB entering the aquifer. AB entering the partially saturated zone are retained temporally and then leach to the aquifer by gravity flow after consecutive precipitation. The concentration of AB in runoff is controlled by the time lapse between the introduction of AB to the field and the event of rainfall. Barrios et al. (2020) reported that longer intervals are associated with a low concentration of chlortetracycline, lincomycin, and tiamulin, resulting in a runoff (Barrios et al., 2020). Heavy rainfall could make the AB reach deeper in the subsurface environment. Various land application methods like broadcasting, incorporation, and injection show different results in terms of mobility of AB to reach groundwater along with leaching and runoff (Barrios et al., 2020; Stacey R. Joy et al., 2013). The accumulation of antibiotics may happen when the dissipation rates slow down compared to the rate of input.

2.4.1.3 *Biodegradation*

Biodegradation is recognized as the main route for the removal pathways, but it is difficult to quantify degradation from field data. Biodegradation is highly influenced by the soil microbiome, soil pH, temperature, and the presence of multiple AB. Biodegradation of AB in the subsurface environment is limited due to the prevalence of restricted aerobic conditions (anoxic and anaerobic), low microbial diversity and activity (Boy-Roura et al., 2018), and less initial concentration of the AB (Pan and Chu, 2016). Change in microbial community structure and composition to promote biodegradation could be achieved only at a higher concentration of pollutants. The degradation in surface soil is quite faster than groundwater due to the aerobic environment at topsoil and anaerobic condition in a deeper zone. Slow degradation in subsurface results increases the abundance and prevalence of AB in the groundwater. The temperature difference may change the biodegradation rate of AB. Change in thermodynamic conditions in groundwater may lead to a redox reaction, which includes electron transfer which ultimately metabolizes the organic pollutant (Banzhaf and Hebig, 2016). Rath et al. (2019) observed enhanced degradation chlortetracycline from 18.8% to 89.8% for a temperature rise from 5°C to 45°C (Rath et al., 2019). Coexistence of multiple pollutants in the subsurface may decrease the degradation of individual AB because certain AB may eliminate the microbiome responsible for the degradation of other specific antibiotics. Plants play a crucial role by uptaking the AB from the subsurface (Azanu et al., 2016; Tian et al., 2019) and the roots of certain plants release enzymes that may degrade the AB and greatly metabolize the antibiotics in the subsurface (Klement et al., 2020). Plants develop the complex enzymatic system for their survival in the contaminated environment, which transforms drugs into metabolites that are stored in vacuoles or cell walls (Bártíková et al., 2015).

2.4.2 Antibiotic-resistant bacteria and antibiotic resistance genes

Many researchers had reported the elevated levels of ARB and ARG existence in the soil and groundwater upon the application of livestock manure for agricultural purposes, discharge of partially treated/untreated municipal wastewater to river streams, and reuse of the reclaimed wastewater for recreational activities. Emergence and dissemination of ARB in the complex environmental matrix could occur through successive enrichment of resistant bacteria and horizontal gene transfer of their resistance genes to the other bacterial communities present in the environment. Prolonged exposure of microbial communities to AB exerts selective pressure to develop resistance toward the single/multiple AB (Guo et al., 2018). Studies on the application of swine manure/biosolids on farmland reveal the development of the resistance to AB such as tetracycline and sulfonamides by soil microorganisms (Fahrenfeld et al., 2014). When ABs are used in livestock animals, commensal gut bacteria of the animals develop resistance toward the AB due to selective pressure through various mechanisms (Chee-Sanford et al., 2009). The excreta of livestock animals act as the source of the occurrence of antibiotic resistance and dissemination. The storage and usage of livestock manure on agricultural land future vary the proportion of ARB. In the cases of reclaimed wastewater and biosolids application to land, the wastewater treatment plants act as the hotspot for the antibiotic resistance development.

Horizontal gene transfer is considered as one of the important mechanisms of dissemination since ARGs can be transferred from the donor ARB, bacteriophages, free DNA available in the environmental matrix and even from dead cells to the other live bacteria. Different mechanisms of horizontal gene transfer are described as follows: (i) conjugation process involves the transfer of DNA containing ARG from the donor cells to the recipient cells through a cell-to-cell contact phenomenon via sexual pilus; (ii) during the transformation process, free ARGs obtained from the dead ARBs are transferred, integrated, and functionalized in the recipient bacterial DNA; (iii) transduction process involves the transfer of ARGs from one bacterium to another through bacteriophages; and (iv) gene transfer agents (GTAs) are bacteriophage-like particles that possess random elements of the host bacterial genetic material rather than the bacteriophage DNA. GTA is transferred through cell lysis or transduced into the recipient bacteria (Bello-López et al., 2019; Von Wintersdorff et al., 2016). The detailed fate and transport of ARB/ARG in the livestock manure applied to land, and the mechanism of horizontal gene transfer, is represented in Figure 2.3a and b. Genetic mobile elements such as plasmids, transposons, and integrons play a crucial role in dissemination of antibiotic resistance. The occurrence of class 1 integron gene (*intI*1) and its correlation with the existence of antibiotic resistance are widely reported for various environmental matrices (Burch et al., 2014). Moreover, heavy metals are constituents of wastewater and part of the natural environment could impose selective pressure for AB and heavy metal resistance to bacteria. Since the mechanisms of selective pressure for heavy metals and antibiotic resistance are similar, many studies have reported the occurrence and correlation for co-selection of heavy metal and antibiotic resistance in the natural environment (Mazhar et al., 2021).

Persistence and transport of ARB in the soil matrix upon introduction is a concern of environmental and health risk factors. The presence of active AB residue in the soil could impact the natural soil bacterial communities (Duan et al., 2019). Amendment of ABs, along with the nutrient-rich manure, could fasten the adaptation of microbial communities to the new soil environment (Chee-Sanford et al., 2009). ABs may exert the selective pressure to acquire antibiotic resistance and speed up the propagation through horizontal gene transfer among the soil bacterial communities (Barrios et al., 2020). Persistence of ARB in the soil depends on the environmental conditions such as moisture, pH, temperature, organic content, adsorption/desorption, hydraulic condition, and competency to other microbial communities (Chee-Sanford et al., 2009). Depending on the subsurface environmental conditions, ARB/ARGs can be trapped in the soil matrix through adsorption or further transported to the confined aquifers leading to the contamination of groundwater. The propagation of ARB in the soil is rapid, with heavily amended manure. Additionally, the levels of ARB in the soil manured to that of non-manured control samples in 5 months were studied. It showed the occurrence of rebound effect among the microbial communities in a specific period (Sengeløv et al., 2003). Increase in ARG abundance was always not maintained as steep in the soil environment. Pig manure applied soil exhibited persistency in ARG (macrolide, lincosamide, and streptogramin B) abundance for <20 days (Zhou et al., 2010); however, the persistence of *sul1* in soil was observed for more than 2 months (Heuer and Smalla, 2007). Bacterial resistance to tetracycline, macrolides,

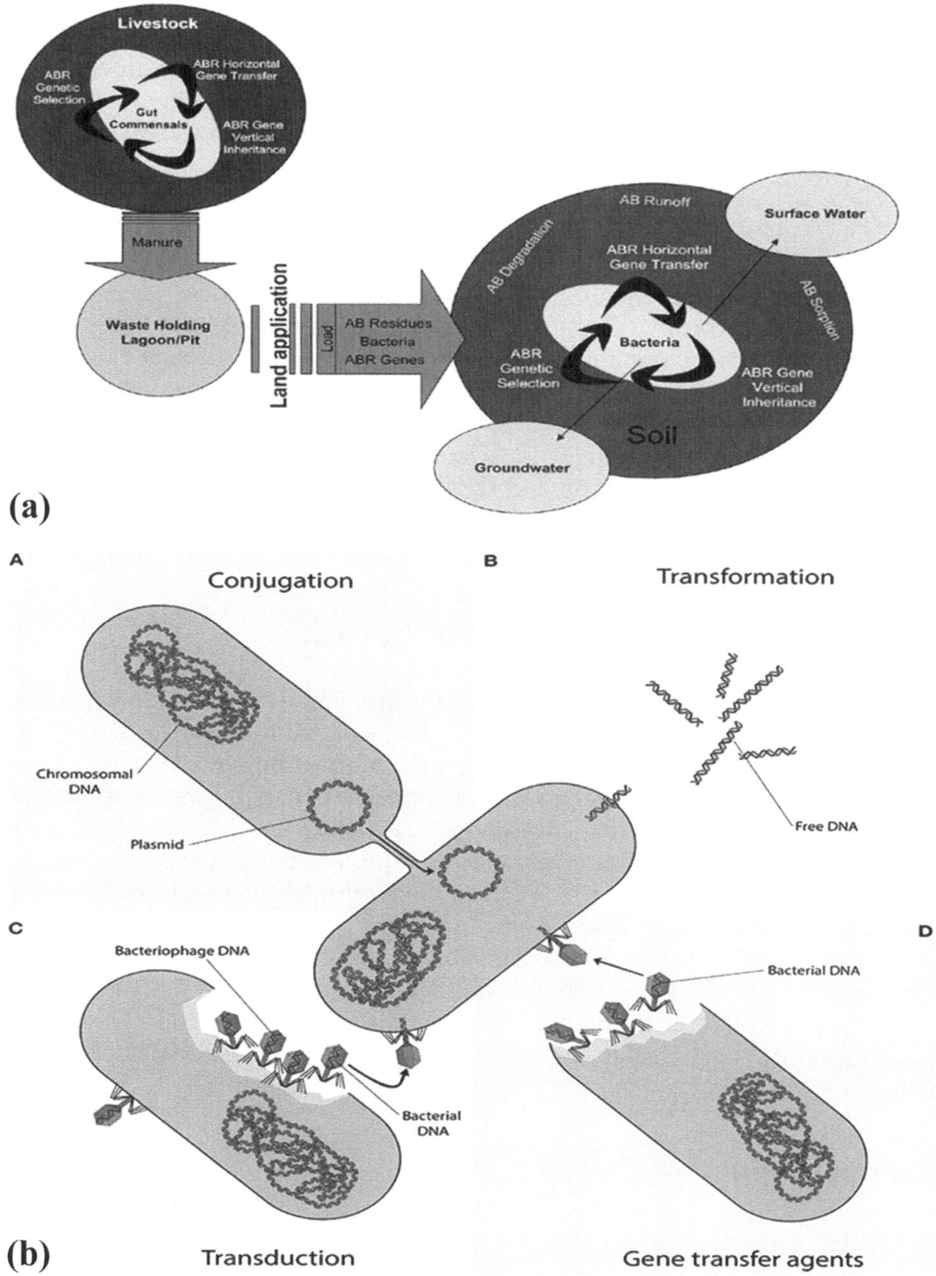

Figure 2.3 (a) Fate and transfer of antibiotics, antibiotic-resistant bacteria, and their resistant genes in the soil environment sourced through the application of livestock waste as manure (Reprinted with permission from Chee-Sanford et al., 2009), and (b) an illustration of mechanisms of horizontal gene transfer via conjugation, transformation, transduction, and gene transfer agents (Reprinted with permission from Von Wintersdorff et al. 2016).

and streptomycin AB was observed for 8 months in the farmland applied with pig manure (Sengeløv et al., 2003). Fate and transport of ARG and ARB in the soil matrix is a complex process. Antibiotic resistance in the soil matrix is not only driven by the presence of AB and the selection pressure, but it also depends on the growth/decay of the ARB/ARG and the tendency of horizontal gene transfer among the microbial communities (Fahrenfeld et al., 2014). The mass balance of ARG in the soil matrix can be estimated as per Equation 2.2.

$$ARG_{acc} = ARG_{load} + ARG_{growth} - ARG_{decay} \pm ARG_{run\ off} - ARG_{infitration} \quad (2.2)$$

where the overall accumulation of ARG (ARG_{acc}) in the soil matrix depends on the initial ARG loading rate (ARG_{load}), growth and expansion of ARGs (ARG_{growth}) due to the selection pressure and horizontal gene transfer between the bacterial communities, decay rate of ARG (ARG_{decay}), runoff of the ARGs within the control volume via surface inflow and outflow (ARG_{runoff}), and infiltration of ARGs in the soil matrix ($ARG_{infiltration}$) (Fahrenfeld et al., 2014). The dissipation rates of ARG in soil matrix depend on various factors, including (i) the transport of ARB and free extracellular ARGs, (ii) adsorption of ARB/ARG to the soil matrix, (iii), the decay of free ARGs, and (iv) death of the host bacteria (Fahrenfeld et al., 2014).

2.5 CONCLUSION AND FUTURE PERSPECTIVES

Studies on the presence of ARB and ARG in the groundwater are limited when compared to AB occurrence. Transport of AB, ARB, and AG to subsurface is not well understood, research is limited, and the available information is inadequate. This research gap must be filled for better understanding of the underlying threat. It is evident from the available literature that anthropogenic activities increase the prevalence of AMR components in groundwater. However, a better understanding of the transport and exchange process is lacking. Contamination of groundwater with AMR components could cause serious health problems and diseases when consumed. The presence of ARB in the groundwater could further worsen the situation resulting in treatment failure. Fate of AB in water is well documented, whereas the fate of ARB and ARG is not. Hence, the fate of ARB and ARG from different sources and its impact on groundwater must be studied to curtail AMR spread in groundwater. Regulation and monitoring of AMR spread in groundwater must also be implemented.

CONTRIBUTION

Sasikaladevi R and Kiruthika Eswari V contributed equally as first authors.

REFERENCES

Al Qarni, H., Collier, P., O'Keeffe, J., Akunna, J., 2016. Investigating the removal of some pharmaceutical compounds in hospital wastewater treatment plants operating in Saudi Arabia. *Environ Sci Pollut Res* 23, 13003–13014. https://doi.org/10.1007/s11356-016-6389-7.

Alexander, J., Bollmann, A., Seitz, W., Schwartz, T., 2015. Microbiological characterization of aquatic microbiomes targeting taxonomical marker genes and antibiotic resistance genes of opportunistic bacteria. *Science of the Total Environment* 512–513, 316–325. https://doi.org/10.1016/j.scitotenv.2015.01.046.

Amarasiri, M., Sano, D., Suzuki, S., 2020. Understanding human health risks caused by antibiotic resistant bacteria (ARB) and antibiotic resistance genes (ARG) in water environments: Current knowledge and questions to be answered. *Critical Reviews in Environmental Science and Technology* 50, 2016–2059. https://doi.org/10.1080/10643389.2019.1692611.

Amato, H.K., Wong, N.M., Pelc, C., Taylor, K., Price, L.B., Altabet, M., Jordan, T.E., Graham, J.P., 2020. Science of the Total Environment Effects of concentrated poultry operations and cropland manure application on antibiotic resistant Escherichia coli and nutrient pollution in Chesapeake Bay watersheds. *Science of the Total Environment* 735, 139401. https://doi.org/10.1016/j.scitotenv.2020.139401.

Araújo, S., A.T. Silva, I., Tacão, M., Patinha, C., Alves, A., Henriques, I., 2017. Characterization of antibiotic resistant and pathogenic Escherichia coli in irrigation water and vegetables in household farms. *International Journal of Food Microbiology* 257, 192–200. https://doi.org/10.1016/j.ijfoodmicro.2017.06.020.

Avisar, D., Lester, Y., Ronen, D., 2009. Sulfamethoxazole contamination of a deep phreatic aquifer. *Science of the Total Environment* 407, 4278–4282. https://doi.org/10.1016/j.scitotenv.2009.03.032.

Azanu, D., Mortey, C., Darko, G., Weisser, J.J., Styrishave, B., Abaidoo, R.C., 2016. Uptake of antibiotics from irrigation water by plants. *Chemosphere* 157, 107–114. https://doi.org/10.1016/j.chemosphere.2016.05.035.

Banzhaf, S., Hebig, K.H., 2016. Use of column experiments to investigate the fate of organic micropollutants - A review. *Hydrology and Earth System Sciences* 20, 3719–3737. https://doi.org/10.5194/hess-20-3719-2016.

Barrios, R.E., Khuntia, H.K., Bartelt-Hunt, S.L., Gilley, J.E., Schmidt, A.M., Snow, D.D., Li, X., 2020. Fate and transport of antibiotics and antibiotic resistance genes in runoff and soil as affected by the timing of swine manure slurry application. *Science of the Total Environment* 712. https://doi.org/10.1016/j.scitotenv.2020.136505.

Bártíková, H., Skálová, L., Stuchlíková, L., Vokřál, I., Vaněk, T., Podlipná, R., 2015. Xenobiotic-metabolizing enzymes in plants and their role in uptake and biotransformation of veterinary drugs in the environment. *Drug Metabolism Reviews* 47, 374–387. https://doi.org/10.3109/03602532.2015.1076437.

Bello-López, J.M., Cabrero-Martínez, O.A., Ibáñez-Cervantes, G., Hernández-Cortez, C., Pelcastre-Rodríguez, L.I., Gonzalez-Avila, L.U., Castro-Escarpulli, G., 2019. Horizontal gene transfer and its association with antibiotic resistance in the genus aeromonas spp. *Microorganisms* 7, 1–11. https://doi.org/10.3390/microorganisms7090363.

Böckelmann, U., Dörries, H.-H., Ayuso-Gabella, M.N., Salgot de Marçay, M., Tandoi, V., Levantesi, C., Masciopinto, C., Van Houtte, E., Szewzyk, U., Wintgens, T., Grohmann, E., 2009. Quantitative PCR monitoring of antibiotic resistance genes and bacterial pathogens in three European artificial groundwater recharge systems. *Applied and Environmental Microbiology* 75, 154–163. https://doi.org/10.1128/AEM.01649-08.

Boy-Roura, M., Mas-Pla, J., Petrovic, M., Gros, M., Soler, D., Brusi, D., Menció, A., 2018. Towards the understanding of antibiotic occurrence and transport in groundwater: Findings from the Baix Fluvià alluvial aquifer (NE Catalonia, Spain). *Science of the Total Environment* 612, 1387–1406. https://doi.org/10.1016/j.scitotenv.2017.09.012.

Bradley, P.M., Barber, L.B., Duris, J.W., Foreman, W.T., Furlong, E.T., Hubbard, L.E., Hutchinson, K.J., Keefe, S.H., Kolpin, D.W., 2014. Riverbank filtration potential of pharmaceuticals in a wastewater-impacted stream. *Environmental Pollution* 193, 173–180. https://doi.org/10.1016/j.envpol.2014.06.028.

Burch, T.R., Sadowsky, M.J., LaPara, T.M., 2014. Fate of antibiotic resistance genes and class 1 integrons in soil microcosms following the application of treated residual municipal wastewater solids. *Environmental Science and Technology* 48, 5620–5627. https://doi.org/10.1021/es501098g.

Burke, V., Richter, D., Greskowiak, J., Mehrtens, A., Schulz, L., Massmann, G., 2016. Occurrence of Antibiotics in Surface and Groundwater of a Drinking Water Catchment Area in Germany. *Water Environment Research* 88, 652–659. https://doi.org/10.2175/106143016x14609975746604.

Chee-Sanford, J.C., Mackie, R.I., Koike, S., Krapac, I.G., Lin, Y.-F., Yannarell, A.C., Maxwell, S., Aminov, R.I., 2009. Fate and transport of antibiotic residues and antibiotic resistance genes following land application of manure waste. *Journal of Environmental Quality* 38, 1086–1108. https://doi.org/10.2134/jeq2008.0128.

Chen, G., Liu, X., Tartakevosky, D., Li, M., 2016. Risk assessment of three fluoroquinolone antibiotics in the groundwater recharge system. *Ecotoxicology and Environmental Safety* 133, 18–24. https://doi.org/10.1016/j.ecoenv.2016.05.030.

Chen, H., Gao, B., Li, H., Ma, L.Q., 2011. Effects of pH and ionic strength on sulfamethoxazole and ciprofloxacin transport in saturated porous media. *Journal of Contaminant Hydrology* 126, 29–36. https://doi.org/10.1016/j.jconhyd.2011.06.002.

Chen, H., Ma, L.Q., Gao, B., Gu, C., 2013. Influence of Cu and Ca cations on ciprofloxacin transport in saturated porous media. *Journal of Hazardous Materials* 262, 805–811. https://doi.org/10.1016/j.jhazmat.2013.09.034.

Chen, K.L., Liu, L.C., Chen, W.R., 2017a. Adsorption of sulfamethoxazole and sulfapyridine antibiotics in high organic content soils. *Environmental Pollution* 231, 1163–1171. https://doi.org/10.1016/j.envpol.2017.08.011.

Chen, Q.L., Li, H., Zhou, X.Y., Zhao, Y., Su, J.Q., Zhang, X., Huang, F.Y., 2017b. An underappreciated hotspot of antibiotic resistance: The groundwater near the municipal solid waste landfill. *Science of the Total Environment* 609, 966–973. https://doi.org/10.1016/j.scitotenv.2017.07.164.

Chung, S.S., Zheng, J.S., Burket, S.R., Brooks, B.W., 2018. Select antibiotics in leachate from closed and active landfills exceed thresholds for antibiotic resistance development. *Environment International* 115, 89–96. https://doi.org/10.1016/j.envint.2018.03.014.

Duan, M., Gu, J., Wang, X., Li, Y., Zhang, R., Hu, T., Zhou, B., 2019. Factors that affect the occurrence and distribution of antibiotic resistance genes in soils from livestock and poultry farms. *Ecotoxicology and Environmental Safety* 180, 114–122. https://doi.org/10.1016/j.ecoenv.2019.05.005.

Elliott, S.M., Erickson, M.L., Krall, A.L., Adams, B.A., 2018. Concentrations of pharmaceuticals and other micropollutants in groundwater downgradient from large on-site wastewater discharges. *PLOS ONE* 13, e0206004. https://doi.org/10.1371/journal.pone.0206004.

Fahrenfeld, N., Knowlton, K., Krometis, L.A., Hession, W.C., Xia, K., Lipscomb, E., Libuit, K., Green, B.L., Pruden, A., 2014. Effect of manure application on abundance of antibiotic resistance genes and their attenuation rates in soil: Field-scale mass balance approach. *Environmental Science and Technology* 48, 2643–2650. https://doi.org/10.1021/es404988k.

Fahrenfeld, N., Ma, Y., O'Brien, M., Pruden, A., 2013. Reclaimed water as a reservoir of antibiotic resistance genes: distribution system and irrigation implications. *Frontiers in Microbiology* 4. https://doi.org/10.3389/fmicb.2013.00130.

Frey, S.K., Topp, E., Khan, I.U.H., Ball, B.R., Edwards, M., Gottschall, N., Sunohara, M., Lapen, D.R., 2015. Quantitative Campylobacter spp., antibiotic resistance genes, and veterinary antibiotics in surface and ground water following manure application: Influence of tile drainage control. *Science of the Total Environment* 532, 138–153. https://doi.org/10.1016/j.scitotenv.2015.03.114.

Gallegos, E., Warren, A., Robles, E., Campoy, E., Calderon, A., Sainz, Ma.G., Bonilla, P., Escolero, O., 1999. The effects of wastewater irrigation on groundwater quality in Mexico.

Water Science and Technology, Water Quality and its Management 40, 45–52. https://doi.org/10.1016/S0273-1223(99)00429-1.

Gao, F.Z., Zou, H.Y., Wu, D.L., Chen, S., He, L.Y., Zhang, M., Bai, H., Ying, G.G., 2020. Swine farming elevated the proliferation of Acinetobacter with the prevalence of antibiotic resistance genes in the groundwater. *Environment International* 136, 105484. https://doi.org/10.1016/j.envint.2020.105484.

Gao, P., Munir, M., Xagoraraki, I., 2012. Correlation of tetracycline and sulfonamide antibiotics with corresponding resistance genes and resistant bacteria in a conventional municipal wastewater treatment plant. *Science of the Total Environment* 421–422, 173–183. https://doi.org/10.1016/j.scitotenv.2012.01.061.

Godfrey, E., Woessner, W.W., Benotti, M.J., 2007. Pharmaceuticals in On-Site Sewage Effluent and Ground Water, Western Montana. *Groundwater* 45, 263–271. https://doi.org/10.1111/j.1745-6584.2006.00288.x.

Gowrisankar, G., Chelliah, R., Ramakrishnan, S.R., Elumalai, V., Dhanamadhavan, S., Brindha, K., Antony, U., Elango, L., 2017. Chemical, microbial and antibiotic susceptibility analyses of groundwater after a major flood event in Chennai. *Scientific Data* 4, 170135. https://doi.org/10.1038/sdata.2017.135.

Guo, T., Lou, C., Zhai, W., Tang, X., Hashmi, M.Z., Murtaza, R., Li, Y., Liu, X., Xu, J., 2018. Increased occurrence of heavy metals, antibiotics and resistance genes in surface soil after long-term application of manure. *Science of the Total Environment* 635, 995–1003. https://doi.org/10.1016/j.scitotenv.2018.04.194.

Haack, S.K., Metge, D.W., Fogarty, L.R., Meyer, M.T., Barber, L.B., Harvey, R.W., LeBlanc, D.R., Kolpin, D.W., 2012. Effects on groundwater microbial communities of an engineered 30-day in situ exposure to the antibiotic sulfamethoxazole. *Environmental Science & Technology* 46, 7478–7486. https://doi.org/10.1021/es3009776.

Harb, M., Wang, P., Zarei-Baygi, A., Plumlee, M.H., Smith, A.L., 2019. Background antibiotic resistance and microbial communities dominate effects of advanced purified water recharge to an urban aquifer. *Environmental Science & Technology Letters* 6, 578–584. https://doi.org/10.1021/acs.estlett.9b00521.

Heuer, H., Smalla, K., 2007. Manure and sulfadiazine synergistically increased bacterial antibiotic resistance in soil over at least two months. *Environmental Microbiology* 9, 657–666. https://doi.org/10.1111/j.1462-2920.2006.01185.x.

Huang, J.-J., Hu, H.-Y., Tang, F., Li, Y., Lu, S.-Q., Lu, Y., 2011. Inactivation and reactivation of antibiotic-resistant bacteria by chlorination in secondary effluents of a municipal wastewater treatment plant. *Water Research* 45, 2775–2781. https://doi.org/10.1016/j.watres.2011.02.026.

Joy, S.R., Bartelt-Hunt, S.L., Snow, D.D., Gilley, J.E., Woodbury, B.L., Parker, D.B., Marx, D.B., Li, X., 2013. Fate and transport of antimicrobials and antimicrobial resistance genes in soil and runoff following land application of swine manure slurry. *Environmental Science & Technology* 47, 12081–12088. https://doi.org/10.1021/es4026358.

Jueschke, E., Marschner, B., Tarchitzky, J., Chen, Y., 2008. Effects of treated wastewater irrigation on the dissolved and soil organic carbon in Israeli soils. *Water Science and Technology* 57, 727–733. https://doi.org/10.2166/wst.2008.173.

Katz, B.G., Griffin, D.W., 2008. Using chemical and microbiological indicators to track the impacts from the land application of treated municipal wastewater and other sources on groundwater quality in a karstic springs basin. *Environmental Geology* 55, 801–821. https://doi.org/10.1007/s00254-007-1033-y.

Kibuye, F.A., Gall, H.E., Elkin, K.R., Ayers, B., Veith, T.L., Miller, M., Jacob, S., Hayden, K.R., Watson, J.E., Elliott, H.A., 2019. Fate of pharmaceuticals in a spray-irrigation system: From wastewater to groundwater. *Science of the Total Environment* 654, 197–208. https://doi.org/10.1016/j.scitotenv.2018.10.442.

Kivits, T., Peter, H., Beeltje, H., Grif, J., 2018. Presence and fate of veterinary antibiotics in age-dated groundwater in areas with intensive livestock farming. *Environmental Pollution* 241, 988–998. https://doi.org/10.1016/j.envpol.2018.05.085.

Klement, A., Kodešová, R., Golovko, O., Fér, M., Nikodem, A., Kočárek, M., Grabic, R., 2020. Uptake, translocation and transformation of three pharmaceuticals in green pea plants. *Journal of Hydrology and Hydromechanics* 68, 1–11. https://doi.org/10.2478/johh-2020-0001.

Kruć, R., Dragon, K., Górski, J., 2019. Migration of pharmaceuticals from the warta river to the aquifer at a riverbank filtration site in Krajkowo (Poland). *Water* 11, 2238. https://doi.org/10.3390/w11112238.

Kuroda, K., Murakami, M., Oguma, K., Muramatsu, Y., Takada, H., Takizawa, S., 2012. Assessment of groundwater pollution in Tokyo using PPCPs as sewage markers. *Environmental Science & Technology* 46, 1455–1464. https://doi.org/10.1021/es202059g.

Lewandowski, J., Putschew, A., Schwesig, D., Neumann, C., Radke, M., 2011. Fate of organic micropollutants in the hyporheic zone of a eutrophic lowland stream: Results of a preliminary field study. *Science of the Total Environment* 409, 1824–1835. https://doi.org/10.1016/j.scitotenv.2011.01.028.

Li, X., Atwill, E.R., Antaki, E., Applegate, O., Bergamaschi, B., Bond, R.F., Chase, J., Ransom, K.M., Samuels, W., Watanabe, N., Harter, T., 2015. Fecal indicator and pathogenic bacteria and their antibiotic resistance in alluvial groundwater of an irrigated agricultural region with dairies. https://doi.org/10.2134/jeq2015.03.0139.

Li, X., Liu, C., Chen, Y., Huang, H., Ren, T., 2018. Antibiotic residues in liquid manure from swine feedlot and their effects on nearby groundwater in regions of North China. *Environmental Science and Pollution Research International* 25, 11565–11575.

Li, X., Watanabe, N., Xiao, C., Harter, T., Mccowan, B., Liu, Y., Atwill, E.R., 2014. Antibiotic-resistant *E. coli* in surface water and groundwater in dairy operations in Northern California. *Environmental Monitoring and Assessment* 186, 1253–1260. https://doi.org/10.1007/s10661-013-3454-2.

Liu, Z., Han, Y., Jing, M., Chen, J., 2017. Sorption and transport of sulfonamides in soils amended with wheat straw-derived biochar: Effects of water pH, coexistence copper ion, and dissolved organic matter. *Journal of Soils and Sediments* 17, 771–779. https://doi.org/10.1007/s11368-015-1319-8.

Ly, D.K., Chui, T.F.M., 2012. Modeling sewage leakage to surrounding groundwater and stormwater drains. *Water Science and Technology* 66, 2659–2665. https://doi.org/10.2166/wst.2012.496.

Matlou, D.P., Bissong, M.E.A.T., Tchatchouang, C.D.K., Adem, M.R., Foka, F.E.T., Kumar, A., Ateba, C.N., 2019. Virulence profiles of vancomycin-resistant enterococci isolated from surface and ground water utilized by humans in the North West Province, South Africa: a public health perspective. *Environmental Science and Pollution Research* 26, 15105–15114. https://doi.org/10.1007/s11356-019-04836-5.

Mazhar, S.H., Li, X., Rashid, A., Su, J.M., Xu, J., Brejnrod, A.D., Su, J.Q., Wu, Y., Zhu, Y.G., Zhou, S.G., Feng, R., Rensing, C., 2021. Co-selection of antibiotic resistance genes, and mobile genetic elements in the presence of heavy metals in poultry farm environments. *Science of the Total Environment* 755, 142702. https://doi.org/10.1016/j.scitotenv.2020.142702.

Navon, R., Hernandez-Ruiz, S., Chorover, J., Chefetz, B., 2011. Interactions of carbamazepine in soil: Effects of dissolved organic matter. *Journal of Environmental Quality* 40, 942–948. https://doi.org/10.2134/jeq2010.0446.

Negreanu, Y., Pasternak, Z., Jurkevitch, E., Cytryn, E., 2012. Impact of treated wastewater irrigation on antibiotic resistance in agricultural soils. *Environmental Science & Technology* 46, 4800–4808. https://doi.org/10.1021/es204665b.

O'Dwyer, J., Hynds, P., Pot, M., Adley, C.C., Ryan, M.P., 2017. Evaluation des niveaux d'antibiorésistance des isolats d'E.coli en provenance des eaux souterraines du Centre Ouest Irlandais et explication des indicateurs potentiels de résistance. *Hydrogeology Journal* 25, 939–951. https://doi.org/10.1007/s10040-017-1546-8.

Pan, M., Chu, L.M., 2017. Fate of antibiotics in soil and their uptake by edible crops. *Science of the Total Environment* 599–600, 500–512. https://doi.org/10.1016/j.scitotenv.2017.04.214.

Pan, M., Chu, L.M., 2016. Adsorption and degradation of five selected antibiotics in agricultural soil. *Science of the Total Environment* 545–546, 48–56. https://doi.org/10.1016/j.scitotenv.2015.12.040.

Peng, X., Ou, W., Wang, C., Wang, Z., Huang, Q., Jin, J., Tan, J., 2014. Occurrence and ecological potential of pharmaceuticals and personal care products in groundwater and reservoirs in the vicinity of municipal landfills in China. *Science of the Total Environment* 490, 889–898. https://doi.org/10.1016/j.scitotenv.2014.05.068.

Powell, K., Taylor, R., Barrett, M., Pedley, S., Sellwood, J., Trowsdale, S., Lerner, D., 2003. Microbial contamination of two urban sandstone aquifers in the UK. *Water Research* 37, 339–352. https://doi.org/10.1016/s0043-1354(02)00280-4.

Rahube, T.O., Marti, R., Scott, A., Tien, Y.-C., Murray, R., Sabourin, L., Zhang, Y., Duenk, P., Lapen, D.R., Topp, E., 2014. Impact of fertilizing with raw or anaerobically digested sewage sludge on the abundance of antibiotic-resistant coliforms, antibiotic resistance genes, and pathogenic bacteria in soil and on vegetables at harvest. *Applied and Environmental Microbiology* 80, 6898–6907. https://doi.org/10.1128/AEM.02389-14.

Rath, S., Fostier, A.H., Pereira, L.A., Dioniso, A.C., de Oliveira Ferreira, F., Doretto, K.M., Maniero Peruchi, L., Viera, A., de Oliveira Neto, O.F., Dal Bosco, S.M., Martínez-Mejía, M.J., 2019. Sorption behaviors of antimicrobial and antiparasitic veterinary drugs on subtropical soils. *Chemosphere* 214, 111–122. https://doi.org/10.1016/j.chemosphere.2018.09.083.

Reynolds, J.H., Barrett, M.H., 2003. A review of the effects of sewer leakage on groundwater quality. *Water and Environment Journal* 17, 34–39. https://doi.org/10.1111/j.1747-6593.2003.tb00428.x.

Rodriguez-Mozaz, S., Chamorro, S., Marti, E., Huerta, B., Gros, M., Sànchez-Melsió, A., Borrego, C.M., Barceló, D., Balcázar, J.L., 2015. Occurrence of antibiotics and antibiotic resistance genes in hospital and urban wastewaters and their impact on the receiving river. *Water Research* 69, 234–242. https://doi.org/10.1016/j.watres.2014.11.021.

Rozman, D., Hrkal, Z., Váňa, M., Vymazal, J., Boukalová, Z., 2017. Occurrence of pharmaceuticals in wastewater and their interaction with shallow aquifers: A case study of Horní Beřkovice, Czech Republic. *Water* 9, 218. https://doi.org/10.3390/w9030218.

Sasikaladevi, R., Eswari, V.K., Nambi, I.M., 2020. Antibiotic resistance and sanitation in India: Current situation and future perspectives. In: *The Handbook of Environmental Chemistry*. Springer, Berlin, pp. 1–28.

Scandura, J.E., Sobsey, M.D., 1997. Viral and bacterial contamination of groundwater from on-site sewage treatment systems. *Water Science and Technology, Health-Related Water Microbiology* 35, 141–146. https://doi.org/10.1016/S0273-1223(97)00249-7.

Sengeløv, G., Agersø, Y., Halling-Sørensen, B., Baloda, S.B., Andersen, J.S., Jensen, L.B., 2003. Bacterial antibiotic resistance levels in Danish farmland as a result of treatment with pig manure slurry. *Environment International* 28, 587–595. https://doi.org/10.1016/S0160-4120(02)00084-3.

Sidhu, J. P.S., Gupta, V.V.S.R., Stange, C., Ho, J., Harris, N., Barry, K., Gonzalez, D., Van Nostrand, J.D., Zhou, J., Page, D., Tiehm, A., Toze, S., 2020. Prevalence of antibiotic resistance and virulence genes in the biofilms from an aquifer recharged with stormwater. *Water Research* 185, 116269. https://doi.org/10.1016/j.watres.2020.116269.

Song, L., Li, L., Yang, S., Lan, J., He, H., McElmurry, S.P., Zhao, Y., 2016. Sulfamethoxazole, tetracycline and oxytetracycline and related antibiotic resistance genes in a large-scale landfill, China. *Science of the Total Environment* 551–552, 9–15. https://doi.org/10.1016/j.scitotenv.2016.02.007.

Spielmeyer, A., Höper, H., Hamscher, G., 2017. Long-term monitoring of sulfonamide leaching from manure amended soil into groundwater. *Chemosphere* 177, 232–238. https://doi.org/10.1016/j.chemosphere.2017.03.020.

Srinivasan, P., Sarmah, A.K., Manley-Harris, M., 2014. Sorption of selected veterinary antibiotics onto dairy farming soils of contrasting nature. *Science of the Total Environment* 472, 695–703. https://doi.org/10.1016/j.scitotenv.2013.11.104.

Sui, Q., Cao, X., Lu, S., Zhao, W., Qiu, Z., Yu, G., 2015. Occurrence, sources and fate of pharmaceuticals and personal care products in the groundwater: A review. *Emerging Contaminants* 1, 14–24. https://doi.org/10.1016/j.emcon.2015.07.001.

Szekeres, E., Chiriac, C.M., Baricz, A., Szőke-Nagy, T., Lung, I., Soran, M.L., Rudi, K., Dragos, N., Coman, C., 2018. Investigating antibiotics, antibiotic resistance genes, and microbial contaminants in groundwater in relation to the proximity of urban areas. *Environmental Pollution* 236, 734–744. https://doi.org/10.1016/j.envpol.2018.01.107.

Threedeach, S., Chiemchaisri, W., Watanabe, T., Chiemchaisri, C., Honda, R., Yamamoto, K., 2012. Antibiotic resistance of Escherichia coli in leachates from municipal solid waste landfills: Comparison between semi-aerobic and anaerobic operations. *Bioresource Technology* 113, 253–258. https://doi.org/10.1016/j.biortech.2012.01.086.

Tian, R., Zhang, R., Uddin, M., Qiao, X., Chen, J., Gu, G., 2019. Uptake and metabolism of clarithromycin and sulfadiazine in lettuce. *Environmental Pollution* 247, 1134–1142. https://doi.org/10.1016/j.envpol.2019.02.009.

Tong, L., Qin, L., Guan, C., Wilson, M.E., Li, X., Cheng, D., Ma, J., Liu, H., Gong, F., 2020. Antibiotic resistance gene profiling in response to antibiotic usage and environmental factors in the surface water and groundwater of Honghu Lake, China. *Environmental Science and Pollution Research* 27, 31995–32005. https://doi.org/10.1007/s11356-020-09487-5.

Von Wintersdorff, C.J.H., Penders, J., Van Niekerk, J.M., Mills, N.D., Majumder, S., Van Alphen, L.B., Savelkoul, P.H.M., Wolffs, P.F.G., 2016. Dissemination of antimicrobial resistance in microbial ecosystems through horizontal gene transfer. *Frontiers in Microbiology* 7, 1–10. https://doi.org/10.3389/fmicb.2016.00173.

Walia, K., Sharma, M., Vijay, S., Shome, B.R., 2019. Understanding policy dilemmas around antibiotic use in food animals & offering potential solutions. *Indian Journal of Medical Research* 149, 107–118. https://doi.org/10.4103/ijmr.IJMR_2_18.

Wang, P., Wu, D., You, X., Li, W., Xie, B., 2019. Distribution of antibiotics, metals and antibiotic resistance genes during landfilling process in major municipal solid waste landfills. *Environmental Pollution* 255, 113222. https://doi.org/10.1016/j.envpol.2019.113222.

Wang, S., Wang, H., 2015. Adsorption behavior of antibiotic in soil environment: A critical review. *Frontiers of Environmental Science and Engineering* 9, 565–574. https://doi.org/10.1007/s11783-015-0801-2.

WHO 2017. WHO | Global priority list of antibiotic-resistant bacteria to guide research, discovery, and development of new antibiotics. World Health Organisation.

Wolters, B., Fornefeld, E., Jechalke, S., Su, J.-Q., Zhu, Y.-G., Sørensen, S.J., Smalla, K., Jacquiod, S., 2019. Soil amendment with sewage sludge affects soil prokaryotic community composition, mobilome and resistome. *FEMS Microbiology Ecology* 95. https://doi.org/10.1093/femsec/fiy193.

Wu, D.L., Zhang, M., He, L.X., Zou, H.Y., Liu, Y.S., Li, B.B., Yang, Y.Y., Liu, C., He, L.Y., Ying, G.G., 2020. Contamination profile of antibiotic resistance genes in ground water in comparison with surface water. *Science of the Total Environment* 715. https://doi.org/10.1016/j.scitotenv.2020.136975.

Xie, W.-Y., McGrath, S.P., Su, J.-Q., Hirsch, P.R., Clark, I.M., Shen, Q., Zhu, Y.-G., Zhao, F.-J., 2016. Long-Term impact of field applications of sewage sludge on soil antibiotic resistome. *Environmental Science & Technology* 50, 12602–12611. https://doi.org/10.1021/acs.est.6b02138.

Yao, L., Wang, Y., Tong, L., Deng, Y., Li, Y., Gan, Y., Guo, W., Dong, C., Duan, Y., Zhao, K., 2017. Occurrence and risk assessment of antibiotics in surface water and groundwater from different depths of aquifers: A case study at Jianghan Plain, central China. *Ecotoxicology and Environmental Safety* 135, 236–242. https://doi.org/10.1016/j.ecoenv.2016.10.006.

You, X., Wu, D., Wei, H., Xie, B., Lu, J., 2018. Fluoroquinolones and B-lactam antibiotics and antibiotic resistance genes in autumn leachates of seven major municipal solid waste landfills in China. *Environment International* 113, 162–169. https://doi.org/10.1016/j.envint.2018.02.002.

Yu, Z., He, P., Shao, L., Zhang, H., Lü, F., 2016. Co-occurrence of mobile genetic elements and antibiotic resistance genes in municipal solid waste landfill leachates: A preliminary insight into the role of landfill age. *Water Research* 106, 583–592. https://doi.org/10.1016/j.watres.2016.10.042.

Yuan, Q.-B., Guo, M.-T., Yang, J., 2015. Fate of antibiotic resistant bacteria and genes during wastewater chlorination: implication for antibiotic resistance control. *PLoS ONE* 10, e0119403. https://doi.org/10.1371/journal.pone.0119403.

Zhang, H., Zhou, Y., Huang, Y., Wu, L., Liu, X., Luo, Y., 2016. Residues and risks of veterinary antibiotics in protected vegetable soils following application of different manures. *Chemosphere* 152, 229–237. https://doi.org/10.1016/j.chemosphere.2016.02.111.

Zhou, Z., Raskin, L., Zilles, J.L., 2010. Effects of swine manure on macrolide, lincosamide, and streptogramin B antimicrobial resistance in soils. *Applied and Environmental Microbiology* 76, 2218–2224. https://doi.org/10.1128/AEM.02183-09.

Chapter 3

Fate and transport of micro- and nanoplastics in graywater, fresh and marine water systems

G. Sruthi, S. Aishwarya, Samarshi Chakraborty, and S. Karthika
Vellore Institute of Technology

G. Divyapriya
Virginia Polytechnic Institute and State University

K. Sivagami
Vellore Institute of Technology

CONTENTS

DOI: 10.1201/9781003157465-4

3.1 INTRODUCTION

Due to the widespread application of plastics nowadays, microplastic and nanoplastics are omnipresent. These tiny particles of plastics are found from Pacific Ocean to Arctic Sea ice and from the farms to urban areas. Microplastics are often detected and accumulated as debris in river and marine systems (Buragohain et al., 2020; La Daana et al., 2018, Patel et al., 2020). The micro- and nanoplastics (MNPs) can easily be ingested by human and aquatic life causing severe health hazards. Fishes often consume these plastics, which then find its way into our body through the water we drink or the food that we eat. Based on the chemical composition of the polymer, associated health hazards can be determined, and polymers can be ranked accordingly. The density of the polymer would determine whether the plastic waste would remain floating or sink to the bottom of the river and ocean systems. Polymer density greater than 1 g/cm^3 (polyethylene terephthalate density is 1.32) implies that plastic debris will settle down at the bottom of the river/ocean considering water density to be 1 g/cm^3; when the density is lower than 1 g/cm^3 (polyethylene density is 0.91–0.94), plastic debris will remain in a suspended state (Paul-Pont et al., 2018). Therefore, polymer composition and its density provide researchers the key information on the environmental and health impacts of plastic pollution and help in determining the methods of plastic removal. Microplastics and nanoplastics are derived from various sources such as single-use plastics, fishing net, textiles, cosmetics, agricultural runoff, paints, tyres, urban dust, packaging etc. The MNPs sourced from household wastewater are transported to the wastewater treatment plant (WWTP) facilities. Inefficient separation of microplastic and nanoplastics from wastewater results in the transport of these particles into the river and ocean. The MNPs are hard to be captured from the riverine and ocean systems which further complicate the plastic pollution in the aquatic environment. This chapter discusses the overview of MNP sources, fate and transportation into the environment, municipal wastewater, river, and marine water systems. It also discusses the removal methodologies of MNP in WWTP and coarse plastic separation in river and marine systems.

3.2 MACRO-, MICRO-, AND NANOPLASTICS IN THE ENVIRONMENT – SOURCES AND EFFECTS

The microplastics are mainly classified into two categories, namely, primary microplastics and secondary microplastics (Germanov et al., 2018; Andrady, 2017). Primary microplastics are those plastics which are already less than 5.0mm in size before entering the environment. These can include microfibers from clothes, plastic pellets, and microbeads. The primary microplastics often come from the preparation of facial cleansers, cosmetics, and textiles. Secondary microplastics are produced by physical, chemical, biological, human activity, and photodegradation of larger plastic debris (Cai et al., 2018). Secondary microplastics are created by the degradation of larger plastics like water bottles and plastic bags in the environment through natural weathering processes. The biggest problem with plastics pollution is that the large plastic debris disintegrates with time into MNPs under the exposure of UV radiation and mechanical erosion. Based on size, plastic debris can be classified into several categories: (i) macroplastic (greater than 5 cm), (ii) mesoplastic (5 cm–5 mm), (iii) microplastic (100 µm–5 mm), and (iv) nanoplastic (1–100 nm) (Van Emmerik and Schwarz, 2020). Plastics are mostly broken down by mechanical forces (Christopher et al., 2016). Therefore, macroplastics degrade into microplastics, which will further breakdown into nanoplastics.

The surface features, composition, and size of MNPs vary with respect to their source. For example, primary MNPs found in personal care products tend to be smaller than 0.3 mm, and they contain some additives such as polymethyl methacrylate, Teflon, polyethylene, polypropylene, and polyethylene terephthalate (Storck et al., 2015). The shape of primary MNPs may be spherical or cylindrical with the size around 5 mm. The low-density polymers are used in packaging, while the high-density polymers like polyvinyl chloride and polyester are used in construction and textiles. These plastics become the sources of secondary MNP fragments and fibers from sewage effluent or surface runoff (Sadri et al., 2014).

MNP affects aquatic life in water bodies. Microplastics are consumed by aquatic life more than they are perceived to be. Scientists have observed the physical damage as well as inflammation of organs due to the accumulation of microplastics caused by jabbing and rubbing of plastics to the cell walls or organ walls (Xu et al., 2020). Animals exposed to microplastics have had lesser chances of reproduction of offspring, which has widely affected the subsequent generations too. In human beings, the ingested microplastic particles and hazardous chemicals such as bisphenol A (BPA) and pesticides which are added during the production can affect the immune system, growth, and reproduction (Wang et al., 2020). Bisphenols are known to interfere with hormones, and there are studies linking bisphenol exposure to reduced fertility in men and women. The plastic will be able to accumulate in certain organs causing blockages in the bloodstream or near the walls of the organ and then can lead to inflammation or carcinogenic diseases (Thompson, 2018).

The plastic wastes will further become home for various microorganisms and act as a spot of origin for various diseases and other disturbances in the environment. It is thus necessary to ensure their proper management and disposal. Plastics like polyethylene terephthalate, polycarbonate etc. act as a support and become the perfect environment on which the bacteria form biofilms (Lobelle and Cunliffe, 2011). The source of plastics present in the water bodies in India are drinking water bottles, utensils and household items, the discarded pipelines, and plastic bags. Due to sunlight and higher atmospheric temperature, these plastics tend to degrade into smaller particles and eventually break. The composition sometimes gets altered and leads to off-gassing and increases the carbon content also. These released chemicals tend to affect the life forms, disrupting their normal functionality (Kale et al., 2015).

The microstructure and molecular arrangement of these plastics, especially the plastic fibers used in clothing materials like nylon, silk, rayon, acrylic etc. provides larger surface area for growth and leads to the development of bacterial colonies which continue to multiply and possess the potential to cause diseases. The various organisms present in the water bodies may also indirectly assist in the growth of these microorganisms and their colonies and subcolonies (Lobelle and Cunliffe, 2011). Further, the plastics containers like shampoo, medicine (antibiotics), perfume bottles, and tubes may have the contents still left in them. The microorganisms react with chemicals and get modified genetically sometimes. Due to prolonged exposure, the bacteria develop resistance against the antibiotics (Xu et al., 2020). This results in the development of diseases in the aquatic organisms, which would be eventually passed down the food chains and in turn cause widespread chronic diseases in humans (Di Cesare et al., 2016).

3.3 OCCURRENCE, FATE, AND TRANSPORT OF MICRO- AND NANOPLASTICS IN WWTP

Sewage treatment plants receive the wastes from different areas. The composition may vary depending on the place and surrounding regions. After the process of treatment, the harmful wastes like primary, secondary microplastics, and other organic matter are removed and then sent for composting or landfilling etc. The treated and untreated wastewater effluents like those domestic and industrial waste effluents are discharged into septic tank or sterilization pond and have not undergone treatment processes. This untreated sewage effluent contains heavy metals such as sulfur, zinc, copper, and other toxic metals that are released into the rivers; they act as source of microplastics to be entered into the environment (Gahlot et al., 2020, Kumar et al., 2021). This section describes the occurrence, fate, and transport of MNPs in WWTP including the interaction between MNPs and pharmaceuticals, the role of microplastics in the emergence of antibiotic resistance, and MNP removal methodologies.

3.3.1 Occurrence and detection of MNPs

The MNPs enter the WWTPs along with domestic, industrial, and commercial discharges. The microplastics loading in the wastewater entering the treatment system tend to vary based on the location, population density, and the industrial/economic activity in a location. The sources that contribute to the MNP loading in WWTPs are the plastic beads used in pharmaceutical and personal care products, and synthetic fibers used in clothing. It has been reported in multiple studies that the major portion of the microplastics constituting the effluent stream is found to contain microfibers with microbeads or pellets forming a smaller portion. About 700,000 microplastic fibers are released into wastewater in a single wash from washing machines (Napper and Thompson, 2016). Conley et al. (2019) have reported that nearly 10,000 microplastics/person/day contribute to the WWTP. The particles resulting from road and tyre wear also contribute to microplastics load in wastewater.

A typical WWTP treats the sludge containing fat, oil, and grease (FOG) with other impurities in sludge treatment facilities. The sludge along with FOG is then incinerated and discarded to a landfill. It is likely that most of the microplastics would be removed during the waste treatment accumulated in sludge. Magnusson and Norén (2014) noted that the amount of microplastic effluent entering a Swedish WWTP with a load of 12,000 population equivalent was 15,000 microplastic particles per m^3. More than 99% of this amount was retained in the sewage sludge and only eight microplastics per m^3 were found in the treated effluent (Iyare et al., 2020). Li et al., (2018) pointed out that the sources of microplastics in sewage sludge are laundry and plastic-industry wastewater (Corradini et al., 2019). The number of plastic particles present in 10 million tonnes of sewage sludge generated each year in the EU is approximately 7.2×10^{12}–1.49×10^{14} per year (Li et al., 2018). As per 2010 data, on an average, 1.56×10^{14} microplastic particles per year enter the natural environment in China.

It is often difficult to quantify the nanosized plastics in the environmental samples. There are some practical difficulties associated with the sampling techniques. The analytical tools used for data collection such as microscopy and other spectroscopy are time-consuming techniques; they tend to disturb the characteristics of particles. In addition, some techniques such as dynamic light scattering for tracking nanoparticles will not be able to distinguish the natural and artificial nanoparticles. In a recent study conducted at bench scale for the removal of nanoparticles in the range <400 nm, nearly 92%±3% of nanoparticles are removed during filtration with 0.22-um filters (Murray and Örmeci, 2020). The particle size distribution after the filtration process indicates that the removal is fairly uniform in all the size ranges. This is especially important because if the smaller particles have not been removed and if filtration shows the preferential removal of larger particles only, then the removal may not be efficient with respect to smaller particles that might have a larger impact. They also observed that centrifugation of the samples at 10,000 rpm for 10 min showed 99%±1%, which is excellent removal efficiency, but again the smaller particles are not efficiently removed. Even if only a few particles escape the treatment, it would contribute largely to the increase in the concentration of NPs in the discharged effluent. Figure 3.1 shows some examples of microplastics collected from the wastewater effluents and their removal and transport to surface water.

3.3.2 Interaction between microplastics and pharmaceuticals

Even though antibiotics are usually used in the pharmaceutical industry, sewage effluents are also known to have antibiotics in them. Due to the physicochemical properties of the MNP such as hydrophobicity and surface area, it affects the fate of soluble organic contaminants present in the wastewater, including pesticides, pharmaceuticals, petroleum hydrocarbons, halogenated organic compounds, phthalates, surfactants etc. Often the occurrence of antibiotics such as tetracycline, sulfamethoxazole, ciprofloxacin, norfloxacin, trimethoprim, and ofloxacin is widely reported in the wastewaters and sewage sludge (Barancheshme and Munir, 2019). Adsorption of pharmaceutical compounds on the polyethylene microplastic is studied by Razanajatovo et al. (2018). Adsorption of compounds such as sulfamethoxazole, propranolol, and sertraline on microplastics was found to be 15.31%, 21.69%, and 28.61%, respectively. Properties of pharmaceutical compounds such as $\log_{Kow}$ and pK_a and microplastics properties such as the surface area, age, and charge greatly influence the adsorption mechanism (Razanajatovo et al., 2018). pH_{pzc} (point of zero charge) value of the PE microplastics is 7.12, and pH of the wastewater influences the sorption. Moreover, mechanisms such as electrostatic attraction and the hydrogen binding can play a role in adsorption phenomenon (Wu et al., 2016).

3.3.3 Influence of microplastics on the emergence of antibiotic resistance

Antibiotic resistance is referred to the state of bacteria to resist the effects of antibiotics. The overuse and misuse of antibiotics leads to the emergence of the antibiotic-resistant bacteria (ARB) in the environment (Lu et al., 2020; Xu et al., 2020). WWTPs act as the hotspot for the emergence and dissemination of antibiotic resistance. Since the microplastics act as the supporting platform for the attachment of antibiotics and

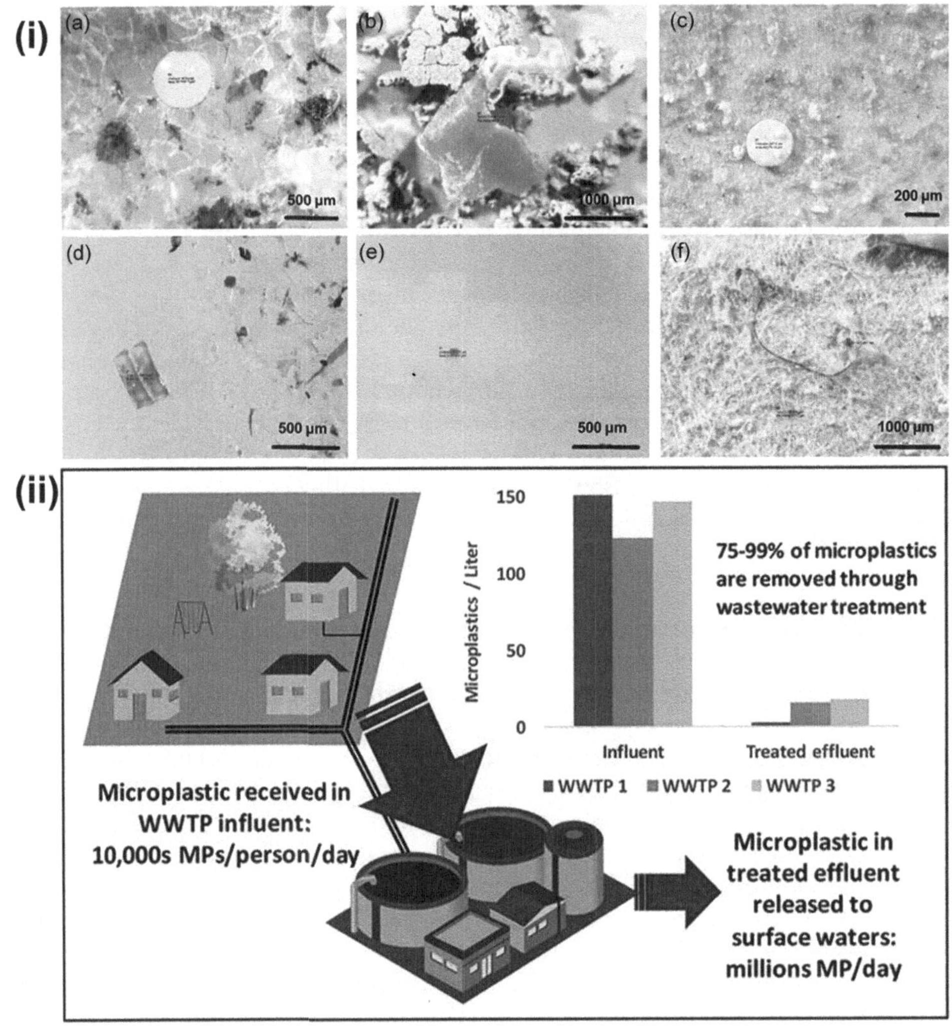

Figure 3.1 (i) Microplastics collected at wastewater treatment plants: (a) pellet, (b) fragment, (c) foam, (d) sheet, (e) granule, and (f) fiber (adapted with permission from Gies et al., 2018) and (ii) removal and transport of microplastics from wastewater treatment plant to surface waters. (Adapted with permission from Conley et al., 2019.)

bacteria, the chance for the development of antibiotic resistance is high. The resistance may be developed due to the presence of a single antibiotic or a range of antibiotics (Li et al., 2018). Antibiotic resistance is developed due to genetic mutation, and the mutated microbes, especially bacteria, transfer the genes to other bacteria and continue the cycle. The bacteria can interchange their DNA, and they also transfer the DNA to the other bacteria via reproduction (Sivagami et al., 2020).

Various studies reported the ARB resistance toward antibiotics such as tetracycline and sulfonamide in the domestic wastewater (Wang et al., 2014). Adsorption of heavy metals onto the microplastics can also trigger the antibiotic resistance in the wastewater (Kedzierski et al., 2018). High concentrations of metals in the environment can apply co-relation pressure and generate resistance through genetic coupling. Co-resistance, co-regulation, and cross-resistance are the mechanisms involved. A cross-linkage formed between two or more same or diverse resistance genes is called as co-relation. Cross-resistance provides resistance to more than one antimicrobial agent like antibiotic and heavy metal. ARS can also be activated by metal exposure (Li et al., 2017). When exposed to metals like Cu and Zn, the microbes acquire resistance to the metal, and due to this process, they acquire both metal- and antibiotic-resistant genes (Poole, 2017; Di Cesare et al., 2016). ARB can be transported along with wastewater from WWTPs, gets washed into larger water bodies where come in contact with the other pollutants (organic wastes, micro- and nanoplastics, micro- and nanofibers etc.) in the same water body, which provide them the medium to grow. ARBs are more disseminated to the environment through sewage treatment plants as effluents to rivers, wastewater reuse, irrigation etc.

3.3.4 MNP removal at different processes

The conventional WWTPs employing primary, secondary, and tertiary treatment technologies are specifically designed for treating municipal wastes. Since the MNP enter the same treatment facilities through various sources, the performances of the WWTP in handling MNP loads are compared and analyzed in this section.

Preliminary studies conducted at WWTP showed that a significant amount of fibers and particles are able to escape the treatment process, and on an average, 4.9 (±1.4) fibers and 8.6 (±2.5) particles were found per liter of treated wastewater. The effluent is found to contain the fiber concentration of 25 times higher than that of the receiving water body (Talvitie et al., 2015). Hence, it appears that even WWTPs can be considered as one of the point sources of microplastics entering the water bodies (Murphy et al., 2016). To evaluate the removal efficiency of municipal WWTP, a study is conducted in Kenkaveronniemi WWTP located near the city center of Mikkeli, Finland. The treated wastewater and the sludge samples are taken every 15 days for 3 months. The treatment facility consists of conventional activated sludge and a pilot-scale membrane reactor. Advanced tools such as Fourier transform infrared spectroscopy and Raman microscopy are used to characterize the samples. The treated water from membrane bioreactor (MBR) is found to contain 0.4 microplastics/L, whereas the effluent from the activated sludge process is found to contain 1.0 microplastic/L before discharge. It shows that the microplastics escape through the treatment steps and enter into the ocean sources (Lares et al., 2018). In addition, these particles tend to adsorb a variety of pathogens and other chemicals that are present in the WWTPs, since their surface area is high. In such cases, the wastewater discharged into the environment introduces an additional risk for the oceanic systems.

Although conventional WWTPs are efficient in the removal of MPs, their loading in the aquatic systems is quite high since larger volumes of wastewater are discharged as effluents per day. Hence, the efficiencies of some advanced final treatment technologies such as MBR, disk filter, rapid sand filtration, and dissolved air floatation have also been studied (Talvitie et al., 2017). MBR systems eliminate the need for secondary clarifiers and tertiary treatment. For the better removal of organic and suspended solids, the novel MBR that combines the conventional activated sludge process with membrane filtration can be used. The MBR contains an aerated biological reactor into which the membranes are suspended. The membrane porosities ranging from 0.035 to 0.4 μm result in a high-quality effluent. Thus, the use of MBR eliminates the subsequent processes such as sedimentation and filtration used in typical WWTPs. According to a study conducted in four different municipal WWTPs employing different final treatment methods, the MBR that treats the primary effluent removed 99.9% from 6.9 to 0.005 microplastic/L (Talvitie et al., 2017). The other advanced techniques such as disk filter, rapid sand filtration, and dissolved air floatation treating secondary effluents are found to remove 40%–98.5% (from 0.5–2.0 to 0.03–0.3 microplastics/L), rapid sand filter 97% (from 0.7 to 0.02 microplastics/L), dissolved air flotation 95% (from 2.0 to 0.1 microplastics/L). Hence, the advanced techniques are necessary to enhance the quality of effluent before discharge (Talvitie et al., 2017).

The microplastics are found in untreated effluents, treated as well as heat-dried sludge. The microbeads of polyethylene, polypropylene used in personal care products such as exfoliants, toothpastes, and polystyrene particles, which function as a scrubbing agents in facial scrubs, find their way in the WWTPs and they filter out the larger particulates. They are transported into terrestrial and oceanic systems through treated effluent discharge and sludge disposal. Another main contribution of microplastics to sludge comes from synthetic textile fibers. Based on a study, the concentration of microplastics in sewage sludge was approximately 720–14,900 particles per kg (wet weight) and from 1,000 to 170,900 particles per kg (dry weight). The majority of microplastics in sludge about 63%–90% were fibers. Since the sludge would be used as a fertilizer, the microplastic content in the sludge has also become a source of soil contamination (Corradini et al., 2019). Since the WWTPs are not designed for treating plastics, a new question also arises as to how the presence of MNPs affect the efficiency of the existing facility. Almost 72% of the total amount of primary microplastic and 88% of secondary microplastics present in the sewage can be removed during the primary treatment itself. The coarse and fine screens of size range 6–150 mm and less than 6 mm, respectively, remove larger microplastics (Iyare et al., 2020). It is also noted that the primary treatment process removes microfibers significantly (Talvitie et al., 2015). The secondary treatment that removes the suspended and dissolved organic solids using microorganisms has shown further 17% removal of microplastics from the primary effluent. The tertiary treatment involving the use of sand or activated carbon filtration has shown further removal ranging from 5% to 20%. Studies show that for the microplastics in the range of 400 to 45 μm, the removal can be complete in the facilities having tertiary treatment (Carr et al., 2016).

3.4 OCCURRENCE, FATE, AND TRANSPORT OF MACRO-, MICRO-, AND NANOPLASTICS IN RIVERINE AND MARINE SYSTEMS

Scientists believe that microplastics are present in oceans around the world, but unlike larger plastics floating around, microplastics are really hard to be identified and get removed. Moreover, ocean currents play a large role in the situation of microplastics, making them very difficult to be cleaned. A 2015 study predicted that around 8 million tons of plastics enters the oceans every year (Jambeck et al., 2015). Out of this, only 1% is found floating on the surface of water. This leads us to the conclusion that bulk of plastics in the oceans are settled or buried in sediments. Microplastics in oceans can pose a huge threat to marine life, whether it is a sea turtle nibbling onto a plastic cover or a dolphin entangled in a fishing gear. Organisms like plankton and fish larva are known to consume MP, and filter-feeding animals like oysters and scallops that human also consume, ingest particles as they filter seawater. Although larger plastics like bottles or polythene bags contribute to the formation of microparticles in oceans, many believe that the occurrence of MPs is due to plastic products found in facial cleansers and makeup. According to the European Chemicals Agency, 36,000 tons of microplastics is released into the environment from beauty products alone (Imogen et al., 2015). Nadella Beilzair commented, "When you're washing products off your face, where are you washing it to? Down the drain and into the oceans and what lives in those? All the fish. They have no choice but to eat that have those in them," and this is where bioaccumulation comes in (Kneafsey, 2017).

3.4.1 Rivers as transport pathways to ocean

Although the percentage of pollutants in freshwater is less compared to that of marine bodies, the accumulation of microplastics has only increased as the years pass by. A survey was conducted to compare the concentration of MPs in Laurentian Great Lakes and St. Lawrence River between 2 years 2011 and 2016 in USA, which shows that microplastics were present in both sediment and surface waters with a higher concentration of pellets and fragrance, indicating contributions from both primary and secondary sources (Ballent et al., 2016). Moreover, the microplastics found in St. Lawrence River were comparable in size, shape, and composition to those found in Laurentian Great Lakes, indicating that the microplastics could have flown downstream along the various rivers and lakes (Castaneda et al., 2014).

According to another resource, a study in Lake Hovsogl in Mongolia by Free et al. (2014) was reported to have microplastics, resulting in a total density of 20,264 items per sq km; but the place where the lake is situated is in a rural area, without much access to such quantities of plastics. This could be concluded as the microplastics from urban areas, along with water current, have settled out to this lake. Rivers can retain or transport MNPs based on its discharge rates and flow of current. Rivers are considered to be temporary sinks which eventually a delay out the diffusion of MNPs to sea water. Materials travel faster during monsoons due to the increased flow rate. Danube River is known to be the largest and most important transport route for plastic products from Germany and Austria to the Black Sea. But export patterns are not always clear. In Thames River, although no major particle movement was noticed from up to downstream sites, high abundance of microplastics were found in sites near the sewage treatments discharges (Lechner et al., 2014).

3.4.2 Partitioning and mass flow behavior of these compounds in marine and subsurface systems

The transport of MNPs in the marine system are mainly through the gravity-driven transport followed by settling in the sediments. The conveyance through biological

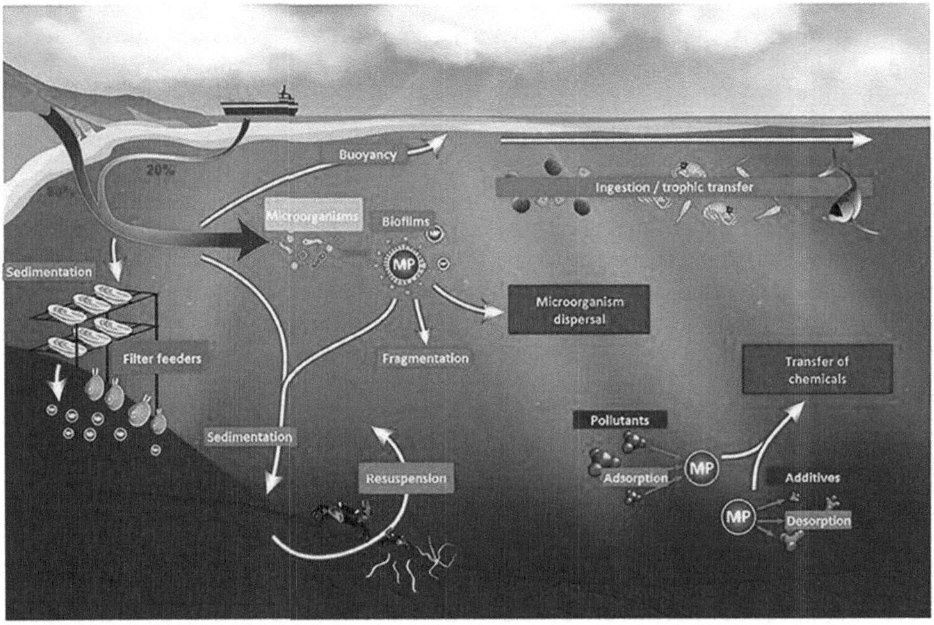

Figure 3.2 An illustration of fate of microplastics in the marine system. (Adapted with permission from Paul-Pont et al., 2018).

processes and thermohaline currents, the deposited microplastics distributes in deep marine flow. The sedimentological and stratigraphic knowledge of marine system will be useful in understanding the fate and distribution of microplastics. The photooxidative degradation and disintegration of polymers were accelerated by wave action and turbulences, increased temperatures, physical abrasion, and higher UV irradiation on the surface. There is a clear evidence that the microplastics provide large surface-to-volume ratio and enrich the sorption of hydrophobic organic pollutants like polycyclic aromatic hydrocarbons, chlorinated compounds etc. (Napper et al., 2015). The adsorption coefficient (KOC) and distribution coefficient of polymers play a major role in the sorption and desorption processes. The sorption of PAH compounds like phenanthrene in polymers is 2–4 times higher than the natural sediments since microplastics can be transported to long distance. Velzeboer et al. (2014) also found a similar magnitude of sorption to microplastics of PE matter and sediment organic matter. Fate of microplastics in the marine environment is illustrated in Figure 3.2.

3.4.3 MNPs and hydro geochemical processes

Another major problem caused by the plastics accumulation is its interference with the hydrogeochemical processes. Groundwater and aquifers are an important source of portable water. Thus, managing and maintaining the proper quality of water is a prime concern for our survival. There are various processes that affect the quality and quantity of the water available, although it may change based on the location and regional factors. The origin of the water, the path of flow, the soil composition of that region and the surrounding regions, the porosity of soil, the presence of vegetation and different organisms in the soil and their activity are few of the main factors that affect the water quality in groundwater and aquifers. Hydrogeochemical characteristics and the quality of groundwater present in different aquifers are important in solving the problems that are related to water conservation and management. Evaporation, ion exchange, silicate weathering, and dissolution of carbonate minerals are the important hydrogeochemical processes that affect the groundwater chemistry (Sajil et al., 2016). Major ion chemistry of groundwater and the aquifers is used to study the subsurface hydrogeochemical processes (Liu et al, 2020). Resistivity profiling between two interfaces or formations is another way of understanding the subsurface systems. Resistivity sounding determines the thickness and resistivity of different horizontal or low dipping subsurface layers, including the aquifer zone.

There are certain steps that need to be followed for the process, the ignorance of which will lead to leakage of the waste into the soil and result in pollution. Each layer of the landfill needs to be set properly. In case of thin outer lining or wrong profiling, the MNPs and other chemicals tend to seep through the porous layers of the soil and eventually mix and contaminate the water table. This contamination over time increases, making the water unusable. Further, if the top layer is not covered and sealed properly, then the rain water tends to wash away these wastes into the water bodies, leading to pollution and other ill effects (Ganesan, 2017; Abdel-Shafy et al., 2018).

The landfills contain both organic and inorganic wastes. Similar to the situation in oceans, the wastes tend to mix up and this affects the life cycle of the microorganisms and the soil-dwelling living beings. Studies have shown that chemicals in soil force earthworms and other soil creatures to behave differently, which in turn affects the fertility of the soil also (Jiang et al, 2017). The chemicals and the degraded plastic compounds seep through the porous soil and contaminate the water table. This water when consumed will be life-threatening. The proper ratio of calcium, magnesium, sodium, chlorides, and sulfates present in the soil and rock is essential for the smooth hydrogeological processes. This ratio establishes a concentration gradient, which sets up the path for the flow of water, which in turn assists the other processes. This helps in maintaining proper pH of the water also, as water with too high or low pH cannot be consumed. Other processes like mining and other activities also affect pH of water (Liu et al., 2019). When leaching occurs due to improper landfills, the chemicals that are released from the degraded plastics and other wastes, especially pharmaceuticals, tend to disturb the component balance both in soil and in the water, which in turn affects the rock-water interactions like evaporation, precipitation, weathering etc. (Kirstein et al., 2018). It is thus a matter of utmost importance to control the pollution and other disastrous effects caused by the plastics and other associated wastes.

3.4.4 Occurrence of macroplastic contamination in Indian marine, riverine, and subsurface water system

India is still backward when it comes to maintaining microplastics in environment. According to laboratory results of 2017, it is seen that 50% of the face wash products and 67% of the face masks consist of plastics. Several attempts have been made globally to control the usage of plastics. The beaches were known to be the most common source of pollution. Samples that are collected at high tide are significantly higher than the samples collected during low tides. Out of the total detected debris, plastic fragments were known to be 24%–27%, with polyethylene, polystyrene, and polypropylene being the main types. Marine animals found dead due to the ingestion of plastic were known to be 10.1% (Karthik et al., 2018). A report was made in 2017, in which the microplastic pollution in Vembanad Lake in Kerala was noted. According to the survey, the abundance of MP from the samples ranged from 96 to 496 particles/m sq. with a mean abundance of 252.80 particles /m sq. Low-density polyethylene was known to be the most common, and this was the first report from India that focused on MPs in lake sediments, which helped India to research more on the distribution and impact of this emerging pollutant on the species of marine life (Sruthy and Ramasamy, 2017).

In the following section, the authors have discussed three main polluted rivers in the country, which attributes significantly to plastic waste present in the Indian marine systems. The absence of a formal plastic recycling framework in many parts of India has led to the unregulated dumping of plastic waste into the river systems, combined with a lack of public knowledge, causing major water pollution in the river system. The impact of such devastating pollution is not restricted within the inland river systems. When a river or its tributaries flow into the ocean or sea, contaminants are often brought over. Similarly, all the plastic and nonplastic contaminants are transported to the Bay of Bengal, the Arabian Sea, and the Indian Ocean by Indus, Ganges, and Brahmaputra, thereby raising the total plastic contamination of the ocean. Indus, Brahmaputra, and Ganges are ranked among the top ten plastic-polluting rivers in the world that contribute to plastic fragments contained in the ocean. The River Indus is considered the second most plastic-polluted river in the world, whereas the Ganges ranked sixth under the same category. The world's oceans can be divided or categorized into six zones/garbage patches in the following order, based on their geographical location and the quantity of plastic fragments found in them: (i) North Pacific garbage patch (Pieces: 2 trillion, Weight: 192.8 million lbs), (ii) Indian Ocean garbage patch (Pieces: 1.3 trillion, Weight: 188.3 million lbs), (iii) North Atlantic garbage patch (Pieces: 930 billion, Weight: 112.9 million lbs), (iv) South Pacific garbage patch (Pieces: 491 billion, Weight: 42 million lbs), (v) South Atlantic garbage patch (Pieces: 297 billion, Weight: 25.6 million lbs), and (vi) Mediterranean Sea garbage patch (Pieces: 247 billion, Weight: 46.3 million lbs)[1].

The writers have therefore limited their topic to the Indian Ocean garbage patch, as this chapter focuses on the Indian subcontinent. To understand the fate of microplastics and nano-plastics in the Indian riverine and marine ecosystem, it is important to understand the formation process of the Indian Ocean garbage patch. When floating plastics are collected at the center of subtropical circulating ocean currents, the ocean garbage patch is usually created. Over time, tiny microplastics and nanoplastics that can be ingested by fish and enter our food cycles break down into large pieces of floating plastics, posing significant health risks. Therefore, to mitigate the above impacts, before it reaches the ocean, plastic should be eliminated from river systems. The only way to control plastic waste in the ocean is to adopt such a policy where plastics carried by inland waterways are prevented from entering the ocean. The spectrum of plastic contamination in Indian riverine and marine environments will be addressed in the current subsection. The adverse health and environmental risks associated with plastic waste will also be addressed. The section will conclude with a debate on state-of-the-art technologies for river and oceans clean-up to reduce and mitigate the problem.

3.4.5 Extent of plastic pollution in Indian riverine systems

The River Ganges should be the starting point to address the level of plastic contamination in the Indian River system. While it is considered the "Holiest River" in India, the Ganges and the Brahmaputra River are considered to be the sixth most polluted-plastic rivers in the world. The Ganges River is heavily polluted with industrial waste, untreated sewage, domestic waste, as well as 1.2 billion pounds of plastic waste each year, further exacerbating the river's already-deteriorating water quality. According to the study conducted by Lebreton et al. (2017), 1.15–2.41 million tonnes of plastic waste reaches the ocean every year through various riverine systems worldwide. Among these rivers, the majority come from the continent of Asia. However, owing to a lack of significant data, such an estimate is widely debated (Geraeds et al., 2019; Van Emmerik and Schwarz, 2020). In the recent past, the studies on river-based plastic pollution have gained impetus with several investigations on Beijiang River (Tan et al., 2019), Rhine (Mani et al., 2015), Danube, Yangtze Estuary (Zhao et al., 2014), Selenga River (Battulga et al., 2019), etc. It should also be noted that, except for limited studies on the River Ganges (Sarkar et al., 2019), certain studies on Indian riverine systems are mostly absent, restricting the in-depth research from the Indian context.

It is important to understand the scale of plastic waste produced in India to comprehend the depth of this issue. It is also important to consider the status of the country's current recycling scenario. The country produces 25,940 tonnes of plastic waste according to the Central Pollution Control Board (CPCB). When put together on an annual basis, it accounts for 9.5 million tonnes of waste per year. The study also revealed that about 60% of produced plastics are collected for recycling, which can be considered significantly high compared to many developed countries like the USA as a global recycling rate, which stands at 20%. However, as 40% of produced plastics remained uncollected, the majority of these unprocessed plastics end up in Indian riverine, ocean systems, or landfill sites, approximately 9400 tonnes per day which is a massive concern for CPCB. Post-implementation of plastic waste management rules (2016), the latest CPCB report published in 2018 revealed that the total plastic waste generated was 660,787.85 tonnes per annum, which is significant, but underestimates the actual plastic pollution scenario as it only takes into account data submitted by only 13 State Pollution Control Boards (SPCBs). In a nutshell, it can be inferred that unless all the states strictly enforce the plastic waste guidelines of the central government and limit the use of single-use plastic, restrict, monitor, and remove floating plastic waste from rivers within their state territory, it is almost impossible to reverse the current scenario.

To give a nationwide perspective on the current plastic waste problem in Indian riverine systems, the authors have identified the major polluted rivers in the country and divided them into four regions. In northern India, Ganges, Yamuna, and Indus contribute significantly to the overall floating plastic waste problem in the country. In addition to the above, the Gomti River is also massively polluted with floating plastic waste. The Indus (164,332 tonnes) and Meghna (river in Bangladesh)-Brahmaputra-Ganges (72,845 tonnes) bear some of the world's largest volumes of plastic waste to the oceans, according to the United Nations. In terms of plastic pollution, the River Indus that passes through India and Pakistan is ranked second, whereas Ganges-Brahmaputra together ranked sixth. Rivers such as Mithi, Sabarmati, Ulhas, Vaitarna, Godavari, Bhima, Krishna, Tapi, Kundalika, Panchganga, Mula-Mutha, Pelhar, etc. are considered to be the most polluted rivers in western India that bring a large amount of plastic waste, industrial waste, and domestic waste into the ocean. In South India, rivers like Godavari, Krishna, and Cooum are also severely affected due to plastic, municipal, and industrial wastes, thus making these rivers toxic not only to humans but also for aquatic life forms. In eastern India, along with the Ganges, Hooghly (Tributary river of Ganges) and Damodar rivers are massively impacted by plastic pollution. In the majority of the aquatic fish that are consumed by the humans, guts are generally removed before cooking. It is also to be mentioned that the effect of microplastics on other body tissues of these marine organisms depends on the amount of microplastic ingestion and subsequent transfer, distribution, and accumulation within the body. Similarly, the presence of microplastics in the human gut can not only create blockages but also affect the food digestion process. Due to the large surface area of microplastics, it can absorb different hydrophobic pollutants and transfer them to the human body, which may have several negative impacts (Lusher, 2015).

3.4.6 Modern river and ocean clean-up technology

To solve marine plastic problems and reduce the scale of the Indian Ocean garbage patch, the first step is to cut off the main source of plastic waste (PET bottles, plastic toys, and plastic bags, etc.) entering the ocean (Geraeds et al., 2019). To achieve those results, drastic measures need to be taken to clean-up our riverine networks. There are few new technologies capable of cleaning up floating plastic waste from rivers and oceans. The following technologies are discussed: (i) interceptor, (ii) bubble barrier, (iii) trash wheel, (iv) floating boom, (v) System 001 or Wilson, and, (vi) floating seabin trash collector. The technology can be classified according to where it can be applied, i.e., (i) river-based clean-up strategy and (ii) ocean-based clean-up strategy. Interceptor, bubble barrier, trash wheel, and floating sea bin trash collector (effective in static water bodies, and with low flow rate) are effective for river clean-up, whereas for ocean-based clean-up, System 001/Wilson and floating boom technology are the best possible options. However, it is also to be pointed out that among the above-mentioned techniques, interceptor and trash wheel utilize similar methods for river clean-up, whereas the floating boom and System 001 are same strategies with minor modifications. Brief details on the above-mentioned strategies are given below.

3.4.6.1 Interceptor

A company named "The Ocean Cleanup" has created an ideal interceptor for river clean-up purposes. Four main components are included in the setup: (i) barrier,

(ii) conveyor belt, (iii) shuttle, and (iv) dumpster. In the overall clean-up process, each part of the setup plays a critical role. The interceptor automatically sends a text message to the local operators to come and collect the waste when it is almost full. Then, operators remove the barge, carry it to the riverside, clear the dumpsters, send the garbage to local waste collection facilities, and take the barge back to the interceptor.

3.4.6.2 Bubble barrier

In the bubble barrier, a bubble screen is created by pumping air through a submerged air nozzle that is placed at the bottom of river/waterways. Due to buoyancy effect, bubbles generate an upward thrust and help to bring submerged plastic debris to the water surface. This strategy has several advantages over the "Interceptor" strategy. Waterway vessel and aquatic life can easily pass through the canal or river without any obstruction. It improves the dissolved oxygen level in water, which is beneficial for aquatic life. The technology is easily scalable and does not require high investment on infrastructure development and is capable of removing heavier plastics (density >1 g/cm^3) from river without the need of expensive setup. The bubble barrier infrastructure can be established without impacting fish migration, shipping, and natural river functioning. Therefore, it is ideal for any rivers and canals having high concentration of plastic debris.

3.4.6.3 Trash wheel

The trash wheel is a solar and hydro-powered trash collector that operates with the same working principle as the "Interceptor." The trash wheel has five main components like an interceptor, i.e., (i) floating boom/barrier, (ii) conveyor belt, (iii) dumpster, (iv) river current-powered water wheel for belt operation, and (v) solar panel for additional electricity. Since the system uses the river current for conveyor belt operation, it is a very eco-friendly and energy-efficient trash removal option. The trash wheel is currently under operation in the Harbor area of Baltimore, Maryland, USA. However, the trash wheel's collection capacity is lower than that of the interceptor. The trash wheel was designed and developed by Mr. John Kellet in 2008.

3.4.6.4 Floating seabin trash collector

Floating seabin trash collector is a trash skimmer, which works effectively to clean public marinas, harbors, ponds, lakes from debris including macroplastics and microplastics. The device works best in a static water pool. Floating seabins are ideal for floating marinas and pontoons. When the Seabin is secured with a floating dock, it can effectively remove a wide variety of trash (microplastics, fishing net, food wrappers, plastic packaging, polystyrene foam, plastic bottles) and hydrocarbon from water via an absorbent pad, which works together with a water pump. The device also consists of steel bracket, which improves its durability for application in area dominated with marine life. The device requires very low maintenance other than the removal of algae and other marine life attached to container wall and removal of trash from seabin depending on the trash storage capacity and amount of plastic debris where the device has been employed.

3.4.6.5 Floating boom

Floating boom is also an effective way to contain and remove plastic debris from rivers and ocean. System 001, interceptor, and trash wheel use different forms of floating boom to contain and direct plastic debris for effective removal. Floating seabin trash collector can also effectively remove plastic from marinas and harbors in conjunction with a floating boom.

3.4.7 Recommendations to reduce MNPs pollution in the Indian context

The disintegration of microplastics into nanoplastics into the environmental compartments is difficult to measure. The nanofibers and particles spread in the air and are carried away by the action of wind. The traces are even found on the snow in the areas like Arctic and Antarctic regions. Hence, the threat is visible ahead of us, and it is not in the distant future. From the available literature, it appears that there is only a limited study on the accumulation and the consequences of microplastics and nanoplastics in the rivers, coastal area, and oceanic resources. Only few studies have quantified the microplastic concentration along the coastal areas of Karnataka, Goa, and Maharashtra. It is known that most of the microplastic release into the marine environment comes from the mismanaged waste generated from the nearby land. Hence, the critical locations where the microplastic emissions are high into the water bodies have to be identified and the possible sources of release, its seasonal variations, the portioning of microplastics, economic importance of the area, and the public awareness all have to be assessed.

From the perspective of treatment technologies, the studies on the source of pollutants and their partitioning in WWTPs have to be developed. The effect of portioning in the treatment facility on the various compartments of the environment must be studied thoroughly. In addition, we have no data on the characteristics of MPs and their interaction with other pollutants in WWTPs. This domain has a great scope for research. From the practical perspective, single-use plastics need to be avoided at any cost, since it is the main source of MPs in the environment, which have undergone fragmentation over time. In addition, when a product is designed to be recycled easily, it will reduce the problem to an extent. Even though the recycle is quite high in our country, tonnes of plastics ends up either as a land fill or as contaminated water bodies. More efficient recycle system has to be designed and responsible recycling has to be adopted, so that the chances of accidental littering which would carry plastics everywhere due to the wind will be avoided.

From the population perspective, the current knowledge of nanoplastics and their possible effects in the human body is very less when compared to what we know about microplastics. Since the traces of microplastics and nanoplastics are even found in drinking water, there is an urgent need of preventing the increase in the pollution level. The WHO has now understood the importance of pollution due to microplastic level in drinking water. The organization has raised its concern on this issue, since most of the population in the world are not benefitted by adequate treated drinking water supply especially like our country. Hence, the data on the pollution due to nanoplastics in drinking water, their sources, and the efficacy of different treatment methods adopted in the country need to be studied thoroughly. The effect of microplastic entry into the human body is dependent on the concentration and the mode of entry into the body, via oral consumption or inhalation. The effect of their accumulation in the body system and its effect on various organs may be quite complex to study but we need further studies to ascertain the facts. The task appears to be very huge in terms of complexity and the level of effort needed.

3.5 CONCLUSION

The inefficient removal of microplastics and nanoplastics from WWTP results in the transport of these particles into river and ocean. MNPs are hard to be captured from the riverine and ocean systems, which further complicate the plastic pollution in the aquatic environment. With the limited knowledge on the effects of the plastics of any form in India, it appears that reducing the generation of plastics at source should be a key mantra to reduce plastic pollution. From the discussion, it is evident that the journey of plastics from its source through various compartments of environment and their effects on various life forms until the end of the life cycle is quite complex. The main source of the oceanic and riverine pollution is the excessive human activity on the land. Hence, the public awareness and the cooperation from the major plastic producers could play a key role to drastically reduce the consumption. The hesitation of the plastic generating companies and their intention to bypass or delay the government regulation has resulted in the stagnation of the efforts of government and the public. Even though the major plastics producers are initiating the recycling efforts as a means to show their social responsibility, most of the programs are insufficient to address the huge crisis. In a huge populated country like India, though some efforts are taken to ban single-use plastics, it appears that we are nowhere near the solution. The expenditure incurred in collecting, segregating, and treating the wastes and their disposal would be cost-prohibitive. Hence, the efforts taken by major academic and research institutions to develop biodegradable plastics for packaging, function specific bioalternates for micro-/nanoplastics in cosmetics, and other applications have to be promoted. Public participation in governmental regulations to fight against the threats caused due to plastics, the environmental consciousness, and collective responsibility of citizens in handling the trash would help the planet to breathe.

REFERENCES

Abdel-Shafy, H.I. and Mansour, M.S., 2018. Solid waste issue: Sources, composition, disposal, recycling, and valorization. *Egyptian Journal of Petroleum*, 27(4), pp. 1275–1290. https://doi.org/10.1016/j.ejpe.2018.07.003.

Andrady, A.L., 2017. The plastic in microplastics: A review. *Marine Pollution Bulletin*, 119(1), pp. 12–22. https://doi.org/10.1016/j.marpolbul.2017.01.082.

Ballent, A., Corcoran, P.L., Madden, O., Helm, P.A. and Longstaffe, F.J., 2016. Sources and sinks of microplastics in Canadian Lake Ontario nearshore, tributary and beach sediments. *Marine Pollution Bulletin*, 110(1), pp. 383–95. https://doi.org/10.1016/j.marpolbul.2016.06.037.

Barancheshme, F. and Munir, M., 2019. Development of antibiotic resistance in wastewater treatment plants. In Antimicrobial Resistance-A Global Threat. IntechOpen.

Battulga, B., Kawahigashi, M. and Oyuntsetseg, B. 2019. Distribution and composition of plastic debris along the river shore in the selenga river basin in Mongolia. *Environmental Science and Pollution Research*, 26(14), 14059–14072. https://doi.org/10.1007/s11356-019-04632-1.

Buragohain, P., Nath, V. and Sharma, H.K., 2020. Microbial degradation of waste: A review. *Current Trends in Pharmaceutical Research*, 7(1). www.dibru.ac.in/ctpr.

Cai, M., He, H., Liu, M., Li, S., Tang, G., Wang, W., Huang, P., Wei, G., Lin, Y., Chen, B. and Hu, J., 2018. Lost but can't be neglected: huge quantities of small microplastics hide in the South China Sea. *Science of the Total Environment*, 633, pp. 1206–1216. https://doi.org/10.1016/j.scitotenv.2018.03.197.

Carr, S.A., Liu, J. and Tesoro, A.G., 2016. Transport and fate of microplastic particles in wastewater treatment plants. *Water Research*, 91, pp. 174–182. https://doi.org/10.1016/j.watres.2016.01.002.

Castaneda, R.A., Avlijas, S., Simard, M.A. and Ricciardi, A., 2014. Microplastic pollution in St. Lawrence river sediments. *Canadian Journal of Fisheries and Aquatic Sciences*, 71(12), pp. 1767–1771. https://doi.org/10.1139/cjfas-2014-0281.

Corradini, F., Meza, P., Eguiluz, R., Casado, F., Huerta-Lwanga, E. and Geissen, V., 2019. Evidence of microplastic accumulation in agricultural soils from sewage sludge disposal. *Science of the Total Environment*, 671, pp. 411–420. https://doi.org/10.1016/j.scitotenv.2019.03.368.

Conley, K., Clum, A., Deepe, J., Lane, H. and Beckingham, B., 2019. Wastewater treatment plants as a source of microplastics to an urban estuary: Removal efficiencies and loading per capita over one year. *Water Research X*, 3, p. 100030. https://doi.org/10.1016/j.wroa.2019.100030.

Crawford, C.B., Russell, M., Ewins, C. and Quinn, B., 2016. Chemical adsorption of hydrophobic polycyclic aromatic hydrocarbons in the marine environment onto microplastic polymers and subsequent desorption in a simulated gut. In *SETAC Europe 26th Annual Meeting*. http://nantes.setac.eu/?contentid=851.

Di Cesare, A., Eckert, E.M., D'Urso, S., Bertoni, R., Gillan, D.C., Wattiez, R. and Corno, G., 2016. Co-occurrence of integrase 1, antibiotic and heavy metal resistance genes in municipal wastewater treatment plants. *Water Research*, 94, pp. 208–214. https://doi.org/10.1016/j.watres.2016.02.049.

Danthurebandara, M., Van Passel, S., Nelen, D., Tielemans, Y. and Van Acker, K., 2012. Environmental and socio-economic impacts of landfills. *Linnaeus Eco-Tech*, pp. 40–52.

Engler, R.E., 2012. The complex interaction between marine debris and toxic chemicals in the ocean. *Environmental Science & Technology*, 46(22), pp. 12302–12315. https://doi.org/10.1021/es3027105.

Free, C. M., et al. 2014. High-levels of microplastic pollution in a large, remote, mountain lake. Marine Pollution Bulletin 85(1), pp. 156–163. https://doi.org/10.1016/j.marpolbul.2014.06.001.

Ganesan, P., 2017. Landfill sites, solid waste management and people's resistance: A study of two municipal corporations in Kerala. *International Journal of Environmental Studies*, 74(6), pp. 958–978. https://doi.org/10.1080/00207233.2017.1374076.

Gahlot, R., Taki, K., & Kumar, M. (2020). Efficacy of nanoclays as the potential adsorbent for dyes and metal removal from the wastewater: A review. *Environmental Nanotechnology, Monitoring & Management*, 14, p. 100339.

Gies, E.A., LeNoble, J.L., Noël, M., Etemadifar, A., Bishay, F., Hall, E.R. and Ross, P.S., 2018. Retention of microplastics in a major secondary wastewater treatment plant in Vancouver, Canada. *Marine Pollution Bulletin*, 133, pp. 553–561. https://doi.org/10.1016/j.marpolbul.2018.06.006.

Geraeds, M., van Emmerik, T., de Vries, R. and bin Ab Razak, M.S., 2019. Riverine plastic litter monitoring using unmanned aerial vehicles (UAVs). *Remote Sensing*, 11(17), p. 2045. https://doi.org/10.3390/rs11172045.

Germanov, E.S., Marshall, A.D., Bejder, L., Fossi, M.C. and Loneragan, N.R., 2018. Microplastics: no small problem for filter-feeding megafauna. *Trends in Ecology & Evolution*, 33(4), pp. 227–232. https://doi.org/10.1016/j.tree.2018.01.005.

Iyare, P.U., Ouki, S.K. and Bond, T., 2020. Microplastics removal in wastewater treatment plants: A critical review. *Environmental Science: Water Research & Technology*, 6(10), pp. 2664–2675. https://doi.org/10.1039/d0ew00397b.

Jambeck, J. R., et al. 2015. Plastic waste inputs from land into the ocean. Science 347(6223), pp. 768–771.

Jiang, X.J., Liu, W., Wang, E., Zhou, T. and Xin, P., 2017. Residual plastic mulch fragments effects on soil physical properties and water flow behavior in the Minqin Oasis, northwestern China. *Soil and Tillage Research*, 166, pp. 100–107. https://doi.org/10.1016/j.still.2016.10.011.

Kale, S.K., et al. 2015. Microbial degradation of plastic: a review. *Journal of Biochemical Technology* 6(2), pp. 952–961.

Karthik, R., Robin, R.S., Purvaja, R., Ganguly, D., Anandavelu, I., Raghuraman, R., Hariharan, G., Ramakrishna, A. and Ramesh, R., 2018. Microplastics along the beaches of southeast coast of India. *Science of the Total Environment*, 645, pp. 1388–1399. https://doi.org/10.1016/j.scitotenv.2018.07.242.

Kedzierski, M., d'Almeida, M., Magueresse, A., Le Grand, A., Duval, H., César, G., Sire, O., Bruzaud, S. and Le Tilly, V., 2018. Threat of plastic ageing in marine environment. Adsorption/desorption of micropollutants. *Marine Pollution Bulletin*, 127, pp. 684–694. https://doi.org/10.1016/j.marpolbul.2017.12.059.

Kirstein, I.V., Wichels, A., Krohne, G. and Gerdts, G., 2018. Mature biofilm communities on synthetic polymers in seawater-Specific or general? *Marine Environmental Research*, 142, pp. 147–154. https://doi.org/10.1016/j.marenvres.2018.09.028.

Kneafsey, M., 2017. *Geographies of Rural Cultures and Societies*. Abingdon: Routledge.

Lares, M., Ncibi, M.C., Sillanpää, M. and Sillanpää, M., 2018. Occurrence, identification and removal of microplastic particles and fibers in conventional activated sludge process and advanced MBR technology. *Water Research*, 133, pp. 236–246. https://doi.org/10.1016/j.watres.2018.01.049.

La Daana, K.K., Gårdfeldt, K., Lyashevska, O., Hassellöv, M., Thompson, R.C. and O'Connor, I., 2018. Microplastics in sub-surface waters of the Arctic Central Basin. *Marine Pollution Bulletin*, 130, pp. 8–18. https://doi.org/10.1016/j.marpolbul.2018.03.011.

Lechner, A., Keckeis, H., Lumesberger-Loisl, F., Zens, B., Krusch, R., Tritthart, M., Glas, M. and Schludermann, E., 2014. The Danube so colourful: A potpourri of plastic litter outnumbers fish larvae in Europe's second largest river. *Environmental Pollution*, 188, pp. 177–181. https://doi.org/10.1016/j.envpol.2014.02.006.

Li, D., Shi, Y., Yang, L., Xiao, L., Kehoe, D.K., Gun'ko, Y.K., Boland, J.J. and Wang, J.J., 2020. Microplastic release from the degradation of polypropylene feeding bottles during infant formula preparation. *Nature Food*, 1(11), pp. 746–754. https://doi.org/10.1038/s43016-020-00171-y.

Li, L.G., Xia, Y. and Zhang, T., 2017. Co-occurrence of antibiotic and metal resistance genes revealed in complete genome collection. *The ISME Journal*, 11(3), pp. 651–662. https://doi.org/10.1038/ismej.2016.155.

Li, X., Chen, L., Mei, Q., Dong, B., Dai, X., Ding, G. and Zeng, E.Y., 2018. Microplastics in sewage sludge from the wastewater treatment plants in China. *Water Research*, 142, pp. 75–85. https://doi.org/10.1016/j.watres.2018.05.034.

Liu, J., Gao, Z., Wang, M., Li, Y., Shi, M., Zhang, H. and Ma, Y., 2019. Hydrochemical characteristics and possible controls in the groundwater of the Yarlung Zangbo River Valley, China. *Environmental Earth Sciences*, 78(3), p. 76. https://doi.org/10.1007/s12665-019-8101-y.

Liu, J., Gao, Z., Wang, Z., Xu, X., Su, Q., Wang, S., Qu, W. and Xing, T., 2020. Hydrogeochemical processes and suitability assessment of groundwater in the Jiaodong Peninsula, China. *Environmental Monitoring and Assessment*, 192, pp. 1–17. https://doi.org/10.1007/s10661-020-08356-5.

Lobelle, D. and Cunliffe, M., 2011. Early microbial biofilm formation on marine plastic debris. *Marine Pollution Bulletin*, 62(1), pp. 197–200. https://doi.org/10.1016/j.marpolbul.2010.10.013.

Lu, J., Zhang, Y., Wu, J., Wang, J. and Cai, Y., 2020. Fate of antibiotic resistance genes in reclaimed water reuse system with integrated membrane process. *Journal of Hazardous Materials*, 382, p. 121025. https://doi.org/10.1016/j.jhazmat.2019.12102.

Lusher, A., 2015. Microplastics in the marine environment: Distribution, interactions and effects. In: Bergmann M., Gutow L., Klages M. (eds.) *Marine Anthropogenic Litter* (pp. 245–307). Springer, Cham. https://doi.org/10.1007/978.

Magnusson, K. and Norén, F., 2014. Screening of microplastic particles in and down-stream a wastewater treatment plant.

Mao, D., Yu, S., Rysz, M., Luo, Y., Yang, F., Li, F., Hou, J., Mu, Q. and Alvarez, P.J.J., 2015. Prevalence and proliferation of antibiotic resistance genes in two municipal wastewater treatment plants. *Water Research*, 85, pp. 458–466. https://doi.org/10.1016/j.watres.2015.09.010.

Mani, T., Hauk, A., Walter, U. and Burkhardt-Holm, P., 2015. Microplastics profile along the Rhine River. *Scientific Reports*, 5(1), pp. 1–7. https://doi.org/10.1038/srep17988.

Mato, Y., Isobe, T., Takada, H., Kanehiro, H., Ohtake, C. and Kaminuma, T., 2001. Plastic resin pellets as a transport medium for toxic chemicals in the marine environment. *Environmental Science & Technology*, 35(2), pp. 318–324. https://doi.org/10.1021/es0010498.

Murray, A. and Örmeci, B., 2020. Removal effectiveness of nanoplastics (<400 nm) with separation processes used for water and wastewater treatment. *Water*, 12(3), p. 635. Https://Doi.Org/10.3390/W12030635.

Murphy, F., Ewins, C., Carbonnier, F. and Quinn, B., 2016. Wastewater treatment works as a source of microplastics in the aquatic environment. *Environmental Science & Technology*, 50(11), pp. 5800–5808. https://doi.org/10.1021/acs.est.5b05416.

Kumar, M., Kuroda, K., A. K., Patel, N., Bhattacharya, P., Joshi, M. and Joshi, C. G. (2021). Decay of SARS-CoV-2 RNA along the wastewater treatment outfitted with Upflow Anaerobic Sludge Blanket (UASB) system evaluated through two sample concentration techniques. *Science of the Total Environment*, 754, p. 142329.

Napper, I.E., et al. 2015. Characterisation, quantity and sorptive properties of microplastics extracted from cosmetics. *Marine Pollution Bulletin* 99(1-2), pp. 178–185. https://doi.org/10.1016/j.marpolbul.2015.07.029.

Napper, I.E. and Thompson, R.C., 2016. Release of synthetic microplastic plastic fibres from domestic washing machines: Effects of fabric type and washing conditions. *Marine Pollution Bulletin*, 112(1–2), pp. 39–45. https://doi.org/10.1016/j.marpolbul.2016.09.025.

Paul-Pont, I., Tallec, K., Gonzalez-Fernandez, C., Lambert, C., Vincent, D., Mazurais, D., Zambonino-Infante, J.-L., Brotons, G., Lagarde, F., Fabioux, C., Soudant, P. and Huvet, A. 2018. Constraints and priorities for conducting experimental exposures of marine organisms to microplastics, *Frontiers in Marine Science*, 5, p. 252.

Poole, K., 2017. At the nexus of antibiotics and metals: The impact of Cu and Zn on antibiotic activity and resistance. *Trends in Microbiology*, 25(10), pp. 820–832. https://doi.org/10.1016/j.tim.2017.04.010.

Patel, A. K., Bhagat, C., Taki, K. and Kumar, M. (2020). Microplastic vulnerability in the sediments of the Sabarmati river of India. In: *Resilience, Response, and Risk in Water Systems* (pp. 127–138). Springer, Singapore.

Razanajatovo, R.M., Ding, J., Zhang, S., Jiang, H. and Zou, H., 2018. Sorption and desorption of selected pharmaceuticals by polyethylene microplastics. *Marine Pollution Bulletin*, 136, pp. 516–523. https://doi.org/10.1016/j.marpolbul.2018.09.048.

Sadri, S.S., and Richard C.T. 2014. On the quantity and composition of floating plastic debris entering and leaving the Tamar Estuary, Southwest England. *Marine Pollution Bulletin* 81(1), pp. 55-60. https://doi.org/10.1016/j.marpolbul.2014.02.020.

Sarkar, D.J., Sarkar, S.D., Das, B.K., Manna, R.K., Behera, B.K. and Samanta, S., 2019. Spatial distribution of meso and microplastics in the sediments of river Ganga at eastern India. *Science of the Total Environment*, 694, p. 133712. https://doi.org/10.1016/j.scitotenv.2019.133712.

Sajil Kumar, P.J. and James, E.J., 2016. Identification of hydrogeochemical processes in the Coimbatore district, Tamil Nadu, India. *Hydrological Sciences Journal*, 61(4), pp. 719–731. https://doi.org/10.1080/02626667.2015.1022551.

Sivagami, K., Vignesh, V.J., Srinivasan, R., Divyapriya, G., Nambi, I.M., 2020. Antibiotic usage, residues and resistance genes from food animals to human and environment: An Indian scenario. *Journal of Environmental Chemical Engineering*, 8(1), p. 102221. https://doi.org/10.1016/j.jece.2018.02.029.

Sruthy, S. and Ramasamy, E.V., 2017. Microplastic pollution in Vembanad Lake, Kerala, India: the first report of microplastics in lake and estuarine sediments in India. *Environmental Pollution*, 222, pp. 315–322. https://doi.org/10.1016/j.envpol.2016.12.038.

Storck, F.R., Kools, S.A. and Rinck-Pfeiffer, S., 2015. *Microplastics in fresh water resources.* Global Water Research Coalition, Stirling, South Australia, Australia.

Talvitie, J., Mikola, A., Koistinen, A. and Setälä, O., 2017. Solutions to microplastic pollution–Removal of microplastics from wastewater effluent with advanced wastewater treatment technologies. *Water Research*, 123, pp. 401–407. https://doi.org/10.1016/j.watres.2017.07.005.

Talvitie, J., Heinonen, M., Pääkkönen, J.P., Vahtera, E., Mikola, A., Setälä, O. and Vahala, R., 2015. Do wastewater treatment plants act as a potential point source of microplastics? Preliminary study in the coastal Gulf of Finland, Baltic Sea. *Water Science and Technology*, 72(9), pp. 1495–1504. https://doi.org/10.2166/wst.2015.360.

Tan, X., Yu, X., Cai, L., Wang, J. and Peng, J., 2019. Microplastics and associated PAHs in surface water from the Feilaixia Reservoir in the Beijiang River, China. *Chemosphere*, 221, pp. 834–840. https://doi.org/10.1016/j.chemosphere.2019.01.022.

Thompson, A., 2018. From fish to humans, a microplastic invasion may be taking a toll. Scientific American, 4.

Van Emmerik, T. and Schwarz, A., 2020. Plastic debris in rivers. *Wiley Interdisciplinary Reviews: Water*, 7(1), p. e1398. https://doi.org/10.1002/wat2.1398.

Velzeboer, I., Kwadijk, C.J.A.F. and Koelmans, A.A., 2014. Strong sorption of PCBs to nanoplastics, microplastics, carbon nanotubes, and fullerenes. *Environmental Science & Technology*, 48(9), pp. 4869–4876. https://doi.org/10.1021/es405721v.

Wang, F.H., Qiao, M., Lv, Z.E., Guo, G.X., Jia, Y., Su, Y.H. and Zhu, Y.G., 2014. Impact of reclaimed water irrigation on antibiotic resistance in public parks, Beijing, China. *Environmental Pollution*, 184, pp. 247–253. https://doi.org/10.1016/j.envpol.2013.08.038.

Wang, Q., Bai, J., Ning, B., Fan, L., Sun, T., Fang, Y., Wu, J., Li, S., Duan, C., Zhang, Y. and Liang, J., 2020. Effects of bisphenol A and nanoscale and microscale polystyrene plastic exposure on particle uptake and toxicity in human Caco-2 cells. *Chemosphere*, 254, p. 126788. https://doi.org/10.1016/j.chemosphere.2020.126788.

Wu, C., Zhang, K., Huang, X. and Liu, J., 2016. Sorption of pharmaceuticals and personal care products to polyethylene debris. *Environmental Science and Pollution Research*, 23(9), pp. 8819–8826. https://doi.org/10.1007/s11356-016-6121-7.

Xu, S., Ma, J., Ji, R., Pan, K. and Miao, A.J., 2020. Microplastics in aquatic environments: Occurrence, accumulation, and biological effects. *Science of the Total Environment*, 703, p. 134699. https://doi.org/10.1016/j.scitotenv.2019.134699.

Zarfl, C. and Matthies, M., 2010. Are marine plastic particles transport vectors for organic pollutants to the Arctic? *Marine Pollution Bulletin*, 60(10), pp. 1810–1814. https://doi.org/10.1016/j.marpolbul.2010.05.026.

Zhao, S., Zhu, L., Wang, T. and Li, D., 2014. Suspended microplastics in the surface water of the Yangtze estuary system, China: First observations on occurrence, distribution. *Marine Pollution Bulletin*, 86(1–2), pp. 562–568. https://doi.org/10.1016/j.marpolbul.2014.06.032.

Chapter 4

Solid waste and landfill leachate

The transient sources of emerging microbes and legacy contaminants for groundwater pollution

Trinath Biswal and Krushna Prasad Shadangi
Veer Surendra Sai University of Technology

CONTENTS

DOI: 10.1201/9781003157465-5

4.1 INTRODUCTION

Rapid industrial activity is the cause of the economic growth of a country, and it is related to globalization, liberalization, and also the rapid increase in the population growth of the world. Hence, a billion tons of industrial and municipal solid wastes (MSWs) are produced every day throughout the world. The MSW significantly increased every year due to the rapid increase in urbanization and human wants. In most of the developed and developing countries, especially in the high populated cities, huge amounts of MSW are generated, but the lack of proper management and disposal of these huge quantities of solid wastes and landfill sites create problems (Eggen et al., 2010). Landfilling is considered as the simplest, most beneficial, and popular method mainly preferred by the developing and developed countries throughout the globe for disposal of MSW without adequate precautions. The generation and disposal of MSW with the population density is indistinctly interlinked with each other. The open dumping of MSW in the landfill site is the cause of many environmental issues, including water contamination, air pollution, soil contamination, and many health hazard problems. The anaerobic biodegradation of the solid wastes in the landfill area is the cause of the generation of a black thick liquid, which is highly toxic, acidic in nature, and commonly termed as leachate (Nagarajan et al., 2012). Leachate passes through various solid wastes, suspended solids, extracts solutes along with some toxic materials from the landfill site, and finally percolates inside the surface of the earth and mixes with groundwater. The composition and toxicity of the leachate depend on the age of the landfill site, nature of the solid wastes dumped in that landfill site, and climatic condition design of the landfill (Kulikowska & Klimiuk, 2008, Kumar, 2012, Singh et al., 2020). People nearby the landfill area are continuously affected by a number of serious chronic diseases due to the use of groundwater contaminated by leachate. The leachate addition is becoming worse because many of the landfills have no boundary and suitable leachate collection system, which is the cause of an increase in the rate of percolation by the layers of the landfill and contaminated groundwater of the surrounding locations (Luo et al., 2019). Leachate is not only the cause of groundwater contamination but also the cause of surface water pollution in that nearby area. The wastes present in the landfill sites undergo infiltration from precipitation and are then percolated through the solid wastes and take a number of organic and inorganic materials (Renou et al., 2008). Then it flows out from the wastes of the landfill as a black-colored thick viscous toxic liquid and is accumulated at the lowermost part of the landfill site, commonly called as leachate, and then on infiltration in the soil finally mixed with the nearest sources of groundwater. The leachate generated from municipal landfill sites is a highly toxic complex viscous liquid and mostly contains organic materials; inorganic compounds such as salts of NH_4^+, Mg^{2+}, Ca^{2+}, Fe^{2+}, Na^+, K^+, SO_4^{2-}, and Cl^- along with many toxic heavy metals (HMs), including Cr, Cu, Cd, Pb, Zn, and Ni; and some xenobiotic organic materials. The composition of leachate varies widely on the basis of some interacting aspects such as the depth of the waste deposited, the nature of the waste, availability of moisture, atmospheric oxygen, the design of the landfill, landfill operation, and age (Pujiindiyati et al., 2019).

4.2 LANDFILL SITES

4.2.1 Landfill and it's stabilization

Landfill is otherwise termed as a tip, rubbish dump, dump, dumping ground, garbage dump, or the site for the disposal or dumping of waste materials. The landfill is the simple, cost-effective, and oldest method for the disposal of solid wastes. The burial of solid wastes by soil daily, at intermediate time interval and final cover by soil, started in the 1940s. In the past ages, the refuses or wastes are simply dumped or thrown in heaps or dumped into the pits in archeology and commonly termed as a maiden. Landfills are neither some holes on the earth's surface nor open dumps. A simple and basic landfill is generally a dumping yard of solid waste covered with soil and popularly known as a sanitary landfill. The most important gaseous pollutants generated in the landfill sites are CH_4, NH, and CO_2; these gases, in addition to troposphere, cause the harmful effect to the biotic community (Przydatek & Kanownik, 2019).

The large, planned, and sophisticated landfills are designed in a properly planned manner by creating a trench on the earth's surface. The special design of the landfill protects the mixing of the leachates to percolate inside the surface of the earth and mixed with the aquifer causing the groundwater contamination. Nowadays, in some countries, sensors are used in landfill sites to identify emissions of landfill gas (LFG) and leachate concentration. The term "landfill" means the sanitary landfill or municipal landfill. The technique of landfill was first of all introduced before the 20th century in order to minimize the bad practices of open dumping and other practices of unsanitary waste disposal (Cuartas et al., 2018).

The sanitary landfill is considered as an engineered process projected as a bioreactor or biological reactor, where the complex organic wastes are broken down into nontoxic or comparatively less toxic, simple products by the microbial community. The reactor in the landfill must be intended and operated on the basis of standards and guidelines of the proper regulatory agency. In sanitary landfills, the wastes first undergo aerobic decomposition followed by different stages such as acetogenesis, acidogenesis, and methanogenesis. Normally, the solid waste enriched with organic materials will decompose at a greater rate and be converted into smaller nontoxic or less toxic substances, which dissolve in water and gets converted into a liquid phase. These liquid forms of the smaller organic products undergo hydrolysis, and the hydrolyzed substance obtained undergoes transformation and volatilization into CO_2 and CH_4, and a fraction of it remains as a liquid phase called leachate (Lombardi et al., 2017).

Since the wastes of the organic compounds undergo a quick decrease in the volume, a small fraction of volume of these materials is converted into leachate. In the meantime, the chemical oxygen demand (COD) of the leachate enhanced with the increase in the number of recalcitrant materials relative to high reactive organic material present in the leachate. The proper stabilization and conversion of the waste materials mainly depend upon the population and activity of the different kinds of microbes in the landfill and the nutrients necessary for microbial growth and reproduction. There are some landfill areas, which are only used for the purpose of waste management such as temporary storage of the wastes, for consolidation and transfer or for the different stages of processing the wastes including treatment, sorting, recycling, etc. (Abdel-Shafy & Mansour, 2018).

The degradation of the fraction of the organic matter present in the wastes in addition to the percolating rainwater is the cause of the production of a highly toxic and contaminated black thick liquid known as leachate. The leachate generated possesses a high concentration of toxic metallic contaminants, biological oxygen demand (BOD), COD, ammonium cations (NH_4^+), nitrogen, total solids, and turbidity and comparatively many times more than the usual urban wastewater effluents. The leachate produced because of the release of moisture, infiltration, and precipitation is the major cause of the increase in suspended or soluble organic matters, NH_4^+ nitrogen, and inorganic ions; mainly, HMs might be the cause of serious environmental hazards if the leachate generated is not properly collected, stored, treated, and disposed.

The most significant danger to the human health is due to leachate, the use of groundwater containing leachate. The leachate created from the landfill sites normally collected in the low-lying areas, where it percolates inside the soil surface causing contamination of the groundwater and sometimes flows on the land surface and pollutes surface water.

The chemical composition of the leachate is influenced by the kind of wastes, % of moisture, composition of wastes, degree of compaction, size of the particle, hydrology of the landfill site, age of the landfill, climatic condition, and any other specific conditions at the site of the landfill such as the design of the landfill and the kinds of liners used. The properties of leachate also depend on the pretreatment technology of the solid wastes such as separation of recyclable material including plastics, paper, metals, and glass and shredding and bailing off the waste materials (Ağdağ, 2011; Vaverková et al., 2020).

4.2.1.1 Urbanization and population growth

Now urbanization is becoming an important global issue and cause of extensive population growth, and its consequences are found to be more pronounced in developed and developing countries. The population growth and rapid increase in urbanization are the major cause of the requirement of more landfill. The more the growth of population is, especially in urban areas, the more are the quantity or volume of solid wastes generated and the number of landfill sites required. In urban areas, many toxin products, plastic materials, are present in the generated solid wastes, whose management in the landfill requires special techniques.

The rapid rate of increase in population decreases the opportunities, especially in the agricultural sector, because of more occupancy in urban areas and more contribute toward urbanization. The unanticipated immigration is the cause of the rapid growth of slum residential areas and informal unplanned housing in many areas throughout the city in the developing countries of the world.

In most of the major cities of the world, the rapid increase in population in some cases overcomes the capacity of the authorities to deliver the basic facilities, where the waste management is a major factor. Urbanization is directly proportional to the volume of the generation of solid wastes, haphazard waste dumping, handling of wastes,

Figure 4.1 Impact of landfill on environment.

and, ultimately, the cause of degradation of the environment. Currently, the uncontrolled and unplanned dumping of solid wastes due to financial limitations or the adoption of improper technology or institutional dimness made serious public health and ecological problem (Swati et al., 2018; Xue et al., 2015). The landfill is the cause of degradation of the three basic components of the globe such as air, soil, and water and the natural environment of the globe (Figure 4.1).

4.2.1.2 Agricultural waste

Many farmers use animal manure, especially, cow dung composts, as the source of nutrients and fertilizers in different agricultural practices. The compost of agricultural or farm waste is highly toxic, and the haphazard dumping of compost in the agricultural field is the cause of contamination of both soil and water. In 2014, it was established that agricultural waste can be utilized for the stabilizing of soil in the landfill area. The major components of agricultural waste materials are palm oil ash and rice husk ash (RHA), which can be used as a sustainable substitute for soil stabilization instead of using in the manufacturing of conventional Portland cement. Landfill soil is stabilized by itself or on mixing with laterite clay soil by using palm oil fuel ash (POFA). The combination of landfill soil and laterite clay in equal proportion can be stabilized by using 20% of RHA and POFA of 50:50 compositions and provides the highest compressive strength relative to that of other stabilizers (Ayilara et al., 2020; NidzamRahmat et al., 2014).

4.2.1.3 Industrial waste

The industrial wastes include the wastes from power plants, chemical manufacturing units, printing industry, dye and textile industry, metal processing industries, steel making industries, and oil refineries. Although the majority fraction of industrial waste is biodegradable, it is toxic in nature and generally added to landfill sites to stabilize the landfill. The industrial hazardous wastes are of different kinds and added due to different industrial activities which are as follows:

- **Chemical manufacturing units**: The wastes ejected from chemical manufacturing plants cantinas, some toxic bases, and acids (present in the used solvents). It contains some reactive waste materials along with complex organic substances.
- **HM solutions**: The waste materials of steel plants, manufacturing industries, metal processing industries, printing industries, and sludge inks solvents containing toxic HMs either in the pure form or in the combined state.
- **Petroleum refining industry**: The wastewater effluent of petroleum refining industries contains benzene, toluene, and some other hydrocarbons as in the form of sludge.
- **Leather products**: The industries manufacturing leather products ejected as solid wastes to the landfill, which are rich in benzene and toluene.
- **Paper industry**: The injectable solvents from paint and paper industries are particularly enriched in toxic metals.
- **Metal product manufacturing units**: The solid industrial wastes and sludge rejected from metal product manufacturing units contain a higher % of HMs, cyanide waste products, painting wastes, etc.

In addition to hazardous wastes, the landfill containing industrial wastes also contains some nonhazardous materials besides MSW (Kan, 2009 & Gworek et al., 2015).

The landfill rich in industrial wastes is the cause of many negative impacts on our environment as follows:

- **Contamination of air, water, and land resources:** The industrial wastes of the landfill sites are the cause of air, water, and soil contamination. The air pollution is due to industrial dust particles and the release of CH_4 and CO_2 from the landfill sites, which are potent greenhouse gases. The odor produced from landfill sites is also a cause of air pollution. Water pollution is due to the addition of direct effluents and leachate to the nearest water sources, which are enriched in toxic metals and acidic in nature. Industrial wastes contain an extremely high % of toxic metals and, therefore, leads to the cause of soil contamination.
- **Harms related to hazardous industrial wastes:** The wastes generated due to the industrial operation from various industries contain many hazardous materials, including metallic contaminates. Because of the inappropriate disposal of this hazardous waste to the landfill sites and in combination with rainwater, stormwater finally mixes with underground water and nearest sources of surface water, causing potential harm or injury not only to human health but also for other organisms and degrade our ecosystem.
- **Impact on various communities:** The addition of excessive industrial wastes to the landfill site is the cause of the reduction of the values of the properties adjacent to the landfill. Since industrial wastes contain a comparatively higher % of flammable materials, it may be the cause of fire hazard risk. It affects the community, due to the production of dust, pests, and odor (Fazzo et al., 2017; Ferronato & Torretta, 2019; Giusti, 2009).

4.2.1.4 Municipal solid waste

The MSWs are normally termed as garbage or trash and mainly consist of the products thrown after use and packaging materials, broken unused furniture, clothes after use, plastic home utensils, food scraps, unused home appliances, newspapers, e-waste, batteries, etc. The major source of these wastes is homes, hospitals, schools, and businesses areas.

Most of the fraction of MSW both in developing and developed countries are finally settled in the landfills. In many of the developed countries of the world, suitable designed sanitary landfills are mainly used, which mainly consist of bottom liners, leachate collection systems, gas collection systems, leachate treatment facility, final covers, gas treatment facility, water, and air monitoring systems. There is a complex sequence of the chemical, biological, and physical process of the MSW carried out within the landfill, leading to the reduction of the wastes. Although all these processes are naturally occurring processes, their efficiency can be controlled by regulating the internal condition of the landfill (Moya et al., 2017; El-Fadel & Khoury, 2000).

The degradation of MSW is the cause of the production of biogas or LFG and leaching the waste dumped materials from the landfill. The production of LFG and leachate generation can be controlled by adopting suitable scientific procedures. After closure, the landfill must have to be properly controlled and managed in order to avoid a negative impact on human being, domestic animals, wildlife, and our environment. If the municipal authority feels that the landfill has no longer threat to human and animal health after the postclosure care, then it is finally brought to an end (Chugh et al., 1999).

The landfill used for dumping of MSW is normally a distinct region or a particular archeological site, which receives the household wastes, commercial wastes, and biomedical wastes generated from cities or towns. For the construction of a proper MSW landfill, we have to follow the following:

- **Restriction of location:** The landfill must have to be built in an appropriate geological area far away from wetlands, flood plains, city area, and locality or other restricted areas.
- **Requirement of composite liners:** It is normally the flexible geomembrane that covers about 2 feet of compressed clay soil lining the bottom and all the sides of the landfill area. This flexible membrane liner protects the underlying soil and groundwater from the leachate generated from the landfill.
- **System for leachate collection and removal:** A suitable system must have to be kept in order to collect the leachate at the top of the liners site and from the landfill for proper treatment and disposal.
- **Operating procedures:** In this step, the wastes of the landfill are frequently covered by the suitable depth of soil every day in order to avoid bad odor; control litter, pests, insects, and rodents; and protect public health (Abd et al., 2015; Tałałaj et al., 2019).

The landfills of MSW are popular and widely used as a waste management device and landfill microbiology, which serves as an important tool for the degradation of solid wastes in the local ecosystem. Bioprocessing is considered as an important and the primary microbiology of MSW landfills. The bioprocessing of wastes in the landfills

Figure 4.2 Various sources of MSW.

originates from the substrate metabolism and cometabolism through a number of syntrophic microorganisms, which is the cause of partial transformation of complex organic substrates into simpler polymeric materials, and completes the process of mineralization into the water, inorganic salts, and some gases such as biofuel gas, particularly, CH_4.

The decomposition of the waste is characterized by the interactions and evolution of various groups of bacteria, fungi, and archaea because of favorable abiotic and biotic conditions in the landfills. This favorable condition is the cause of the occurrence of hydrolytic, acetogenic, methanogenic, and fermentative processes. But, however, bioprocess dynamics and functional diversity play a key role in influencing the factor for *in situ* bioprocesses that occur in the landfills and cause of degradation of solid wastes in a scientific way and regulate the efficiency of landfills (Meyer-Dombard et al., 2020; Sekhohola-Dlamini & Tekere, 2019). The different major sources of MSW are shown in Figure 4.2.

4.2.2 Advantages or benefits of landfills

4.2.2.1 As an outstanding source of energy

In the landfill, due to anaerobic decomposition of the solid wastes of complex forms of organic or biological matter, CO_2 and CH_4 are produced. In the proper management technique, the gaseous components are collected, properly filtered, purified, and finally used as a source of energy. In the United States, the third source of CH_4 production is from landfill sites. If the gas produced from the landfill is large enough, then it can also be used for the operation of the internal combustion engine, gas turbine, and also the production of electricity. Almost about 70% of the projects of landfill electricity use reciprocating piston (RP) engines (a particular type of internal combustion engine) because of high efficiency, cost-effectiveness, and proper size compatible with most of the landfills. The RP engines normally attain an efficiency of 25%–35% by using LFG (Malinauskaite et al., 2017; Kalyani & Pandey, 2014).

4.2.2.2 *Eco-friendly modern landfills*

The old landfills are generally open-air dump where everything is thrown at it without any care and sanitation or planned manner. In the sanitary landfill, the wastes are covered with soil to minimize the exposure, block LFG, reduce the stench, and decrease the leachate generation. Leachate is a black thick toxic liquid, which drains from the landfill and is properly treated before addition to the soil or used in some other purposes and prevents groundwater contamination. There are many municipal authorities who are focusing on a system of zero-leachate discharge in order to eliminate the risk of underground water contamination. Proper leachate management and soil lining generally prevent damage and seepage. The Sudokwon Landfill houses are one of the largest LFG power plants in the world that convert LFG into electricity. The biogas is generated from the landfill on mixing with the compressed natural gas and used successfully in a different automobile (Kerdsuwan et al., 2015).

4.2.2.3 *Keep the cities clean*

There are many small cities in the world, where there are no landfills or no effective waste management technology for reduction and minimization of solid wastes; therefore, people dumped the solid wastes haphazardly in different vacant spaces near the locality, which is the cause of the health hazard effect of animals and human beings. If wastes are disposed of the landfill sites, then local garbage will be gathered in a particular area without the discrete dumping creating any harm to human beings and degradation of the environment. Hyper consumption is a swear word of our modern eras, which is perfectly applicable to human beings. People in the modern world generate massive amounts of wastes, among which a major fraction of it is disposed into landfills. In modern landfills, these huge quantities of wastes are converted into energy by using incinerators. But, however, billions of tons of wastes or garbage, including e-waste and plastic wastes, finally enters into the ocean and not dumped into the landfill, leading to dangerous marine pollution. These nonbiodegradable wastes, especially, plastics, get converted into microplastic and choke marine life, which is the key factor for disturbing the population of marine zooplankton. Marine zooplankton plays a vital role in eliminating CO_2 from the oceanic environment and troposphere. CH_4 generated from the landfill is a potent greenhouse gas; before it gets ejected into the atmosphere, it can be used as a source of energy. The garbage can be composted in the landfill, and efficient landfill can be done by adopting the optimum population of microbes and temperature to create a nutrient-dense soil conditioner (Brand et al., 2017).

4.2.2.4 *Segregation of hazardous wastes*

A landfill is a platform that provides ultimate dumping of nonrecyclable, recyclable, hazardous, and nonhazardous wastes, and these wastes are separated before dumping in the landfill. The landfill site provides a platform of hazardous materials; therefore, it can be segregated from the public. Household hazardous wastes are defined as any toxic wastes produced from the domestic source and finally disposed in the landfill area. It normally includes the wastes such as electronic, electrical, and computer parts. Household hazardous wastes are corrosive, ignitable, toxic, and reactive ingredients. On the basis of the term of Europe countries, household hazardous wastes may be defined as any waste products that when dumped in the landfill, on composting or on incineration. According to the National Household Hazardous Waste Forum (UK), any waste materials rejected through household sources, which is very difficult to dispose of or treat and is the cause of risk toward human health and our environment due to its biological and chemical properties which are considered as household hazardous wastes (Inglezakis & Moustakas, 2015). The landfill enriched with containerized waste or hazardous solid wastes must have to be regulated and managed more seriously as compared to the landfill of MSW. The landfills where hazardous wastes are disposed of are called secure landfills, where there is 3 m of the distance of separation between the bottom part of the landfill and the groundwater table or underlying bedrock. The secure landfill containing hazardous wastes must have to contain suitable leachate collection systems and two impermeable liners. The upper part of the system checks the accumulation of leachate trapped in the landfill, and the lower part function as a backup; then the leachate collected is pumped into the treatment plant. For minimizing or reducing the leachate from the landfill, a resistant cover or cap has to be kept over the exhausted landfill (Joshi & Ahmed, 2016).

4.2.2.5 Cheap management of solid wastes

The transportation of solid waste or garbage is a problem for us, but if it is transported to the nearest landfill sites, then it becomes cost-effective. This is the cause of a decrease in environmental pollution and prevents damage, due to haphazard dumping of garbage waste. The major problem relating to the MSW is its improper technique of collection, disposal, and management. The lack of disposal and management of MSW is a significant environmental issue throughout the world and leads to contamination of air, soil, and water. This environmental issue is normally associated with the health hazard effects of human beings, and if it is properly managed at the landfill site level, then it becomes cost-effective. Although the hazardous wastes originated from households are in small quantities, their separate disposal and management are a problem and expensive, but their disposal along with other MSW is cost-effective. The separate disposal of MSW, industrial wastes, and medical wastes is a risk and costly, whereas if all these are disposed of collectively at the landfill site, then it is cost-effective and less harmful (Ma et al., 2016).

4.2.2.6 Landfill promotes local business and jobs

The management of landfill sites and segregation of biodegradable and nonbiodegradable wastes and hazardous and nonhazardous wastes of any country of the world requires a huge workforce. The government or authority of the municipality spends million dollar for this work and many job opportunities for local people are created in different stages of management in landfill starting from the collection, storage, disposal, transportation, treatment, and management in the landfill sites. Technical manpower is required in the long-term time horizon, where land filling can convert the wastes into the short-term gas generation and also offer a long-term supply of carbon. After the end of the landfill action, people are engaged for its aftercare management and proper renovation (Wath et al., 2010).

4.3 DISADVANTAGE OF LANDFILLS

The wastes put in the landfill have many advantages in various ways, but there are still many disadvantages associated with it, which are as follows:

4.3.1 Landfills are partly responsible for the change in climate

The MSW landfill is filled with both biodegradable and nonbiodegradable solid wastes. It was estimated that one ton of biodegradable solid waste can able to produce 400–500 cubic meters of LFG due to anaerobic biodegradation. The major constituents of LFG are CO_2 and CH_4, and in addition to these, the other different gases produced from the landfill sites are N_2O, perfluorocarbons, SF6, and hydrofluorocarbons. CH_4 is a significant greenhouse gas and 25 times more potent than CO_2. Hence, there are thousands and thousands of exhausted and active landfills throughout the world, which emit huge quantities of greenhouse gases and provide a significant contribution toward global climate change. The consequence of global climate change is hurricanes, cyclones, and devastating droughts along with many diseases and ultimately affects the life process of human beings. The different negative impacts of LFGs on human beings and the environment are as follows:

- Loss of biodiversity
- Human beings and other organisms suffering unfavorable situation and struggle for their survival
- Alternation of the food chain and food web
- Change in inhabitants of human being, wildlife, and other biological community
- Cause of instability of the ecosystem.
- Promotes deforestation and insect infestation
- The environment becomes more conductive for breeding
- Different epidemic situations are born because of the increase in insect-borne diseases
- Increase in raise in the sea level affecting marine life along with territories and Pacific Island countries (Zhang et al., 2019).

4.3.2 Methane easily catches fire

Methane is an extremely inflammable gas; if the landfill site is in a closet space with a poor ventilation facility, then CH_4 generated from the landfill can easily catch fire, especially in summer, and the fire can spread throughout the landfill within a few seconds causing hazards to the native organisms. Since CH_4 is highly combustible in nature, it sometimes spontaneously catches fire especially in winter and causes toxic emissions. This toxic gaseous emission due to fire is higher in winter and causes serious health hazard risk. The emission of toxic fumes contains a colloidal form of tiny particles, particulate matter, along with some toxic contaminant gases, which has a short-term negative impact on our respiratory system, and its long-term effect is cancer and different metabolic diseases. There are two kinds of landfill fires normally identified. The first one is surface fires, and the other one is seated fires. The surface fire generally occurs in the landfill of underdeveloped countries, where there is a lack of coverage of the wastes either daily or in some intermediate interval. If the landfill is not properly covered by soil daily, then aerobic biodegradation occurs, air interruption supplies the required oxygen, resulting in the increase in the decomposition of wastes by microbes and produces more heat leading to a spontaneous fire. If it is unimpeded, then the fire, in specific, tends to burn deeper into the waste products, ensuing in deep installed fires. The landfill fire is particularly hazardous, if it emits toxic fumes, due to the combustion of a wide range of nonbiodegradable materials. The vital components that cause health hazard risk due to landfill fire are CO, H_2S, volatile organic compounds (VOCs), furans, dioxins, etc. (Pecorini & Iannelli, 2020).

4.3.3 Cause of water and soil contamination

The impact of landfills is devastating because different toxic chemicals, leachates, gaseous components, fumes, and toxins leak from landfills and ultimately mix with groundwater, surface water, and soil causing a health risk to both animals and human beings. Normally, the MSW of the landfill penetrates metals, carbon compounds, ammonium salts, and other toxins into the nearby soil and changes the soil texture, decreases the soil fertility, and kills the essential fauna and flora. The MSW in landfills is the cause of a number of environmental problems, including the combustion of LFG, foul smells, and leakage of leachate, but among these, the leakage of leachate is the most serious impact because it is the cause of both surface water and groundwater contamination. Leachate contains a high % of HMs, complex organic materials, and some toxins and is subsequently acidic in nature. The dumping of HMs such as Cu, Zn, Hg, Cr, Cd, and Ni is continuously increasing in landfill wastes and ultimately causes significant damage to the soil (Vodyanitskii, 2016).

4.3.4 Wildlife and birds

Many animals and certain birds collect their food from the open landfill sites and consume waste plastics, woods, aluminium foil, HM products, drywall, and other hazardous materials as their food source. These hazardous materials are nonbiodegradable products and ingested in our intestine, causing serious health effects, and sometimes may cause death. The garbage dump is responsible to change in the food habits of some wild animals and birds. The major culprit in the landfill site is nondegradable plastic wastes; these wastes are by mistake taken by the wild animals and birds, and they suffer complications in health and disruptive reproductive patterns. There are many toxic chemicals that are leached from plastic materials in the landfill causing environmental contamination. On the basis of the observed behavior of animals nearer to the site of landfills, researchers classified them into three various groups. The first group is peckers, which consists of 19 species of birds. The second group is handlers; these are species of the animals having hand-like body parts on their front, and these animals are able to segregate food from other wastes. The last recognized group is gulpers, which have no hand-like organs and mouthparts and, therefore, unable to separate food from plastic wastes. The last group recognized is gulpers, who have no hand-like organs or mouthparts in their body, hence, unable to separate food from plastic wastes from the landfill. Among the different kinds of animals, large-billed crow exhibits the highest rate of contact with the plastic wastes, whereas Sambar has the lowest rate of contact. Among all these three kinds of animals, peckers and handlers are found to be two times more in contact with the plastic wastes relative to gulpers. Hence, it was concluded that the landfill sites are responsible for both toxicological and physical impacts on animal life and birds. The plastic wastes ingested are the cause of lethal injuries and damage to the digestive tract of the animals, leading to starvation, reduced fitness, ulceration of the stomach, growth problems, and premature death (Plaza & Lambertucci, 2017).

4.3.5 Negative health impact on human beings

People living nearby the landfill or on prolonged exposure to landfills are more prone to respiratory disorders, cancer, fertility loss, and other developmental defects in children. The short-term exposure to LFGs, NH_3 and H_2S, is the cause of eye, throat, and nose irritation and asthma. In addition to that, there arise many health complaints due to open landfill, which includes weight loss, sleep disturbance, and chest pain. The gases such as CH_4, CO_2, and H_2S produced from the landfill decrease O_2 availability to the living tissues, disturbing coordination and causing fatigue, vomiting, nausea, and unconsciousness. Leachate is an important component generated in the landfill; the composition and toxicity of it depend upon many factors, which include the depth and composition of the wastes, accessibility of oxygen and moisture, the design of the landfill, age, and operational technique. The composition of the leachate generated from the landfill is basically the function of the degree of stabilization of the waste and age of the landfill. Groundwater is the potable drinking water in both rural and urban areas, and leachate is a key factor for the contamination of groundwater. There are some toxic chemicals such as cyanides, polychlorinated biphenyls, and mercury released in the untreated form, and its long exposure is the cause of many chronic diseases and even death. The colored plastic products are harmful to the biotic community because of the pigments, and HMs present in it are highly toxic. The different toxic metals present in the colored plastics include Cu, Pb, Se, Cd, and Co, which are highly toxic to human health. Hence, many industrialized countries in the world legally banned the production and use of colored plastic materials. The direct contact of plastic wastes is the cause of blood and skin infections. The incineration operation at the landfill is the cause of dust hazards and chronic respiratory diseases (Verma et al., 2016).

4.4 LANDFILL LEACHATES

Landfill leachate is defined as a thick black liquid effluent that is generated due to the percolation of rainwater through the solid wastes deposited in the landfill along with the presence of moisture in the wastes and the residues formed after the degradation of wastes. Leachate drained from the landfill varies widely in composition depending upon the age and kinds of wastes present in the landfill. Leachate contains both suspended and dissolved substances. Rainwater is the major influencing factor for the generation of leachate in the landfill site. The rainwater precipitation percolates into the deposited wastes of the landfill and binds to particles of nondissolved and dissolved wastes through the means of numerous chemical and physical reactions (Peixoto et al., 2018).

The tributaries of groundwater, biological decomposition of the wastes, and surface runoff also contribute a major role in the production of leachate. The liquid fraction of these wastes and moisture content of the soil cover of the landfill also contribute to some extent toward the generation of landfill. The rate of leachate discharge is closely related to the intensity of rainfalls, infiltration capacity of the soil, and surface runoff. The leachate generated is also greatly influenced by the climatic condition of the local environment because it affects the precipitation input into the landfill and losses owing to evaporation. Again, the rate of leachate production is also affected due to the degree of waste compaction at the top layer, properties of the disposed waste material, and water content. The technology for the treatment of leachate was applied to the landfills in European countries in the 20th century based on the system of "dilute and disperse." Normally, landfills were not properly sealed as a result of which the leachate produced from it undergoes leakage into the nearby surroundings and flows over the land surface, and a major fraction of it percolates inside the soil and mixes with groundwater. Such facilities for dispersion and dilution of leachate are no more adopted today. In the landfills, the waste decomposition basically happens through a number of chemical and biological processes mainly in four different pathways as follows:

- Aerobic phase
- Anaerobic acidogenic phase
- Methanogenic phase
- Waste stabilization phase

The quantity of leachates produced is mainly affected by the intensity of rainfall, precipitation, surface runoff, groundwater infiltration, degree of compaction within the landfill, and evapotranspiration. In order to minimize the leachate generation in the landfill, a number of techniques are adopted, including preventing the entry of water inside the landfill, lining surrounding the landfill, covering layers of soil, and waterproofing layers. The characteristics of the leachate can be determined by numerous chemical and physical parameters such as pH, suspended solids, COD, BOD, ammoniacal nitrogen (NH_4^+-N), total nitrogen, chloride, HM concentration, phosphorus, and alkalinity (Miao et al., 2019; Costa et al., 2019).

4.4.1 Composition of leachates

In many of the studies, it was found that the leachate generated from landfills is a potential source of pollution of soil and groundwater because of the leaching of hazardous substances. Leachate generally contains four different major components along with some minor toxic materials. The major components of the leachate are as follows:

Table 4.1 Classification of leachate on the basis of age (Li et al., 2010)

Sl. no.	*Leachate parameters*	*Age < 5 years*	*Age 5–10 years*	*Age > 10 years*
1	pH	<6.5	6.5–7.5	>7.5
2	COD/BOD_5	0.5–1.0	0.1–0.5	<0.1
3	COD in mg/L	>10,000	4,000–10,000	<4,000
4	Organic materials	80% VFAs	Humic and fulvic acid +5%–30% VFA	Fulvic acids and humic acids
5	Ammonia nitrogen in mg/L	<400	NA	>400
6	Nitrogen in g/L	0.1–0.2	NA	NA
7	TOC/COD	<0.3	0.3–0.5	>0.5
8	Biodegradability	High	Medium	Low
9	Heavy metals in mg/L	Low to medium	Low	Low

- Nutrients mainly nitrogen
- VOCs
- HMs, which are highly toxic to human health
- Toxic organic compounds and complex organic materials

There are various factors that affect the leachate composition, quality, and toxicity, which include the age of the landfill, quantity of total precipitation, seasonal fluctuations of weather condition, amounts of wastes, structural aspects, and kinds of wastes. The composition of leachate normally changes significantly on the basis of the age of the landfill (Abbas et al., 2009). Table 4.1 reflects the variation of the concentration of different parameters on the basis of age, and on the basis of age of the landfill, leachate produced is divided into three categories.

The concentration of COD matters, or organic materials decrease with the increase in the age of the landfill, whereas the concentration of NH_3 increases. The major impact of landfill leachate is the eutrophication of the surface water bodies and toxic effect on fauna, due to the presence of variability of contaminants. A typical landfill leachate has the constituent concentrations which includes 1,000 mg/L BOD5, 30 mg/L TP, 1,000 mg/L Ca, 18,000 mg/L COD, 225 mg/L TN, 300 mg/L K, 250 mg/L Mg, 500 mg/L Na, 300 mg/L SO_4^{2-}, 60 mg/L Fe, and 500 mg/L Cl^-. The composition of the leachate of the landfill mainly depends upon the chemical composition of the wastes, age, climatic conditions, and the rate of degradation of the solid wastes. Because of extreme toxicity, leachate is considered as a major threat to aquatic organisms and the health condition of surface water. The major issue in the treatment of leachate is due to very high concentrations of organic nitrogen and NH_3 and is found to be extremely toxic to aquatic organisms both chronically and acutely.

- The leachate composition is highly variable and heterogeneous in nature and depends upon the different stages of development of the wastes such as aerobic, methanogenic, anaerobic acid, and stabilization stages. There are about 200 hazardous pollutants found in the leachate, which can be divided into four different groups as follows:
- The organic matter in dissolved state (refractory organic materials, mainly fulvic and humic acids, volatile fatty acids).

- The macro inorganic components such as ammonia (NH_4^+ -N), Na^+, Ca^{2+}, K^+, Mn^{2+}, Mg^{2+}, Fe^{2+}, SO_4^{2-}, Cl^-, and HCO_3^-.
- The toxic HMs including Cr^{3+}, Cu^{2+}, Cd^{2+}, Ni^{2+}, Zn^{2+}, Pb^{2+}, and Hg^{2+}.
- The xenobiotic organic substances such as aromatic hydrocarbons, chlorinated aliphatic, phenols, plasticizers, and pesticides (Kjeldsen et al., 2002).

4.4.2 Toxicity of the landfill leachates

The landfill leachate is a complex mixture of various materials such as organic matter in dissolved form, HMs, inorganic macrocomponents, and a wide variety of xenobiotic organic materials. The presence of a higher % of these materials in the landfill leachates is dangerous and extremely toxic to human health, animals, and the environment. These toxic chemicals, if entered inside the bodies of animals and birds, undergo bioaccumulation or biomagnification and simultaneously passed into the food chain and food web and ultimately enter inside the human body.

The assessment of health hazard risks from the leachates is conventionally based on the assessment of individual chemical compounds through the process of chemical analysis. There are some toxic contaminants substances present in the landfill leachate, which is below the detection limits through chemical analysis, and the detection of these toxic substances is very problematic owing to the limitations of analytical techniques. But these chemical techniques of analysis cannot allow estimating the effects of contaminants on ecosystems. The study of ecotoxicology of the responses of the model for biological organisms shows the impact of toxic or poisonous substances and also provides some vital information completing the standard of the chemical analyses.

Although the landfill is a popular solid waste management process, there is a potential risk to the environment due to the addition of leachates from the landfill to the groundwater and surface water. The toxic effect of the leachates can be assessed by using the rapid, cost-effective, and a sensitive biological assay, which is valuable in assessing the risks associated and measures the overall toxicity of the leachates due to the presence of various kinds of chemicals. But still, it has to develop a suitable matrix of bioassays based on the sensitivity of various toxicants evaluating the toxicity of the leachate.

The landfill is considered as a popular option for disposal of wastes, but it was found that most of the landfill in the world is comparatively old and not built in a properly planned manner in order to protect the toxic leachate to contaminate soil and groundwater. The different toxic pollutants present in the leachates have a damaging effect and are accumulative on our ecosystem, food chain, and food web and cause of acute toxicity, genotoxicity, and carcinogenic effects among the biological community including human beings. The treatment of the generated highly toxic leachate from landfill is a great challenge for authorities before its disposal into the proper location to confirm the minimal environmental impact (Baderna et al., 2018).

The leachate management is mainly divided into four different stages as follows:

- Monitoring and regulating the formation of leachate
- Flow of leachate into the environment
- Identification of hazards or damaging effect associated with it
- Treatment prior to its disposal to minimize its impact on the biotic community

There is another method of assessing toxicity of the leachate called biological analysis, which provides only a few or incomplete information regarding the environmental fate of the complex refluxes of leachates. The biological analysis of toxicity delivers the biological impact of all the toxic substances present in the leachate in addition to their biological availability. The ecotoxicological analysis shows the required relationship between the conventional chemical analysis and complete field studies with the application of biological organisms in exclusive controlled or defined environmental conditions (Mukherjee et al., 2014).

4.4.3 Landfill leachate treatment technologies

There are three important stages required for the removal of toxic contaminants from leachates, which include the collection, storage, and appropriate treatment. There are a number of different kinds of technologies adopted for the treatment of leachates; the major technologies that attained the standards fixed by legislation are as follows:

- The biological process of treatment, which includes activated sludge process, anaerobic and aerobic stabilization process, and biological filters or trickling filters.
- The physical–chemical processes, including flotation, flocculation or coagulation, adsorption, air stripping, chemical precipitation, chemical oxidation, pH adjustment, ion exchange, and electrochemical treatment.
- Membrane filtration technologies such as microfiltration, nanofiltration, ultrafiltration, and reverse osmosis.
- Advanced oxidative treatments called as ozonization.

Regardless of many advantages, there are many limitations such as the generation of extensively contaminated leachates, chemical composition, volumetric flow, and constitutes present in the leachate are the key drawbacks of landfills. Hence, before the release of the toxic leachate, it has to be properly treated. Hence, a new advanced modified technology known as membrane technologies (reverse osmosis) is used as the major step of the treatment of landfill leachate for reducing pollutants from the leachates (Peng, 2017).

4.4.4 Factors affecting characteristics of leachate

The properties of landfill leachate of MSW can be influenced by a number of factors, which includes the kinds of wastes, composition of wastes, age of the landfill, hydrogeology of the landfill site, seasonal variation in weather conditions, and intensity of dilution by rainwater, precipitation, and rate of decomposition of the wastes. Among all these factors, the age of the landfill is a significant determining factor in controlling and regulating the composition of the leachate because there are a number of parameters, which change unpredictably during the stabilization of the landfill. On the basis of age, the traditional landfill leachates are broadly classified into three different categories.

- Young landfills (age < 5 years)
- Intermediate landfill (age 5–10 years)
- Mature landfill (age more than 10 years)

With an increase in the age of the landfill, the parameters of leachate such as pH, COD, BOD, and BOD/COD ratio, etc. change significantly (Montusiewicz, 2014).
For example, the concentration of COD and BOD gradually decreases with the increase in the age of the landfill because of the decomposition of organic materials present in the leachates. It was observed that most of the biodegradable organic substances (BOD matters) are degraded in the stabilization stage, whereas the nonbiodegradable organic substances (COD matters) remain as such in this stage. Again, the biodegradability index or BOD/COD ratio decreases with the passage of time. In the leachates of young landfill, exclusively higher biodegradability index (0.5–1) and COD concentration (>10,000 mg/L) were observed, whereas in the mature or old landfill, COD concentration is <4,000 mg/L, and the biodegradability index is <0.1. The pH value of the leachate increases with the increase in age, whereas the HM concentration exhibits a decreasing trend with the increase in age. The % of ammonia nitrogen in leachate does not exhibit a spontaneous decreasing or increasing trend with time (Naveen et al., 2016; Adhikari et al., 2014).

The composition and characteristics of the leachate not only change with age but also differ from place to place. Therefore, the different characteristic parameters fluctuate from place to place. Hence, the quality of leachate is extremely variable in most of the locations, and the variation in the quality of the leachate depends upon a number of interacting factors, mostly composed of the leachate, dissolved oxygen, depth of the landfill, age of waste's moisture content, and design and operation of landfill sites. Temperature is an uncontrollable factor that greatly influences the quality of leachate in the landfill. The seasonal variation of temperature in the landfill influences the chemical reactions carried out within the landfill and bacterial growth. The change in the optimum temperature is the cause of a decrease in the growth rate of microbes because of the deactivation of enzymes and the rupture of cell walls. The increase in the temperature decreases the solubility of many salts in leachate such as $CaSO_4$ and $CaCO_3$. Age is an important factor that affects the quality of leachate generated in landfills. The quality of leachate reaches a peak value over 2–3 years and then gradually declines with the subsequent increase in years because of changes in chemical reactions of the wastes within the landfill at various stages. It was observed that the leachate generated from new landfills has higher values of COD and BOD, which subsequently decrease with an increase in age during ten years, but, however, all the pollutants have not attained the peak stage at the same time. For example, % of organic materials gradually decreases due to biodegradability, whereas the % of inorganic matter subsequently increases with an increase in age because of nonbiodegradability. In anaerobic conditions, SO_4^{2-} present in the leachate is converted into S^{2-}, and its concentration gradually decreases (Dadrasnia et al., 2017; Çoruh et al., 2012).

4.5 LANDFILL GAS

There is a mixture of around hundreds of different gaseous substances produced from the landfill, among which CH_4 and CO_2 are the major components and are called as

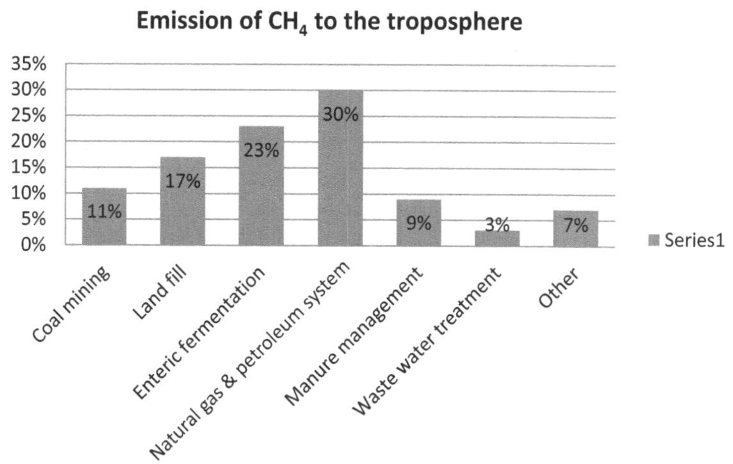

Figure 4.3 Various sources of emission of CH_4 to the troposphere.

LFGs. LFG is normally a natural byproduct obtained due to the degradation of organic wastes in landfill sites. In terms of volume, the LFG mainly consists of 45%–60% of CH_4, 40%–60% of CO_2, and in addition to these two, some other gases are present in the landfill in small amounts, which include N_2, NH_3, O_2, CO, H_2, H_2S, CS_2 along with some nonmethane organic substances such as acrylonitrile, 1,1-dichloroethane, benzene, dichloromethane, carbonyl sulfide, 1,2-cis dichloroethylene, ethyl-benzene, hexane, ethyl ketone, trichloroethylene, methyl tetrachloroethylene, toluene, xylenes, and vinyl chloride. CH_4 is a potent greenhouse gas, which is almost 26 times more potent than CO_2 in climate change. LFG is the 3rd source of CH_4 emission to the atmosphere. Figure 4.3 shows the various sources of production of CH_4.

The rate of production, composition, and toxicity of LFG change with change in the age of the landfill. The production of LFG is due to the anaerobic decomposition of organic or biological substances present in MSW. A ton of MSW contains an average of 150–250 kg of organic carbon; the microorganisms present in the landfill sites decompose the organic matter and transform it into some simple products with the generation of LFG. There are many factors which affect the quantity of production of LFG such as the composition of the wastes, the content of moisture, landfill age, temperature, etc. Normally, the emission of LFG begins one to two years after the disposal of solid wastes and continues up to 15–25 years or till the exhaustion of the landfill. The collected LFG may be used as a source of energy and production of electricity (Fei et al., 2019; Abbas et al., 2009). The LFG is mainly produced in three processes such as bacterial decomposition, chemical reactions, and volatilization.

Bacterial decomposition: In this process, the organic wastes decompose by the bacteria present in the landfill in anaerobic condition, which was created due to soil covering. The wastes rich in organic matters include food wastes, textiles, garden wastes, street sweepings, and discarded paper and wood products, and the decomposition of these wastes is generally carried out in four different phases, and at each phase, the composition of the LFG changes

Volatilization: At the landfill sites, the solid and liquid form of some wastes of rich inorganic compounds change into the vapor phase, which is known as volatilization. Some LFGs are produced due to volatilization of some definite chemical substances dumped into the landfill.

Chemical reactions: The different LFGs, including the nonmethane organic substances, are produced by different chemical reactions among the wastes and environmental components

For example, the bleach of NH_3 and chlorine gas if come in contact with each other in the landfill area, it forms different harmful gases (Abbas et al., 2009; Gautam et al., 2019).

4.6 LANDFILL MINING

The landfill mining (LFM) is now a recent strategy worldwide and is considered as a pioneering strategy to mitigate risks associated with the environment in the landfill sites in order to recover secondary materials and energy from the dumped wastes and permits a well-managed land use at the site of the landfill. The different landfills are mined in order to obtain different raw materials such as metals or energy source (plastic products). The LFM is a new and innovative concept, which is designed in order to close the material loops to a circular economy, improving landfill waste. The secured LFM operations require an average of 103 MJ diesel and 1.9 MJ electricity per ton of mined wastes, which produces about 12 kg of CO_2 equivalent specially in EU countries (Hölzle, 2018). In some cases, it was found that LFM is the cause of reduction of climate impacts relative to business as usual, and it was some contribution toward the reduction in global warming. LFM is mainly essential in the countries where there is a scarcity of the land for the establishment of new landfills. In this process, the space for landfill can be reclaimed and reutilized by the extraction of biodegradable and nonbiodegradable waste products by filling with the solid wastes necessary for disposal. The mining landfill is considered as the simplest and basic form of LFM, and its construction mainly requires three basic constituents such as wood, gypsum, and scrap metal along with fewer quantities of some other construction materials. The wood collected from it can be used as fuel in the coal-fired power plants, and the scrap metals can be again reprocessed and reused (more popular in the United States). The secured landfill is a best useful process for the remediation of hazardous mining wastes; however, the major disadvantages are the proper construction with linear and the requirement of more land area. The EPA of the United States needed closed landfills to have to be monitored and regulated for at least 30 years after the end of the landfill operation. Mining the landfill requires a safe liner at the end of the landfill activity. Although the LFM is beneficial for us because of the recovery of useful resources from wastes, it is the cause of higher risk for human health hazard due to continuous exposure to toxic landfill wastes and leachates. The appropriately stabilized advanced sanitary landfills offer their own benefits. It was observed that almost 80% of the created LFM scenarios exhibit negative results. Hence, the development of treatment plants permits maximum recovery of resources, economically consistent and environmentally sustainable, so that it creates a technological challenge for development and modification. Therefore, for a suitable LFM, it should be operated with full potential, planned policy decisions, and custom-made support systems, and the combination of all these are necessary for material recycling, nature restoration, and energy utilization (Krook et al., 2012; Weng et al., 2015).

4.7 THE FUTURE OF LANDFILLS

In European countries, waste management increases recycling and reuse of the wastes than dumping into the landfill sites. Since 2016, a member of countries throughout the world, specially developed countries, did not permit 35% of biodegradable MSW, but the major objective of the future landfill is the generation of minimal wastes, but practically, it cannot be archived as per expectation; therefore, the major objective of the strategy of waste management, the proper separation of the wastes with proper recycled and reused techniques, must have to be adopted. If this policy is not possible, then the wastes from the landfill may be used for the generation of energy by using the process of thermovalorization. The wastes which have no alternative or possible way of managing should be disposed to the landfill sites. Besides these, the landfill also plays a key role in the management system of MSW. Furthermore, it was found that the disposal of wastes in the landfill is higher than prevention, recycling, and reuse of the wastes. The waste minimization strategy of recycling and reuse of solid wastes in some cases is not feasible and economically acceptable, and in that condition, the only landfill is the only alternative. Although reusing and recycling are not economical, if it is modified by using some modern technology, then it may be beneficial in future. The volume of waste generated is changing with the change in season throughout the year, and sometimes, the quantity of wastes selected for recycling, incinerating, or reusing goes beyond the capacity. Not all the kinds of wastes are incinerated or recycled, and in this case, the only choice is disposal in the landfill. Sometimes, the incinerator or the recycling facility is not functioning well out of order because of maintenance, breakdown, or lack of repair, so in that case, the landfill is the only choice for operation. Hence, landfill has to be established far from the residential areas in a planned and scientific way to maintain sustainability and reduction of pollution load for future generations.

Insulation of the landfills by using impermeable membranes is now becoming a standard in European countries. The membrane used has an average lifetime of 50 years, but in many cases, it can sustain up to 500 years. However, these methods unavoidably fail at any moment, and the process of emission generation starts again. In most of the countries, the emphasis is given to the aftercare landfill management and regulated for at least 30–60 years after the closing of the landfill. Hence, the society that struggles for long-term sustainable development requires long-standing sustainable landfills. According to international recognition, there is no suitable definition of sustainable landfilling. But the sustainability term in landfill is frequently associated with stability, end technologies, threat, and completion to the environment. According to the Anglo-Welsh agency for environmental protection, the closer of the landfill is defined as the situation, when the landfill is physically, biologically, and chemically stabilized to such an extent that the materials present in it are not associated with any risk toward the environment pollution in the surrounding environment of the landfill area. Hence, after the closing of the landfill, no active pollution controls such as leachate and LFG management or any kind of monitoring systems is necessary. According to the Technical University Hamburg, the different stages of the aftercare management have to be never stopped until the potential emission of LFG and leachate is so slow that it does not create any harm to the environment. Although all the definitions are slightly different, there is a general agreement among these are for the sustainable landfill, is that, the landfill which after closer not threat to the environment or human health. Hence, finally we can say that the aftercare effect of the landfill is stopped at the moment of the completion of the landfill (Ismail, 2013).

4.8 ROLE OF MICROBES IN LANDFILL

The MSW landfills are most common and widely used as a potential tool for the management of solid wastes, and microbial community plays a key role in the degradation of the solid wastes in the landfill ecosystem.

The leachate generated from landfill contains high % of NH_4^+ ions and many complex organic materials such as volatile fatty acids (VFAs), various kinds of microscale xenobiotic organic chemicals (XOCs), and refractory aquatic humic matter, which stimulates the microbial community of diverse categories in the treatment of landfill leachate in the bioreactors. These various groups of microorganisms can remove biodegrade organic materials, N_2, and toxic effects of XOCs and generate energy. The advantages of using diverse microbes are that they exhibit dominant properties in the bioreactor and widespread in the treatment of leachate (Meyer-Dombard et al., 2020).

The anaerobic methanogenesis (hydrogenotrophic or acetoclastic) and the aerobic methanotrophy within the landfills are identical to the biodegradation of organic material in a natural environment. Methane gas generated from the landfill diffuses through the cover soil and undergoes oxidation by methanotrophs and produces water vapor and CO_2. The different chemical reactions carried out in the landfill sites are as follows (Meyer-Dombard et al., 2020):

Due to carbon recycling

$$2CH_4 + 3O_2 \rightarrow 2CO + 4H_2O \tag{4.1}$$

$$CH_4 + NO_3^- + 2H^+ \rightarrow NH_4^+ + CO_2 + H_2O \tag{4.2}$$

$$CH_4 + 2O_2 \rightarrow CO_2 + 2H_2O \tag{4.3}$$

$$CH_3COOH \rightarrow 2CO_2 + 4H_2 \tag{4.4}$$

$$CH_4 + NO_2^- + 2H^+ \rightarrow NH_4^+ + CO + H_2O \tag{4.5}$$

$$3N_2 + 5CO + 13H_2O \rightarrow 6NO^{3-} + 5CH_4 + 6H^+ \tag{4.6}$$

$$3H_2 + CO \rightarrow CH_4 + H_2O \tag{4.7}$$

$$2NH_4^+ + CO \rightarrow N_2 + CH_4 + 2H^+ + H_2O \tag{4.8}$$

$$4H_2 + CO_2 \rightarrow CH_4 + 2H_2O \tag{4.9}$$

Due to Nitrogen cycling

$$2NH_4^+ + 3O_2 \rightarrow 2NO_2^- + 4H^+ + 2H_2O \quad (4.10)$$

$$NH_4^+ + NO_2^- \rightarrow N_2 + 2H_2O \quad (4.11)$$

$$2NO_2^- + O_2 \rightarrow 2NO_3^- \quad (4.12)$$

$$H_2S + NO_3^- + H_2O \rightarrow SO_4^{2-} + NH_4^+ \quad (4.13)$$

$$H_2 + NO_3^- \rightarrow NO_2^- + H_2O \quad (4.14)$$

$$3H_2 + N_2 + 2H^+ \rightarrow 2NH_4^+ \quad (4.15)$$

$$3H_2 + NO_2^- + 2H^+ \rightarrow NH_4^+ + 2H_2O \quad (4.16)$$

Sulphur, Iron, carbon coupled cycling

$$H_2S + 2O_2 \rightarrow SO_4^{2-} + 2H^+ \quad (4.17)$$

$$4H_2S + O_2 + 2Fe^{2+} \rightarrow 2FeS_{2\,\mathrm{Pyrite}} + 4H^+ + 2H_2O \quad (4.18)$$

$$CH_4 + Fe^{3+} + 2H_2O \rightarrow CO_2 + 2Fe^{2+} + 3H_2 + 2H^+ \quad (4.19)$$

$$2H_2S + Fe_2O_{3\,\mathrm{Hematite}} + 2H^+ \rightarrow FeS_{2\,\mathrm{Pyrite}} + Fe^{2+} + 3H_2O \quad (4.20)$$

$$12Fe^{2+} + CO_2 + 14H_2O \rightarrow 4Fe_3O_{4\,\mathrm{Magnetite}} + CH_4 + 24H^+ \quad (4.21)$$

$$4FeS_{2\,\mathrm{Pyrite}} + 7CO_2 + 18H_2O \rightarrow 8SO_4^{2-} + 4Fe^{2+} + 7CH_4 + 8H^+ \quad (4.22)$$

$$FeS_{2\,\mathrm{Pyrite}} + 2H^+ \rightarrow 2S + Fe^{2+} + H_2 \quad (4.23)$$

$$2H_2S + Fe^{2+} \rightarrow H_2 + FeS_{2\,\mathrm{Pyrite}} + 2H^+ \quad (4.24)$$

$$4FeS_{2\,\mathrm{Pyrite}} + CO_2 + 8H^+ \rightarrow 8S + 4Fe^{2+} + CH_4 + 2H_2O \quad (4.25)$$

$$H_2S + Fe_3O_{4\,\mathrm{Magnetite}} + 6H^+ \rightarrow S + 3Fe^{2+} + 4H_2O \quad (4.26)$$

The bioprocessing of the wastes in landfills originates from the metabolism and co-metabolism of the wastes by a number of syntrophic microorganisms leading to the limited transformation of complex material into simple polymeric materials and fully mineralization into the water, inorganic salts, and LFG including CH_4. The decomposition of the wastes in the landfill permits the occurrence of hydrolytic, acetogenic, fermentative, and methanogenic processes. The different operational parameters significantly influence the presence of the microbial communities in the landfill and negative impact on the performance of the bioreactor. The high loading of bacteria, low DO, and optimal pH is accumulating the necessary ammonium oxidation bacteria (AOB) and the cause of partial nitrification. The organic matters and nitrite prevent the anaerobic AOB (AnAOB). In the process of treatment of anaerobic LFL in bioreactors, *Methanosarcina* and *Methanosaeta, sulfur-reducing* bacteria, and homoacetogens are mainly the dominant archaea. The heterotrophic denitrifying bacteria, AnAOB, and nitrite-oxidizing bacteria interfere with each other in the landfill site (Lee & Pandey, 2012).

4.9 CONCLUSIONS

Although the landfill is a popular widespread process for the management of solid wastes in a cost-effective method, the uncontrolled or haphazard disposal of solid wastes in the landfill affects the biotic community, our ecosystem, water, air, and soil in many ways. The major ecological problem of landfills is the production of leachate and LFG. The addition of the leachate to the groundwater is the cause of a high health hazard effect for the human being. Hence, in many of the developing and developed countries, landfills are established scientifically in a planned way, having liner at the bottom part along with a leachate collection system. Now, the leachate problem is worsened gradually because of the generation of vast amounts of MSW and its frequent deviation in composition and characteristics. The HMs and toxic chemicals present in the leachate are the cause of severe groundwater contamination and lead to be a highly negative impact on human health and the environment. The HMs such as Cd, Ni, As, and Cr found in the landfill leachate are carcinogenic in nature and cause of many adverse effects to human health depending on the level and duration of exposure. It significantly affects our liver, kidneys, central nervous system, skin, heart, and lungs and subsequently damages our reproductive system. The composition and toxicity of the leachate generated depend upon the age of the landfill, kind of wastes deposited, moisture, and rainfall intensity, design of the landfill, temperature, and population of microbes. Despite so many disadvantages and greater potential of environmental contamination, still, the landfill is a leading method for the management of MSW. The new sustainable landfill is that in which minimal pollutant generation occurs in a cost-effective and eco-friendly way. Hence, in order to avoid environmental degradation and contamination of groundwater, frequent monitoring is required for identifying the hazards created by landfills. The other improved technology that can be used instead of landfill includes ecoutilization of solid wastes, biodrying, gasification, plasma gasification, hydrothermal carbonization, and biological treatment, conversion of wastes into refuse-derived fuel, hydrothermal liquefaction, pyrolysis and molten salt oxidation.

REFERENCES

Abbas, A. A., Guo, J. S., Ping, L.Z., Ya, P. Y., 2009. Review on landfill leachate treatments. *American Journal of Applied Sciences*, 6(4), 672–684.

Abd El-Salam, M. M., Abu-Zuid, G. I., 2015. Impact of landfill leachate on the groundwater quality: A case study in Egypt. *Journal of Advanced Research*, 6(4), 579–586.

Abdel-Shafy, H. I., Mansour, M. S. M., 2018. Solid waste issue: Sources, composition, disposal, recycling, and valorization. *Egyptian Journal of Petroleum*, 27(4), 1275–1290.

Adhikari, B., Khanal, S., Lamichhane, J., Giri, D., 2014. Seasonal Variation of pH, BOD, COD and BOD/COD Ratio in Different Ages of Landfill Leachate in Nepal. *Journal of Biomolecule Reconstruction*, 11(2), 89–99.

Ağdağ, O. N., 2011. Characterization and treatment of Denizli landfill leachate using anaerobic hybrid/aerobic CSTR systems. *Environmental Technology*, 32(7), 699–711.

Ayilara, M., Olanrewaju, O., Babalola, O., Odeyemi, O., 2020. Waste management through composting: Challenges and potentials. *Sustainability*, 12(11), 4456–4478.

Baderna, D., Caloni, F., Benfenati, E., 2018. Investigating landfill leachate toxicity in vitro: A review of cell models and endpoints. *Environment International*, 122, 21–30.

Brand, J. H., Spencer, K. L., O'shea, F. T., Lindsay, J. E., 2017. Potential pollution risks of historic landfills on low-lying coasts and estuaries. Wiley Interdisciplinary Reviews: *Water*, 5(1), 1–12, e1264.

Chugh, S., Chynoweth, D. P., Clarke, W., Pullammanappallil, P., Rudolph, V., 1999. Degradation of unsorted municipal solid waste by a leach-bed process. *Bioresource Technology*, 69(2), 103–115.

Çoruh, S., Elevli, S., Geyikçi, F., 2012. Statistical evaluation and optimization of factors affecting the leaching performance of copper flotation waste. *The Scientific World Journal*, 1–8.

Costa, A. M., Alfaia, R. G. de S. M., Campos, J. C., 2019. Landfill leachate treatment in Brazil – An overview. *Journal of Environmental Management*, 232, 110–116.

Cuartas, M., López, A., Pérez, F., Lobo, A., 2018. Analysis of landfill design variables based on scientific computing. *Waste Management*, 71, 287–300.

Dadrasnia, A., Azirun, M. S., Ismail, S. B., 2017. Optimal reduction of chemical oxygen demand and NH_3–N from landfill leachate using a strongly resistant novel Bacillus salmalaya strain. *BMC Biotechnology*, 17(1). 85–94.

Eggen, T., Moeder, M., Arukwe, A., 2010. Municipal landfill leachates: A significant source for new and emerging pollutants. *Science of the Total Environment*, 408(21), 5147–5157.

El-Fadel, M., Khoury, R., 2000. Modeling settlement in MSW landfills: A critical review. *Critical Reviews in Environmental Science and Technology*, 30(3), 327–361.

Fazzo, L., Minichilli, F., Santoro, M., 2017. Hazardous waste and health impact: A systematic review of the scientific literature. *Environmental Health*, 16, 107–117.

Fei, F., Wen, Z., De Clercq, D., 2019. Spatio-temporal estimation of landfill gas energy potential: A case study in China. *Renewable and Sustainable Energy Reviews*, 103, 217–226.

Ferronato, N., Torretta, V., 2019. Waste mismanagement in developing countries: A review of global issues. *International Journal of Environmental Research and Public Health*, 16(6), 1060–1087.

Gautam, P., Kumar, S., Lokhandwala, S., 2019. Advanced oxidation processes for treatment of leachate from hazardous waste landfill: A critical review. *Journal of Cleaner Production*, 237, 1–14.

Giusti, L., 2009. A review of waste management practices and their impact on human health. *Waste Management*, 29(8), 2227–2239.

Gworek, B., Dmuchowski, W., Gozdowski, D., Koda, E., Osiecka, R., Borzyszkowski, J., 2015. Influence of a municipal waste landfill on the spatial distribution of mercury in the environment. *PLoS One*, 10(7), 1–12, e0133130.

Hölzle, I., 2018. Analysing material flows of landfill mining in a regional context. *Journal of Cleaner Production*, 207, 307–328.

Inglezakis, V. J., Moustakas, K., 2015. Household hazardous waste management: A review. *Journal of Environmental Management*, 150, 310–321.

Ismail, S., 2013. The challenge of future landfill: A case study of Malaysia. *Journal of Toxicology and Environmental Health Sciences*, 5(6), 86–96.

Joshi, R., Ahmed, S., 2016. Status and challenges of municipal solid waste management in India: A review. *Cogent Environmental Science*, 2(1), 1–18.

Kalyani, K. A., Pandey, K. K., 2014. Waste to energy status in India: A short review. *Renewable and Sustainable Energy Reviews*, 31, 113–120.

Kan, A., 2009. General characteristics of waste management: A review. *Energy Education Science and Technology Part A: Energy Science and Research*, 23(1), 55–69.

Kerdsuwan, S., Laohalidanond, K., Jangsawang, W., 2015. Sustainable development and eco-friendly waste disposal technology for the local community. *Energy Procedia*, 79, 119–124.

Kjeldsen, P., Barlaz, M. A., Rooker, A. P., Baun, A., Ledin, A., Christensen, T. H., 2002. Present and long-term composition of MSW landfill leachate: A Review. *Critical Reviews in Environmental Science and Technology*, 32(4), 297–336.

Krook, J., Svensson, N., Eklund, M., 2012. Landfill mining: A critical review of two decades of research. *Waste Management*, 32(3), 513–520.

Kulikowska, D., Klimiuk, E., 2008. The effect of landfill age on municipal leachate composition. *Bioresource Technology*, 99(13), 5981–5985.

Kumar, M., 2012. "Problems, Perspectives and Challenges of Agricultural Water Management" In-Tech international publication, ISBN 979-953-307-706-1 https://www.intechopen.com/books/problems-perspectives-and-challenges-of-agricultural-water-management.

Lee, J., Pandey, B. D., 2012. Bio-processing of solid wastes and secondary resources for metal extraction – A review. *Waste Management*, 32(1), 3–18.

Li, W., Zhou, Q., Hua, T., 2010. Removal of organic matter from landfill leachate by advanced oxidation processes: A review. *International Journal of Chemical Engineering*, 2010, 1–10.

Lombardi, F., Costa, G., Sirini, P., 2017. Analysis of the role of the sanitary landfill in waste management strategies based upon a review of lab leaching tests and new tools to evaluate leachate production. *Ambiente & Agua - An Interdisciplinary Journal of Applied Science*, 12(4), 543–555.

Luo, H., Zeng, Y., Cheng, Y., He, D., Pan, X., 2019. Recent advances in municipal landfill leachate: A review focusing on its characteristics, treatment, and toxicity assessment. *Science of the Total Environment*, 703, 1354–68.

Ma, H., Cao, Y., Lu, X., Ding, Z., Zhou, W., 2016. Review of typical municipal solid waste disposal status and energy technology. *Energy Procedia*, 88, 589–594.

Malinauskaite, J., Jouhara, H., Czajczyńska, D., Stanchev, P., Katsou, E., Rostkowski, P., Spencer, N., 2017. Municipal solid waste management and waste-to-energy in the context of a circular economy and energy recycling in Europe. *Energy*, 141, 2013–2044.

Meyer-Dombard, D. R., Bogner, J. E., Malas, J., 2020. A review of landfill microbiology and ecology: A call for modernization with 'next generation' technology. *Frontier in Microbiology*, 11, 1127–1148.

Miao, L., Yang, G., Tao, T., Peng, Y., 2019. Recent advances in nitrogen removal from landfill leachate using biological treatments – A review. *Journal of Environmental Management*, 235, 178–185.

Montusiewicz, A., 2014. Co-digestion of sewage sludge and mature landfill leachate in pre-bioaugmented system. *Journal of Ecological Engineering*, 15(4), 98–104.

Moya, D., Aldás, C., López, G., Kaparaju, P., 2017. Municipal solid waste as a valuable renewable energy resource: A worldwide opportunity of energy recovery by using Waste-To-Energy Technologies. *Energy Procedia*, 134, 286–295.

Mukherjee, S., Mukhopadhyay, S., Hashim, M. A., Sen Gupta, B., 2014. Contemporary environmental issues of landfill leachate: Assessment and remedies. *Critical Reviews in Environmental Science and Technology*, 45(5), 472–590.

Nagarajan, R., Thirumalaisamy, S., Lakshumanan, E., 2012. Impact of leachate on groundwater pollution due to non-engineered municipal solid waste landfill sites of erode city, Tamil Nadu, India. *Iranian Journal of Environmental Health Science & Engineering*, 9(1), 35–47.

Naveen, B. P., Sivapullaiah, P. V., Sitharam, T. G., 2016. Effect of aging on the leachate characteristics from municipal solid waste landfill. *Japanese Geotechnical Society Special Publication*, 2(56), 1940–1945.

NidzamRahmat, M., RedzwanRaffe, M., Ismail, N., 2014. Utilization of agricultural wastes in stabilization of landfill soil. *MATEC Web of Conferences*, 15(01001), 1–8.

Pecorini, I., Iannelli, R., 2020. Landfill GHG reduction through different microbial methane oxidation biocovers. *Processes*, 8(5), 591–602.

Peixoto, A. L. de C., Salazar, R. F. dos S., Barboza, J. C. de S., IzárioFilho, H. J., 2018. Characterization of controlled landfill leachate from the city of Guaratinguetá- SP, Brazil. *Ambiente & Agua - An Interdisciplinary Journal of Applied Science*, 13(2), 1–16.

Peng, Y., 2017. Perspectives on technology for landfill leachate treatment. *Arabian Journal of Chemistry*, 10, S2567–S2574.

Plaza, P. I., Lambertucci, S. A., 2017. How are garbage dumps impacting vertebrate demography, health, and conservation? *Global Ecology and Conservation*, 12, 9–20.

Przydatek, G., Kanownik, W., 2019. Impact of small municipal solid waste landfill on groundwater quality. *Environmental Monitoring and Assessment*, 191(3), 169–183.

Pujiindiyati, E. R., Satrio, S., Prasetio, R., 2019. Major ions for tracing leachate migration within shallow groundwater in the vicinity of municipal landfill in BantarGebang–Bekasi. *Indonesian Journal of Chemistry*, 19(1), 19–29.

Renou, S., Givaudan, J. G., Poulain, S., Dirassouyan, F., Moulin, P., 2008. Landfill leachate treatment: Review and opportunity. *Journal of Hazardous Materials*, 150(3), 468–493.

Sekhohola-Dlamini, L., Tekere, M., 2019. Microbiology of municipal solid waste landfills: A review of microbial dynamics and ecological influences in waste bioprocessing. *Biodegradation*, 31(1–2), 1–21.

Singh, A., Patel, A. K., Ramanathan, A., Kumar, M., 2020. Climatic influences on arsenic health risk in the metamorphic precambrian deposits of Sri Lanka: A re-analysis-based critical review. *Journal of Climate Change*, *6*(1), 15–24.

Swati, T. I. S., Vijay, V. K., Ghosh, P., 2018. Scenario of landfilling in India: Problems, challenges, and recommendations. In: Hussain, C. (ed.) *Handbook of Environmental Materials Management*. Springer, Cham. pp. 1–16.

Tałałaj, I.A., Biedka, P., Bartkowska, I., 2019. Treatment of landfill leachates with biological pretreatments and reverse osmosis. *Environmental Chemistry Letters*, 17, 1177–1193.

Vaverková, M. D., Elbl, J., Koda, E., Adamcová, D., Bilgin, A., Lukas, V., Zloch, J., 2020. Chemical composition and hazardous effects of leachate from the active municipal solid waste landfill surrounded by farmlands. *Sustainability*, 12(11), 4531–4550.

Verma, R., Vinoda, K. S., Papireddy, M., Gowda, A. N. S., 2016. Toxic pollutants from plastic waste- A review. *Procedia Environmental Sciences*, 35, 701–708.

Vodyanitskii, Y. N., 2016. Biochemical processes in soil and groundwater contaminated by leachates from municipal landfills (Mini review). *Annals of Agrarian Science*, 14(3), 249–256.

Wath, S. B., Vaidya, A. N., Dutt, P. S., Chakrabarti, T., 2010. A roadmap for development of sustainable E-waste management system in India. *Science of the Total Environment*, 409(1), 19–32.

Weng, Y. C., Fujiwara, T., Houng, H. J., Sun, C. H., Li, W. Y., Kuo, Y. W., 2015. Management of landfill reclamation with regard to biodiversity preservation, global warming mitigation and landfill mining: Experiences from the Asia-Pacific region. *Journal of Cleaner Production*, 104, 364–373.

Xue, W., Cao, K., Li, W., 2015. Municipal solid waste collection optimization in Singapore. *Applied Geography*, 62, 182–190.

Zhang, C., Xu, T., Feng, H., Chen, S., 2019. Greenhouse gas emissions from landfills: A review and bibliometric analysis. *Sustainability*, 11(8), 2282–2294.

Chapter 5

Emerging groundwater contaminants

Their manifestation and consequences

Shreeja Datta
Delhi Technological University

Arpita Roy
Sharda University

Suresh Ghotekar
Smt. Devkiba Mohansinhji Chauhan College of Commerce and Science

CONTENTS

DOI: 10.1201/9781003157465-6

5.1 INTRODUCTION

The groundwater sources consist of discharge from sea surfaces majorly via rainwater, rivers, water from lakes, and watercourse. Since most of the freshwater sources are present as icebergs and glacial masses, the water available for drinking and other purposes accounts up to 0.3%, leading to groundwater being a major source of water consumption. In recent years, there has been an increasing attentiveness in prevalence of microorganic contaminants in terrestrial as well as marine surroundings (Kümmerer, 2009), causing groundwater pollution. This pollution of groundwater aided through micro contaminants (organic) has become an issue of rising concern (Pal et al., 2010). Quality of groundwater implies the physical, chemical, and the biological aspects of the intended water use. This quality is often at risk due to the human activities; however, intrusion of some of the harmful substances occurs naturally too. Sustainable management of groundwater should be based on the prevention of contamination, alongside preventing the over exploitation of groundwater resources since this prevention protects all these resources, on contrary to the treatment of the same.

Generation of the waste products includes economic activities which include primary activities (production of commodities such as mining and agriculture), secondary or industrial activities (building, energy generation, and also manufacturing), and tertiary activities (transport which results in the threatening of environment and thus increasing the groundwater pollution) (Postigo et al., 2018, Patel et al., 2019). Two primary origins of groundwater contamination are the point and nonpoint source. With respect to the present scenario, waste sites are able to be extracted from the environment as well. This is not achievable in the case of diffuse contamination sources since they are present in air or are eventually rained out or are of use in agriculture. The current sources for groundwater pollution or contamination are usually the consequences of activities carried out years ago. Along with this, the current waste disposal activities are most likely to affect the quality of groundwater as well as human health in the future. The main contaminants of groundwater arise mainly from land surface via soil zone infiltration. The contaminant plumes, as a result of industrial sources, may cover a large extent of area in due course of time (usually many years), along with the natural flow of groundwater, and further pose a persistent danger for water supply based on the same aquifer. Thus, groundwater contamination is a subject of serious concern for many industrialized countries, and its remediation techniques should be well implemented and executed in every part of the society in order to sustain water management and prevent health hazards.

5.2 MECHANISMS INVOLVED IN GROUNDWATER CONTAMINATION

The contaminants which enter the groundwater system tend to spread throughout the system only when transport mechanisms are available, such as a flowing liquid. This possibility of presence of any compound in groundwater depends on its solubility as well as speciation. As soon as it reaches the unsaturated or saturated zone and comes in contact with subsurface water, numerous physical, biochemical, and geochemical processes decide its providence.

5.2.1 Physical processes

These processes comprise advection, molecular diffusion, hydrodynamic dispersion, filtration, volatilization, and mechanical dispersion. These processes are mainly the prime one responsible for transport of the contaminants in the groundwater and, hence, the reason for the typical contaminant plume generation caused because of point sources of contamination. Delineation of such plumes are able to be carried out with the help of temperature measure or even the pollutant concentration in groundwater.

Advection, primarily being the most significant contaminant movement process, is present in subsurface. Apart from this, it indicates transport of contaminants caused by the prominent and heavy water flow. This advective flow velocity, also known as the average linear velocity of groundwater, is dependent on the bulk or average properties, usually average aquifer material permeability along with the average hydraulic gradient of the two sources involved. This method obeys Darcy's law—the fluid flow occurring in the porous medium. This process enables the contaminant to move according to hydraulic gradient (that is, from high hydraulic potential areas to lower ones). The method of molecular diffusion involves movement of contaminants with respect to the chemical gradient (that is, from the area of high concentration to lower areas). The process of changing of concentrations with respect to the time period is even a function of concentration alteration in space in a one-dimensional system. It is in accordance with Fick's second law, that is described as the rate of depletion or rate of accumulation of concentration with respect to the volume proportional to the curvature of the concentration gradient. Hydrodynamic dispersion is defined as the overlapping effect of two physical processes, the net result being an overall reduction of the concentration of contaminants along with their spread as well, for cleaning the aquifer areas. Physical filtration takes place inside the aquifer, further retaining biological species which are bigger in size than the pores of aquifer in the soil. Volatilization involves partitioning of contaminant's concentration within the vapor and aqueous phases. Besides, it occurs in an aquifer of the vadose zone and in the water table as well. Its volatility is dependent on parameters such as solubility, vapor pressure, and the Henry's law constant—representing volatility of contaminants. Molecular dispersion involves flow path tortuosity, such as size variations of pore as well as path lengths and frictions within pore walls. These make the particles of groundwater and contaminants to travel in a different way in the aquifer porous medium. It also results in dilution effects with mechanical mixing through the flow path. The process of mechanical dispersion is said to be longitudinal when mixing is in the direction of flow of groundwater and is transverse when perpendicular to the direction.

5.2.2 Biochemical processes

Activities of bacteria can efficiently bio-transform and even mineralize pollutants present at the subsurface. Although, this is attained only when bacteria have capacity of using them as a substrate and if those pollutants are available to them. The aquifer's redox potential decides whether the bacteria is present or not. In other words, it decides whether the organic and inorganic contaminant biotransformation takes place or not. Bacteria dependent on oxygen for their survival are able to biotransform or degrade the contaminants using oxygen as the electron acceptor. But when high amounts of contaminants are present, bacterial respiration often reduces the oxygen content in the aquifer and further leads to development of anoxic conditions in it. Biotransformation in the presence of anoxic conditions involves various oxidized compounds for the role of electron acceptors. Firstly, after depletion of oxygen, denitrifying bacteria tends to utilize nitrate (NO_3) (which is a groundwater contaminant itself), as an electron acceptor. After this nitrate depletion, due to the reducing environment in the aquifer, other biological reactions take place consecutively. Although when such species in the aquifer are present, reduction of metals such as manganese (Mn^{4+}), sulfur (SO_4^{2-}), carbon dioxide (CO_2), and iron (Fe^{3+}) takes place. Iron and manganese reduction lead to the breaking of their respective oxides as well as carbonates and also the assemblage of the reduced species in an aqueous phase. Sulfur reduction causes cessation of sulfur resources that are found in contaminated places near mines and, hence, contaminant trace metal release as well. If the environment is highly reducing, methanogenesis can also occur, which utilizes CO_2 as an electron acceptor as well as reduces carbon of +4 state into −4 state for methane production. Apart from these processes, biochemical processes also consist of transpiration, decay, bacterial respiration, and cell synthesis.

5.2.3 Geochemical processes

These reactions comprise small groups of processes such as acid–base reactions, oxidation–reduction, sorption–desorption, surface complexation reactions, ion exchange reactions, precipitation, retardation, and dissolution.

Reactions involving transfer of hydrogen ions between acid and base are acid–base reactions. pH of groundwater and natural buffers of the soil aquifer significantly affect the contaminant and mineral solubility of aquifer. The dissolution of minerals has a crucial role in the mobility of contaminants in groundwater (Kumar et al., 2017, Singh et al., 2020). Firstly, it releases those contaminants which are absorbed onto the minerals and further enhances the concentration of the contaminant. Secondly, this process also leads to alteration in the water reactions and contributes in the contaminant adsorption/precipitation. This happens in carbonate minerals, such as calcite and dolomite, which can efficiently neutralize acid waters like mining leachates. Apart from this, carbonates are even appropriate buffers at ranges of neutral and acid pH. Although at pH more than 9, aluminosilicate minerals show higher buffer efficiency. Hence, the aquifer materials deeply affect the movement of contaminants at subsurface.

Oxidation–reduction (redox) reactions is basically a reaction involving electron transfer among chemical species. Oxidation involves the electron loss or the increase in the oxidation state of the molecule. Lost electrons during oxidation are recovered by the corresponding molecule that was reduced by the same. The contaminants of the groundwater which have been dissolved are found in numerous oxidation states of aquifer, which is directly dependent on the contaminant's speciation, mobility, degradation, and toxicity and is crucial in the cases of heavy metal. Microbes support redox reactions via two ways, at the subsurface: through utilizing the substrate which is organic matter—for building cell mass and through giving rise to required reactants for transfer of electrons during "abiotic redox reactions." The redox potential of water is represented in the form of negative logarithm of its own electron activity (expressed by pE or Eh). Also, it tells the aquifer's redox status, which is responsible for prediction of the movement of elements as well as their biodegradation assessment of organic contaminants.

The sorption process contributes in decreasing the concentration of contaminants in water and is thus very important in contaminant hydrology. An aquifer consists of a matrix, solid in state, where the dissolved substances in the water move freely in their ionic form. Pollutants affinity for sorbing onto the solid phase of aquifer is basically noted by distribution or partition coefficient. Low partition coefficient (K_d) values represent the chemical's preference of staying in the aqueous phase. Along with this, sorption–desorption behavior of a contaminant is also demonstrated using sorption isotherms. K_d is the slope of the isotherm, in a linear model. The above-mentioned model does not take into account the limited and minimal capacity of the solid surface for the process of adsorption and also effects of the contaminant on adsorption. Thus, the Freundlich and Langmuir isotherms were established. The slope of both of the isotherms, at a low concentration of the contaminant, is linear. But these slopes change at high concentration. Since metals with high molecular weight are dissolved as cations, they are removed from groundwater by adsorption. Along with this, the nonpolar (less soluble) compounds are also sorbed onto soils when compared with contaminants with high solubility in water. With a high content of organic material or clay quantity, the sorption efficiency for organic contaminants also increases. The contaminants move slowly in the aquifer as compared to groundwater and is demonstrated by the retardation factor. Contaminants having high K_d values also move slowly in groundwater. This K_d value is used as the source or input or sink term in many such models for contaminant transport. This characteristic also represents various metal affinities and further can be exercised as an indicator for the contamination potential of aquifer (Jakomin et al., 2015).

Reactions involving surface complexation serve a prominent role in dissolution as well as transport/transformation of organic contaminants with precipitation of minerals. Complexes are formed through cation and anion combination. The ligand may possibly be any random inorganic anion, present in groundwater. This ligand also comprises organic molecules such as amino acids. Positively charged or zero charged or negatively charged complexes are possible. Contaminant complexes having zero charge show greater mobility when compared to the other complexes (positively/negatively charged), since they have least interaction with counter or other ions or even aquifer materials. Contaminant complexes involving humic acid or fulvic acid decreases the contaminant's efficiency for movement. On the contrary, the complex of heavy metals such as lead, plutonium, cadmium, and lead and uranium with organic or inorganic ligands enhances their mobility due to their nonavailability for adsorption.

5.3 EMERGING GROUNDWATER CONTAMINANTS

With the increasing growth of population over the last few decades, extensive and intensive exploitation of water resources has taken place to meet the human requirements (drinking and sanitation), economic purposes (tourism, agriculture, industrial use, landscapes and energy production), and significant changes in land use (Kumar et al., 2019). Organic compounds were not known to be prominent in groundwater with respect to their distribution and concentration, but have been currently detected which showed potential for causing serious ecological effects leading to worsening of human health. New chemical synthesis or change in the use and disposal of existing chemicals may form new emerging groundwater contaminants (ECs). Apart from this, emerging contaminants also comprise substances which were present in the environment and yet not discovered for their significance. Their availability in surface waters was reviewed for the purpose of public water supply (Houtman, 2010), for sources with respect to public supplies (Pal et al., 2010), and even for fate of ECs and notable trace pollutants (Murray, 2010). Lapworth et al. (2012) highlighted global spread of pollution of groundwater via various compounds, detected in terms of both current and past historical activities. Carroll et al. (2013) carried out a study which concluded that metropolitan areas (urban areas) put up more pollutants or have a higher contribution, in groundwater contamination when compared to nonurban areas because of their high density. Besides this, over abstraction of groundwater even reduces the level of groundwater table. This leads to reduction of surface water natural discharge flows, groundwater quality impairment, and land subsidence. When the flow pattern of groundwater changes, it displaces saline waters, moves the sea water, and increases surface water infiltration. All of these ultimately lead to contamination of groundwater. This is the reason why studying about such contaminants is significant. Therefore, depending on their origin, the physical properties, along with the chemical characteristics, following are the various kinds of ECs present in the groundwater.

5.3.1 Pharmaceuticals, hormones and sterols, endocrine disruptors

Pharmaceutical chemicals have been of great concern in the aquatic environment since very long (Menon et al., 2020; Mohapatra et al., 2018, 2021; Priyanka and Mohapatra, 2020, Mukherjee et al., 2020). Their major paths are via human excretion, agricultural usage, and disposal of unused products (Barnes et al., 2008). Variety of such products have been discovered on the surface as well as groundwater (Vulliet and Cren-Olivé, 2011). These mainly involve analgesics, anti-inflammatory drugs, antibiotics, antiepileptics, and barbiturates. The prime analgesic and anti-inflammatory sources were from wastewater, such as bank infiltration, urban sewage leakage, artificial recharge, and septic tanks (Carrara et al., 2007). The commonly known antibiotics such as triclosan, sulfamethoxazole, lincomycin, and erythromycin were reported in various studies. The maximum concentration of these antibiotics was in the range of 5.7–2 x 10^3 ng/L in the sites such as animal waste lagoons, septic tanks, wastewater sources, and landfills (Carrara et al., 2007). Less commonly known pharmaceuticals comprised a variety of chemicals related to products improving skin care such as phenoxyethanol, UV blockers such as oxybenzone and drometrizole, lilal, and isopropyl myristate (Stuart et al., 2011). Apart from this, some of the preservatives were also found including paraben compounds such as fungicide or microbiocide such as methylparaben, used in creams and other personal care products. Methylparaben was reported to be frequently present in UK groundwaters (Stuart et al., 2011) at a maximum concentration of 5×10^3 ng/L and might be a useful marker for wastewater pollution in the freshwater environment. Other threats to surface water include tamiflu and chemotherapy drugs such as cyclophosphamide, 5-fluorouracil, or ifosfamide (Johnson, 2008), along with illicit drugs such as amphetamines (Kasprzyk-Hordern et al., 2008). Antibiotics used for veterinary purposes seen in contaminated groundwater were monensin and tylosin (Bartelt-Hunt et al., 2010).

Hormones (mainly sex hormones) such as androgens (testosterone or androstenedione), estrogens (estrone or estriol or 17α- and 17β-estrodiol), and progesterone are also found in groundwater. As shown by Kurwadkar (2014), 17β-estradiol was detected at a concentration of 20,069 ng/L. Few androgens, namely, nandrolone, along with more significant synthetic estrogens or xenoestrogens, namely, 17α-ethinyl estradiol as well as diethylstilbestrol, are broadly utilized in the form of contraceptives. Few of them are also detected in wastewater and treated effluents (Vulliet and Cren-Olivé, 2011). On account with this, a similar group is also present such as cholesterol and its constituent metabolites 5β-coprostanol, along with plant sterols such as β-sitosterol, stigmastanol, and stigmasterol. Phytoestrogens, commonly known as sterols derived from plants and released in wastewater, thus being the biggest source of such chemicals in the environment (Liu et al., 2010).

Endocrine disruptors (EDs) are recognized as the ones which are present inside the body and can block hormones and their effects on regular body functioning. Three prime classes of EDs include estrogenic (in low concentrations 0.1–120 ng/L), which tend to copy or change the natural functioning of estrogen, androgenic (mimics or blocks testosterone) along with thyroidal (responsible for immediate impact on thyroid). On the other hand, natural as well as engineered EDCs are generated in the environment due to human industries, activities, and creatures, mainly by sewage treatment systems prior to the final going in groundwater, soil, silt, and surface water. Around eight compounds are efficient EDS, comprising fragrances; fluoxetine; galaxolide and tonalide, along with various nonsteroidal pharmaceuticals; diclofenac, clofibric acid, naproxen, diclofenac, paracetamol, and ibuprofen (Caliman and Gavrilescu, 2009).

5.3.2 Industrial effluents

Both bisphenol A (used as plasticizer; at a concentration of 0.82–102.00 ng/L) (Osenbruck et al., 2007) and metabolite nonylphenol (known as EDs; at a concentration of 10.30–84.00 ng/L), phenolic in nature; fire retardants; and musk galaxolide, were the noted contaminants. Apart from these, detergents such as plasticizers have also been found. Other groups such as corrosion inhibitors, solvent stabilizers, and surfactants were reported under industrial contaminants (Loos et al., 2010). Another group of fluorinated organic compounds include packaging/protective coatings, refrigerants, and fire retardants, along with their growing impact on the environment and their occurrence as well. There are a lot of industrial compounds which affect the environment, many of which have resulted in well-established problems. The solvents containing chlorine, polyaromatic hydrocarbons as well as fuel oxygenate methyl tertiary-butyl ether (petroleum hydrocarbons), plasticizers, phthalates, bisphenols, and adipates are classified as priority pollutants (Verliefde et al., 2007). 1,4-dioxane (1,1,1-trichloroethane stabilizer) is water soluble and remains the same after exposure to biodegradation; does not bind with soil; and leeches easily. Benzotriazole derivatives found in anthelmintic, antifungal, and antibacterial drugs are resistant and constant in water comprising environment. Dioxins are released due to micropollutant degradation, such as from triclosan (antimicrobial additive).

5.3.3 Everyday use compounds

Compounds such as caffeine, cotinine, and nicotine have been recently discovered in the groundwater. Sweeteners produced artificially such as acesulfame, sucralose, cyclamate along with saccharin are found at higher concentrations in groundwater. It is also interesting to know that Buerge et al. (2009) illustrated how acesulfame due to its persistence, use, and mobility has been found in higher amounts in aquatic environments. Mostly, concentrations of caffeine ranged around $11\text{–}13 \times 10^4$ ng/L, thus proving it serves as a detector for anthropogenic activity and wastewater contamination (Loos et al., 2010).

5.3.4 Food additives

These are mainly either oxidants or can be EDs.. Another example of food additive is triethyl citrate–stabilized foams and used for pharmaceutical coverings, and plasticizers. Hydroxytoluene as well as butylated hydroxyanisole are used for preservation of fat in foods. Others include camphor, citronellal, 1,8-cineole (eucalyptol), heliotropin, cis-3-hexenol, citral, triacetin, phenylethyl alcohol, along with terpineol. Other sweeteners such as sucralose, acesulfame, and saccharin were also detected (Buerge et al., 2009) along with preservatives such as parabens. The above-mentioned groups have not yet studied elaborately, but still have the potential of serving as a marker compound for tracing wastewater sources.

5.3.5 Pesticides

Pesticides have been mostly discovered in lower concentrations in the groundwater globally for a significant time and are now the most common contaminants. A study by Malaguerra et al. (2013) showed how utilization of reactive transport models via three of them, namely, glyphosate and its degradation product aminomethylphosphonic acid along with mecoprop helps in evaluation of vulnerability of groundwater. They further showed that high concentration of mecoprop (>0.1 μg/L) was detected in the wells above surface water. Nearly twenty years ago, it was certain how pesticide degraded groundwater (Kolpin, et al., 2000). Other studies have also shown how pesticide metabolites have higher concentration than the parent compounds of the same in the groundwater (Lapworth and Gooddy, 2006).

5.3.6 Personal care compounds

These are other contaminants which involve both prescribed as well as nonprescribed human and veterinary pharmaceuticals. Some of the other personal care compounds (PCPs) include steroids, cosmetic products, shampoos, engineered hormones, and perfumes. A report showed how UV filters have been proven to display estrogenic activity and the most common PCPs found inside the groundwater (Oviedo and Aga, 2016). Some more PCPs have been mentioned below:

- DEET is N, N-diethyl-meta-toluamide, which is a prime component of insect repellents.
- Parabens are those alkyl esters which contain p-hydroxybenzoic acid; previously exercised as bacteriostatic and fungistatic agents in foods, drugs, as well as cosmetics.
- Bactericide and antifungal agents—triclosan; common ingredient in household products such as antimicrobial sprays, toothpaste, and even soap.
- Polycyclic musks such as tonalide and galoxalide are used as fragrances; present in the agents used for washing and cleaning.
- UV filters—organic filters such as benzophenones and methoxycinnamate.

5.3.7 Surfactants

Different kinds of surfactants such as nonionic, anionic, cationic, and amphoteric, along with antifoaming agents, are usually present. Priority contaminants such as octyl- and nonyl-phenol (OP and NP) are termed as "priority pollutants." They are exercised in alkylphenol ethoxylates generation for nonionic surfactant manufacture. Both of the parent ethoxylates with their constituent metabolites, carboxylic degradation products, and alkyl phenols have been seen to remain persistent in aquatic environments (Soares et al., 2008). The compounds based on nonionic polyethylene glycol are utilized in the form of antifoaming agents. Another kind of such compounds, namely, siloxanes, is utilized in many PCPs in the form of antifoaming agents. Also, they remain a matter regarding their toxicity as well as transport in aquatic bodies (Richardson, 2007). On the contrary, chlorides are availed as homologues, emulsifiers, and antiseptics, which have been further recognized in the form of ECs (Ahrens, 2009). The surfactants which are amphoteric in nature comprise coconut-based products, for example, the commonly used cocamidopropyl betaine. The anionic ones involve perfluorinated chemicals at concentration > 0.3 mg/L (EA, 2007). All of these are quite persistent in surface waters and even wastewaters (Lara-Martín, 2010).

5.3.8 Heavy metals

Arsenic: Groundwater pollution caused by arsenic is affected by variations taking place seasonally. For instance, analysis of groundwater quality of around 45 samples at the time of monsoon and 55 after monsoon, collected from the Nawalparasi District of Nepal showed an increase in arsenic in the premonsoon period samples. Shrestha et al. (2013) concluded that this change is because of the decrease in water table levels in this specific season. Another interesting result in context of arsenic concentration variation is based on the tube well depths. The ones which were drilled into shallow aquifers displayed how the lower Pleistocene aquifers had high concentration of arsenics as compared with the upper Holocene aquifer. Observations from Ghaghara Basin, India, have clearly confirmed arsenic occurrence in groundwater. Shah (2013) carried out a survey in Basti, Faizabad, and Gonda regions of Uttar Pradesh, India, and concluded that 42%, 38%, and 61%, respectively, of tube wells have higher arsenic concentrations than standard values given by the WHO with respect to safety of drinking water. It also reported that around 86%, 69%, and 35% of tube wells of Faizabad, Gonda, and Basti districts, respectively, are present at lower depth (21–45 m), and 47% of such wells are situated at 10–35 m depth in the aquifer named Holocene Newer Alluvium. Such arsenic contamination usually takes place due to anthropogenic activities. Also, arsenic being ubiquitous in the environment due to natural conditions triggers its movement, hence causing the contamination. Another study by Alam et al. (2016) showed a concentration of up to 133 ng/L in Jharkhand. Lastly, the values exceeded the level of 10 ppb in Punjab.

Fluoride: Enhanced fluoride levels present in drinking water cause serious health issues. This enhancement in groundwater has been reported in some of the portions of Rajasthan, India. One of the surveys conducted by Vikas et al. (2013) included 66 numbers of samples and analyzed fluoride's concentration remained to the extent of 0.35–9.3 mg/L along with 62% of the samples surpassing the minimum limit for water meant for consumption (1.5 mg/L). That report also stated that groundwater of the aquifers which are shallow, that is the weathered area of soil–clay complex, contains higher fluoride concentration when compared with deep aquifers, consisting of mainly hard rocks. Another study showed the increased concentrations because of the processes of weathering (the one done chemically) and leaching of fluorine-containing minerals inside hard rocks. Singaraja et al. (2013) also reported such findings where it was found that fluoride occurrence in groundwater was present in hard rock aquifers of coastal areas of South India. Another study involving 135 samples also proved that concentration of 3.2 mg/L exceeded the standard consumable water regulations. Other parts of South India have also reported high incidence of elevated fluoride levels. Hanipha and Hussain (2013) carried out a study involving 23 samples, taken from Dindigul, Tamil Nadu, India, where the concentrations were 2.47–5.26 mg/L, again more than the drinking water standards (1.0 mg/L). Many of the researchers concluded the reason for this high concentration to be the weathering of rocks/soil, along with chemical fertilizers of agricultural lands. They also added another reason to it, that the evapotranspiration as well as fluoride enrichment because of weak alkaline pH conditions (ranging from 7.2 to 8.2) is responsible for the same.

Chromium (valence state—six): This element can be a result of unregulated release of industrial wastewater. Rao et al. (2013) carried out a survey on samples of groundwater from the industrial area of Ranipet, Tamil Nadu, India and observed that chromium (hexavalent) concentration exceeded the standard guidelines of the WHO for drinking water. Alongside this, it was then concluded that this increase was due to the surface release of waste as well as partially treated sewage. A similar study was conducted by Srivastava et al. (2013) where 21 groundwater samples were extracted from various industrial areas in the district of Kanpur Dehat. It was also stated that chromium occurrence in industrial areas of Kanchanpur in Kanpur, Uttar Pradesh, India, was because of inaccurate waste disposal comprising alkaline chromite sulfate. Further, favorable hydrogeological parameters with climatic conditions of a region also facilitates chromium conduction, moving from soil to groundwater. The prevalence of hexavalent forms of chromium in water bodies such as springs and underground sources in some areas of Zanzibar was also reported by Mohamed et al. (2013). An illustrative analysis of 30 such samples showed 96.6% of samples having increased concentrations (above the WHO safe regulation of drinking water-0.05 mg/L) of hexavalent chromium; the concentration ranges between 0.05 and 0.4 mg/L.

5.3.9 Nitrate

Cruz et al. (2013) reported the groundwater pollution and impact of agricultural practices caused by nitrate. Some studies even reported that groundwater samples from agricultural lands enhanced the nitrate levels, when compared with the groundwater from nonagricultural lands. Another experiment by Torok-Oance et al. (2013) involved 17 wells which were measured in order to check the nitrite as well as nitrate concentration. It was seen that both of their concentration levels surpassed the standard regulations for drinking water in 41.17% (nitrate) and 11.76% (nitrite) of the samples. They analyzed those observations which lead to the conclusion that livestock farms, manure deposits, and latrines are responsible for these increased concentration levels. Haque et al. (2013) elucidated the rapid growth of major urban cities and their consequent groundwater contamination. They studied that the urbanization of few of the mega cities of South Asia has a causal relationship with groundwater pollution. They further stated that such increased contamination is due to groundwater over extraction, along with the induced recharge due to intrusion by the seawater. This intrusion is a possible consequence of shipbreaking activities which is further root cause for nitrate contamination. Across India, nitrate concentrations in groundwater are usually more than 45 mg/L as illustrated by Shukla and Saxena (2016).

5.4 SOURCES OF GROUNDWATER CONTAMINATION

Emerging organic contaminant (EOC) sources present in the environment which poses an impact on the groundwater are of two kinds: point and nonpoint or diffuse sources of pollution. The former arises from discrete areas, and their inputs are present in the aquatic bodies which are easily described in a distinct way. Plumes of pollution are usually more inhibited. Some of the examples are industrial residues (food processing plants, manufacturing plants, and hospitals), treatment plants for municipal sewage, merged waste or stormwater congestions, buried septic tanks, and extraction of resources by mining and sites for waste disposal (farm waste lagoons, sites for landfills and industrial confinements). On the other hand, diffuse sources initiate via undefined sources that specifically take place in wide geographical areas such as agricultural runoffs from sources of biosolids and manure, diffuse aerial deposition, stormwater and urban runoffs, and sewage leakage from reticulated urban systems (Terry et al., 2008; Nakada et al., 2008). Such sources of groundwater contamination are also identified as follows:

a. larger spreading of geographical scales
b. lesser environmental loading with respect to point sources
c. presence of large potential for weakening in the soil as well as subsurface via natural means and,
d. poor description with respect to the polluter, still are a real obstacle for monitoring, regulating and even assessing their effect on groundwater assets.

Waste materials extracted via waste lagoons or biosolids, landfills, septic tanks, industrial as well as municipal wastewater are some of the other sources. Landfill sources generally carry a higher median value for all the EOCs as compared with the median values of rest of the sources. Certain contaminants are further source specific such as contrasting agents of hospitals and pharmaceuticals (veterinary purposes) are easily detected in surroundings as compared with the contaminants having various effective sources, for example, caffeine. A major proportion of pharmaceuticals has plenty of sources present in the environment although it is still plausible for their elucidation on a subregional scale (Clara et al., 2004; Buerge et al., 2009).

5.4.1 Point sources

Many experiments have proved EOC's fate in groundwater followed by the wastewater intrusion (both sewage and industrial), along with contaminated sources of surface water and continue to provide larger scopes for research with respect to their sources and fate. Díaz-Cruz and Barceló (2008) showed how artificial recharge technology has the potential of being an efficient EOC source in groundwater especially at the time of shorter residence time, alongside being a matter of threat for the adjoining groundwater resources as well as the surface water resources. It can be seen how this specific source is a significant EOC input in the environment worldwide, particularly in those areas where the treatment of wastewater is rudimentary or nonexistent. Although on the one hand, the treatment systems are becoming advanced, simultaneously, on the other hand, the process of removal still remains incomplete, and thus as a result, these sources remain the major input for the environment. Also, molecules which are released in the environment are dependent on the origin of the treated water. Even the landfill sites have proven to be significant EOC sources in groundwater for several years and still continue to be the same today due to sites for landfill construction, waste material management, and poor choices with respect to the locations of landfill. Along with this, septic tanks are also an important source for contamination of groundwater via EOCs, specifically in the case of shallow groundwater tables with high transmissivity. Point sources are responsible for the higher production of EOCs due to the reasons such as (i) they are easily detectable in the environment (ii) for their detection, engineered solutions are more straightforward and thus can deliver the greatest environmental benefit, (iii) stronger history of regulatory control and easier connection between polluter and pollution, and (iv) easier to predict, conceptualize, and monitor.

5.4.2 Diffuse sources (nonpoint sources)

While treating wastewater, incomplete removal of EOCs results in residual concentration in the biosolids which are appealed for the soil (Clarke and Smith, 2011). Since they have usually high concentrations in wastewater linked biosolids and also high solubilities, they have two kinds of halogenated hydrocarbons, namely, perfluorochemicals and polychlorinated alkane, which are known to be significant contaminants in groundwater (Clarke and Smith, 2011). When these EOCs set foot in the surface water, they tend to remain in the surface water via vertical/lateral hydraulic exchange and are shifted to a groundwater body (Lewandowski et al., 2011). Talking about diffuse leakage of sewerage systems, they are a matter of concern for EOCs in the groundwater as it is responsible for natural weakening processes in subsurface. Aerial sources extracted from agriculture, industries, and transport systems provide a diffuse as well as low loading up to the land surface. Since the subsurface has low concentrations along with high attenuation potential, specifically in the soil area, such kinds of diffuse sources are usually not known to be prominent for contamination in groundwater. They are termed as "nonpoint source pollution" and have been originated from undefined sources which typically are present in the broad geographical scales. For instance, a major proportion of pesticides has been used for agricultural and horticultural purposes. Once these are released, they are reduced via both biotic and abiotic mechanisms. Stuart et al. (2012) studied and assessed approaches and risk factors for pesticide degradants. Even the application of biosolids as well as manure derived from processing of sewage sludge has the scope for enhancement of nutrient levels inside the soil. Although the incomplete EOC removal during such treatment results in release of solid residuals, these are likely to remain potential sources for contamination of groundwater.

5.5 IMPACTS OF GROUNDWATER POLLUTION

Widespread groundwater contaminations have various repercussions, including health hazards, ecological imbalance, environmental deterioration, ozone layer depletion, decreased potency of farm lands, infertility of crop fields, and even scarcity of resources. Various national and even regional studies have shown biases toward the sites of contamination in order to know the real frequency of occurrence as well as the distribution in the environment. More of the regional experiments are required to analyze the EOC incidence in the groundwater and also recognize spatial as well as temporal contamination nature, in sources from leakage of urban sewage, biosolids, transport networks, and manure (Musolff et al., 2009). It has been proven that environments which are more arid are usually susceptible to such contamination because of the intentional or unintentional wastewater discharge (Cabeza et al., 2011). The basic ratio of this natural by artificial recharge shows less potential of natural attenuation as well as dilution. Thus, the impact on groundwater due to such pursuits is higher and mostly less sustainable for longer time (Fram and Belitz, 2011). Groundwater which has been contaminated by EOCs is a matter of concern since they may be hazardous for coming decades, owing to longer residence times as well as persistence because of the decreasing chemistry and reducing rates of degradation by microbes. Specifically, alluvial aquifers and karstic ones are prone to contamination and are the center for various studies. Apart from this, the impact of this pollution can also be seen in human health which can be even severe. This involves serious illnesses such as vomiting, headache, nausea, eye and nose irritation, and respiratory diseases, sometimes leading to death. Along with these, there are also chronic diseases caused by contamination of groundwater such as cancer; anemia; circulatory and nervous system problems; liver and kidney damage; reproductive difficulties; bone diseases; and hair loss. Groundwater pollution also leads to water scarcity since most of the population of the world is dependent on groundwater for their regular chores and use. Assessment of groundwater pollution was conducted in Babati, Manyara, Tanzania by Pantaleo et al. (2018). The overall distribution of contaminants detected, comprised 435 sources in the vicinity of Babati. Different contaminations of contaminants were detected. Nitrates were detected at a concentration of 357.7 mg-NO_3/L, which is not within the WHO recommended levels (70–50.0 mg-NO_3/L for drinking water). The microbiological contamination results included mostly sources polluted by fecal coliforms (FCs) and total coliforms; concentrations exceeded the WHO standard of 0 colony forming unit/100 mL for FC in shallow wells. In addition to this, the total phosphorus (TP) concentrations were 0.34 mg/L with the standard deviation being around 0.1 mg-TP/L (Ulén and Snäll, 2007). Water consumption with high phosphate concentrations generally results in digestive health problems (Domagalski and Johnson, 2012). Along with this, the maximum concentration for groundwater solute content was around 30 mg/L around Babati. The mean values for the same also were above the WHO values of 10 mg/L. The fluoride amounts on the other hand also exceeded the standard readings. Consumption of such concentrations of fluoride also leads to fluorosis in children's teeth (Rao et al., 2015). These above-mentioned studies prove the impact of the groundwater samples in the area. The results of the study also show that such pollution will further increase with population increment, if necessary, measures are not taken, ultimately restricting the community of Babati for using groundwater wells for consumption.

5.6 INDIAN SCENARIO

According to the reports of the National Commission for Integrated Water Resources Development, the overall water usage and utilization in India was around 710 billion cubic meter (BCM) in the year of 2010. Predictions and estimations say that it will rise up to 1,180 BCM by 2050. Annually, the average per capita availability of water in India was recorded to be 1,544 cubic meters in 2011 which is likely to decrease to 1,341 and 1,140 in the years of 2025 and 2050, respectively. In a way, India is the largest user of groundwater around the world and withdraws around 253 BCM of groundwater annually which is roughly 25% of the total, globally. The overall amount of renewable groundwater resources totals 447 BCM, 441 out of which is the average accessible groundwater for the developmental process. Around 62% is the developmental stage of groundwater in India, and around 90% of the net withdrawal of groundwater is used for irrigation purposes (CGWB, 2017). India is at the 120th position with respect to water quality index among 122 countries, with around 70% contaminated water. Talking about groundwater, around 60% of the Indian districts involve water quality issues due to either depletion or contamination. Fifty-four percentage of the wells of groundwater are gradually decreasing, and 21 of the major cities will be short of it by the end of 2020, thus affecting millions of people. Along with the World Bank's help, the Government of India expanded one of its schemes, namely Groundwater Management and Regulation Scheme and came up with the Atal Bhujal Yojana. Initially, in 2013, it was started by the Union government. It was stated that this project will be initiated from June 5, 2018, and will further go on till September 29, 2023. Also, seven of the states including Gujarat, Madhya Pradesh, Maharashtra, Rajasthan, Haryana, Uttar Pradesh, and Karnataka were selected for their participation in the program. This program also includes a provision which permits even the current states to be included at the time of implementation. The yojana mainly revolves around the interventions for improving the quantity of groundwater. Provisions for the quality of groundwater are limited to planning as well as monitoring of groundwater as required. Its activities consist of arrangement of supply mechanisms for groundwater management; state specific institutional framework to achieve sustainable management; increase of recharge of groundwater and improvement in efficiency of water use; and strengthened community-based organizations for foster administration.

5.6.1 Indian industries and groundwater contamination

The contamination caused in the groundwater is due to the water pollution in the water bodies which can often result in severe ill effects on living organisms. This pollution occurs because of the untreated contaminant discharge, mainly industrial sewage, directly or indirectly into water bodies (Matta, 2015). This waste production has increased because of the increase in urbanization. For instance, textile manufacturing units usually need huge amounts of water, thus giving rise to massive amounts of wastewater comprising heavy metals and organic contaminants, and are released into a common sewage drain for industrial areas. These toxic pollutants leach and percolate within the soil profile and are finally settled in groundwater, hence causing health hazards in humans and livestock consumption (Bharti et al., 2007). In order to maintain the contaminant's behavior, it is essential for measuring the contaminants of industrial areas, where wastewater is released in a common industrial drain used for irrigation purposes. Approximate heavy metal concentration of lead, cadmium, and nickel was 42.358, 1.927, and 7.96 mg/kg, respectively, in those soils which were irrigated by contaminated groundwater from textile industries of Panipat (Haryana) (Bharti et al., 2013). Bharti et al. (2013) also noted that the maximum transfer factor values for cadmium and nickel were found to be 321.11 and 83.49, respectively, from the groundwater to agricultural soil. It was also proven that this contamination of groundwater affected the human subjection to metals via food chains (Lokeshwari and Chandrappa, 2006).

A matter of great concern is the rapid industrial growth in the areas of Gujarat, consisting of Ankleshwar and Vapi (known as the golden corridor area). Such industries are responsible for the generation of wastewater filled with toxic metals further contaminating the groundwater, thus leading to serious environmental issues through affecting soil, aquatic organisms, animals, plants, and human beings. Patel and Das (2015) conducted a survey at the edge of Vapi industrial belt (VIB), where pH of 32 samples of groundwater ranged from neutral to alkaline (6.31 to 8.06, to be more specific), and the mean value was 7.34. Around 700 m away from VIB is a village named Eklahare that showed higher salinity in water, thus concluding the water to be unsafe for irrigation. Along with this, it also had higher BOD and COD levels with respect to the Indian Standards, thus indicating the presence of contaminants in groundwater sources (Yadav and Kumar, 2011). Near VIB, toxic elements such as lead and nickel were found in water. Apart from this, in the northern states also, the quality of groundwater has deteriorated because of the increasing agricultural activities, industrialization, and urbanization (CGWB, 2005).

5.7 REMEDIATION

Groundwater is a significant source which cannot be under looked. A majority of people are directly or indirectly dependent on groundwater for their daily chores. This requirement of groundwater and its contamination issues is of great concern. Thus, there has been research going on in order to find out new remedial measures. As each aquifer of groundwater is of a kind, an organic contaminant's mineralization takes place via biotic and abiotic processes, for example, chemical conversion into carbon dioxide and water, is dependent on the spatial heterogeneity of properties of subsurface such as microorganism populations, pore structure, and hydraulic conductivity along with energy and carbon sources. This information is supported by laboratory studies including transport, degradation, and sorption (as displayed by chemical dispersion as well as the hydraulic conductivity). This approach should be turned into plans for designing and engineering the required and necessary conditions for restoration. Scientists and groundwater experts have suggested various methods for remediating the contamination. Such remediation technologies can be either an onsite technique or an in situ technique. Various onsite techniques include pump and treat method, phytoremediation, air sparging, soil water extraction, and permeable reactive barriers. On the other hand, in situ technologies are yet not well developed, but still are effective to some extent. Since, contaminant plume and its migration are often difficult to track and define, artificial recharge and surface water harvesting have come into picture for aquifer system sustainability, along with in situ dilution of groundwater, contaminated by arsenic, fluoride, and other geogenic contaminants.

Majorly, there are three groups for remedial measures:

1. restriction of containment: contaminant in a specific domain for prevention of additional spread;
2. transfer of removal: contaminant from an open environment to a managed one; and
3. transformation of treatment: contaminant into a nonhazardous substance.

5.7.1 Containment technologies

This containment is usually achieved by controlling the fluid flow which has the contaminant or by stopping the motility of the contaminant. It has been described in the following section involving the physical barriers as well as hydraulic ones in order to contain the contaminants.

5.7.1.1 Physical barriers

The main motive of this kind of barrier is to prevent the contamination spread by controlling the flow of water. Generally, a barrier is fitted in front, that is the downgradient part of the contaminant zone, but sometimes also fitted in an upgradient or both the cases. Such hurdles are exercised for the incoherent materials such as sand or soil and also in consolidated media such as rocks if special techniques can be employed. One of the prime reasons for considering this kind of barrier is due to the existence of substantial layers present horizontally with little permeability where it can be placed. In the absence of a seating of lesser permeability zones like this, contaminated water flows beneath such physical barriers. Another reason for the same is the barrier's permeability itself. Since the aim of the physical barrier is to lessen the flow of the fluid via the target zone, its permeability should also certainly be very low. Final reason for the same as contaminant's potential to degrade its performance. There also exist three major kinds of physical barriers: slurry walls, that is, trenches having clay or mixtures of it as well as soil; the grout curtains, represented by hardened matrices produced via cement as well as cement-like materials and injected in the ground; and lastly, sheet piling, that is, large iron sheets driven into the ground. Apart from this, barrier properties should be in accordance with the contaminant properties for least chances of failure.

5.7.1.2 Hydraulic of pressure-driven barriers

The most significant criteria for hydraulic barriers to occur is just like the reason for physical barriers such as manipulation and control of flow of water. Not able to understand and rewrite the sentence. While the physical barriers are made up of solid substances, these kinds of barriers, on the other hand, are based on the water pressure manipulation. Due to the differences in pressure caused by extraction of water using well or drains, hydraulic barriers are generated. The prime performance factor depends on its potential of capturing the contaminant plume, or in other words, to limit the contamination spread. This capture of plume is based on the number and also the positioning of drains and wells, along with the water flow rate via the same. Frequently, efforts have been carried out for optimization of design as well as functioning of the containment system in order to achieve plume capture at greater levels. The basic form of this barrier is formed via a drain system, which is built with the help of installation of a horizontal pipe (perforated), at the subsurface which is fitted such that it allows maximum contaminated water capture. This water is then collected and further removed via gravity pumping (active pumping). Exercising of such systems is limited to small contaminated zones. The large zones contain contaminant plumes stretching up to a certain length. The above-mentioned and other reasons are responsible for these barriers to be the most commonly used containment method. It has some disadvantages, including operational costs, maintenance and disposal needs, and storage and treatment of large quantities of contaminated water.

5.7.2 Removal technologies

5.7.2.1 Pump-and-treat mechanism

One of the commonly used techniques to remove contaminated groundwater involves the use of more than one extraction wells from the subsurface. The pumped water which is contaminated at the subsurface is subjected to certain types of treatment operation consisting of carbon adsorption, air stripping, or biological treatment system. Usually used for saturated subsurface systems as aquifers, this method is also helpful for removing pollutants of vadose area. In this particular instance, it is known as in situ soil washing. This approach involves penetration passages, apart from boreholes or wells, which are used for introduction of water into contaminated zones. While making use of water flushing for the ultimate removal of pollutants along with plume capture, good results are expected. Studies on such systems have proven how this method is quite useful for contaminant plume containing and shrinking processes. However, it is also ineffective for complete removal of contaminants from subsurface. The factors responsible for this include presence of low-permeability zones, rate-limited/nonlinear desorption, and presence of immiscible liquid. One way for overcoming these factors is for containment and removal of contaminant source zones— the areas where they are disposed of. If this source zone is left untreated, it automatically serves as a constant source.

5.7.2.2 Air scattering/sparging

The process of in situ air sparging enhances the mass removal rate from contaminated water systems. It involves injection of air into the targeted contaminated zone, in order to undergo mass transfer (volatilization) of volatile and semivolatile contaminants from groundwater to air bubbles. Due to buoyancy, these bubbles move in the upper direction toward the vadose zone—the region where a soil-venting structure is introduced for capturing the contaminated air. Reports have also proven that the administered air in the porous media (which is water-saturated) generally travels in the form of distinct channels which contain just a fraction of the overall cross section of that specific zone, instead of surpassing the entire channel as bubbles. Such a channeling occurrence decreases the air sparging efficiency. Another similar drawback is the lower permeability zones above the target zone, for targeting the localized contamination zones. In this context, another technology of air sparging consists of in-well aeration. With the help of this procedure, air sparging takes place in the well's water column instead of formation of the subsurface itself.

5.7.3 Treatment (in situ technologies)

5.7.3.1 Bioremediation

This method involves natural microbial activity for cleaning the polluted sites—especially for contaminant degradation. Mainly, two kinds of bioremediation are present: in situ treatment where treatment takes place in place of a contaminated site, and ex situ treatment is the treatment above ground of contaminated soil. These processes basically incorporate a complex system of biological activities. Thus, it becomes difficult at times to perfectly predict its outcome and design efforts. For successful bioremediation to take place, characterization of site and identification of limiting factors are extremely significant. Soil samples and water collected from the site are examined in order to verify the microbial presence and check whether they can degrade the contaminant. Strict monitoring of groundwater at the site is usually carried out for characterizing environmental factors which have an impact on biodegradation such as oxygen concentrations, pH, nutrient concentrations, and temperature. Sorbed contaminants are basically not available directly for microbial cells. Rate as well as magnitude of such contaminant biodegradation are thus decreased. The common limiting factor in case of in situ procedures is oxygen availability in the saturated zone. Oxygen is the utmost requirement for aerobic biodegradation to occur. Although it carries limited water solubility, it is saturated and thus needs replenishment with the help of aqueous diffusion (slow process). Apart from this, phosphorus, nitrogen, and nutrients are basic additives used for increasing the biodegradation rate. Various contaminated sites consist of organic waste materials, which are clearly carbon rich but very poor in nitrogen as well as phosphorus. Another approach is sequential aerobic or anaerobic biodegradation processes, whereby the biodegradation is increased in the presence of anaerobic conditions, and further through aerobic bioremediation. This particular initiative will be of great use in the case of mixed-waste sites.

5.7.3.2 Remediation by microorganisms

Process through which risky and harmful contaminants are broken down or converted into safer ones via microbial activity is termed as "microbial remediation." Their response toward toxic metals (usually heavy metals) shows the relevance of the remediation of groundwater. *Klebsiella pneumoniae* showed remediation of heavy metals at a concentration of 15 mM, slightly higher than the normal range of concentrations (Sharma et al., 2000). Sanyal et al. (2005) conducted a survey and concluded that *Fusarium oxysporum*, a fungus remediates aqueous Pb and Cd ions. Carbon dioxide is produced during metabolic activity, which further reacts with metal ions and produces metal carbonates. When a fluidized reactor was used, 94% chromium was absorbed. *Pseudomonas fluorescens* and *Pseudomonas aeruginosa* were able to remove metals such as lead, zinc, manganese, iron, and copper, at a pH of 7–9 with temperature of 25°C–35°C (Paranthaman and Karthikeyan, 2015). Twelve bacterial isolates were found by Sunil et al. (2015), which were tolerant to mercury and lead from suburban parts of Mysore, Karnataka, India, including *Streptomyces flavomacrosporus*.

5.7.3.3 Phytoremediation

Phytoremediation is an efficient and significant method for achieving the goal of treating contaminants. This process includes the application of living plants for extraction of contaminants from groundwater via their interaction with hydrological systems. This method is implied for the treatment of both kinds of contaminants, that is, organic and inorganic. Interaction of plants and contaminants occurs via various methods. Accumulation involves contaminant intake into plant tissues. This approach has an advantage where the removal of contaminants actually takes place at the site via plant harvesting, post contaminant accumulation. The second approach for the same is stabilization which involves contaminant stabilization via root complexation and root surface in the root zone. This method is beneficial since it avoids any risk like exposure to wildlife, and also, harvestation of plants is not needed. Phytovolatilization is the mechanism wherein uptake of water and contaminants takes place with the help of the xylem, and further vaporization occurs in the atmosphere via foliage. Another process is phytodegradation which involves absorption, assimilation as well as degradation of pollutants by plants into nontoxic or lesser toxic states. This method occurs in the course of metabolic processes supported with few enzymes. Ghosh (2010) illustrated how *Hydrilla verticillata* showed potential of being an accumulator of arsenic and cadmium, along with *Ipomoea aquatica* being of cadmium. Another species reported was *Typha angustifolia* which showed the efficiency for acting as phytoremediators of metals such as nickel, zinc, and copper present in wastewater generated from industries (Chandra and Yadav, 2010). Similarly, Kumar et al. (2012) examined the effect of phytoremediation using five native aquatic plants, *Eichhornia crassipes*—successful removal of copper and nickel; *Hydrilla verticillata* and *Marsilea minuta*—lead and chromium removal, respectively; *Bacopa monnieri*; and *Ipomoea aquatica*. *Garcinia cambogia* (Indian local plant) has proved to be used for removal of arsenic from the groundwater (Kamala et al., 2005). Lastly, *Pistia stratiotes*, that is, water lettuce, showed efficiency of absorbing arsenic with a range of 0.25–5 mg/L at neutral pH within 144-hour time period (Basu et al., 2003).

5.8 FUTURE PERSPECTIVE

EOCs used in the groundwater in the form of wastewater source tracers is currently a developing research area. This research would be very helpful and valuable when evaluating the sources for pollution because of dilution potential of groundwater resources. Although the sources for contamination can be discriminated against, it still remains unclear and requires further research to be done. So far, there are many studies which have illustrated EOCs along with other recognized environmental tracers (such as isotopes or time tracers for groundwater residence) for developing processes for better interpretation of their contaminant's fate along with the transport at subsurface. For this understanding, field as well as modeling studies are required for deciding this fate in the unsaturated areas. Basic and initial experimentation about degradation and sorption of the EOC in soil and aquifers is also needed for developing modeling abilities in the subsurface. A detailed study is also required for assessment of importance of various sources as well as transport process apart from the EOC's chemical properties in controlling their own occurrence. On experimenting their fate and degradation methods, it was found that these processes are facilitated by isotope labeled approaches in laboratory studies as well as field work (Nowak et al., 2011). It was also proved that colloidal transport can be a significant EOC pathway within the subsurface simultaneously allowing transport of EOCs faster in comparison to that in the dissolved phase. EOC relationships with colloids such as iron oxyhydroxides as well as fulvic-like substances, along with natural colloids which are heterogeneous, are a matter of further investigation. Their fate in the zone of surface water interaction with groundwater, namely, hyporheic zone, is still not studied deeply and hence is a significant research aspect for subsequent ventures (Banzhaf et al., 2011). Currently, very few studies are available regarding latest treatment technologies such as chemical and biological treatments, adsorption processes, and membrane filtration. So far, it seems that no method is efficient enough to remove each and every contaminant; therefore, advanced treatment should be applied for successful removal. Some of such techniques are in use and have shown the possibility of quantifying the emerging contaminants. Research should also be done for detection of new contaminants in a complex environment along with finding their efficient removal strategies. Various such emerging contaminants are present which are not detected/reported because of lack of experiments and knowledge. Apart from this, behavior of contaminants varies in natural environments showing several ecotoxicological effects, usually not predicted via conventional lab tests. Hence, it is essential for the development of newer protocols in order to keep a check on these effects via ecotoxicity tests with suitable endpoints. Future research should also focus on treatment techniques for converting contaminants into more pharmacologically active metabolites.

5.9 CONCLUSION

Groundwater is an indispensable resource for societal development involving human health and ecosystem preservation. Degradation in the groundwater quality because of contamination poses a serious threat globally. Several different groundwater sources which lead to its pollution have been identified so far, consisting of anthropogenic origins, such as effluents from industries, PPCPs, life-style compounds, ibuprofen, carbamazepine, bisphenol A, sulfamethoxazole, pharmaceuticals, caffeine, pesticides, fragrances, flame and fire retardants, hormones, nicotine metabolites and also surfactants, along with natural sources. Most of such EOCs are prioritized as chemical substances for the regulation as well as treatment purpose with respect to their respective environments along with human health effects. Research based on site specificities has been focused toward point sources of wastewater in order to lower down the risk of contamination due to diffuse sources such as biologically derived solids as well as leakage from urban sewage. Apart from this, it has also been proven that prevention of such contamination is more significant than its remediation. These methods for prevention comprise appropriate waste disposal, controlling and monitoring of hazardous materials, conduction of environmental audit as well as intensification of health education. If these do not work, then the role of remediation comes into play, consisting of bioremediation, stream stripping, air stripping, oxygen sparging, and chemical and thermal treatment. The process for regulating these chemicals in the groundwater and even in the environment is a very demanding task to perform, requiring understanding and interpretation of prime properties of contaminants, distribution, and behavior. Thus, more novel monitoring methods should be implemented in order to control these emerging contaminants keeping in record the appropriate standards too.

REFERENCES

Ahrens, L., 2009. Polyfluorinated compounds in waste water treatment plant effluents and surface waters along the River Elbe, Germany. *Marine Pollution Bulletin* 58: 1326–1333.

Alam, M.O., Shaikh, W.A., Chakraborty, S., Avishek, K., Bhattacharya, T., 2016. Groundwater arsenic contamination and potential health risk assessment of Gangetic Plains of Jharkhand, India. *Exposure and Health* 8(1): 125–142.

Banzhaf, S., Krein, A., Scheytt, T., 2011. Investigative approaches to determine exchange processes in the hyporheic zone of a low permeability riverbank. *Hydrogeology Journal* 19: 591–601.

Barnes, K.K., Kolpin, D.W., Furlong, E.T., Zaugg, S.D., Meyer, M.T., Barber, L.B., 2008. A national reconnaissance of pharmaceuticals and other organic wastewater contaminants in the United States I. *Groundwater. Science of the Total Environment* 402: 192–200.

Bartelt-Hunt, S., Snow, D.D., Damon-Powell, T., Miesbach, D., 2010. Occurrence of steroid hormones and antibiotics in shallow groundwater impacted by livestock waste control facilities. *Journal of Contaminant Hydrology* 123: 94–103.

Basu, A., Kumar, S., Mukherjee, S., 2003. Arsenic reduction from aqueous environment by water lettuce (Pistia stratiotes L.). *Indian Journal of Environmental Health* 45(2):143–150.

Bharti, P.K., 2007. Effect of textile industrial effluents on ground water and soil quality in Panipat region (Haryana), Ph. D. thesis, Gurukul Kangri University, Hardwar, 191.

Bharti, P.K., Kumar, P., Singh, V., 2013. Impact of industrial effluents on ground water and soil quality in the vicinity of industrial area of Panipat city, India. *Journal of Natural and Applied Sciences* 5(1): 132–136.

Buerge, I.J., Buser, H.R., Kahle, M., Müller, M.D., Poiger, T., 2009. Ubiquitous occurrence of the artificial sweetener acesulfame in the aquatic environment: an ideal chemical marker of domestic wastewater in groundwater. *Environmental Science & Technology* 43: 4381–4385.

Cabeza, Y., Candela, L., Teijón, J., Tamoh, K., 2011. Presence of pharmaceuticals, personal care products, and priority substances (2008/105EU) in groundwater (Llobregat delta, Barcelona, Spain). Geophysical Research Abstracts 13 EGU2011–3524-1, 2011, EGU General Assembly 2011.

Caliman, F.A., Gavrilescu, M., 2009. Pharmaceuticals, personal care products and endocrine disrupting agents in the environment are reviewed. *Clean* 37: 277–303.

Carrara, C., Ptacek, C.J., Robertson, W.D., Blowes, D.W., Moncur, M.C., Sverko, E., Backus, S.1., 2007. Fate of pharmaceutical and trace organic compounds in three septic systems plumes, Ontario, Canada. *Environmental Science and Technology* 42: 2805–2811.

Carroll, S., Liu, A., Dawes, L., Hargreaves, M., Goonetilleke, A., 2013. Role of land use and seasonal factors in water quality degradations. *Water Resource Management* 27: 3433–3440.

CGWB, 2017. Dynamic groundwater resources of India (As on 31st March 2013). Ministry of Water Resources, River Development & Ganga Rejuvenation, Government of India, Faridabad. Central Ground Water Board.

CGWB, 2005. District groundwater management studies of Palghat District, Kerala, Technical report series. Central Ground Water Board.

Chandra, R., Yadav, S., 2010. Potential of Typha angustifolia for phytoremediation of heavy metals from aqueous solution of phenol and melanoidin. *Ecological Engineering* 36: 1277–1284.

Clara, M., Strenn, B., Kreuzinger, N., 2004. Carbamazepine as a possible anthropogenic marker in the aquatic environment: investigation on the behavior of carbamazepine in wastewater treatment during groundwater infiltration. *Water Research* 38: 947–954.

Clarke, B.O., Smith, S.R., 2011. Review of 'emerging' organic contaminants in biosolids and assessment of international research priorities for the agricultural use of biosolids. *Environment International* 37: 226–247.

Cruz, J.V., Silva, M.O., Dias, M.I, Prudencio, M.I., 2013. Groundwater composition and pollution due to agricultural practices at Sete Cidades volcano (Azores, Portugal). *Appllied Geochemistry* 29: 162–173.

Díaz-Cruz, M.S., Barceló, D., 2008. Trace organic chemicals contamination in ground water recharge. *Chemosphere* 72: 333–342.

Domagalski, J., Johnson, H., 2012. Phosphorus and groundwater: establishing links between agricultural use and transport to streams. Fact Sheet 2012–3004. US Geological Survey. doi: 10.3133/fs20123004.

Environment Agency (EA). (2007) *Investigation of PFOS and Other Perfluorochemicals in Groundwater and SurfaceWater in England and Wales.* Environment Agency, Bristol, UK.

Fram, M.S., Belitz, K., 2011. Occurrence and concentrations of pharmaceutical compounds in groundwater used for public drinking-water supply in California. *Science of the Total Environment* 409: 3049–3417.

Ghosh, S., 2010. Wetland macrophytes as toxic metal accumulators. *International Journal of Environmental Science* 1:523–528.

Hanipha, M.M., Hussain, Z.A., 2013. Study of groundwater quality at Dindigul town, Tamil Nadu, India. *International Journal of Environmental Science* 2(1): 68–73

Haque, S.J., Onodera, S., Shimizu, Y., 2013. An overview of the effects of urbanization on the quantity and quality of groundwater in South Asian megacities. *Limnology* 14: 135–145.

Houtman, C., 2010. Emerging contaminants in surface waters and their relevance for the production of drinking water in Europe. *Journal of Integrative Environmental Science* 7: 271–295

Jakomin, L.M., Marban, L., Grondona, S.I., Galli, M.G., Martinez, D.E., 2015. Mobility of heavy metals (Pb, Cd, Zn) in the Pampeano and Puelche aquifers. Argentina. Partition and retardation coefficients. *Bulletin of Environmental Contamination & Toxicology* 95: 325–331.

Johnson, A.C., 2008. Do cytotoxic chemotherapy drugs discharged into rivers pose a risk to the environment and human health? An overview and UK case study. *Journal of Hydrology* 348: 167–175.

Kamala, C.T., Chu, K.H., Chary, N.S., Pandey, P.K., Ramesh, S.L., Sastry, A.R.K., Sekhar, K.C., 2005. Removal of arsenic(III) from aqueous solutions using fresh and immobilized plant biomass. *Water Research* 39(13): 2815–2826.

Kasprzyk-Hordern, B., Dinsdale, R.M., Guwy, A.J., 2008. The occurrence of pharmaceuticals, personal care products, endocrine disruptors and illicit drugs in surface water in South Wales, UK. *Water Research* 42: 3498–3518.

Kolpin, D.W., Thurman, E.M., Linhart, S.M., 2000. Finding minimal herbicide concentrations in groundwater? Try looking for their degradates. *Science of the Total Environment* 248: 115–122.

Kumar, M., Patel, A. K., Das, A., Kumar, P., Goswami, R., Deka, P., Das, N., 2017. Hydrogeochemical controls on mobilization of arsenic and associated health risk in Nagaon district of the central Brahmaputra Plain, India. *Environmental Geochemistry and Health* 39(1): 161–178.

Kumar, M., Snow, D.D., Li, Y., Shea, P.J., 2019. Perchlorate behavior in the context of black carbon and metal cogeneration following fireworks emission at Oak Lake, Lincoln, Nebraska, USA. *Environmental Pollution* 253, 930–938.

Kumar, N., Bauddh, K., Dwivedi, N., Barman, S.C., Singh, D.P., 2012. Accumulation of metals in selected macrophytes grown in a mixture of drain water and tannery effluent and their phytoremediation potential. *Journal of Environmental Biology* 33: 923–927.

Kümmerer, K., 2009. The presence of pharmaceuticals in the environment due to human use— present knowledge and future challenges. *Journal of Environmental Management* 90: 2354–2366.

Kurwadkar, S., 2014. Emerging trends in groundwater pollution and quality. *Water Environment Research* 86. doi: 10.2175/106143014X14031280668290.

Lapworth, D.J., Baran, N., Stuart, M.E., Ward, R.S., 2012. Emerging organic contaminants in groundwater: A review of sources, fate and occurrence. *Environmental Pollution* 163: 287–303.

Lapworth, D.J., Gooddy, D.C., 2006. Source and persistence of pesticides in a semi-confined chalk aquifer of southeast England. *Environmental Pollution* 144: 1031–1044.

Lara-Martín, P.A., 2010. Occurrence of alkyltrimethylammonium compounds in urban estuarine sediments: Behentrimonium as a new emerging contaminant. *Environmental Science & Technology* 44(19): 7569–7575.

Lewandowski, J., Putschew, A., Schwesig, D., Neumann, C., Radke, M., 2011. Fate of organic micropollutants in the hyporheic zone of a eutrophic lowland stream: results of a preliminary field study. *Science of the Total Environment* 409: 1824–1835.

Liu, Z.H., Kanjo, Y., Mizutani, S., 2010. A review of phytoestrogens: Their occurrence and fate in the environment. *Water Research* 44: 567–577.

Lokeshwari, H., Chandrappa, G.T., 2006. Impact of heavy metal contamination of Bellandur Lake on soil and cultivated vegetation. *Current Science* 91(9): 622–627

Loos, R., Locoro, G., Comero, S., Contini, S., Schwesig, D., Werres, F., Balsaa, P., Gans, O., Weiss, S., Blaha, L., Bolchi, M., Gawlik, B.M., 2010. Pan-European survey on the occurrence of selected polar organic persistent pollutants in groundwater. *Water Research* 44: 4115–4126.

Malaguerra, F., Albrechtsen, H., Binning, P.J., 2013. Assessment of the contamination of drinking water supply wells by pesticides from surface water resources using a finite element reactive transport model and global sensitivity analysis techniques. *Journal of Hydrology* 476: 321–331.

Matta, G., 2015. Effect of water quality on phytoplankton ecology of Upper Ganga Canal. *International Journal of Scientific & Engineering Research* 6(2): 762–768.

Menon, N.G., Mohapatra, S., Padhye, L.P., Tatiparti, S.S.V., Mukherji, S. (2020) Review on Occurrence and Toxicity of Pharmaceutical Contamination in Southeast Asia. In: Kumar M., Snow D., Honda R. (eds) *Emerging Issues in the Water Environment during Anthropocene.* Springer Transactions in Civil and Environmental Engineering. Springer, Singapore.

Mohamed, A.A.J., Rahman, I.A., Said, S.A., Lim, L.H., Mchenga, I.S., 2013. Levels of hexavalent chromium, copper, and total hardness in springs and underground water in Zanzibar Island. *Asian Journal of Applied Sciences* 01(05): 212–219.

Mohapatra, S., Menon, N., Padhye, L.P., Tatiparti, S.S.V., Mukherji, S. (2021) Natural Attenuation of Pharmaceuticals in the Aquatic Environment and Role of Phototransformation. In: Kumar M., Snow D., Honda R., Mukherjee S. (eds) *Contaminants in Drinking and Wastewater Sources.* Springer Transactions in Civil and Environmental Engineering. Springer, Singapore.

Mohapatra, S., Padhye, L.P., Mukherji, S. (2018) Challenges in Detection of Antibiotics in Wastewater Matrix. In: Gupta T., Agarwal A., Agarwal R., Labhsetwar N. (eds) *Environmental Contaminants: Energy, Environment, and Sustainability.* Springer, Singapore.

Mukherjee, S., Mazumder, P., Joshi, M., Joshi, C., Dalvi, S.V., Kumar, M., 2020. Biomedical application, drug delivery and metabolic pathway of antiviral nanotherapeutics for combating viral pandemic: A review. *Environmental Research* 191: 110119.

Murray, K.E., 2010. Prioritizing research for trace pollutants and emerging contaminants in the freshwater environment. *Environmental Pollution* 158: 3462–3471.

Musolff, A., Leschik, S., Moder, M., Strauch, G., Reinstorf, F., Schirmer, M., 2009. Temporal and spatial patterns of micropollutants in urban receiving waters. *Environmental Pollution* 157: 3069–3077.

Nakada, N., Kiri, K., Shinohara, H., Harada, A., Kuroda, K., Takizawa, S., Takada, H., 2008. Evaluation of pharmaceuticals and personal care products as water-soluble molecular markers of sewage. *Environmental Science and Technology* 42: 6347–6353.

Nowak, K.M., Miltner, A., Gehre, M., Schäffer, A., Kästner, M., 2011. Formation and fate of bound residues from microbial biomass during 2,4-D degradation in soil. *Environmental Science and Technology* 45: 999–1006.

Osenbruck, K., Gläser, H.R., Knoller, K., Weise, S.M., Moder, M., Wennrich, R., Schirmer, M., Reinstorf, F., Busch, W., Strauch, G., 2007. Sources and transport of selected organic micropollutants in urban groundwater underlying the city of Halle (Saale), Germany. *Water Research* 41: 3259–3270.

Oviedo, K.N., Aga, D.S., 2016. Lessons learned from more than two decades of research on emerging contaminants in the environment. *Journal of Hazardous Materials* 316: 242–251.

Pal, A., Gin, K.Y.H., Lin, A.Y.C., Reinhard, M., 2010. Impacts of emerging organic contaminants on freshwater resources: Review of recent occurrences, sources, fate and effects. *Science of the Total Environment* 408: 6062–6069.

Pantaleo, P.A., Komakech, H.C., Mtei, K.M., Njau, K.N., 2018. Contamination of groundwater sources in emerging African towns: The case of Babati town, Tanzania. *Water Practice & Technology* 13: 980–990. doi:10.2166/wpt.2018.104.

Paranthaman, S.R., Karthikeyan, B., 2015. Bioremediation of heavy metal in paper mill effluent using Pseudomonas spp. *International Journal of Microbiology* 1:1–5.

Patel, A.K., Das, N., Kumar, M., 2019. Multilayer arsenic mobilization and multimetal co-enrichment in the alluvium (Brahmaputra) plains of India: a tale of redox domination along the depth. *Chemosphere* 224: 140–150.

Patel, J.K., Das, A., 2015. Assessing toxic metals contamination in soil, water and plant bodies around an industrial belt. *International Journal of Research in Applied, Natural and Social Sciences* 3(2): 5–20.

Postigo, C., Martinez, D.E., Grondona, S., Miglioranza, K.S.B., 2018. Groundwater Pollution: Sources, Mechanisms, and Prevention. 87–96.

Priyanka, M.S., Mohapatra, S. (2020) Shifts and Trends in Analysis of Contaminants of Emerging Concern: Sulfonamides. In: Kumar M., Munoz-Arriola F., Furumai H., Chaminda T. (eds) *Resilience, Response, and Risk in Water Systems*. Springer Transactions in Civil and Environmental Engineering. Springer, Singapore. doi:10.1007/978-981-15-4668-6_14.

Rao, G.T., Rao, V.V.S.G., Ranganathan, K., 2013. Hydrogeochemistry and groundwater quality assessment of Ranipet industrial area, Tamil Nadu, India. *Journal of Earth System Science* 122(3): 855–867.

Rao, N.S., Rao, S. P., Dinakar, A., Rao, P.V., Marghade, P., 2015. Fluoride occurrence in the groundwater in a coastal region of Andhra Pradesh, India. *Applied Water Science* 7(3): 1467–1478.

Richardson, S.D., 2007. Water analysis: emerging contaminants and current issues. *Analytical Chemistry* 79: 4295–4324.

Sanyal, A., Rautaray, D., Bansal, V., Ahmad, A., Sastry, M., 2005. Heavy metal remediation by a fungus as a means of production of lead and cadmium carbonate crystals. *Langmuir* 21(16): 7220–7224.

Shah, B.A., 2013. Status of groundwater arsenic pollution in Holocene aquifers from parts of the Ghaghara Basin, India: Its relation to geomorphology and hydrogeological setting. *Physics and Chemistry of the Earth* 58–60: 68–76.

Sharma, P.K., Balkwill, D.L., Frenkel, A., Vairavamurthy, M.A., 2000. A new Klebsiella planticolastarin (Cd-1) grows anaerobically at high cadmium concentrations and precipitates cadmium sulfide. *Applied and Environmental Microbiology* 66(7): 3083–3087.

Shrestha, R.K., Regmi, D., Kafle, B.P., 2013. Seasonal variation of arsenic concentration in groundwater of Nawalparasi District of Nepal. *International Journal of Applied Sciences and Biotechnology* 2(1): 59–63.

Shukla, S., Saxena, A., 2016. Nitrate Contamination of Groundwater in India: Its Occurrence, Health Impacts and Mitigation Measures.

Singaraja, C., Chidambaram, S., Anandhan, P., Prasanna, M.V., Thivya, C., Thilagavathi, R., 2013. A study on the status of fluoride ion in groundwater of coastal hard rock aquifers of south India. *Arabian Journal of Geosciences* 6: 4167–4177.

Singh, A., Patel, A. K., Ramanathan, A., Kumar, M., 2020. Climatic influences on arsenic health risk in the metamorphic precambrian deposits of Sri Lanka: a re-analysis-based critical review. *Journal of Climate Change 6*(1), 15–24.

Soares, A., Guieysse, B., Jefferson, B., Cartmell, E., Lester, J.N., 2008. Nonylphenol in the environment: a critical review on occurrence, fate, toxicity and treatment in wastewaters. *Environment International* 34(7): 1033–49.

Srivastava, A., Singh, S., Prakash, R., Srivastava, S.K., 2013. Impact of tanneries on ground water quality in Kanchandpur area, Kanpur dehat district, U.P. *Journal Ins. Pub. Health Engineers* 2: 19–26.

Stuart, M., Lapworth, D., Crane, E., Hart, A., 2012. Review of risk from potential emerging contaminants in UK groundwater. *Science of the Total Environment* 416: 1–21.

Stuart, M.E., Manamsa, K., Talbot, J.C., Crane, E.J., 2011. Emerging contaminants in groundwater. British Geological Survey Open Report, OR/11/013. 1.

Sunil, K.C.R., Swati, K., Bhavya, G., Nandhini, M., Veedashree, M., 2015. Streptomyces flavomacrosporus, a multi-metal tolerant potential bioremediation candidate isolated from paddy field irrigated with industrial effluents. *International Journal of Life Sciences* 3: 9–15.

Terry, A.S., Carter, A.D., Humphrey, R.L., Capri, E., Grua, B., Panagopoulous, A.C., Pulido-Bosch, A., Kennedy, S.H., 2008. A monitoring programme for 1,3-dichloropropene and metabolites in groundwater in five EU countries. *Pesticide Management Science* 64: 923–932.

Torok-Oance, R., Tulhina, D., Lazar, M., 2013. Preliminary study on groundwater pollution with nitrites and nitrates in Gataia (Timis County). Annals of West University of Timişoara, ser. *Biology* 16(2): 89–96.

Ulén, B., Snäll, S., 2007. *Forms and retention of phosphorus in an illite-clay soil profile with a history of fertilisation with pig manure and mineral fertilisers.* Geoderma. Elsevier B.V.

Verliefde, A., Cornelissen, E., Amy, G., Van der Bruggen, B., van Dijk, H., 2007. Priority organic micropollutants in water sources in Flanders and the Netherlands and assessment of removal possibilities with nanofiltration. *Environmental Pollution* 146: 281–289.

Vikas, C., Kushwaha, R., Ahmad, W., Prasannakumar, V., Reghunath, R., 2013. Genesis and geochemistry of high fluoride bearing groundwater from a semi-arid terrain of NW India. *Environmental Earth Sciences* 68: 289–305.

Vulliet, E., Cren-Olivé, C., 2011. Screening of pharmaceuticals and hormones at the regional scale, in surface and groundwaters intended for human consumption. *Environmental Pollution* 159: 2929–2934.

Yadav, S.S., Kumar, R., 2011. Monitoring water quality of Kosi River in Rampur district, Uttar Pradesh, India. *Advances in Applied Science Research* 2(2): 197–201.

Section 2

Techniques to study the fate, transport and removal of microbial contaminants

Chapter 6

Assessment of microbial load removal potential of vertical constructed wetland systems treating dairy farm wastewater

Deepa Minakshi, Swati Singh, Pradeep Kumar Sharma, and Anju Rani
Graphic Era Deemed to be University

Piyush Malaviya
University of Jammu

CONTENTS

DOI: 10.1201/9781003157465-8

6.1 INTRODUCTION

Presently, approximately 2,450 million of the world population is engaged in agricultural sector, out of which, nearly two-thirds of the total population is associated with livestock farming. India supports nearly 57% of the world's buffalo population and 16% of the overall cattle population (Statistics, 2012; www.nddb.org.in). As per the records, India has roughly 70,000 dairy farms with a total population of 113 million cattle (source: Department of Animal Husbandry and Dairying 2018 (http://dahd.nic.in/about-us/divisions/cattle-and-dairy-development)). India produces around 108 million tons of milk annually. In India, dairy farming and related activities have become a principal source of rural economy. In India, the per capita milk production was 130 g/day in 1950–1951, which is likely to increase up to 574 g/day by 2022 (Sudarsan et al. 2012).

6.2 WASTEWATER GENERATION IN DAIRY FARMS

The dairy farms generate wastewater rich in organic and inorganic constituents such as dissolved and suspended solids (SSs), sulphates, phosphates, nitrogen compounds, oil and grease, pathogens, etc. (Minakshi et al. 2018; 2021). In addition to above, the wastewater also contains casein, detergents, and sanitizers used for washing (Barnett et al. 2010). The dairy farm wastewater consists of cattle urine, dung, and disposed milk that results into high organic matter content (i.e., BOD: 1,500 mg/L and COD: 3,000 mg/L) in the wastewater.

Untreated dairy influents, when disposed in surface water bodies, deplete dissolved oxygen (DO) content of receiving water. Prolonged discharge of untreated wastewaters to such water bodies leads to the breeding of flies and insects and causes contamination of pollutants. The discharge of untreated wastewater also causes taste and odor problems, imparts color and turbidity, and hence promotes eutrophication and disturbs the ecological balance by polluting water, air, and biodiversity in the ponds/lakes (Deshpande et al. 2012). The Central Pollution Control Board (CPCB) has issued the permissible limits for discharge of dairy farm wastewater in the surrounding environment (Table 6.1).

Almost all the pollutants in dairy wastewater are biodegradable, resulting into its easy treatment by different biological treatment methods, i.e., aerobic and anaerobic processes (Raghunath et al. 2016). There are various commercial wastewater treatment methodologies, including trickling filters, aerated ponds, moving bed bioreactors, sequential batch reactors, rotating biological contractors, activated sludge processes, etc., available for the treatment of dairy farm wastewater. However, the dairy farm owners still show that their least/no interest is adopting these technologies

Table 6.1 Dairy wastewater discharge standards by CPCB (CPCB, 2000)

Permissible limits			
Parameters	*Inland surface water*	*Public sewers*	*Land for irrigation*
TSS (mg/L)	100	600	200
pH value	5.5–9	5.5–9	5.5–9
Ammonical nitrogen (as N) (mg/L)	50	50	-
NH_3 (mg/L)	5	-	-
BOD [3 days at 27°C] (mg/L)	30	350	100
COD (mg/L)	250	-	-
Phosphorous (mg/L)	5	-	-
NO_3-N (mg/L)	10	-	-

Table 6.2 Economical and technical details of CWs (Sharma et al. 2018)

S. no.	*Parameters*	*Description*
1.	Area required	1–3 m^2/PE
2.	Operation cost (electricity)	Rs. 100–200/month
3.	Cost of construction	Rs. 6,000–7,500/m^2
4.	Treatment performance	BOD: 96% TN: 92% TP: 99% Microorganisms: up to 99%
5.	Benefits	Decentralized technology Cost-effective No sludge disposal problem Eco-friendly Technically feasible Able to tolerate influent load fluctuations

for treatment of wastewater (Sharma et al. 2013a, b). The major reasons for such ignorance by dairy farm owners include complex treatment processes, high installation, operation, and maintenance costs, expensive, sludge disposal problem, lack of skilled efficient labors, etc. Therefore, in order to bring attention of dairy farm owners for the treatment of wastewater in their premises, it is utmost important to introduce a cost-effective, simple yet effective technique for treating dairy wastewater and reduce the risk of infections. One of the technologies that is suitable to be adopted under such circumstances is constructed wetland (CW) technology (Kato et al. 2009). CWs have become popular in the recent years due to their relatively less operation and installation charges (Table 6.2) and few maintenance necessities (Sharma et al. 2018).

6.3 CONSTRUCTED WETLANDS

CWs are artificially designed wastewater purifying systems that mimic natural wetland filters in several ways, i.e., treatment processes, surface vegetation, filter substrates, and microbial communities (Hammer and Bastian 1989). In a CW system, physicochemical and biological processes occur together and lead to removal of pollutants from wastewater (Sharma et al. 2018). These systems utilize aerobic and anaerobic activities of microbial community (Akratos and Tsihrintzis 2007; Kadlec et al. 2000), physicochemical characteristics of filter media, and phytoremediation capabilities of surface vegetation for purifying wastewater. CW technology has been adopted as one of the natural wastewater purification methods due to use of similar processes that occur in natural wetlands. However, CWs are operated under controlled environmental conditions. These systems are the channels of fixed depth (generally 0.5–1 m), packed with porous filter media and with macrophytes grown over the bed surfaces. For maintaining direction of flow, hydraulic retention time, and the water level in the CW system, it is lined with impervious clay/HDPE sheets/concrete liners. For designing a CW system, type of wastewater, organic load, target effluent quality, filter media, microorganisms, and surface vegetation are considered as key factors.

The treatment process takes place when the wastewater moves through the filter media and vegetation, resulting in an interaction among the filter substrates, plant roots, and wastewater. This interaction in the form of rhizofiltration and sedimentation results into degradation of organic matter biologically (Mthembu et al. 2013).

6.4 CONFIGURATIONS OF CONSTRUCTED WETLAND TREATMENT UNITS

Various types of CWs can be distinguished depending on the design characteristics as well as treatment processes that are occur in CW systems during the removal of pollutants (Figure 6.1).

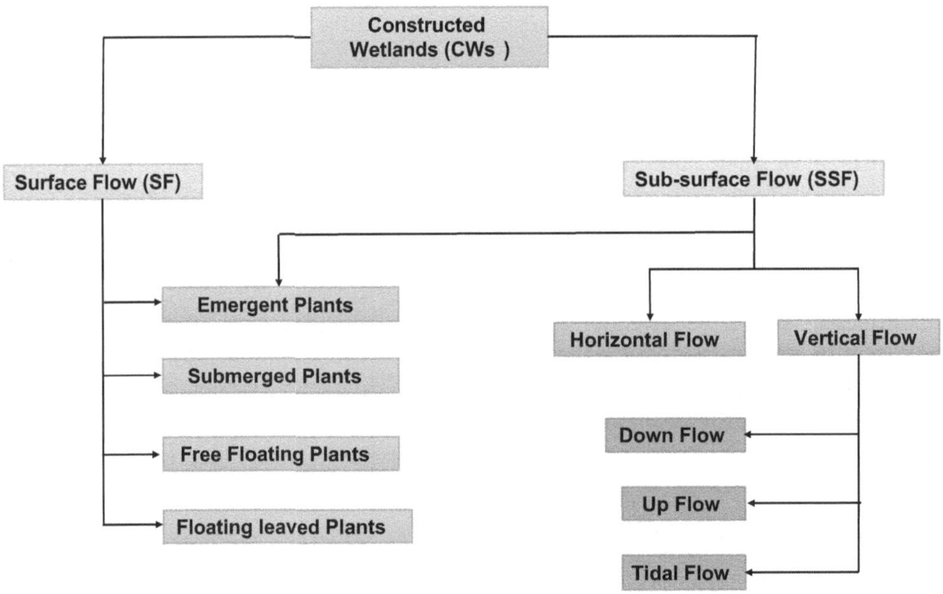

Figure 6.1 Types of CWs. (Adapted from Vymazal and Kröpfelová, 2008.)

Broadly, there are two major categories of CWs that are classified by design, operation, and CW configuration, i.e., surface flow or free water surface and subsurface flow (SSF).

SSF CWs have been further categorized into horizontal SSF (HSSF), vertical SSF (VSSF), and hybrid CW systems. In HSSF CWs, wastewater is allowed to flow horizontally through the filter media substrate, whereas in the case of VSSF CWs, the wastewater is loaded intermittently (batch-feeding process) over the superficial layers of sand or gravel substrates and finally drains at the base of the VSSF bed prior to sample collection (Healy et al. 2007; Kato et al. 2009). Hybrid systems involve different combinations of HSSF and VSSF CWs in one unit.

6.5 MECHANISMS FOR POLLUTANT AND PATHOGEN REMOVAL IN CWs

In a CW system, the treatment processes occur in various filtering zones such as filter media substrate, root and pore water zone, plants and their roots, waste matter, water, air, biomass zone, etc. In the filter media, wastewater purification occurs as a result of complex biochemical reactions that occur at different zones of filter media. The various substrate zones in a CW system are characterized by different levels of dissolved O_2 that facilitate in removal of contaminants from wastewater. The filter media in the CWs serve as mechanical as well as biological filters. SSs present in the wastewater are removed physically, whereas organic substances are readily fixed and absorbed by the biofilms formed around the filter media particles. The biological filtering process is an outcome of aerobic and facultative microbial actions that grow on the surfaces of filter media and plant roots. Eventually, a huge amount of suspended and dissolved organic matter (BOD and COD) is reduced and gets converted to gases such as CO_2, N_2 and NO_3, and H_2O by aerobic and anoxic reactions. Pathogens are filtered out at the surface of the CW bed. The overview of contaminant removal mechanisms in CWs has been discussed in Table 6.3. The treatment processes of various pollutants in CWs have been divided as follows:

Table 6.3 Overview of contaminant removal mechanisms in CWs

Pollutants	*Removal mechanisms*
SSs	Adsorption to biofilms, filtration, and decomposition by native bacteria in soil.
Organic matter (expressed as BOD or COD)	Conversion of insoluble organic matter into soluble form and subsequent removal by bacteria present in biofilm over plants roots and bed matrix. Removal of insoluble organic matter by sedimentation/filtration.
Phosphorous	Adsorption, Fe/Al/Ca precipitation, and minor uptake by plants.
Nitrogen	Removal by microbial nitrification and denitrification processes, and plant uptake (very limited impacts only).
Pathogenic microorganisms	Filtration, adsorption, and natural die-off.

6.5.1 Suspended solids

Settleable or SSs existing in the wastewater are removed by sedimentation and filtration processes (Stowell et al. 1981; Vymazal 2010). These processes possess an important role in reducing the SS content of wastewater. In vertical CWs, removal of SSs is associated with the hydraulic design parameters and microbial diversity of the filter material (Manios et al. 2003). Previous studies on HF CW systems suggest that a CW bed area of 3–5 m^2/PE provides high removal of SSs from wastewater (Vymazal 2010). However, a large fraction of SSs remains in an unstable form in the influent (Bavor and Schulz 1993), and in most of the treatment wetland systems, bulks of settleable solids are reduced by mechanical or pretreatment units (sedimentation tanks or Imhoff tanks) prior to their discharge into the actual wetland systems.

6.5.2 Organic components

Settleable organic components are rapidly reduced by accumulation and percolation under quiescent conditions. In addition, fixed and suspended microbial growth also degrades soluble organic matter under aerobic and anaerobic conditions. For aerobic degradation of organic matter, oxygen is provided by diffusion or seepage through the roots in the rhizospheric zone.

Aerobic decomposition of organic components is accomplished by the actions of aerobic heterotrophic bacteria as per following Equation (6.1):

$$(CH_2O)+O_2=CO_2+H_2O \tag{6.1}$$

Anaerobic degradation is determined by a multistep process that takes place in a CW system in an oxygen-deficient environment (Cooper et al. 1997). This process may also be enhanced by the presence of either facultative or obligate anaerobes. Further organic matter uptake also occurs by the macrophytes, which generally remains less in comparison to biological degradation (Watson et al. 1989; Cooper et al. 1997).

6.5.3 Nitrogen

In CWs, nitrogen removal is accomplished by two major pathways, i.e., physicochemical and biological treatment techniques (Lee et al. 2009). In these systems, nitrogen removal processes are diverse and includes volatilization, ammonification, matrix adsorption, plant uptake, nitrification and denitrification, etc.

6.5.3.1 Volatilization

Ammonium volatilization is a physicochemical mechanism where ammonium nitrogen remains in equilibrium with gaseous and hydroxyl forms as depicted in Equation (6.2):

$$NH_3\,(aq.)+H_2O=NH_4^{\;+}+OH^- \tag{6.2}$$

In a CW system, the volatilization process is affected by many operating factors. Vymazal (1995) reported that the degree of NH_4^+ volatilization in water is affected by NH_4^+ concentration, temperature, wind speed, type, and population of aquatic macrophytes as well as the system's ability to alter the pH value in 24-hour cycles (i.e., lack of CO_2 leads to increase in rate of volatilization).

6.5.3.2 *Ammonification*

Ammonification occurs in wastewaters that are rich in organic nitrogen (Anbin et al. 2018). Ammonification converts organic nitrogen into ammonia, which further gets transformed from organic forms via a series of multistep energy-releasing, biochemical processes. Kinetically, ammonification process occurs more speedily than nitrification in CW systems (Lee et al. 2009). Ammonification rate is greatly reliant on temperature, C/N ratio, pH, accessible nutrients, and soil conditions and is maximum within the aerated zones. The ammonification rate reduces when there is a transition of mineralization between aerobic to facultative anaerobic and obligate anaerobic microflora (Lee et al. 2009; Reddy et al. 1984). Kadlec and Knight (1996) suggested that ammonification takes place faster as compared to nitrification process. Previous literatures have reported a broad range of ammonification rates that have values fluctuating in range of 0.004–0.53 g N/m^2d (Tanner et al. 2002).

6.5.3.3 *Nitrification*

Nitrification may be described as the biological oxidation of ammonium into nitrate through an intermediary step of nitrite formation. It is related with chemoautotrophic bacteria, and it can be acknowledged that the heterotrophic nitrification process takes place and has extreme importance (Sharma et al. 2015). In the mechanism, ammonium oxidation and/or nitrite provides energy to the nitrifying bacteria. CO_2 serves as a carbon source for production of new cells. Nitrification is a dual-step process carried out by different microbial assemblages (Schmidt et al. 2003).

The initial step of nitrification deals with the conversion of ammonium to nitrite which is firmly accomplished by chemolithotrophic bacteria. Microbial species that belong to genera *Nitrosospira*, *Nitrosovibrio*, *Nitrosolobus*, *Nitrosococcus*, and *Nitrosomonas* have been recognized to carry out the reaction process (Paul and Clark 1996). The second step involves the nitrite (NO_2^-) oxidation into nitrate (NO_3^-), which is carried out by facultative chemolithotrophic bacteria which consume organic components for the energy production required for their growth. In comparison to the ammonia-oxidizing bacteria, a single nitrite-oxidizing bacterial species is present inside the soil as well as freshwater viz. *Nitrobacter winogradskyi*.

Several bacterial species (*Arthrobacter globiformis*, *Aerobacter aerogenes*, *Mycobacterium phlei*, *Streptomyces griseus*, *Thiosphaera*, and *Pseudomonas spp.*) have been studied for their nitrification potential. Major fungal species (*Aspergillus flavus*, *Penicillium*, or *Cephalosporium*) are also involved in the process (Paul and Clark 1996). Nitrification rates in treatment wetlands were specified to fluctuate between 0.01 and 2.15 g N/m^2d with an average value of 0.048 g N/m^2d (Tanner et al. 2002).

6.5.3.4 Denitrification

Denitrification process refers to conversion of nitrate into nitrogen molecules or dinitrogen via intermediates such as NO_2^-, N_2O, and NO (Paul and Clark 1996). From the biochemical point of view, it is a microbial process where nitrogen oxide functions in the form of a terminal electron acceptor for respiratory electron transport system. Electron transport is supported by an electron-donating substrate through various transporters toward a more oxidized form of nitrogen (Sharma et al. 2015). The consequential (resultant) free energy is preserved in ATP, followed by phosphorylation which can further be utilized by denitrifying microorganisms to sustain respiration. The following Equation (6.3) depicts the denitrification process (Hauck 1984):

$$6(CH_2O) + 4NO_3^- = 6CO_2 + 2N_2 + 6H_2O \quad (6.3)$$

Most of the chemoautotrophs are denitrifers, i.e., they gain energy merely from the chemical reactions that utilize organic components in the form of electron donors and cellular carbon source (Hauck 1984). Denitrifying bacteria in soil belong to genera *Bacillus*, *Micrococcus*, and *Pseudomonas*, whereas in aquatic systems, *Pseudomonas*, *Aeromonas*, and *Vibrio* dominate. Other genera that are responsible for denitrification include *Achromobacter*, *Aerobacter*, *Alcaligenes*, *Azospirillum*, *Brevibacterium*, *Flavobacterium*, *Spirillum*, or *Thiobacillus*. Vymazal (1995) studied the impact of various environmental factors, i.e., lack of O_2, soil moisture, redox potential, temperature, pH, denitrifer presence, organic compounds and existing overlying water, etc., on the denitrification process. The amount of N_2O evolved depends on the quantity of nitrogen denitrified as well as on the proportion of N_2 to N_2O formed.

The estimated rate of denitrification differs largely in the previous works and was reported to be between 0.003 and 1.02 g N/m^2d (Reddy and D'angelo 1997). Previous investigations related to conversion of nitrogen under anaerobic and anoxic environmental conditions revealed that NH_3 is transformed into N_2 by mechanisms along with conventional nitrification of NH_3 followed by denitrification process (Hunt et al. 2005).

6.5.3.5 Matrix adsorption

In its reduced form, ammoniacal nitrogen is quite constant and can easily be adsorbed on the active sites of the CW bed medium. The ionic exchange of ammonical nitrogen on the matrix exchange site does not provide NH_4-N removal for a long period. However, NH_4-N sorption is assumed to be changeable. In case NH_4-N is vanished from the wetland through the nitrification process, the exchange equilibria are likely to reallocate itself. Hence, the adsorbed NH_4-N, in a continuous flow system, will be in balance with NH_4-N solution. Batch loading of water shows rapid removal of NH_4-N by adsorption mechanisms due to NH_4-N depletion on the adsorption sites throughout its rest phases (Cooper et al. 1997).

6.5.3.6 Plant uptake

Nitrogen assimilation denotes a diversity of biological mechanisms which is related to the inorganic nitrogen transformation into organic components that function as a structural block for cells. NH_3 and NO_3-N are the two major nitrogen forms that are generally utilized for assimilation. Since ammonium-nitrogen is more actively reduced in comparison to nitrate, it is a better nitrogen uptake source (Kadlec and Knight 1996). Biological assimilation of nitrogen involves a role of macrophyte, microbes, and algae in a CW system. Ammonia is easily combined into amino acids due to activities of various autotrophs and microbial heterotrophs. Macrophyte species vary in their ideal nitrogen forms absorbed, on the basis of forms accessible in the soil (Lambers et al. 1998). Most macrophyte species are, yet, able to absorb any form of solvable nitrogen, particularly if acclimatized to its existence (Atkin 1996).

6.5.4 Phosphorous

Phosphorus removal mechanisms in CWs include peat/soil deposition, adsorption/desorption on media, precipitation/dissolution, uptake by plants and microbes, leaching, mineralization, and burial. Adsorption and retaining capacity of phosphorous by CW substrates is regulated by the interrelation of redox potential, pH, minerals, and concentrations of native soil phosphorous. Richardson and Marshall (1986) studied that soil adsorption and peat accretion regulate long-lasting phosphorus sequestration in CWs. Sorption and biomass accumulation are both stable procedures, which means they possess limited capability and therefore might not collaborate to continuing sustainable deduction (Dunne and Reddy 2005).

6.5.4.1 Adsorption and precipitation

In a CW system, major mechanisms for P removal include precipitation and/or adsorption on filter material and minor uptake by microbes and plants (Minakshi et al. 2019). Phosphorus adsorption ability of soil usually gets increased with the clay/mineral proportion of the soil. The steadiness among P adsorption and vice versa retains the P equilibrium between the solid phase and soil pore water. In CWs, P adsorption is associated with either high Al, Fe in acidic conditions, or Ca levels in alkaline conditions (Richardson 1985). The adsorption of P by soil is usually regulated by the phosphate concentration within the soil pore water as well as the capability of the solid phase to refill phosphate into the soil pore water.

Precipitation is often referred to the interaction of phosphate ions with metallic cations, thus forming amorphous or poorly crystalline solids. These chemical reactions classically take place at higher phosphate or metalloid cation concentrations (Pallavi et al. 2015). Various cations precipitate phosphate under certain circumstances. There are some vital minerals that are capable of precipitating phosphate in the wetlands are (Reddy and D'angelo 1997, Singh et al. 2020): apatite, hydroxylapatite, variscite, strengite, vivianite, and wavellite. Additionally, phosphorus coprecipitates with other minerals, for instance, ferric oxy hydroxide and the carbonate minerals, including calcite ($CaCO_3$).

6.5.4.2 Plant uptake

CWs are more or less saturated at least periodically. The macrophytes growing on the wetland bed surface adapt to grow in water-saturated soils. It is, therefore, an inherent property of wetlands that they are vegetated by wetland plants, but many times, it has been a matter of question if macrophytes are required in CWs. Wetland plants are morphologically adjusted for developing in water-saturated sediments by taking advantage of larger inner pores to allow passage for oxygen into the roots and rhizomes. The internal O_2 passage fulfills the respiratory demands of the buried tissues and provides oxygen to the rhizosphere by its seepage from the roots. The seepage of O_2 from the roots generates aerobic environmental conditions and enhances both aerobic breakdown of organic matter and growth of nitrifying bacteria. Microbes residing within the soil and roots of plants are thus capable to gain oxygen directly from their surrounding environments (Brix 1994).

Plants absorb nutrients for growth mainly through their root system. The quantity of nutrients absorbed is generally removed by harvesting and is usually irrelevant in contrast to the loadings with wastewater (Kim and Geary 2001). Tanner (2001) stated a yearly net reduction of 2%–8% for N and 1.9%–5.3% for P. For low loaded systems and particularly SSF and surface flow CWs, the quantity of nutrients reduced by harvesting can be noticeably greater (Lin et al. 2002; Stottmeister et al. 2003; Kyambadde et al. 2004). In case the wetland plants remain unharvested, the minerals and nutrients assimilated in the plant gets reverted back to water during decomposition of the plants.

6.5.5 Pathogens

The presence of coliform bacteria and viruses along with protozoan pathogens (mainly in tropical and subtropical nations) are considered water quality indicators (Rivera et al. 1995). In dairy farms, cattle manure is mainly associated with pathogenic bacterial species such as *Escherichia sp.*, *Salmonella* sp., *Clostridium* sp., *Mycoplasma* sp., *Campylobacter* sp., and *Mycobacterium* sp. (Ibekwe et al. 2003).

CWs are often recognized to offer appropriate arrangement of physicochemical and biological aspects for the pathogen reduction in wastewater. In physical mechanisms, pathogen removal occurs by filtration (mechanical purification) and sedimentation while chemical mechanisms include oxidation, UV radiation, exposure to biocides, and adsorption to organic matter. The biological removal mechanisms involve antibiosis, predation by nematodes, protists, and zooplanktons, lytic bacterial and viral attack, as well as natural die-off (Seidel et al. 1978; Gersberg et al. 1989; Cooper et al. 1996, Kumar et al. 2020).

6.6 CASE STUDY: ASSESSMENT OF MICROBIAL REMOVAL EFFICIENCIES OF VERTICAL CW SYSTEMS FOR DAIRY FARM WASTEWATER TREATMENT

6.6.1 Aims and background

Vertical subsurface flow CWs have been considered as more efficient as compared to other CWs since it provides high aeration that favors removal of microorganisms (Zurita et al. 2009). Filter materials also a play significant role as some processes such as adsorption, filtration, and sedimentation are influenced by filter material characteristics

(Redder et al. 2010; Tanner et al. 2012). Finer filter substrates (soil, sand, etc.) attain efficient removal of microbes as compared to coarser materials (Headley et al. 2013).

The present case study deals with the assessment of removal of microbial population in dairy farm wastewater by three vertical subsurface flow CW (VSSF CW) systems.

6.6.2 Experimental setup

6.6.2.1 Vertical flow CW Systems

In the current study, three VSSF CW systems (CW1, CW2, and CW3) were constructed near a dairy farm in Dehradun, Uttarakhand, India (Latitude 30.3165°N; Longitude 78.0322°E). CW1, CW2, and CW3 systems were filled with 10 mm gravels, 20 mm gravels, and sand, respectively, and were planted with *Arundo donax* (giant reed). The total area of each CW system was 4 m^2 (Figure 6.2). All the three units were intermittently loaded with 220 L (hydraulic loading rate: 55 mm/m^2) (Sharma et al. 2018).

6.6.2.2 Sampling of dairy farm wastewater and analysis methods

The sampling and analysis were conducted for one year with a frequency of once in a week. Samples were taken (1 Lplastic bottles) and stored in a refrigerator for further analysis. The water samples were analyzed for parameters such as pH, oxidation

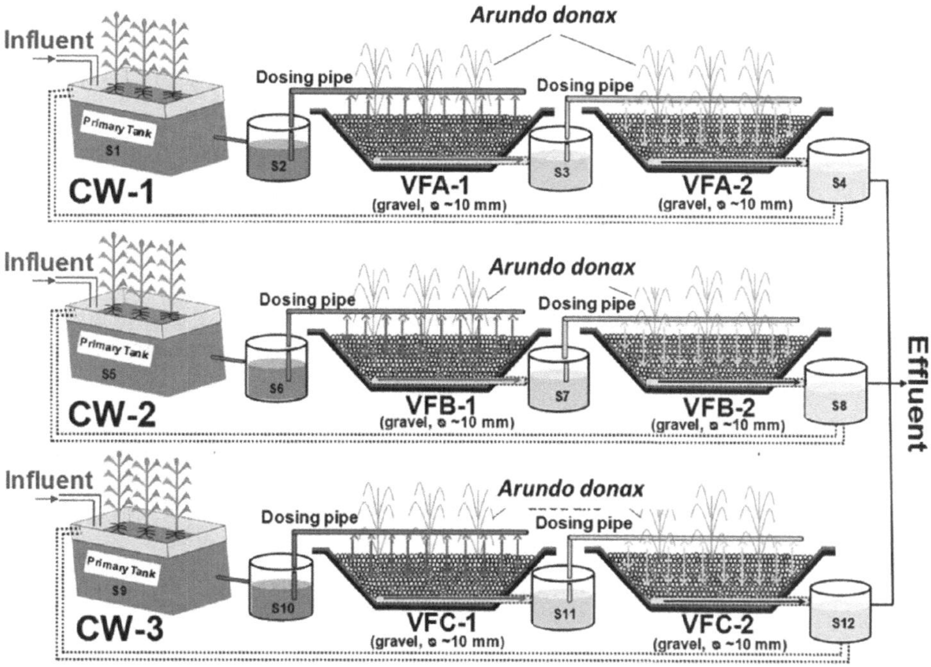

Figure 6.2 Schematic layout of CW1, CW2, and CW3 for the treatment of dairy farm wastewater.

reduction potential (ORP), water temperature, DO, total SSs (TSSs), and microbial load. Parameters such as pH, ORP, and water temperature were analyzed by multiparameter system (HachSensION + MM150) whereas DO was analyzed by DO meter and sensor (Hach).

For microbial load analysis, dairy influent and effluent samples (1 ml) were added to 10 ml distilled water, and serial dilutions were prepared (10^{-3}, 10^{-4}, 10^{-5} and 10^{-6}), according to the standard methods. Standard dilution plate method and spread plate technique was used for enumeration and isolation of microbes from dairy wastewater sample on nutrient agar medium plates at pH 7.2. After 24 hours incubation at 37°C, discrete microbial colonies were selected for counting, and data was expressed according to the colony forming units (CFUs). Serial dilutions of 10^{-3}, 10^{-4}, 10^{-5}, and 10^{-6} of each sample were inoculated on nutrient agar medium plates for total microbial load, with three replicates of each dilution. (CFUs)/mL was calculated using following formulae (6.4).

$$\text{CFU/mL} = \{(\text{No. of colonies} * \text{Dilution factor}) / \text{Volume of culture plate}\} \quad (6.4)$$

6.7 RESULTS AND DISCUSSION

Dairy wastewater possesses a huge number of microbes due to presence of organic carbon nitrogen and phosphorus. These microbes may be pathogens which in turn can contaminate soil and water if not removed from wastewater before its discharge.

6.7.1 pH, temperature, and ORP in dairy farm wastewater

The dairy wastewater characteristics have been shown in Table 6.4. In the present study, the average temperature of the influent and effluents of CW1, CW2, and CW3 was recorded as 12.1 ± 1.5, 15.3°C ± 2.5°C, 15.1°C ± 2.9°C, and 14.9°C ± 2.7°C, respectively. The average influent pH was recorded as 7.3 ± 0.5 during the experimental period. The treated dairy effluent showed pH variation between 7.6 and 8.1 in all the three CW units with an average pH range of 7.8 ± 0.2 in CW1, 7.9 ± 0.2 in CW2, and 7.8 ± 0.2 in CW3 (Table 6.4).

In a study by Smith et al. (1989), it was reported that higher microbial activity in water is observed at temperature of either 15°C or above. In this study, since the

Table 6.4 Dairy influent and effluent characteristics

	Influent	*CW1*	*CW2*	*CW3*
pH	7.3 ± 0.5	7.8 ± 0.2	7.9 ± 0.2	7.8 ± 0.2
Temperature (°C)	17.1 ± 1.5	15.3 ± 2.5	15.1 ± 2.9	14.9 ± 2.7
ORP (mV)	−40.1 ± 15.1	−67 ± 5.3	−67.3 ± 5.2	−62.5 ± 6.5
DO (mg/L)	4 ± 0.4	4.8 ± 0.3	5.3 ± 0.4	5.7 ± 0.5
DO increase (%)	-	22.3 ± 13.4	35.1 ± 19.7	45.9 ± 13.6
TSS (mg/L)	152.7 ± 28.9	103.8 ± 42	116.8 ± 50	10 ± 2.9
TSS removal rate (%)	-	33.1 ± 19.1	25.0 ± 22.7	93.4 ± 1.6
CFU per mL	219 ± 74.6	182 ± 61.7	140.5 ± 39.2	124 ± 14.2

temperature of the dairy wastewater was between 16 and 18, the occurrence and density of the microbes were noticeably higher. However, pH and temperature data recorded during the current assessment period depicted that there is no such visible association between microbial density with pH and temperature. This may be further explained from the fact that they are able to withstand minor variation in pH and temperature. In a similar study, it was observed that there was no apparent relationship between pH and temperature of wastewater which was perhaps due to the fact that most of the microbial populations are tolerant to slight variations in pH and temperature (Olutiola et al. 2010).

The average ORP in the dairy influent was observed as −40 ± 15.1 mV while the effluent in all the CW units showed average ORP in the range of −60.8 ± 6.9 to −67.8 ± 6.6 mV (Table 6.4). The maximum change in ORP was recorded in CW1 and CW2 while sand filled system CW3 showed least change in ORP during the assessment period.

6.7.2 Dissolved oxygen and TSS

DO reflects the amount of oxygen dissolved in the water which is required by the microorganisms for degrading organic components of wastewater. Average influent DO concentration was recorded as 4.0 ± 0.4 mg/L. The average DO was increased in all CW systems and was found to be 4.8 ± 0.3 mg/L in CW1, 5.3 ± 0.4 mg/L in CW2, and 5.7 ± 0.5 mg/L in CW3 (Figure 6.3). Minimum DO increase of 22.3% was recorded in the CW1 (10 mm gravel-filled) system while maximum increase of 45.9% in DO was noticed in the CW3 (sand-filled) system. The roots of the plants also release oxygen through their roots which is used by microorganisms for degradation of organic matter as well as for nitrification of NH_4-N from wastewater. The water within the CW can increase the rate of released oxygen by allowing movement of oxygen away from rhizosphere (Dong et al. 2011) altering the DO content of dairy wastewater.

Dissolved oxygen content of wastewater is an important factor during microbial load removal study (Smith et al. 1989; Olutiola et al. 2010). DO content of the treated effluent was observed to be higher than that of the influent, showing inverse relationship between DO and microbial load (CFU). Percent DO increase is a major criterion for determining the efficiency of CWs.

The presence of animal dung in the dairy wastewater remains in the form of SSs. The TSS content of the dairy influent was recorded between 135 and 186 mg/L with an average concentration of 152.7 ± 28.9 mg/L. After the treatment process, decrease in the TSS concentration of the effluents was observed which was recorded as 103.8 ± 42 mg/L (CW1), 116.8 ± 50 mg/L (CW2), and 10 ± 2.9 mg/L (CW3). The highest TSS removal (93.4%) was recorded in sand-filled CW bed (CW3) (Figure 6.3 and Table 6.4).

6.7.3 Microbial load estimation

Microbial load per mL was estimated in the dairy influent along with three CW systems. Serial dilution of highly concentrated wastewater sample was done in order to reduce the dense culture of cells to a more usable concentration. Each single dilution reduces the concentration of microorganisms by a specific amount. Serial dilution 10^{-6}

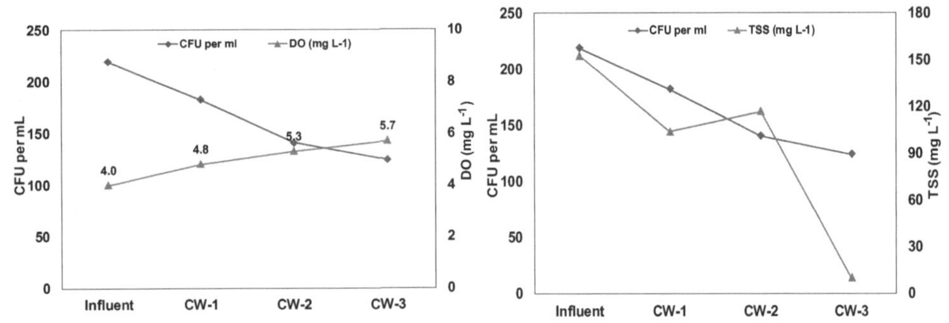

Figure 6.3 Relationship between CFU per mL with DO and TSS concentration.

showed satisfactory results due to less microbial colonies on the plates as compared to other dilutions (10^{-1}, 10^{-2}, 10^{-3}, 10^{-4} and 10^{-5}), and thus was considered for further studying the microbial load in each CW system. Microbial load was determined in terms of colony forming units per milliliter (CFU per mL).

In 10^{-6} serial dilution, the average microbial count in the influent, CW1, CW2, and CW3 outlet was 219 ± 74.6, 182 ± 61.7, 140.5 ± 39.1, and 124 ± 14.2 CFU/mL, respectively (Table 6.4). The average microbial load removal rate of CW1, CW2, and CW3 was recorded as 16.8%, 35.8%, and 43.4%, respectively. Microbial load reduction is highly dependent on the CW type, for instance, VSSF CWs favor good removal of microbes due to higher DO concentrations. Constructed wetlands remove microbes from wastewater through adsorption and filtration mechanisms at the bed surfaces. Sand being smallest in size offers a maximum removal rate of microbes.

6.7.4 Correlation studies between microbial load removal and wastewater parameters

The removal processes in CWs are accomplished by the combined interaction of substrates used, aeration capacity, mode of feeding wastewater (batch/continuous), plant species, microbial communities of the CW systems, etc. Efficiency of the wetland system is affected by multiple components, and therefore, correlation analysis is significant in knowing the ecological, physicochemical, and biological processes in the CW systems. Figure 6.4 illustrates the correlation studies of the wastewater parameters with microbial load per mL in CW3 (sand filled) system.

The percent increase in DO concentration at the CW3 outlet was inversely correlated with microbial load count ($r^2 = 0.806$). The levels of microbial communities showed strong positive correlation ($r^2 = 0.9085$) with dissolved oxygen concentration at the CW3 outlet. From the correlation studies of the current data, it can be inferred that in the influent, DO was low at high microbial load while it increased when the wastewater was subjected to treatment process leading to decrease in the microbial load. Similarly, correlation of TSS concentration and removal rate was performed with microbial load, and the r^2 values obtained was 0.5544 and 0.3953 (Figure 6.3) which depicted a moderate positive correlation. The significant positive correlation

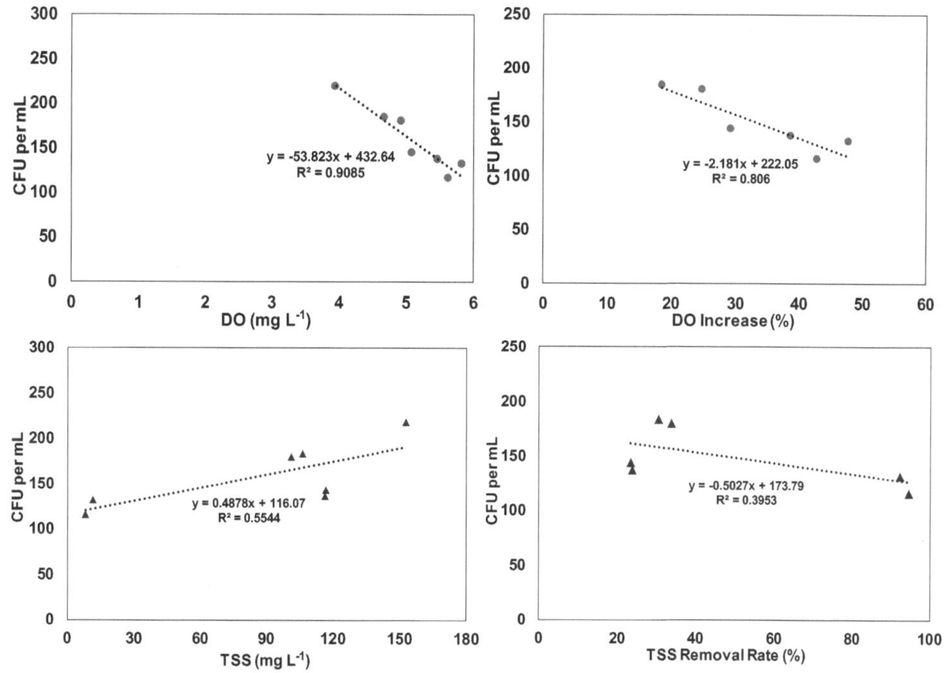

Figure 6.4 Correlation between microbial removal and wastewater parameters.

($r^2=0.5544$) between microbes and TSS might be indicative of the fact that microbes are not only associated with SSs in dairy wastewater. Hence, it is rational to speculate that microbes and TSS may originate from different sources and their discharge by dairy wastewater is governed by different processes (He et al. 2010).

Different unconventional wetland substrates have been used for treating different types of wastewaters, and their removal efficiencies have been monitored in various studies. Each substrate is different in physical and chemical characteristics accounts for varied removal efficiency (Patra et al. 2006). The finer substrate particles (such as sand) achieve higher microbial load removal rates as compared to coarser (e.g., gravels) particles (Headley et al. 2013).

6.8 CONCLUSIONS

In the present study, average microbial load in the influent and effluents of CW1, CW2, and CW3 systems were observed as 219 ± 74.6, 182 ± 61.7, 140.5 ± 39.2, and 124 ± 14.2 CFU/mL. The average microbial load removal rate of CW1, CW2, and CW3 was recorded as 16.8%, 35.8%, and 43.4%, respectively. Although sand-filled CW systems provide maximum removal of microbial load from dairy wastewater, overall microbial load reduction was not found significant. A significant correlation was observed between DO concentration and removal of microbial population from wastewater. The outcomes of the present study suggested that VSSF constructed wetland specifically sand-filled systems have good potential for the removal of TSS from dairy farm wastewater. The outcome of the present study showed average removal of pathogens with this CW design and suggests incorporation of other auxiliary filter materials such as activated charcoal layer at the top of the bed.

ACKNOWLEDGMENTS

Authors wish to thank DBT, Ministry of Science and Technology, GOI for financial support by research grant (BT/PR7545/BCE/8/999/2013) and Graphic Era Deemed to be University for providing necessary infrastructure.

REFERENCES

Akratos, C. S., & Tsihrintzis, V. A. 2007. Effect of temperature, HRT, vegetation and porous media on removal efficiency of pilot-scale horizontal subsurface flow constructed wetlands. *Ecol. Eng.* 29(2), 173–191.

Anbin Xie, A., Chen, H., & You, S. 2018. Advance of nitrogen removal in constructed wetland. *Mater. Sci. Eng.* 301, 1–4. doi:10.1088/1757-899X/301/1/012120.

Atkin, O. K. 1996. Reassessing the nitrogen relations of Arctic plants: A mini-review. *Plant, Cell & Environ.* 19(6), 695–704.

Barnett, J. W., Robertson, S. L., & Russell, J. M. 2010. Environmental issues in dairy processing. Environment Portfolio, New Zealand Dairy Research Institute, Private Bag, 11029, Palmerston North.

Bavor, H. J., & Schulz, T. J. 1993. *Sustainable Suspended Solids and Nutrient Removal in Large-Scale, Solid Matrix, Constructed Wetland Systems*, Lewis Publishers, Boca Raton, FL, pp. 646–656.

Brix, H. 1994. Functions of macrophytes in constructed wetlands. *Water Sci. Technol.* 29(4), 71–78.

Cooper, P. F., Job, G. D., Green, M. B., & Shutes, R. B. E. 1996. Reed beds and constructed wetlands for wastewater treatment. *Eur. Water Pollut. Control.* 6(7), 49.

CPCB. 2020. Environmental Standards for Ambient Air, Automobiles, Fuels, Industries and Noise, Pollution control law, Series: PCLS/4/2000-2001. Central Pollution control board. 1-119.

Cooper, P. F., Job, G. D., Green, M. B., & Shutes, R. B. E. 1997. Reed beds and constructed wetlands for wastewater treatment. *Eur. Water Pollut. Control* 6(7), 49.

Department of Animal Husbandry and Dairying. 2018. http://dahd.nic.in/about-us/divisions/cattle-and-dairy development.

Deshpande, D. P., Patil, P. J., & Anekar, S. V. 2012. Biomethanation of dairy waste. *Res. J. Chem. Sci.* 2(4), 35–39.

Dong, C., Zhu, W., Gao, M., Zhao, L. F., Huang, J. Y., Zhao, & Y. Q. 2011. Diurnal fluctuations in oxygen release from roots of Acoruscalamus Linn in a modeled constructed wetland. *J. Environ. Sci. Health, Part A* 46 (3), 224–229.

Dunne, E. J., & Reddy, K. R. 2005. Phosphorus biogeochemistry of wetlands in agricultural watersheds. In: *Nutrient Management in Agricultural Watersheds: A Wetland Solution*, Wageningen Academic Publishers, Wageningen, The Netherlands, pp. 105–119.

Gersberg, R. M., Lyon, S. R., Brenner, R., & Elkins, B. V. 1989. Integrated wastewater treatment using artificial wetlands: A gravel marsh case study. In: *Constructed Wetlands for Wastewater Treatment: Municipal, Industrial and Agricultural*, pp. 145–152.

Hammer, D. A. & Bastian, R. K. 1989. Wetland ecosystems: Natural water purifiers? In: *Constructed Wetlands for Wastewater Treatment*, D.A. Hammer, ed., Lewis Publishers, Chelsea, MI, pp. 5–19.

Hauck, R. D. 1984. Atmospheric nitrogen. Chemistry, nitrification, denitrification, and their interrelationships. In: *The Natural Environment and the Biogeochemical Cycles*, Springer, Berlin, Heidelberg, pp. 105–125.

He, J., Valeo, C., Chu, A., & Neumann, N. F. 2010. Characteristics of suspended solids, microorganisms, and chemical water quality in event-based stormwater runoff from an urban residential area. *Water Environ. Res.* 82(12), 2333–2345.

Headley, T., Nivala, J., Kassa, K., Olsson, L., Wallace, S., Brix, H., & Müller, R. 2013. Escherichia coli removal and internal dynamics in subsurface flow ecotechnologies: Effects of design and plants. *Ecol. Eng.* 61, 564–574.

Healy, M. G., Rodgers, M., & Mulqueen, J. 2007. Treatment of dairy wastewater using constructed wetlands and intermittent sand filters. *Bioresour. Technol.* 98(12), 2268–2281.

Hunt, P. G., Poach, M. E., & Liehr, S. K. 2005. Nitrogen cycling in wetland systems. In: *Nutrient Management of Agricultural Watersheds: A Wetlands Solution*, Wageningen Academic Publishers, pp. 93–104.

Ibekwe, A. M., Grieve, C. M., & Lyon, S. R. 2003. Characterization of microbial communities and composition in constructed dairy wetland wastewater effluent. *App. Environ. Microbiol.* 69(9), 5060–5069.

Kadlec, R. H., & Knight, R. L. 1996. *Treatment Wetlands*, Lewis Publisher, Boca Raton, FL.

Kadlec, R., Knight, R., Vymazal, J., Brix, H., Cooper, P., & Haberl, R. 2000. Constructed wetlands for pollution control: Processes, performance, design and operation. IWA publishing. Scientific and Technical Report No. 8, London.

Kato, K., Sharma, P. K., Inoue, T., Ietsugu, H., Koba, T., Kitagawa, K., Tomita, K., Matsumoto, T., & Nagasawa, T. 2009. New design of hybrid reed bed constructed wetland systems for the treatment of milking parlor waste water in cold climate conditions of Hokkaido, Northern Japan, In: IWA newsletter: *Specialist Group of Macrophytes in Water Pollution Control*, 34, pp. 8–12.

Kim, S. Y., & Geary, P. M. 2001. The impact of biomass harvesting on phosphorus uptake by wetland plants. *Water Sci. Technol.* 44(11–12), 61–67.

Kumar, M., Chaminda, T., Patel, A. K., Sewwandi, H., Mazumder, P., Joshi, M., & Honda, R. 2020. Prevalence of antibiotic resistance in the tropical rivers of Sri Lanka and India. *Environ. Res.* 188, 109765.

Kyambadde, J., Kansiime, F., Gumaelius, L., & Dalhammar, G. 2004. A comparative study of Cyperus papyrus and Miscanthidiumviolaceum-based constructed wetlands for wastewater treatment in a tropical climate. *Water res.* 38(2), 475–485.

Lambers, H., Chapin, F. S., & Pons, T. L. 1998. Biotic influences. In: *Plant Physiological Ecology*, Springer, New York, pp. 378–494.

Lee, CG, Fletcher, T.D., Sun, G. 2009. Nitrogen removal in constructed wetland systems: Review. *Eng. Life Sci.* 9(1), 11–22.

Lin, Y. F., Jing, S. R., Lee, D. Y., & Wang, T. W. 2002. Nutrient removal from aquaculture wastewater using a constructed wetlands system. *Aquaculture* 209(1–4), 169–184.

Manios, T., Stentiford, E. I., Millner, P. 2003. Removal of total suspended solids from wastewater in constructed horizontal flow subsurface wetlands. *J. Environ. Sci. Health, Part A* 36, 1073–1085.

Minakshi, D., Sharma, P.K., Rani, A., & Malaviya, P. 2018. Treatment of dairy farm effluent using recirculating constructed wetland units. In: *Advances in Health and Environment Safety*, N. A. Siddiqui, S. M. Tauseef, K. Bansal, eds., Springer, Singapore, pp. 57–66.

Minakshi, M., Sharma, P.K., Rani, A., & Malaviya, P. 2019. Phosphorous removal potential of vertical constructed wetlands filled with different filter materials. *J. Graph. Era Univ.*, 7(1), 53–63.

Minakshi, M., Sharma, P.K., & Rani, A. 2021. Effect of filter media and hydraulic retention time on the performance of vertical constructed wetland system treating dairy farm wastewater. *Environ. Eng. Res.* doi:10.4491/eer.2020.436.

Mthembu, M. S., Odinga, C. A., Swalaha, F. M., & Bux, F. 2013. Constructed wetlands: A future alternative wastewater treatment technology. *Afr. J. Biotechnol.* 12(29).

Olutiola, P. O., Awojobi, K. O., Oyedeji, O., Ayansina, A. D. V., & Cole, O. O. 2010. Relationship between bacterial density and chemical composition of a tropical sewage oxidation pond. *Afr. J. Environ. Sci. Technol.* 4(9), 595–602.

Patra, A. K., Abbadie, L., Clays-Josserand, A., Degrange, V., Grayston, S. J., Guillaumaud, N., & Philippot, L. 2006. Effects of management regime and plant species on the enzyme activity and genetic structure of N-fixing, denitrifying and nitrifying bacterial communities in grassland soils. *Environ. Microbiol.* 8(6), 1005–1016.

Paul, E. A., & Clark, F. E. 1996. *Soil Microbiology and Biochemistry*. 2nd ed., Academic Press, San Diego, CA, pp. 340.

Pallavi, D., Kumar, M., & Sarma, K. P. 2015. Speciation of heavy metals in surface sediment of the Brahmaputra river, Assam, India. *J. Environ. Res. Dev.* 9, 944–952.

Raghunath, B. V., Punnagaiarasi, A., Rajarajan, G., Irshad, A., & Elango, A. 2016. Impact of dairy effluent on environment—A review. In: *Integrated Waste Management in India*, Springer, Cham, pp. 239–249.

Rivera, F., Warren, A., Ramirez, E., Decamp, O., Bonilla, P., Gallegos, E., & Sánchez, J. T. 1995. Removal of pathogens from wastewaters by the root zone method (RZM). *Water Sci. Technol.* 32(3), 211–218.

Redder, A., Dürr, M., Daeschlein, G., Baeder-Bederski, O., Koch, C., Müller, R., & Borneff-Lipp, M. 2010. Constructed wetlands–Are they safe in reducing protozoan parasites? *Int. J. Hygiene Environ. Health.* 213(1), 72–77.

Reddy, K. R., Patrick, W. H., & Broadbent, F. E. 1984. Nitrogen transformations and loss in flooded soils and sediments. *Crit. Rev. Environ. Sci. Technol.* 13(4), 273–309.

Reddy, K. R., & D'angelo, E. M. 1997. Biogeochemical indicators to evaluate pollutant removal efficiency in constructed wetlands. *Water Sci. Technol.* 35(5), 1–10.

Richardson, C. J. 1985. Mechanisms controlling phosphorus retention capacity in freshwater wetlands. *Science* 228(4706), 1424–1427.

Richardson, C. J., & Marshall, P. E. 1986. Processes controlling movement, storage, and export of phosphorus in a fen peatland. *Ecol. Monogr.* 56(4), 279–302.

Schmidt, I., Sliekers, O., Schmid, M., Bock, E., Fuerst, J., Kuenen, J. G., & Strous, M. 2003. New concepts of microbial treatment processes for the nitrogen removal in wastewater. *FEMS Microbial. Rev.* 27(4), 481–492.

Seidel, K., Happel, H., & Graue, G. 1978. Contributions to revitalisation of waters. Stiftung Limnologische Arbeitsgruppe.

Sharma, P. K., Takashi, I., Kato, K., Ietsugu, H., Tomita, K., & Nagasawa, T. 2013a. Effects of load fluctuations on treatment potential of a hybrid sub-surface flow constructed wetland treating milking parlor waste water. *Ecol. Eng.* 57, 216–225.

Sharma, P. K., Takashi, I., Kato, K., Ietsugu, H., Tomita, K., & Nagasawa, N. 2013b. Seasonal efficiency of a hybrid sub-surface flow constructed wetland system in treating milking parlor wastewater at northern Hokkaido. *Ecol. Eng.* 53, 257–266.

Sharma, P. K., Rani, A., & Palni, L. M. S. 2015. Constructed Wetlands: A clean technology for treatment of wastewater using microbes and macrophytes. *ENVIS Bull. Himal. Ecol.* 23, 79.

Sharma, P. K., Minakshi, D., Rani, A., & Malaviya, P. 2018. Treatment efficiency of vertical flow constructed wetland systems operated under different recirculation rates. *Ecol. Eng.* 120, 474–480.

Smith, D. B., Hess, A. F., & Opheim, D. 1989. Control of distribution system coliform regrowth. In *Proceedings of the AWWA Water Quality Technology Conference*. American Water Works Association, Denver, pp. 1009–1029.
Statistics, B. B. A. H. 2012. Department of Animal Husbandry, Dairying & Fisheries, Ministry of Agriculture, Government of India. Krishi Bhavan, New Delhi.
Stottmeister, U., Wießner, A., Kuschk, P., Kappelmeyer, U., Kästner, M., Bederski, O., & Moormann, H. 2003. Effects of plants and microorganisms in constructed wetlands for wastewater treatment. *Biotechnol. Adv.* 22(1–2), 93–117.
Stowell, R., Tchoba, G., Colt, J., & Ludwig, R. 1981. Concepts in aquatic treatment system design. *J. Environ. Eng. Div.* 107(5), 919–940.
Sudarsan, J. S., Thattai, D., & Das, A. 2012. Phyto-remediation of dairy-waste water using constructed wetland. *Int. Pharm. Bio. Sci.* 3(3).
Singh, A., Patel, A. K., Deka, J. P., & Kumar, M. 2020. Natural recharge transcends anthropogenic forcing that influences arsenic vulnerability of the quaternary alluviums of the Mid-Gangetic Plain. *NPJ Clean Water* 3(1), 1–12.
Tanner, C. C. 2001. Plants as ecosystem engineers in subsurface-flow treatment wetlands. *Water Sci. Technol.* 44(11–12), 9–17.
Tanner, C. C., Kadlec, R. H., Gibbs, M. M., Sukias, J. P., & Nguyen, M. L. 2002. Nitrogen processing gradients in subsurface-flow treatment wetlands—influence of wastewater characteristics. *Ecol. Eng.* 18(4), 499–520.
Tanner, C. C., Sukias, J. P., Headley, T. R., Yates, C. R., & Stott, R. 2012. Constructed wetlands and denitrifying bioreactors for on-site and decentralised wastewater treatment: Comparison of five alternative configurations. *Ecol. Eng.* 42, 112–123.
Vymazal, J. 1995. *Algae and Element Cycling in Wetlands*. Lewis Publishers Inc.
Vymazal, J., & Kröpfelová, L. 2008. Is concentration of dissolved oxygen a good indicator of processes in filtration beds of horizontal-flow constructed wetlands? In: *Wastewater Treatment, Plant Dynamics and Management in Constructed and Natural Wetlands*, Springer, Dordrecht, pp. 311–317.
Vymazal, J. 2010. Constructed wetlands for wastewater treatment: Review. *Water* 2, 530–549; doi:10.3390/w2030530.
Watson, J. T., Reed, S. C., Kadlec, R. H., & Knight, R. L. 1989. Performance expectations and loading rates. In: *Constructed Wetlands for Wastewater Treatment: Municipal, Industrial and Agricultural*, p. 319.
Zurita, F., De Anda, J., & Belmont, M. A. 2009. Treatment of domestic wastewater and production of commercial flowers in vertical and horizontal subsurface-flow constructed wetlands. *Ecol. Eng.* 35(5), 861–869.

Chapter 7

Occurrence and fate of viruses in water

Muskan Chakraborty
Deccan Education Society's Fergusson College

Saumya Vyas
Kristu Jayanti College

Unnati Bhalerao and Abhranil Gangopadhayya
National Institute of Virology

Manjita Srivastava
National Institute of Virology

Muneesh Kumar Barman and Kailash Chand
National Centre for Cell Sciences

Sai Tejaswi Lavuri
Chalmeda Ananda Rao Institute of Medical Sciences

Meenakshi Singh and Prudhvilal Bhukya
ACTREC, Tata Memorial Centre and National Institute of Virology

CONTENTS

ABBREVIATIONS AND ACRONYMS

AGE:	Acute Gastroenteritis
BKV:	BK in BKV is an abbreviation for the first patient from whom this virus was isolated
CCL:	Contaminant Candidate List
CoV:	Coronavirus

DOI: 10.1201/9781003157465-9

COVID-19: Coronavirus Disease 2019
CPE: Cytopathic effect
CPV: Canine Parvovirus
DFA: Direct Fluorescent Antibody Staining
DNA: Deoxyribonucleic acid
EIA: Enzyme immunoassay
ELISA: Enzyme-linked immunosorbent assay
EPA: Environmental Protection Agency
FIPV: Feline Infectious Peritonitis Virus
HAdv: Human Adenovirus
HAstV: Human Astrovirus
HAV: Hepatitis A virus
HBoV: Human Bocavirus
HBuV: Human Bufavirus
HCoV: Human Coronavirus
HEV: Hepatitis E virus
HIV/AIDS: Human Immunodeficiency Virus/Acquired Immunodeficiency Syndrome
HPV: Human Papillomavirus
HPyV: Human Polyomavirus
HuCv: Human Calicivirus
ICTV: International Committee on Taxonomy of Viruses
JCV: John Cunningham virus
LAMP: Loop-mediated isothermal amplification
LNV: Liao Ning Virus
MERS: Middle Eastern Respiratory Syndrome
MPN: Most Probable Number
NASBA: Nucleic acid sequence-based amplification
PBV: Picobirnavirus
PCR: Polymerase chain reaction
PCV: Porcine Cyclovirus
PFU: Plaque Forming Units
PVC: Polyvinyl chloride
RIA: Radio-immunoassay
RIFA: Radioimmuno-focus-assay
RNA: Ribonucleic acid
RT-PCR: Reverse-transcriptase polymerase chain reaction
RT-qPCR: Reverse Transcriptase- quantitative Polymerase Chain Reaction
SARS: Severe Acute Respiratory Syndrome
SARS CoV-2: Severe Acute Respiratory Syndrome Coronavirus-2
TMLV: Torque teno-like mini virus
TTMDV: Torque teno midi virus
TTMV: Torque teno mini virus
TTV: Torque teno virus
UV: Ultraviolet
VLP: Virus-like particles
WHO: World Health Organization

7.1 INTRODUCTION

Access to clean potable water is a basic human right and is a prerequisite for good health. Poor quality water, whether used for drinking, domestic purposes, or recreational use, is hazardous and can cause disease outbreaks that are a major threat not only to public health but also to the socioeconomic development and well-being of people. Unclean and unsafe water is due to the activities of humans in regions of high population density where there is intensive land use. This serves as a transmission mechanism for a plethora of communicable diseases such as amoebiasis, cholera, bacillary dysentery, typhoid, giardiasis as well as several viral gastrointestinal diseases (Fongaro et al. 2015; Kumar et al. 2020a). The World Health Organization (WHO) records that approximately 3.4 million people die from water-related diseases annually (Guo et al. 2011; Kumar et al. 2021). It has also been postulated that this burden is essentially greater in the pediatric populations, particularly in the under-five age group (Liu et al. 2012; WHO | Water Quality and Health Strategy 2013–2020 2016).

Predictions indicate that by 2025, a good portion of the world's population will be living in regions with scarce water. To reduce this burden, people in different parts of the world have started utilizing wastewater for irrigation purposes and as a nutrient source for crop production (WHO | Guidelines for the Safe Use of Wastewater, Excreta and Greywater - Volume 2 2016; WHO | Water Quality and Health Strategy 2013–2020 2016). This practice poses health risks for workers, communities, and consumers, due to the possible existence of several infectious pathogens in wastewater. Dangerous levels of biological and chemical contaminants such as bacteria, fungi, viruses, protozoans and parasites, and arsenic and fluoride, respectively, in drinking water originates from natural sources as a result of inadequate and improper treatment of urban, industrial, or agricultural wastewater, and this poses a serious threat to the population (WHO | Water Quality and Health Strategy 2013–2020 2016; Flanagan, Johnston, and Zheng 2012). In Pakistan, approximately 64% of the total sewage generated is disposed of in rivers or directly into the Arabian Sea (Tabraiz et al. 2016). Biological contaminants in wastewater include high levels of pathogenic bacteria such as *Salmonella, E. coli, and Campylobacter*, protozoans such as *Amoeba*, *Giardia*, and *Cryptosporidium*, and enteric viruses, for example, human adenoviruses, influenza viruses, hepatitis viruses, rotaviruses, and circoviruses. These biological contaminants can cause serious infections in not only humans but also other animals and can lead to substantial health-related and economic losses (Fongaro et al. 2015; Silvia Bofill-Mas et al. 2006; Mazumder et al. 2020). Pathogens can find their way into the rivers from sources such as pathogen-contaminated groundwater, excreta from livestock, terrestrial or aquatic fauna, discharge of sewage, as well as lateral inputs from pastures, riparian zones, and wastewater treatment plant effluents (Pandey et al. 2014). Investigators have identified myriads of such pathological microorganisms responsible for outbreaks, and the field of research has made efforts to study their sources, possible methods of disinfection, and how these organisms gain resistance to water disinfection (Craun et al. 2006).

Viruses are found to be the most abundant and genetically diverse microorganisms in water systems including the ocean (Breitbart and Rohwer 2005; Suttle 2005). Up to 10 million virus-like particles (VLPs) per ml of seawater have been detected microscopically. A majority of these VLPs are bacteriophages, which are viruses that infect bacteria (Breitbart and Rohwer 2005). Some of the VLPs are also enteric viruses that are commonly found to infect humans as well as other animals are highly resistant to extreme environmental conditions such as highly acidic or alkaline pH, high temperature, salinity, and ultraviolet radiation that are otherwise lethal to microorganisms, and they can therefore survive for comparatively longer periods in water (Fongaro et al. 2015; Griffin, Plummer, and Long 2008; Fong and Lipp 2005). During 1991–2002 in the United States, there were 12 reported outbreaks of norovirus and 2 of hepatitis A (Craun et al. 2006). During 2000 and 2007, there were 136 outbreaks of overall waterborne viral diseases in Europe (Fongaro et al. 2015). In the United States, at least 1870 outbreaks of waterborne diseases occurred between 1920 and 2002 (23 per year), and in Europe, 39% of all the diarrhea cases were because of unsafe water (Fongaro et al. 2015; Craun et al. 2006). However, outbreaks in surface water systems have decreased recently, essentially due to stringent regulations and better treatment protocols (Craun et al. 2006).

A pathogen is said to be "emerging" when it crosses a species barrier to infect a new host population or when it becomes increasingly prevalent in a given population (Cleaveland, Haydon, and Taylor 2007). Changes in and/or between the disease-affected host species and the reservoir population and/or changes of vector species are what result in the emergence of such pathogens (Pandey et al. 2014). Viruses are more often associated with emerging and reemerging infectious diseases because of the possibility of cross-species transmission and their ability to adapt to new environments by mutation as well as recombination. The risk of emerging waterborne viral diseases is found to increase where the quality of water, sanitation, and personal hygiene is low (Guo et al. 2011; WHO | Water Quality and Health Strategy 2013–2020 2016; Rusiñol and Girones 2019). The sewage drainage infrastructure and management in developing countries is usually very poor, due to which the groundwater is at a greater risk of contamination (Tabraiz et al. 2016; Singh et al. 2019). Most foodborne outbreaks are caused by waterborne viruses which high-risk pathogens and these include hepatitis A virus (HAV), hepatitis E virus (HEV), and norovirus, which are transmitted via the consumption of aquatic life forms such as bivalve shellfish, or via fresh produce, and prepared foods (Rusiñol and Girones 2019). This chapter focuses on the various viruses that are transmitted by different sources of water, their occurrence, and persistence in aquatic environments. It also provides an overview of the techniques for detection of viruses in water and the prevention of waterborne viral diseases.

7.2 VIRAL CONTAMINATION OF WATER FROM DIFFERENT SOURCES

The presence of any microbial pathogen in water poses a public health risk and can also prove hazardous to the other several species of this planet. Their occurrence in water is aggravated by increased turbidity due to sewage discharge, surface runoffs, industrial effluents, and human activities (Opere, John, and Ombori 2020). Narrowing down to viruses in water, enteroviruses, adenoviruses, noroviruses, rotaviruses, astroviruses, human polyomaviruses, HAVs, and HEVs are common pathogens of humans that have been detected in wastewaters, and their concentration varies with location and with season. These viruses are commonly found in different sources of water including surface water, groundwater, and drinking water, as has been described in Figure 7.1. (Masachessi et al. 2018).

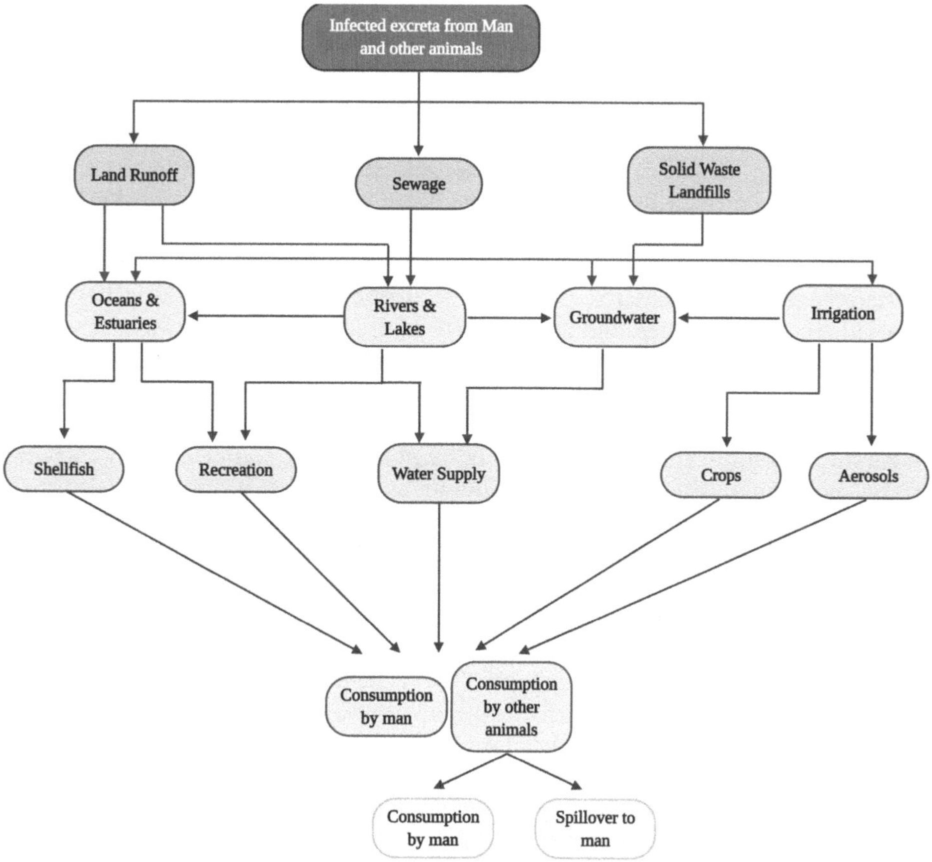

Figure 7.1 Potential sources viruses in water. (Created with biorender.com.)

Surface water is an essential source of potable water, and hence, its quality is a huge concern worldwide (Opere, John, and Ombori 2020). Viruses can occur in any surface water that has been exposed to fecal wastes, surface runoffs, leaching septic systems, or poorly treated sewage water (C. P. Gerba and Rose 1990; Xagoraraki, Yin, and Svambayev 2014; Mukherjee et al. 2020). They may be released into environmental waters by wastewater treatment plants via biosolids that are applied to the land as soil conditioner and fertilizer (Xagoraraki, Yin, and Svambayev 2014). During rainfall, unprecedented entry of untreated sewage and wastewater into surface waters may occur because of sewer overflow, and the ingestion of such contaminated water can expose individuals to several pathogens (Xagoraraki, Yin, and Svambayev 2014).

The prevalence of some viruses in surface waters has been affected by seasonal changes (Opere, John, and Ombori 2020). Viruses are believed to survive for longer periods in significantly low temperature conditions. Enteroviruses occurrence peaks during early autumn or late summer, and concentrations of human caliciviruses are increased during the colder months of the year. However, it has been reported that adenoviruses occur in surface water bodies all year round, irrespective of the season (Opere, John, and Ombori 2020).

Viruses are also found in sediments. Upon entry into natural water, they adsorb on to surfaces of sediment particles that settle, or they may also get resuspended into the water column. Researchers have detected viruses primarily in water and the topmost layer of sediment, whereas none in the bottom sediment (Xagoraraki, Yin, and Svambayev 2014). Elmahdy et al. showed the occurrence of rotaviruses and human adenoviruses in surface water sediments (Elmahdy et al. 2016). Garcia-Aljaro et al. (2017) reported the presence of enteroviruses on river water sediments. Epilithic biofilms are those which occur on stones, and these may form on stones in surface waters as a result of virus uptake from the water and can then subsequently get released after a specific residence time. However, the contribution of these films to the risk of transference of viruses is yet to be deduced.

Runoff and infiltration during rainfall can be another source of viral entry into surface and groundwater. Sewage overflow or seepage from septic tanks is a major source of contamination of groundwater. Viruses in groundwater may also be a result of using sewage and sewage sludges for disposal or irrigation, in domestic waste landfills and waste stabilization ponds, or deep well injection of sewage (C. P. Gerba and Rose 1990). Animal wastes and manure contain high concentrations of infectious zoonotic viruses and are important sources of groundwater microbiological contamination (Xagoraraki, Yin, and Svambayev 2014). The disposal of used diapers in sanitary landfills is another means by which untreated sewage can be introduced into water systems. Infected fecal matter has been reported to contain 106 plaque-forming units (PFUs) of enteroviruses per gram and 1010 of rotaviruses per gram, and these viruses can persist longer in groundwater than in surface waters incubated at similar temperatures (Yates, Gerba, and Kelley 1985). Human noroviruses can remain infectious in groundwater at room temperature in the dark for at least 61 days, and this long-term persistence of the human noroviruses is a result of their highly stable capsid protein coats, which is further facilitated by the formation of virus aggregates which is a common occurrence in natural environments (Seitz et al. 2011). Hepatitis A, Norwalk, and rotaviruses have been the most prevailing etiological agents for groundwater disease outbreaks (C. P. Gerba and Rose 1990). Under viable conditions, viruses are capable of traveling even hundreds of meters through soil and groundwater, owing to their small size (C. Gerba et al. 2020). The most likely route of viral contamination of groundwater is through permeable soils via flow routes that are caused by plant roots, cracks and fissures in rocks, and other natural phenomena (Xagoraraki, Yin, and Svambayev 2014). The concentration of contaminants varies with different soil types and through different soil layers (Tabraiz et al. 2016). A limiting factor for the movement of viruses in groundwater is their ability to adsorb to soil particles. Polioviruses can adsorb readily to soil, while the coliphage MS-2 does so poorly (Goyal and Gerba 1979). Viruses can move easily in sandy soils, as well as in regions of high rainfall.

Inadequate treatment of source water can lead to the recovery of viruses from drinking water. Viruses have been detected in drinking water that met the apparently acceptable water quality measurements of turbidity, total residual chlorine, and total coliform bacteria (Keswick et al. 1984). Viruses from treated drinking water were first isolated and reported in the 1960s. Between 1981 and 1985, 14 recoveries were reported, including that of rotavirus and HAV (Keswick, et al. 1984; Hejkal, Gerba, and Rao 1982). The development of more rapid, sensitive, and accurate methods of concentration and laboratory detection of viruses was a prime reason for the increased occurrence of enteric viruses in drinking water, and these numbers have been observed to increase over the years (Rose et al. 1986). An increased curiosity in the prevalence of viruses in drinking water, too, has unquestionably contributed to this rise (C. P. Gerba and Rose 1990). Estimating the concentration of viruses in source waters, along with the estimation of the efficiencies of drinking water treatment processes can help us deduce the concentration of viruses in drinking water. (Skraber et al. 2005). The assessment of water quality should not only be done after its treatment but also within the distribution system and individual households (Gibson et al. 2011). Operational difficulties at treatment plants such as inadequate floc formation, floc breakdown, and filter overloading can cause a surge in the quantity of particulate matter in treated water, rendering the disinfection protocols ineffective and enabling the viruses to survive such disinfection treatments (Keswick et al. 1984; Mohapatra et al. 2014; Sharma et al. 2021).

Biofilms present in the water entering water distribution systems may also accumulate pathogenic viruses, and these can colonize the drinking water pipes despite the presence of free chlorine (Lehtola et al. 2004; Hallam et al. 2001). Biofilms may also be considered to be a shelter for viruses from inactivating substances utilized in drinking water treatments. Problems can arise if these infectious viruses detach from the biofilms and are released into the water supply. In a case study in Nokia, Finland, bulk water was found to be constantly recontaminated by the presence of adenoviruses in biofilms in the water distribution network (Miettinen et al. 2012). Biofilms comprise sorption sites that can promote the binding and accumulation of other biocolloidal matter such as viral particles, which can then interact with the natural biofilms (Leclerc and Moreau 2020). The consequences of these interactions are dependent on the initial virus concentration in the drinking water that enters the distribution system, as well as on the attachment and detachment rate of these viruses to biofilms, and lastly, on the rates at which the viruses can get inactivated in the water and the biofilms. Hence, it may be possible to hypothesize that pathogenic viruses, including emerging viruses, can occur in natural biofilms (Skraber et al. 2005). However, it is necessary to note that as more advanced virus detection methods are developed, more viral contamination of drinking water will be reported (C. P. Gerba and Rose 1990).

7.3 OCCURRENCE AND FATE OF SPECIFIC VIRUS FAMILIES IN WATER

Adenoviridae

Belonging to the genus *Mastadenovirus*, adenoviruses are icosahedral viruses that are nonenveloped with a nucleocapsid having a helical symmetry. The genomic content is double-stranded, contributing to their longer survival, linear, and of around 26–45 kb (La Rosa et al. 2012). These are important human pathogens that cause respiratory and eye infection, including fever, cough, and sore throat, with children and the immunocompromised being more susceptible. They have a global distribution with respect to causing acute respiratory infections. Poor sanitary conditions contribute to a higher incidence of human adenovirus since the high titer of viral particles in sputum or oral secretions can cause the transmission of the virus. Prevalence is seen in drinking water, rivers, dam water, and coastal waters. In addition, adenoviruses are enlisted in the EPA's Drinking Water Contaminant Candidate List, a list of unregulated contaminants of water that are a serious threat to public health (Mena and Gerba 2008; Mena and Gerba 2009). Consumption of contaminated water and inhalation of aerosolized droplets are the major routes of spread of the infection. Outbreaks linked to pools, ponds, and lakes describe human adenovirus as a major etiological agent for conjunctivitis fever (Sinclair, Jones, and Gerba 2009). The viral load in feces of infected individuals is high (La Rosa et al. 2012). Several serotypes, around 51, that cause unique infection with clinical manifestations associated with gastrointestinal, respiratory, and urinary tract, with morbidity or mortality being uncommon, have been isolated (Heerden et al. 2005). The incubation period is around 2 days to 2 weeks, which depends on the viral type and acquisition mechanism. Influent and effluent sewage samples contain a high load of the virus. Since adenoviruses are resistant to disinfection, they have also been detected in wastewater effluents. Inactivation of adenovirus can be done by the use of monochromatic UV sources (Eischeid, Thurston, and Linden 2011). Human adenovirus is found in groundwater with large survival rates which increases the risk of public health. Adenoviruses have been known to be used as a virological marker since they are easily detectable by polymerase chain reaction (PCR) techniques. Evaluation of encapsidated genome, nucleic acid, and viral particles showed virus persistence is at lower temperatures of around 50°C (Silva, García-Zapata, and Anunciação 2011).

Anelloviridae

Anelloviruses are single-stranded, unenveloped circular DNA genome viruses that comprise the *Torque teno virus* (TTV). They have high genomic variability and are found at a high prevalence in human populations (Biagini and de Micco 2008). The first representative of the TTV was first discovered in 1997, after the discovery of the *Torque teno mini virus* and *Torque teno midi virus*. They include 11 genera and are ubiquitous. They have been known to replicate in high titers and cause diseases including hepatitis and multiple sclerosis ("Circoviridae and Anelloviridae" 2017). Water is the secondary enteric transmission route for TTV. TTVs are used as an indicator of excreta-contaminated water. It was proposed by Griffen that international standards of water quality testing must include detection and quantification of TTVs (Griffin, Plummer, and Long 2008). Metagenomic sequencing studies are used to detect the presence of this virus. A study in Japan shows that of 5% of its surface, waters are infected with the TTV. Monitoring of sludge treatment plants detected the presence of TTV genetic material in effluent and influent water (Haramoto et al. 2005). In India, TTV DNA is very much prevalent in untreated wastewater, which indicates that fecal contamination of water is a serious problem in the country (Vaidya, Chitambar, and Arankalle 2002).

Astroviridae

Astroviruses, shed in feces, come after *rotavirus* and *norovirus* as a major cause of hospitalization due to infantile acute gastroenteritis (AGE) (Bosch, Pintó, and Guix 2014; Abad et al. 2001). Although not common in normal healthy adults, these viruses are also found to infect the geriatric and immunocompromised populations (Vu et al.

2017; Le Guyader et al. 2000). They are members of the family *Astroviridae* and are nonenveloped, icosahedral, positive-sense RNA viruses with a smooth margin that were first described in 1975 as viruses with a characteristic five- or six-pointed star-like appearance and a diameter of ≈28–41 nm (Bosch, Pintó, and Guix 2014).

Although astrovirus infections are prevalent year-round, a higher incidence is observed during winters with a decline during the warmer periods and maximum incidence during monsoons (Bosch, Pintó, and Guix 2014). They are ubiquitous in nature, and infections are prevalent worldwide with a higher incidence in developing countries (Donato and Vijaykrishna 2017). Transmissions essentially occur via the fecal–oral route with contaminated food and water serving as vehicles and are able to survive on utensils and other inert surfaces suggesting fomites as vehicles for secondary transmission of the virus (Bosch, Pintó, and Guix 2014). Astroviruses shed in feces enter sewage water, and treatment protocols are known to be incapable of eliminating viral pathogens in general due to which astroviruses released into the environment contaminate different water bodies (Vu et al. 2017). Foodborne outbreaks of human astrovirus infections arise due to contaminated food including bivalve mollusks grown in sewage polluted water as well as fruits and vegetables irrigated with contaminated water (Yates 2013). Poor hygiene and improper food handling after harvest may also result in transmission of the virus (Jain and Jain 2014; Bosch, Pintó, and Guix 2014).

In addition to humans, astroviruses within the genus *Mamastrovirus* are known to infect other mammals, including pigs, sheep, cattle, deer, monkeys, cats, dogs, rodents, minks, bats, rabbits, foxes, marmots, yaks, camels, and cheetahs among many others. They have also been characterized from marine animals including sea lions, whales, and bottlenose dolphins (Donato and Vijaykrishna 2017). Lastly, avian astroviruses, classified under the genus *Avastrovirus*, are also of significance because of their high economic implications to the poultry industry. Turkeys, chicken, ducks, and certain wild aquatic and migratory birds are known to be infected by these viruses (Bosch, Pintó, and Guix 2014). Astroviruses are known to show a high genetic variability due to recombination events that can occur during simultaneous infections with multiple strains which presents these viruses with a zoonotic potential. Advances in next-generation sequencing and other diagnostic techniques have also shown that cross-species transmission of astroviruses may occur (Vu et al. 2017). Thus, astroviruses are of great ecological importance since they can cross the species barrier and show a great amount of divergence with respect to the hosts they can infect. This warrants their surveillance in the environment since they pose a threat not only to mankind but also to several other host species as well.

In order to prevent astrovirus infections, one essentially needs to break the chain of transmission of these viruses. This requires detection and subsequent inactivation of astroviruses that are commonly found to contaminate food and water (Yates 2013). Astroviruses in water are inactivated by chlorine. A chlorine concentration of 1 mg/mL with a holding time of 1 hour is said to produce a 3-log reduction in astrovirus titers (Jain and Jain 2014; Bosch, Pintó, and Guix 2014).

Caliciviridae

Caliciviridae is a family of viruses to which one of the most important waterborne enteric viruses belongs (La Rosa, Bonadonna et al. 2020). Viruses in this family cause asymptomatic infections in humans as well as animals. They are associated with both outbreaks and sporadic illnesses, with symptoms that include mild to AGE, and even meningitis, eye infections, respiratory infections, hepatitis, myocarditis, vesicular lesions, or hemorrhagic disease (Rusiñol and Girones 2019). The genera *Norovirus* and *Sapovirus* are important etiologies of severe gastroenteritis in humans and are hence referred to as human caliciviruses. Caliciviruses are unenveloped viruses with icosahedral symmetry, which are small in size (approximately 27 nm) and possess a single-stranded, positive-sense, polyadenylated RNA genome of about 7.5–8.5 kb. Being nonculturable, large-scale production of these viruses is not possible (Farkas et al. 2008). They are highly contagious even at extremely low concentrations (Rusiñol and Girones 2019).

Noroviruses and sapoviruses are found worldwide, with infections occurring round the year, peaking during the colder months in temperate climates. Sapoviruses are also primarily associated with pediatric gastroenteritis, rarely needing hospitalization. Noroviruses account for over 90% of viral gastroenteritis outbreaks worldwide (Patel et al. 2009). Incubation periods range from 18 to 72 hours, with the shedding in stools being maximal over the first 24–48 hours after infection (Jain and Jain 2014). According to previous studies, 1,011 genome copies/g stool for *Sapovirus* and 1010 genome copies/g of *Norovirus* have been detected from feces of infected individuals (McCall et al. 2020). These viruses occur in a range of water environments such as rivers, sewage, recreational waters, and groundwater, with a persistence period of about 1–3 months. Noroviruses are also unsusceptible to chlorination in concentrations up to 10 ppm and can also tolerate temperature extremes of below 0°C and above 60°C (La Rosa et al. 2012). Outbreaks were recorded in Finland in 2016 and 2018 due to broken drinking water pipes and subsequent wastewater intrusion into the distribution system, with sapoviruses and noroviruses as the main etiology (Kauppinen et al. 2019). Strict surveillance, decontamination, and personal hygiene are of the utmost importance to prevent gastroenteritis associated with caliciviruses.

Circoviridae

The *Circoviridae* family includes the genera *Circovirus* and *Cyclovirus*, which predominantly consist of viruses seldom identified in human disease conditions (Abbas et al. 2019). These are unenveloped, single-stranded DNA viruses with a circular genome of ~2 kb in size (Smits et al. 2013). Metagenomic analyses of sewage water in the United States, Spain, Uganda, and Ethiopia detected several viruses similar to those of the *Circoviridae* family (O'Brien et al. 2017; Rusiñol and Girones 2019). Circoviruses are stable and resistant to environmental degradation (Todd 2000). High concentrations of circoviruses have been detected in wastewater (Farkas et al. 2020). Cycloviruses, on the other hand, have been found to infect humans and chimpanzees, and they have been detected in feces and tissues of domestic animals, bats, as well as dragonflies. Other than feces, *Cyclovirus* genomes have been reported from human samples such as CSF, blood serum, and respiratory secretions (Rusiñol and Girones 2019). Porcine circovirus (PCV) commonly infects pigs, but has also been occasionally isolated from nonporcine animals such as rodents, insects, canines, and mussels. In Brazil, PCV was found in conventionally treated and chlorinated water and in untreated surface water which was ingested by pigs and in treated and chlorinated tap water meant for human consumption. The virus was also present in the water from the Jacutinga River, which is redirected and treated for domestic supply, and in the Pinhal River (Zhai et al. 2019).

A study by Allan and Ellis (2000) found that PCV1 can withstand incubation at 70°C for 15 minutes and treatment with organic solvents such as chloroform and can be inactivated by treatment at pH 3 (Allan and Ellis 2000).

Coronaviridae

Coronaviruses are enveloped viruses with single-stranded, positive-sense RNA genomes and they belong to the order *Nidovirales*, family *Coronaviridae*, and subfamily *Coronavirinae* (La Rosa, Bonadonna et al. 2020). This family consists of several viruses that can cause a wide spectrum of diseases such as the mild common cold as well as the more serious diseases such as severe acute respiratory Syndrome (SARS) and Middle East respiratory syndrome (MERS) (La Rosa, Bonadonna et al. 2020). In addition to these is the coronavirus disease-2019 (COVID-19) caused by the severe acute respiratory syndrome coronavirus-2 (SARS-CoV-2) which has become a public health emergency around the globe. *Coronavirinae* includes the genera *Alphacoronavirus, Betacoronavirus, Gammacoronavirus, and Deltacoronavirus.* Coronaviruses that commonly infect humans, human coronaviruses (HCoVs), include HCoV-229E and HCoV-NL63 both of which belong to the genus *Alphacoronavirus* and HCoV-HKU1, HCoV-OC43, SARS-CoV, SARS-CoV-2, and MERS-CoV belonging to the genus *Betacoronavirus.* In addition to being pathogens of humans, many coronaviruses are also known to infect different host species, including bats, pigs, cats, cattle, and other wildlife as well (La Rosa, Bonadonna et al. 2020).

HCoVs are essentially pathogens of the respiratory system with direct contact as the primary mode of transmission. Other reported and common modes of transmission include aerosolization and transmission via fomites. Waterborne transmission has not yet been reported, but these viruses have been detected in feces of the infected individuals (Esper, Ou, and Huang 2010; Jevšnik et al. 2013; Risku et al. 2010; Vabret et al. 2006; Isakbaeva et al. 2004; La Rosa, Bonadonna et al. 2020). Thus, feco-oral transmission is also a major possibility. This mode of transmission was hypothesized for a coronavirus-associated outbreak in Amoy Garden, Hong Kong, which probably broke out due to aerosolization of fecally contaminated wastewater (McKinney, Gong, and Lewis 2006). The presence of such large clusters of cases is suggestive of contamination of the sewage or the ventilation system (La Rosa et al. 2012).

A study demonstrated that persistence of SARS-CoV in domestic sewage and hospital wastewater is strongly influenced by temperature and that higher temperatures can rapidly inactivate the virus as a result of increased protein denaturation and activity of extracellular enzymes (Pinon and Vialette 2018; La Rosa, Bonadonna et al. 2020). Furthermore, a more rapid decline in viral titers at 23°C–25°C than at 4°C proves the temperature sensitive nature of these viruses (La Rosa, Bonadonna et al. 2020). Similarly, SARS-CoV-2 gets thermally inactivated if subjected to 56°C for 30 minutes (Kumar et al. 2020b; Mohapatra et al., 2020). The virus has not only been detected in domestic and hospital sewage but also in dechlorinated tap water and urine and feces. However, what they also observed was that SARS-CoV is less resistant to chlorine disinfection than bacteria, which is indicative that disinfection methods are effective toward enveloped viruses including coronaviruses (Wang et al. 2005a; La Rosa, Bonadonna et al. 2020). Gundy and coworkers studied the survival of HCoV-229E and also that of the feline infectious peritonitis virus (a feline coronavirus) in filtered and nonfiltered tap water and sludge effluents of wastewater. They observed that CoVs got inactivated faster in filtered tap water than unfiltered tap water, indicating the protective nature of suspended solids in water to which the viruses could adsorb on (Gundy, Gerba, and Pepper 2009; La Rosa, Bonadonna et al. 2020). In another study by Wang et al., SARS-CoV genome was repeatedly detected by reverse-transcriptase PCR (RT-PCR) from the sewage of a hospital hosting patients of SARS. However, they also observed a drop in detection levels after disinfection of the sewage, and they concluded that this reduction could be a result of viral inactivation by disinfectants, a lesser virus concentration, or reduction in infectivity during the concentration process (Wang et al. 2005b; La Rosa, Bonadonna et al. 2020). Another study investigating the occurrence of alpha- and beta-coronaviruses in the surface waters of Wadi Hanifa, Riyadh, detected only one RT-PCR-positive out of 21 total water samples (La Rosa, Bonadonna et al. 2020; Blanco et al. 2019). Another study that checked the viral diversity in sewage sludge showed that coronaviruses were present in 83% of the sample and that the coronavirus HQU1 is the second most prevalent RNA virus in sewage sludge (Bibby and Peccia 2013; La Rosa, Bonadonna et al. 2020). Lastly, the newly discovered SARS-CoV-2, which has afflicted people of varied ages, ethnicities, and nationalities these past few months, has also been found to occur in municipal wastewaters worldwide (La Rosa, Iaconelli et al. 2020; Wu et al. 2020; Ahmed et al. 2020; Medema et al. 2020; La Rosa, Bonadonna et al. 2020). Although, there has been no exacting evidence of SARS-CoV-2 transmission via contaminated water. Nonetheless, these crown-shaped viruses are known to cause severe diseases in humans and animals alike, and this warrants their constant surveillance in not just water, but in the environment in general.

Hepeviridae

Hepeviridae includes viruses that are unenveloped and have a positive-sense, single-stranded RNA genome of a 7.2kb in size. They are considered to be one of the prime causes of acute viral hepatitis of enteric transmission (Baez et al. 2017). Waterborne transmission is associated with the HAV and HEV, their presence in water bodies being a major public concern. The WHO has estimated around 20 million HEV infections covering over 3 million symptomatic cases and a mortality rate of approximately 56,600 each year ("Hepatitis E" 2020). The transmission is found to be via the fecal–oral route and due to the consumption of contaminated seawater or freshwater (van der Poel 2014). Shedding of the HEV in the feces of infected individuals can contaminate water sources. Regions that have poor sanitation are at a higher risk of large waterborne outbreaks of hepatitis E. The virus can survive in tap water for 60days, 6weeks in river water, and 30weeks in seawater with a higher probability of HEV detection during the summer months (Rodríguez-Lázaro et al. 2012; Ippagunta et al. 2007). HEV occurs sporadically and as epidemic forms in India (Kaur et al. 2017). Most of the epidemics of waterborne hepatitis in the country occur due to HEV (Naik et al. 1992). Several cases of detection of HEV in urban sewage have been reported in Spain and Italy. Monitoring the prevalence of HEV in wastewater can be performed by testing influent and effluent water, and it can be characterized using molecular methods. Disease prevention requires strict surveillance of water standards at regular intervals and personal hygiene ("Hepatitis E" 2020).

Orthomyxoviridae

Orthomyxoviridae is a family of single-stranded, negative-sense, enveloped viruses. The genus *Isavirus* is transmitted through water and can cause hemagglutination in several fish species, but not in aves and mammals. The virus infects cells at a low pH and can show maximal replication in vitro at temperatures between 10°C and 15°C. Replication reduces above 15°C and comes to a total halt at 25°C. In addition, viral replication is inhibited by actinomycin D and α-amanitin (Orthomyxoviridae - Negative Sense RNA Viruses - Negative Sense RNA Viruses (2011) - International Committee on Taxonomy of Viruses (ICTV) 2020).

Influenza viruses cause respiratory disease and are constantly reemerging due to reassortments that can occur between the segmented genomes of two virus strains which coinfect the same cell (La Rosa et al. 2012). Influenza A viruses can infect mammals (including humans), and a variety of avian species, with most infected birds acting as "silent" (asymptomatic) reservoirs (La Rosa et al. 2012; Orthomyxoviridae - Negative Sense RNA Viruses - Negative Sense RNA Viruses (2011) - International Committee on Taxonomy of Viruses (ICTV) 2020). H1 to H16 and N1 to N9 subtype of influenza A viruses have been detected in wild waterfowls (La Rosa et al. 2012). Although rare, the interspecies transmission has been well documented. Influenza B virus, first isolated in 1940, is mainly a pathogen of humans, causes respiratory diseases, and is known to be responsible for epidemics occurring every few years, whereas the influenza C virus causes fewer outbreaks in humans, essentially infects the pediatric populations, and may also infect pigs (Orthomyxoviridae - Negative Sense RNA Viruses - Negative Sense RNA Viruses (2011) - International Committee on Taxonomy of Viruses (ICTV) 2020). Antigenic drift is a slower process in influenza B viruses than in influenza A viruses. Viral transmission occurs mainly by aerosolization (human and most nonaquatic hosts), via fomites, direct contact, or fecally contaminated water (Dalziel et al. 2016; Orthomyxoviridae - Negative Sense RNA Viruses - Negative Sense RNA Viruses (2011) - International Committee on Taxonomy of Viruses (ICTV) 2020; La Rosa et al. 2012). The inactivation of the virus can be influenced by using detergents, OH– radicals, higher temperatures, or by higher levels of salinity (Pinon and Vialette 2018; Wigginton, Ye, and Ellenberg 2015). Studies on the H5N1 virus demonstrate the effects of acidic pH on the hemagglutinin (HA) surface protein, which indicates that subtypes carrying different HA have different thresholds of pH (Pinon and Vialette 2018; Wigginton, Ye, and Ellenberg 2015). In addition, the virus was readily inactivated by ultrafiltration, chlorination, and UV treatment, but the same did not hold for the coagulation-flocculation-settling treatment of drinking water (Wigginton, Ye, and Ellenberg 2015). Ribonucleoproteins may protect the viral RNA from environmental stresses, making it more stable. Its survival is not affected by freezing and thawing cycles as the lipid envelope can solidify to a hardened gel. In addition, viruses from mammalian cells have higher survival rates as compared to viruses released from avian cells in the water at 35°C (Pinon and Vialette 2018). In more biologically active water, virus particles may adhere to particulate matter and become less or no longer infectious (Dalziel et al. 2016). Human cases of avian influenza have recently become more recurrent (Abad et al. 2001). Avian influenza viruses have been detected in lakes and ponds, where they can persist for long periods, and hence, they play an important role as environmental virus reservoirs (La Rosa et al. 2012). Case studies in Cambodia and Vietnam suggested that contaminated water could be

Table 7.1 Occurrence of influenza A virus

Strain	*Year*	*Host*
H1N1	1918–1920 (Spanish flu)	Pigs
	1977 (Russian flu)	Poultry
	2009–2010 (swine flu)	Humans
H2N2	1957–1958	Poultry
"Asian influenza"		Humans
H3N2	1968–69	Poultry
"Hong Kong influenza"		Pigs
(New H3 hemagglutinin, N2 neuraminidase from the 1957 H2N2)		Humans
H1N2	2001–2004	Wild birds (Natural host) Aw et al. (2004)
(reassortant virus between H1N1 and H3N2)		
H5N1	1997, 2003	Poultry
	Hong Kong	Human (spill over)
H9N2	1998	Poultry
		Human (spill over)

Source: ICTV & CDC.

a potential risk factor for H5N1 infection (Vong et al. 2009; de Jong et al. 2005; Vong et al. 2008). Viral RNA has also been detected in human stool samples, with its levels ranging from 4.9×103 to 8.0×107 PCR copies per gram of stool, and in municipal wastewater (Table 7.1).

Papillomaviridae

The family *Papillomaviridae* includes double-stranded DNA viruses that are small and nonenveloped. These genotypes infect the skin and mucosa and are associated with conditions such as dysplastic lesions and can lead to invasive carcinoma. Human papillomaviruses (HPVs) cause cervical cancer, including the high-risk HPV 16 and other low-risk HPV types. These are not considered to be classic waterborne pathogens. The first evidence of the presence of HPVs came from a study that described the prevalence in raw sewage (Symonds 2008). Sequencing has also identified the new HPV112 strain from raw sewage (Fratini, di Bonito, and La Rosa 2014; Cantalupo et al. 2011). Metagenomic studies have also identified 11 new strains of HPV from sewage water (Bibby and Peccia 2013). Recovery of HPVs by nested PCR has detected their presence in raw sewage that can enter water through the washing of skin and mucous membranes or through urine (Ahmed et al. 2009; La Rosa et al. 2013). A majority of sewage samples contain alpha- and beta-type genera of the virus. These epitheliotropic viruses have been detected in urban wastewaters and surface waters. Shedding of feces is a proposed source of transmission. The global prevalence of HPV in water bodies is 11.7%. Susceptibility of HPV to disinfection is difficult. This virus is found to be resistant to heat and desiccation and is thermostable, surviving a treatment at 56°C for an hour (Bonnez et al. 1994; Roden, Lowy, and Schiller 1997). The possibility of waterborne HPV infection accounts for oral HPV infection among sexually inactive people.

Parvoviridae

Members of the family *Parvoviridae* are small and nonenveloped DNA viruses, with linear, single-stranded genomes of the size 4–6 kb. The ruggedness of the virions confers prolonged protection to the genome. They may remain infectious for months or even years in the environment. However, a larger concentration of parvoviruses is required to cause sustained infections in humans. *Parvovirinae* and *Densovirinae* are subfamilies which consist of viruses that infect vertebrate and invertebrate hosts, respectively. Parvovirus infections in humans can manifest with a broad range of pathologies that are rarely life-threatening. However, mortality rates for canine parvovirus (genus *Protoparvovirus*), which replicates in dogs, can reach around 90% if untreated. Densoviruses are highly infectious and are known to pose severe economic or ecological risks, for example, in shrimp aquaculture. The human bocaviruses (HBoV) 1–4 and human bufavirus also belong to this family and cause respiratory and gastrointestinal illnesses in humans (Rusiñol and Girones 2019). HBoV has been detected in fecal samples from children. HBoV was also detected in untreated wastewater samples from Germany, Egypt, Nepal, Italy, Norway, and the United States. According to a study, the time required for 99% inactivation of H-1 parvovirus, presumably infecting humans, at 20°C, pH 7, and a chlorine dose of 0.2 mg/L free chlorine was 3.2 minutes (Churn, Boardman, and Bates 1984).

Picobirnaviridae

Picobirnaviridae is a family of unenveloped, linear, dsRNA viruses with genomes of 4.1–4.6 kbp size, which includes only one genus, the *Picobirnavirus* (PBV). They have a wide geographical distribution among humans as well as other mammals and have also been reported in birds, reptiles, and invertebrates. They are opportunistic pathogens that have been mainly identified in untreated sewage samples, wastewater samples, stool samples from children with diarrhea, in immunocompromised patients, and even in individuals, mammals, and birds lacking signs of gastroenteritis (Rusiñol and Girones 2019). They may be shed for a long period in the feces of infected individuals, which may be due to persistence of infection (Malik et al. 2018, Malik et al. 2014). Symonds and coworkers detected human PBV in raw sewage samples and final effluent samples they tested (Rusiñol and Girones 2019). In India, it was first detected in fecal specimens of children from a slum in Kolkata, and around 3.57% of the diarrheic and 1.64% of the nondiarrheic samples were found to be PBV-positive (van Leeuwen et al. 2010). Picobirnaviruses are capable of cross-species transmission and are among the potentially emerging waterborne viruses (Rusiñol and Girones 2019). These viruses are not culturable in laboratory settings, due to which their pathogenicity has not been described. The seasonality in infection does not seem to play a major role. PBV infections occur in concordance with a number of other diarrheal causes such as rotaviruses, astroviruses, caliciviruses, *Escherichia coli*, and *Salmonella* (Malik et al. 2018). With further research, it may be possible to determine if this virus can be utilized as a microbial water quality indicator (Rusiñol and Girones 2019).

Picornaviridae

Enterovirus is a genus that causes the most disease among *Picornaviridae* members. They have been found in various aquatic environments across the world, such as river water, seawater, and water from wells (Abbaszadegan, Stewart, and LeChevallier 1999;

Lodder and de Roda Husman 2005). Parsley plants irrigated by contaminated river water analyzed by metagenomics revealed the presence of several picornaviruses, such as Aichivirus, Ampivirus, rat hunnivirus, rodent hepatovirus, and enterovirus species A and B (Fernandez-Cassi et al. 2017). The sequence identities of the isolates found in this study were rather low, indicative of new Picornaviridae members to be found in the future. A Korean study from 1997 to 1998 showed the presence of enterovirus clones in tap water samples from various sites, most of them having sequence similarity to Type 1 poliovirus, others being similar to coxsackievirus B or echovirus type 6 (Lee and Kim 2002). The occurrence and outbreaks caused by viruses in recreational water such as lakes and pools have been reviewed in the past (Sinclair, Jones, and Gerba 2009). Polluted freshwater lakes, sewage contaminated seawater, and swimming areas exposed to lake, pond, reservoir, or ocean water were found to be sources of infections by coxsackievirus. HAV infections seemed to have occurred at a comparatively low morbidity rate. Treatment of drinking water can remove picornaviral agents, as proved in a Chinese study, recording a 97% removal rate of enteroviruses from source water (46% of samples) to treated drinking water (21% of samples) (Ye et al. 2012). In 2005, a Netherlands study showed that culturable enteroviruses were present at low concentrations of 1–6 PFU/l in the river Maas, whereas the river Waal had enteroviruses at 0.3–10 PFU/l (Lodder and de Roda Husman 2005). Raw sewage seemed to harbor enteroviruses at about 100 PFU/ml, whereas treated sewage seemed to remove 1.4 log10 units, on average, of the virus. HAV was found to persist in water for 21 days without any loss of infectivity at 4°C or −20°C, but with a reduction of 3.09 log10 PFU/ml at 21°C (Leblanc et al. 2019). Viability losses of 1.99 log10, 0.58 log10, and 0.49 log10 were seen at 21°C, 4°C, and −20°C, respectively, for HAV in bottled spring water incubated over 21 days, and so were seen decreasing in genomic copies per milliliter. Other reports describe the low occurrence of viruses such as HAV in the Sinos River watershed, only in 1 out of 84 samples harboring HAV at a concentration of 1.5 * 104 cg/ liter (de Souza et al. 2018). The *Salivirus* genus is yet another *Picornaviridae* member which can be found in stool and contaminated water such as sewage and transmits through the fecal–oral route (Reuter, Pankovics, and Boros 2017). SalV was found to be prevalent throughout the study period of November 2016 to December 2018, in 32.6% of environmental water samples collected from six locations across Thailand (Badru et al. 2018). Another study from Japan, published in 2013, revealed the presence of SalV in 93% raw sewage samples, secondary-treated sewage samples, and 16% river water samples (Haramoto, Kitajima, and Otagiri 2013). Clinical and sewage samples analysis in an Indian study showed a close relation between enteroviruses detected from both, but also some that could not be detected in the other, perhaps due to cell culture–based detection method drawbacks (Tiwari and Dhole 2018).

Polyomaviridae

Human polyomaviruses are double-stranded DNA viruses that are small and unenveloped. They are known to be used as sewage tracking viruses due to their presence in urine, feces, and wastewaters. They have been known to be present in sewage and environmental samples like surface waters, groundwater, drinking water, seawaters, and sludge (Sílvia Bofill-Mas, Pina, and Girones 2000). The first detection of JCV and BKV was done in sewage samples from Europe and Africa. Human polyomavirus 6 has also been detected and sequenced from sewage samples by next generation sequencing

(Cantalupo et al. 2011). Serious illnesses including leukoencephalopathy and nephropathy have been recorded in immunocompromised patients (Silvia Bofill-Mas 2019). Human polyomavirus 2 (JC and BK polyomavirus) infects humans persistently. These groups of viruses are resistant to inactivation (Nims and Plavsic 2013). High temperatures (>70°C) are needed to effectively inactivate these viruses thermally. HPyV2 is known to be present in high concentrations in the environment and is used as an indicator of contamination of water because of its ability to persist in wastewater and surface waters due to the high stability of the virus and its specificity to the human host (Silvia Bofill-Mas 2019). Environmental surveillance of these viruses needs to be conducted globally.

Reoviridae

One of the leading causes of AGE in children, the *Rotavirus*, has been detected in wastewater, surface water, drinking water, and groundwater (Gerba et al. 1996). The *Rotavirus* has a winter seasonality and can transmit via person-to-person fecal–oral or respiratory routes, as well as foodborne routes, in addition to being waterborne (Abbaszadegan and Alum 2015). Rotaviruses are capable of surviving secondary sewage treatment and show more resistance to chloramines and ultraviolet light disinfection than other enteroviruses. The attack rate of waterborne rotaviruses was found to vary from 4% to more than 70%. Up to 10^9 virus particles per gram of stool may occur (Carter 2005). About 10^9 particles can occur per milliliter and comprise up to 2 mg per gram of stool. However, rotavirus can also be inactivated efficiently by chlorine at a CT (concentration X contact time) value of 112.5 at 3.75 mg/L ("AJPH.74.3.263," n.d.). Case fatality rates of 0.01%, 1%, and 50% for the general population, the elderly, and the immunocompromised were values estimated for the US population in 1996 (Gerba I et al. 1996). In Malatya, Turkey, a group A rotavirus outbreak was caused due to low chlorine levels (0–0.05 mg/L) in drinking water, due to the contamination of a large water depository from a water well, and the outbreak was stopped by effective chlorination and blocking the passage of water between the well and the depository (Koroglu et al. 2011). Despite viruses of human health importance occurring in wastewater, the distribution of rotaviruses, among other kinds, can vary widely, which needs to be detected using advanced methods such as metagenomics (Aw, Howe, and Rose 2014). Water for food preparation as a virus-commodity combination should be considered for prevention and control measures of rotaviruses (Radin 2014). There are variations to the reports on the viral concentrations in stool, like one which mentions that patients shed about 10^{11} virus particles per gram of feces for a period of 8 days (Bridle 2013). While the Rotavirus is an important human health concern, other reoviruses also infect humans, causing disease. In Lake Needwood, a freshwater body in Maryland, USA, a *Seadornavirus*, with sequence homology to *Banna virus*, another seadornavirus, was found (Djikeng et al. 2009). It was inferred that the lake has a *Seadornavirus* with sequence similarity to the *Banna virus*. It may have been that the virus was introduced to the United States or is endemic to North America. A 2018 publication noted the presence of *Liao ning virus* in mosquitoes from Australia, and yet another, speculated that the finding of viral RNA in water-containing pans for mosquito breeding could mean that the virus uses water as a means to infect larvae (Prow et al. 2018; Rames 2020). Infected feces and vomit can disseminate viruses into drinking water and recreational water, highlighting an important fecal–oral route that such viruses can take (Ganesh and Lin 2013). Therefore, strict surveillance and decontamination measures are the needful for eliminating waterborne viruses, thereby reducing their disease burden throughout the globe.

7.4 DETECTION OF VIRUSES FROM WATER SAMPLES

The occurrence and behavior, possible pathogenicity of different viruses in water bodies, as well as their impact on public health, are all factors that mandate the virological assessment and analysis of water. Detection of viruses in water have several plausible applications such as the direct detection of viral pathogens for the screening of water that is intended for public consumption; the investigation of possible waterborne illness outbreaks; and the study of virus persistence in aquatic environments over time, especially for uncultivable viruses (Soule et al. 2000; Rutjes et al. 2006). This occurrence of viruses in water may be a result of inadequacies and inefficiencies in wastewater and drinking water treatment.

Recovery of the virus from a large sample is an essential step. An ideal concentration method should be simple, easy to perform, rapid, affordable, suitable to recover various types of viruses, and with maximal virus recoveries (Green and Lewis 1999). Several methods, for example, ultrafiltration, lyophilization, flocculation, and adsorption-elution, can be employed for virus concentration (Kopecka et al. 1993). Virus adsorption-elution microporous filter method (Viradel) is the most widely used technique for virus concentration from water samples. Dependence on the electrostatic interactions, van der Waals forces, and hydrophobic interactions between electronegative viruses and the negatively or positively charged filters in aqueous systems is seen (C. P. Gerba and Rose 1990). Viral elution from these filter surfaces is based on the disruption of these interactions. Ultrafiltration is another method of concentration that requires no additional recovery steps (Collins et al. 2003). Soluble ultrafilters made of sodium alginate can be used for the concentration of polioviruses, followed by their dissolution in sodium citrate for the virus recovery. However, it can only be used for small sample volumes, requires prefiltration for severely turbid samples, and may lead to loss of solids-associated viruses (Lanciotti and Kerst 2001).

Cell culture and plaque assays have been used for the detection of viruses. Appropriate animal or plant cell lines are infected with the desired virus, which, after a particular period, shows cytopathic effects that can be observed microscopically. However, this method is labor-intensive, time-consuming, requires a specific type of cell culture for a virus or group of viruses, and not all viruses are cultivable in vitro and exhibit cytopathic effects (Lanciotti and Kerst 2001; Hang et al. 2005; Yang et al. 2014). Formation of clear zones or plaques is seen in the areas (monolayer of host cells) infected with the virus. PFUs determine the concentration of viruses. Several viruses, such as enteroviruses, reoviruses, adenoviruses, and even astroviruses or rotaviruses, can be detected using these methods (Bosch et al. 2008; C. P. Gerba and Rose 1990) (Table 7.2).

For the viruses which cannot grow in cell cultures, other methods such as immunofluorescence enzyme-linked immunosorbent assay (ELISA), radioimmuno-focus assay, and molecular diagnostic techniques such as PCR have proven to be better and less time-consuming (Xagoraraki, Yin, and Svambayev 2014; C. P. Gerba and

Table 7.2 Common viruses that are associated with occurrence in water systems and are of importance to human health

Family	*Notable examples*	*Transmission modes*	*Laboratory diagnosis*	*Therapy*	*Prophylaxis*
Adenoviridae	Human adenovirus A-G	Fecal–oral route, Airborne transmission	ELISA, PCR, virus isolation	No specific treatment except supportive care.	No available vaccine. maintenance of proper hygiene of hands and eyes
Anelloviridae	TTV, Torque teno mini virus, Torque teno midi virus.	Fecal–oral route, via exchange of bodily fluids like blood	PCR	Use of antiviral drug valganciclovir in small amounts.	No vaccine available for humans yet, maintenance of hygiene is key
Astroviridae	*Mamastrovirus* *Avastrovirus*	Waterborne, foodborne, fecal–oral route, fomite transmission	Electron microscopy, radioimmunoassay, enzyme immunoassay, RT-PCR, real-time RT-quantitative PCR	No specific antiviral therapy available, Only supportive treatment to combat dehydration (using oral rehydration solutions or intravenous fluid replacement) and pyrexia (using antipyretic drugs)	No available vaccination, maintenance of hygiene is key
Caliciviridae	*Sapovirus* *Norovirus*	Direct contact with an infected host, foodborne, waterborne fecal–oral route, fomite transmission, via vomitus or respiratory secretions	Electron microscopy, RT-PCR, real-time RT-PCR, nucleic acid sequence-based amplification and enzyme immunoassay, radioimmunoassay	No specific antiviral therapy available, Only supportive treatment to combat dehydration (using oral rehydration solutions or intravenous fluid replacement) and pyrexia (using antipyretic drugs)	No available vaccination, maintenance of hygiene is key

(*Continued*)

Table 7.2 (Continued) Common viruses that are associated with occurrence in water systems and are of importance to human health

Family	*Notable examples*	*Transmission modes*	*Laboratory diagnosis*	*Therapy*	*Prophylaxis*
Circoviridae	*Porcine circovirus (PCV1 and PCV2)*	Fecal–oral route, direct contact with an infected host	PCR, in situ hybridization, dot blot hybridization, ELISA	No specific antiviral therapy, antimicrobials to prevent secondary infections	No vaccine available for humans yet, maintenance of hygiene is key
Coronaviridae	*HCoV-229E, HCoV-NL63, HCoV-HKU1, HCoV-OC43, SARS-CoV, SARS-CoV-2, MERS-CoV*	Direct contact, aerosolization, fomite transmission, probable waterborne transmission, feco–oral transmission	RT-PCR, enzyme-based immunoassays, electron microscopy	No specific antiviral therapy. Therapy also varies from virus to virus. Supplementary nutrients are administered, and supportive therapy also suggested	No vaccines yet. Vaccine candidate undergoing trials.
Hepeviridae	*Orthohepevirus A;* HEV-1 and HEV-2	Fecal–oral route, Waterborne transmission	Anti-HEV-IgM antibody detection, enzyme immunoassay, Western blot, immunofluorescent antibody blocking assay, RT-PCR	No specific antiviral therapy available. Ribavirin is an off-label drug that may be administered. Oral rehydration therapy.	No available vaccine. Improved public hygiene
Orthomyxoviridae	*Influenza A virus (H1N1, H3N2), Influenza B Virus*	Direct contact, aerosolization, fomite transmission	Rapid ELISA, immunofluorescence, DFA, indirect fluorescent antibody staining, RT-PCR, virus isolation by tissue culture, hemagglutination and hemagglutination inhibition	Oseltamivir, Zanamivir, Peramivir, Baloxavir marboxil, Amantadine, Rimantadine	Vaccination, maintenance of hygiene is key

(Continued)

Table 7.2 (Continued) Common viruses that are associated with occurrence in water systems and are of importance to human health

Family	*Notable examples*	*Transmission modes*	*Laboratory diagnosis*	*Therapy*	*Prophylaxis*
Papillomaviridae	Human papillomavirus 16,18 HPV 66	Sexually transmitted infection	Colposcopy, acetic acid test, biopsy, PCR, southern blot hybridization, in situ hybridization	CIN treatment includes a loop electrosurgical excision procedure (LEEP), cold knife cone biopsy, cold-coagulation and cryotherapy.	Primary (vaccination) or secondary prevention programs (cervical screening)
Parvoviridae	Human *Parvovirus B19*	Direct contact, aerosolization, fomite transmission, parenteral (blood transfusion), mother to fetus	Radioimmunoassay, enzyme immunoassay, nucleic acid hybridization, light and electron microscopy, Southern blot, Western blot	Treatment in humans usually involves relieving symptoms, such as fever, itching, and joint pain and swelling	No vaccine available for humans yet, maintenance of hygiene is key
Picobirnaviridae	Human *Picobirnavirus*	Fecal–oral.	PAGE, RT-PCR	No specific antiviral therapy available, Only supportive treatment available	No vaccine available yet, maintenance of hygiene is key
Picornaviridae	*Enterovirus, Aichivirus, Aphthovirus, Hepatovirus, Cardiovirus*	Waterborne, foodborne, orofecal route, saliva, respiratory droplets, conjunctival secretions, skin lesion exudates	PCR, neutralization assay, sequencing, metagenomics, cell culture, electron microscopy	Ruprintrivir, Pyridazinyl oxime ethers, Pleconaril, Milrinone	Vaccination
Reoviridae	*Orbivirus, Rotavirus, Coltivirus, Oryzavirus, Seadornavirus, Phytoreovirus*	Contaminated food, hands, water, orofecal route, arthropods, aerosol	ELISA, complement fixation test, immunoelectron microscopy, PCR, neutralization assay, immunofluorescence, cell culture, polyacrylamide gel electrophoresis	Symptomatic treatment of dehydration by oral rehydration solution, intravenous fluid injection, Ribavirin, Antipyretics	Vaccination

Rose 1990). PCR has emerged to be an extremely reliable, rapid, and accurate detection technique. As opposed to cell culture, the molecular methods provide quicker results, high specificity, and sensitivity even for viruses that are difficult to culture (Xagoraraki, Yin, and Svambayev 2014). They indicate the presence of enteroviruses in water since free nucleic acids get degraded in sewage and cannot exist freely in water beyond a particular period. Thus, one can infer from positive PCRs that the source of these in nucleic acids is the complete viral particles. PCRs cannot be conclusive regarding the infectivity of these viral particles that may have been inactivated by water disinfection protocols (Kopecka et al. 1993). Conventional PCR amplifies and detects viral DNA despite the presence of other DNA contaminants. After the amplification, electrophoretic techniques are employed to visualize the results. Quantitation is done using the most probable number which yields units comparable to PFUs (Xagoraraki, Yin, and Svambayev 2014; Chang et al. 1958). RT-PCR detects enteroviruses and noroviruses by a complementary strand production. The sensitivity of RT-PCR is affected due to RNases which tends to lower the RNA concentration. (Steininger et al. 2002). Sensitization of transcriptase and polymerase due to presence of humic acids and metals as inhibitors is seen (Soule et al. 2000; Rutjes et al. 2006). It is used for detection of viruses which are difficult to identify, such as rotaviruses and HAVs (Green and Lewis 1999). Modifications of conventional PCR are nested PCR, multiplex PCR, and real-time PCR. Nested PCR enhances the specificity and sensitivity (Kopecka et al. 1993). Nested PCR is a method that employs two sets of primers, one set nested within the nucleic acid defined by the second primer pair. The outer primers generate an amplicon, while the inner primers amplify the target sequence of DNA. Multiplex PCR targets multiple DNA sequences concurrently, whereas, in real-time PCR, the amplification, as well as quantification of the target sequences, happens simultaneously. Probes with attached dyes are used in real-time PCR, which fluoresce upon release during amplification. The fluorescence signal is detected using a standard curve, and the viral genome copies can be quantified. When merged with the cell culture, the PCR can even determine the infectivity of viruses using a procedure called integrated cell culture PCR.

Immunoassays such as immunofluorescence, RIA, and ELISA are more rapid and cost-effective techniques that are based on the production of a specific antibody against the desired virus. RIA and ELISA need a high viral load (10,000–100,000 virus particles), making them unsuitable for samples with lower viral loads (C. P. Gerba and Rose 1990). Nucleic acid sequence-based amplification and loop-mediated isothermal amplification techniques are also very sensitive and specific. Nucleic acid sequence-based amplification involves the rapid amplification of particular genome sequences and is especially suitable for the amplifying RNAs (Collins et al. 2003). It utilizes three enzymes, reverse transcriptase, RNA polymerase, and RNase. The product is a single-stranded RNA which is complementary to the target RNA (Lanciotti and Kerst 2001). Loop-mediated isothermal amplification requires no complicated equipment and is simple to perform. It is a valuable tool for the diagnosis of viral infectious diseases and relies on auto-cycling strand displacement DNA synthesis performed by using the Bst DNA polymerase large fragment. It is rapid, sensitive, and highly specific for the target sequence, requires low reaction temperatures (63°C–65°C), and is affordable (Lanciotti and Kerst 2001; Hang et al. 2005; Yang et al. 2014).

7.5 METHODS OF DISINFECTION AND PREVENTION OF THE SPREAD OF WATERBORNE VIRUSES

Water supplies contain several viruses leading to the spread of various diseases that can be sporadic or epidemic (Chang 1968). Disinfection of water reduces the virus concentration to an acceptable level. The genome of waterborne viruses including the protein capsid contributes to the resistance to specific disinfectants (Gall et al. 2015). Conventional treatment of virally contaminated sewage effluents using chlorination is able to reduce viral concentrations to a factor of 10^8 (Chang 1968). Turbid viral samples can be filtered through membrane filtration. For soft waters, about 100 ppm of calcium is added facilitating viral adsorption. The development of the enterovirus index is essential for circumstantial evidence for the transmission of viral infection in the community. Survival time upon storage gives an idea about virus survival (Shuval, Cymbalista, and Goldblum 1969). Salts (polyvalent ions) increase the survival time of enterovirus in the high-temperature range (Wallis and Melnick 1962). Data from Chang et al. indicate the disinfection of the virus by the process of flocculation. The lower efficiency of flocculation under field conditions removes around 95% of the waterborne virus (Chang 1968).

Seasonal changes in temperature and turbidity influence viral removal efficiency. Sand filtration has been known to remove about 1%–50% of poliovirus. Rapid sand filters impregnated with performed alum flock are better for virus removal. The viral removal efficiency of flocculation and filtration combined is better. Viral destruction with chlorination methods includes chloramines which are effective disinfectants since they hydrolyze to form HOCl (Johnson, Fields, and Darlington 1967). Ozone as a disinfectant is a better virucidal. Coin reported that 0.4ppm of ozone has known to destroy poliovirus in tap water in around 30 minutes (7). An amount of 0.2ppm of ozone can destroy enterovirus. Ozonation of water is advantageous due to the absence of odor and taste whereas the disadvantage being that it is hard to disperse ozone uniformly and maintenance of a residual concentration. Iodination has been used for the disinfection of water samples in rural/underdeveloped areas. Elementary iodine can be added to water at a constant dose and calculation of the residual iodine can be done by seen by color change. The ease with the treatment of iodine makes it usable to treat small water supplies (Chang 1968). Chlorine dioxide and bromine at low concentrations can also be used. Table 7.3 highlights various disinfection methods for different viruses, their merits, and demerits.

Prevention of viral borne diseases involves various measures including consumption of purified water and practicing hygienic practices. Preventive vaccination for various viruses is also an approach. Populations that are vulnerable to unsafe drinking waters should have protocols and sensors that are cost-effective and easy for use (Gall et al. 2015). For the prevention of waterborne viral disease outbreaks, the investigation by molecular characterization methods, identification of risk factors, and application of control measures are essential. Environmental surveillance on sewage plants and drinking water helps to monitor the shedding of these viral particles. Assessment of vaccine programs provides a rapid and economical way for prevention (Ruggeri and Fiore 2012).

Table 7.3 Disinfection methods for various viruses, their merits, and demerits

Method of disinfection	*Viruses disinfected*	*Merits*	*Demerits*
Chlorination	Poliovirus, enterovirus	High virucidal efficiency. Dichloramines are effective—more hydrolysis.	Enteric viruses are less resistant. Not suitable for turbid waters.
Ozonation	Poliovirus (0.4 ppm in 30 minutes), Enterovirus (0.2 ppm)	Absence of odor and taste.	Hard to disperse uniformly. Difficult to maintain a residual concentration.
Iodination	Enterovirus	Applied on field waters (rural areas). Not complicated. Estimation of residual iodine is easy.	Less virucidal efficiency. High dosage causes physiological effects.
Chlorine dioxide	Retrovirus	Not affected by pH. No taste/odor	Less virucidal efficiency.

Source: Chang 1968.

7.6 CONCLUSION

The unquestionable occurrence of pathogenic viruses has a health hazard due to presence in the environment, to mankind and animals alike, and it is highly imperative to tackle this problem. The most important as well as the relatively uncommon viruses found in water bodies have therefore been highlighted upon in this chapter. Waterborne viruses that are classified by the WHO under the categories of high- and medium-health significance include rotaviruses, noroviruses, adenovirus, astrovirus, HAVs and HEVs, other caliciviruses, and enteroviruses, as well as coxsackieviruses and polioviruses. Other viruses that could potentially be transmitted through water include cytomegaloviruses and polyomaviruses. Most of these pathogens cause gastrointestinal infections and can also manifest as systemic infections that may have dire consequences to the health of an individual. These viruses have low die-off rates and low infectivity doses (Xagoraraki, Yin, and Svambayev 2014). Thus, the desired control of these viruses is with good reason (Gall et al. 2015). The occurrence of these viruses can be monitored by a wide variety of detection techniques as is mentioned in this chapter, and these methods can vary from virus to virus. However, precise detection, quantification and proper evaluation of infectivity doses remain challenging (Xagoraraki, Yin, and Svambayev 2014).

As we know, virologically safe water is not just myth, but can indeed be obtained by utilizing the galactic range of technologies which are currently available and ever-evolving. These technologies allow for the reuse of wastewater by appropriately treating it so as to yield virus-free effluent. However, sometimes viruses maybe detected in treated water as well. This could be a result of improper treatment or even due to secondary pollution of distribution systems. The effectiveness of the disinfection or treatment processes and occurrence of these viral pollutants can also be measured by a range of indicator systems. In conclusion, it is essential to study the fate and occurrence of viruses in water. In order to obtain virologically safe water, it is essential to improve treatment techniques by improving our knowledge about the effect of individual disinfection techniques on various viruses, which can be understood by studying these viruses at a molecular level. Such studies can be conducted by using high-end technologies such as mass spectrometry which allows to study effects of disinfectants on the amino acid patterns of viral capsids. Recent years have also seen the use of PCR techniques to understand how disinfectants alter regions of the viral genome in order to inactivate them. Similar and more advanced techniques may be implemented in the future to understand the effect of disinfection on these viruses, and this may help improve the scenario of occurrence of viruses in water systems. Lastly, several novel, emerging, and reemerging viruses are being detected in the environment from time to time, and with their detection comes the responsibility of studying their occurrence patterns and implementing appropriate interventional strategies.

CONTRIBUTION

Meenakshi Singh and Prudhvilal Bhukya are the corresponding authors. Muskan Chakraborty and Saumya Vas have contributed equally.

REFERENCES

Abad, F. X., C. Villena, S. Guix, S. Caballero, R. M. Pintó, and A. Bosch. 2001. "Potential role of fomites in the vehicular transmission of human astroviruses." *Applied and Environmental Microbiology* 67 (9). American Society for Microbiology: 3904–7. doi:10.1128/AEM.67.9.3904-3907.2001.

Abbas, A. A., L. J. Taylor, M. I. Dothard, J. S. Leiby, A. S. Fitzgerald, L. A. Khatib, R. G. Collman, and F. D. Bushman. 2019. "Redondoviridae, a family of small, circular DNA viruses of the human oro-respiratory tract associated with periodontitis and critical illness." *Cell Host and Microbe* 25 (5). Cell Press: 719–729. doi:10.1016/j.chom.2019.04.001.

Abbaszadegan, M., and A. Alum. 2015. "Waterborne enteric viruses: Diversity, distribution, and detection." In: *Manual of Environmental Microbiology*, pp. 3.1.5-1–3.1.5-13. ASM Press. doi:10.1128/9781555818821.ch3.1.5.

Abbaszadegan, M., P. Stewart, and M. LeChevallier. 1999. "A strategy for detection of viruses in groundwater by PCR." *Applied and Environmental Microbiology* 65 (2). American Society for Microbiology: 444–49. doi:10.1128/aem.65.2.444-449.1999.

Ahmed, W., A. Goonetilleke, D. Powell, K. Chauhan, and T. Gardner. 2009. "Comparison of Molecular Markers to Detect Fresh Sewage in Environmental Waters." *Water Research* 43 (19). Elsevier Ltd: 4908–17. doi:10.1016/j.watres.2009.09.047.

Ahmed, W., N. Angel, J. Edson, K. Bibby, Aaron Bivins, J. W. O'Brien, P. M. Choi, et al. 2020. "First confirmed detection of SARS-CoV-2 in untreated wastewater in australia: A proof of concept for the wastewater surveillance of COVID-19 in the community." *Science of the Total Environment* 728: 138764. doi:10.1016/j.scitotenv.2020.138764.

"AJPH.74.3.263." n.d.

Allan, G. M., and J. A. Ellis. 2000. "Porcine circoviruses: A review." *Journal of Veterinary Diagnostic Investigation* 12 (1). American Assoc. of Veterinary Laboratory Diagnosticians: 3–14. doi:10.1177/104063870001200102.

Aw, T. G., A. Howe, and J. B. Rose. 2014. "Metagenomic approaches for direct and cell culture evaluation of the virological quality of wastewater." *Journal of Virological Methods* 210. Elsevier: 15–21. doi:10.1016/j.jviromet.2014.09.017.

Badru, S., P. Khamrin, K. Kumthip, A. Yodmeeklin, S. Surajinda, K. Supadej, S. Sirilert, et al. 2018. "Molecular detection and genetic characterization of salivirus in environmental water in Thailand." *Infection, Genetics and Evolution* 65. Elsevier B.V.: 352–56. doi:10.1016/j.meegid.2018.08.023.

Baez, P. A., M. C. Lopez, A. Duque-Jaramillo, D. Pelaez, F. Molina, and M.-C. Navas. 2017. "First evidence of the Hepatitis E virus in environmental waters in Colombia." *PLOS ONE* 12 (5): e0177525. doi:10.1371/journal.pone.0177525.

Biagini, P., and P. de Micco. 2008. "Anellovirus." In *Encyclopedia of Virology*, pp. 104–10. doi:10.1016/B978–012374410-4.00694-4.

Bibby, K., and J. Peccia. 2013. "Identification of viral pathogen diversity in sewage sludge by metagenome analysis." *Environmental Science and Technology* 47 (4). American Chemical Society: 1945–51. doi:10.1021/es305181x.

Blanco, A., I. Abid, N. Al-Otaibi, F. J. Pérez-Rodríguez, C. Fuentes, S. Guix, R. M. Pintó, and A. Bosch. 2019. "Glass wool concentration optimization for the detection of enveloped and non-enveloped waterborne viruses." *Food and Environmental Virology* 11 (2). Springer New York LLC: 184–92. doi:10.1007/s12560-019-09378-0.

Bofill-Mas, S. 2019. "Polyomavirus." In *Global Water Pathogen Project*. Michigan State University. doi:10.14321/waterpathogens.16.

Bofill-Mas, S., N. Albinana-Gimenez, P. Clemente-Casares, A. Hundesa, J. Rodriguez-Manzano, A. Allard, M. Calvo, and R. Girones. 2006. "Quantification and stability of human adenoviruses and polyomavirus JCPyV in wastewater matrices." *Applied and Environmental Microbiology* 72 (12). American Society for Microbiology: 7894–96. doi:10.1128/AEM.00965-06.

Bofill-Mas, S., S. Pina, and R. Girones. 2000. "Documenting the epidemiologic patterns of polyomaviruses in human populations by studying their presence in Urban Sewage." *Applied and Environmental Microbiology* 66 (1): 238–45. doi:10.1128/AEM.66.1.238-245.2000.

Bonnez, W., R. C. Rose, C. Borkhuis, C. da Rin, and R. C. Reichman. 1994. "Evaluation of temperature sensitivity of human papillomavirus type 11 by using the human xenograft severe combined immunodeficiency mouse model." *Journal of Clinical Microbiology*. American Society for Microbiology. doi:10.1128/jcm.32.6.1575-1577.1994.

Bosch, A., S. Guix, D. Sano, and R. M. Pintó. 2008. "New tools for the study and direct surveillance of viral pathogens in water." *Current Opinion in Biotechnology*. Elsevier. doi:10.1016/j.copbio.2008.04.006.

Bosch, A., R. M. Pintó, and S. Guix. 2014. "Human Astroviruses." *Clinical Microbiology Reviews* 27 (4). American Society for Microbiology: 1048–74. doi:10.1128/CMR.00013-14.

Breitbart, M., and F. Rohwer. 2005. "Here a virus, there a virus, everywhere the same virus?" *Trends in Microbiology*. Elsevier Current Trends. doi:10.1016/j.tim.2005.04.003.

Bridle, H. 2013. "Overview of waterborne pathogens." In *Waterborne Pathogens: Detection Methods and Applications*, pp. 9–40. Elsevier B.V. doi:10.1016/B978-0-444-59543-0.00002-5.

Cantalupo, P. G., B. Calgua, G. Zhao, A. Hundesa, A. D. Wier, J. P. Katz, M. Grabe, et al. 2011. "Raw sewage harbors diverse viral populations." *MBio* 2 (5). American Society for Microbiology. doi:10.1128/mBio.00180-11.

Carter, M. J. 2005. "Enterically infecting viruses: Pathogenicity, transmission and significance for food and waterborne infection." *Journal of Applied Microbiology*, 98:1354–80. doi:10.1111/j.1365-2672.2005.02635.x.

Chang, S. L. 1968. "Water borne viral infections and their prevention." *Bulletin of the World Health Organization* 38 (3). World Health Organization: 401–14. /pmc/articles/PMC2554483/?report=abstract.

Chang, S. L., G. Berg, K. A. Busch, R. E. Stevenson, N. A. Clarke, and P. W. Kabler. 1958. "Application of the 'Most Probable Number' method for estimating concentrations of animal viruses by the tissue culture technique." *Virology* 6 (1). Academic Press: 27–42. doi:10.1016/0042-6822(58)90057-6.

Churn, C. C., G. D. Boardman, and R. C. Bates. 1984. "The inactivation kinetics of H-1 parvovirus by chlorine." *Water Research* 18 (2): 195–203. doi:10.1016/0043-1354(84)90069-1.

"Circoviridae and Anelloviridae." 2017. In *Fenner's Veterinary Virology*, 259–68. doi:10.1016/B978-0-12-800946-8.00013-1.

Cleaveland, S., D. T. Haydon, and L. Taylor. 2007. "Overviews of pathogen emergence: Which pathogens emerge, when and why?" *Current Topics in Microbiology and Immunology*. doi:10.1007/978-3-540-70962-6_5.

Collins, R. A., L. S. Ko, K. L. So, T. Ellis, L. T. Lau, and A. C. Hoi Yu. 2003. "A NASBA method to detect high- and low-pathogenicity h5 avian influenza viruses." In *Avian Diseases*, 47, pp. 1069–74. American Association of Avian Pathologists. doi:10.1637/0005-2086-47.s3.1069.

Craun, M. F., G. F. Craun, R. L. Calderon, and M. J. Beach. 2006. "Waterborne outbreaks reported in the United States." *Journal of Water and Health* 4 (S2): 19–30. doi:10.2166/wh.2006.016.

Dalziel, A. E., S. Delean, S. Heinrich, and P. Cassey. 2016. "Persistence of low pathogenic influenza a virus in water: A systematic review and quantitative meta-analysis." Edited by Jonas Waldenström. *PLOS ONE* 11 (10). Public Library of Science: e0161929. doi:10.1371/journal.pone.0161929.

de Jong, M. D., B. van Cam, P. T. Qui, V. M. Hien, T. T. Thanh, N. B. Hue, M. Beld, et al. 2005. "Fatal avian influenza A (H5N1) in a child presenting with diarrhea followed by coma." *New England Journal of Medicine* 352 (7). Massachussetts Medical Society: 686–91. doi:10.1056/NEJMoa044307.

de Souza, F. G., F. P. da Silva, R. Staggemeier, C. Rigotto, and F. R. Spilki. 2018. "Low occurrence of hepatitis A virus in water samples from an urban area of Southern Brazil." *Revista Do Instituto de Medicina Tropical de Sao Paulo* 60. Instituto de Medicina Tropical de Sao Paulo. doi:10.1590/s1678-9946201860069.

Djikeng, A., R. Kuzmickas, N. G. Anderson, and D. J. Spiro. 2009. "Metagenomic analysis of RNA viruses in a freshwater lake." *PLoS ONE* 4 (9): e7264. doi:10.1371/journal.pone.0007264.

Donato, C., and D. Vijaykrishna. 2017. "The Broad Host Range and Genetic Diversity of Mammalian and Avian Astroviruses." *Viruses*. MDPI AG. doi:10.3390/v9050102.

Eischeid, A. C., J. A. Thurston, and K. G. Linden. 2011. "UV disinfection of adenovirus: Present state of the research and future directions." *Critical Reviews in Environmental Science and Technology* 41 (15): 1375–96. doi:10.1080/10643381003608268.

Elmahdy, E. M., G. Fongaro, C. D. Schissi, M. M. Petrucio, and C. R.M. Barardi. 2016. "Enteric viruses in surface water and sediment samples from the catchment area of Peri Lagoon, Santa Catarina State, Brazil." *Journal of Water and Health* 14 (1). IWA Publishing: 142–54. doi:10.2166/wh.2015.295.

Esper, F., Z. Ou, and Y. T. Huang. 2010. "Human coronaviruses are uncommon in patients with gastrointestinal illness." *Journal of Clinical Virology* 48 (2): 131–33. doi:10.1016/j.jcv.2010.03.007.

Farkas, K., D. I. Walker, E. M. Adriaenssens, J. E. McDonald, L. S. Hillary, S. K. Malham, and D. L. Jones. 2020. "Viral indicators for tracking domestic wastewater contamination in the aquatic environment." *Water Research* Elsevier Ltd. doi:10.1016/j.watres.2020.115926.

Farkas, T., K. Sestak, C. Wei, and X. Jiang. 2008. "Characterization of a rhesus monkey calicivirus representing a new genus of caliciviridae." *Journal of Virology* 82 (11). American Society for Microbiology: 5408–16. doi:10.1128/jvi.00070-08.

Fernandez-Cassi, X., N. Timoneda, E. Gonzales-Gustavson, J. F. Abril, S. Bofill-Mas, and R. Girones. 2017. "A metagenomic assessment of viral contamination on fresh parsley plants irrigated with fecally tainted river water." *International Journal of Food Microbiology* 257. Elsevier B.V.: 80–90. doi:10.1016/j.ijfoodmicro.2017.06.001.

Flanagan, S. V., R. B. Johnston, and Y.Zheng. 2012. "Aarsenic Dans l'eau Des Puits Tubulaires Au Bangladesh: Impacts Sanitaires et Économiques, et Implications En Vue de Sa Réduction." *Bulletin of the World Health Organization* 90 (11). World Health Organization: 839–46. doi:10.2471/BLT.11.101253.

Fong, T.-T., and E. K. Lipp. 2005. "Enteric viruses of humans and animals in aquatic environments: Health risks, detection, and potential water quality assessment tools." *Microbiology and Molecular Biology Reviews* 69 (2). American Society for Microbiology: 357–71. doi:10.1128/mmbr.69.2.357-371.2005.

Fongaro, G., J. Padilha, C. D. Schissi, M. A. Nascimento, G. B. Bampi, A. Viancelli, and C. R. M. Barardi. 2015. "Human and animal enteric virus in groundwater from deep wells, and recreational and network water." *Environmental Science and Pollution Research* 22 (24): 20060–66. doi:10.1007/s11356-015-5196-x.

Fratini, M., P. di Bonito, and G. La Rosa. 2014. "Oncogenic papillomavirus and polyomavirus in water environments: Is there a potential for waterborne transmission?" *Food and Environmental Virology* 6 (1): 1–12. doi:10.1007/s12560-013-9134-0.

Gall, A. M., B. J. Mariñas, Y. Lu, and J. L. Shisler. 2015. "Waterborne viruses: A barrier to safe drinking water." *PLoS Pathogens* 11 (6). Public Library of Science: e1004867. doi:10.1371/journal.ppat.1004867.

Ganesh, A., and J. Lin. 2013. "Waterborne human pathogenic viruses of public health concern." *International Journal of Environmental Health Research* 23 (6): 544–64. doi:10.1080/09603123.2013.769205.

García-Aljaro, C., E. Ballesté, M. Muniesa, and J. Jofre. 2017. "Determination of CrAssphage in water samples and applicability for tracking human faecal pollution." *Microbial Biotechnology* 10 (6). John Wiley and Sons Ltd: 1775–80. doi:10.1111/1751-7915.12841.

Gerba, C. P., and J. B. Rose. 1990. "Viruses in Source and Drinking Water." pp. 380–96. Springer, New York. doi:10.1007/978-1-4612-4464-6_18.

Gerba, C. P., G. Bitton - Groundwater Pollution Microbiology, John Wiley and, and undefined 1984. 2020. "Microbial Pollutants: Their Survival and Transport Pattern to Groundwater." Accessed October 1.

Gerba, C. P., J. B. Rose, C. N. Haas, and K. D. Crabtree. 1996. "Waterborne rotavirus: A risk assessment." *Water Research* 30: 2929–2940.

Gibson, K. E., M. C. Opryszko, J. T. Schissler, Y. Guo, and K. J. Schwab. 2011. "Evaluation of human enteric viruses in surface water and drinking water resources in Southern Ghana." *American Journal of Tropical Medicine and Hygiene* 84 (1). The American Society of Tropical Medicine and Hygiene: 20–29. doi:10.4269/ajtmh.2011.10-0389.

Goyal, S. M., and C. P. Gerba. 1979. "Comparative adsorption of human enteroviruses, simian rotavirus, and selected bacteriophages to soils." *Applied and Environmental Microbiology* 38 (2): 241–247.

Green, D. H., and G. D. Lewis. 1999. "Comparative detection of enteric viruses in wastewaters, sediments and oysters by reverse transcription-PCR and cell culture." *Water Research* 33 (5). Elsevier Sci Ltd: 1195–1200. doi:10.1016/S0043-1354(98)00313-3.

Griffin, J. S., J. D. Plummer, and S. C. Long. 2008. "Torque teno virus: An improved indicator for viral pathogens in drinking waters." *Virology Journal* 5 (1): 112.

Gundy, P. M., C. P. Gerba, and I. L. Pepper. 2009. "Survival of coronaviruses in water and wastewater." *Food and Environmental Virology* 1 (1). Springer: 10–14. doi:10.1007/s12560-008-9001–6.

Guo, Y., J. T. Schissler, M. C. Opryszko, K. E. Gibson, and K. J. Schwab. 2011. "Evaluation of human enteric viruses in surface water and drinking water resources in Southern Ghana." *The American Journal of Tropical Medicine and Hygiene* 84 (1): 20–29. doi:10.4269/ajtmh.2011.10-0389.

Hallam, N. B., J. R. West, C. F. Forster, and J. Simms. 2001. "The potential for biofilm growth in water distribution systems." *Water Research* 35 (17). Elsevier Ltd: 4063–71. doi:10.1016/S0043-1354(01)00248-2.

Hang, M. P., C. Nakajima, K. Ohashi, and M. Onuma. 2005. "Loop-mediated isothermal amplification for rapid detection of newcastle disease virus." *Journal of Clinical Microbiology* 43 (4). American Society for Microbiology Journals: 1646–50. doi:10.1128/JCM.43.4.1646-1650.2005.

Haramoto, E., H. Katayama, K. Oguma, and S. Ohgaki. 2005. "Application of cation-coated filter method to detection of noroviruses, enteroviruses, adenoviruses, and torque teno viruses in the Tamagawa river in Japan." *Applied and Environmental Microbiology* 71 (5): 2403–11. doi:10.1128/AEM.71.5.2403-2411.2005.

Haramoto, E., M. Kitajima, and M. Otagiri. 2013. "development of a reverse transcription-quantitative PCR assay for detection of salivirus/klassevirus." *Applied and Environmental Microbiology* 79 (11). American Society for Microbiology (ASM): 3529–32. doi:10.1128/AEM.00132-13.

Heerden, J., M. M. Ehlers, J. C. Vivier, and W. O. K. Grabow. 2005. "Risk assessment of adenoviruses detected in treated drinking water and recreational water." *Journal of Applied Microbiology* 99 (4): 926–33. doi:10.1111/j.1365-2672.2005.02650.x.

Hejkal, T. W., C. P. Gerba, and V. C. Rao. 1982. "Reduction of cytotoxicity in virus concentrates from environmental samples." *Applied and Environmental Microbiology* 43 (3). American Society for Microbiology (ASM): 731–33. doi:10.1128/aem.43.3.731-733.1982.

"Hepatitis E." 2020. Accessed October 17. https://www.who.int/en/news-room/fact-sheets/detail/hepatitis-e.

Ippagunta, S. K., S. Naik, B. Sharma, and R. Aggarwal. 2007. "Presence of hepatitis E virus in sewage in Northern India: Frequency and seasonal pattern." *Journal of Medical Virology* 79 (12): 1827–31. doi:10.1002/jmv.21017.

Isakbaeva, E. T., N. Khetsuriani, R. S. Beard, A. Peck, D. Erdman, S. S. Monroe, S. Tong, et al. 2004. "SARS-associated coronavirus transmission, United States." *Emerging Infectious Diseases* 10 (2). Centers for Disease Control and Prevention (CDC): 225–31. doi:10.3201/eid1002.030734.

Jain, P., and A. Jain. 2014. "Waterborne viral gastroenteritis: An introduction to common agents." In *Water and Health*, pp. 53–74. Springer India. doi:10.1007/978-81-322-1029-0_4.

Jevšnik, M., A. Steyer, T. Zrim, M. Pokorn, T. Mrvič, Š. Grosek, F. Strle, L. Lusa, and M. Petrovec. 2013. "Detection of human coronaviruses in simultaneously collected stool samples and nasopharyngeal swabs from hospitalized children with acute gastroenteritis." *Virology Journal* 10 (1): 46. doi:10.1186/1743-422X-10-46.

Johnson, J. H., J. E. Fields, and W. A. Darlington. 1967. "Removing viruses from water by polyelectrolytes." *Nature* 213 (5077): 665–67. doi:10.1038/213665a0.

Kauppinen, A., T. Pitkänen, H. Al-Hello, L. Maunula, A.-M. Hokajärvi, R. Rimhanen-Finne, and I. T. Miettinen. 2019. "Two drinking water outbreaks caused by wastewater intrusion including sapovirus in Finland." *International Journal of Environmental Research and Public Health* 16 (22). MDPI AG: 4376. doi:10.3390/ijerph16224376.

Kaur, M., S. K. Sidhu, K. Singh, P. Devi, M. Kaur, and N. J. Singh. 2017. "Hepatitis E virus: A leading cause of waterborne viral hepatitis in Northwest Districts of Punjab, India." *Journal of Laboratory Physicians* 9 (02). Georg Thieme Verlag KG: 121–24. doi:10.4103/0974-2727.199636.

Keswick, B. H., C. P. Gerba, H. L. Dupont, and J. B. Rose. 1984. "detection of enteric viruses in treated drinking water." *Applied and Environmental Microbiology* 47. http://aem.asm.org/.

Kopecka, H., S. Dubrou, J. Prevot, J. Marechal, and J. M. Lopez-Pila. 1993. "Detection of naturally occurring enteroviruses in waters by reverse transcription, polymerase chain reaction, and hybridization." *Applied and Environmental Microbiology* 59 (4). American Society for Microbiology (ASM): 1213–19. doi:10.1128/aem.59.4.1213-1219.1993.

Koroglu, M., Y. Yakupogullari, B. Otlu, S. Ozturk, M. Ozden, A. Ozer, K. Sener, and R. Durmaz. 2011. "A waterborne outbreak of epidemic diarrhea due to group a rotavirus in Malatya, Turkey." *New Microbiologica* 34.

Kumar, M., A. K. Thakur, P. Mazumder, K. Kuroda, S. Mohapatra, J. Rinklebe, ..., and P. Bhattacharya. 2020a. Frontier review on the propensity and repercussion of SARS-CoV-2 migration to aquatic environment. *Journal of Hazardous Materials Letters*, 1, 100001.

Kumar, M., K. Kuroda, A. K. Patel, N. Patel, P. Bhattacharya, M. Joshi, and C. G. Joshi. 2021. Decay of SARS-CoV-2 RNA along the wastewater treatment outfitted with Upflow Anaerobic Sludge Blanket (UASB) system evaluated through two sample concentration techniques. *Science of the Total Environment*, 754, 142329.

Kumar, M., P. Mazumder, S. Mohapatra, A. K. Thakur, K. Dhangar, K. Taki, S. Mukherjee, et al. 2020b. "A Chronicle of SARS-CoV-2: Seasonality, environmental fate, transport, inactivation, and antiviral drug resistance." *Journal of Hazardous Materials*, October. Elsevier B.V.: 124043. doi:10.1016/j.jhazmat.2020.124043.

La Rosa, G., L. Bonadonna, L. Lucentini, S. Kenmoe, and E. Suffredini. 2020. "Coronavirus in water environments: Occurrence, persistence and concentration methods - A scoping review." *Water Research* Elsevier Ltd. doi:10.1016/j.watres.2020.115899.

La Rosa, G., M. Fratini, L. Accardi, G. D'Oro, S. della Libera, M. Muscillo, and P. di Bonito. 2013. "Mucosal and cutaneous human papillomaviruses detected in raw sewages." *PLoS ONE* 8 (1). Public Library of Science: 52391. doi:10.1371/journal.pone.0052391.

La Rosa, G., M. Fratini, S. della Libera, M. Iaconelli, and M. Muscillo. 2012. "Emerging and potentially emerging viruses in water environments." *Annali Dell'Istituto Superiore Di Sanita* 48 (4). Istituto Superiore di Sanita: 397–406. doi:10.4415/ANN_12_04_07.

La Rosa, G., M. Iaconelli, P. Mancini, G. B. Ferraro, C. Veneri, L. Bonadonna, L. Lucentini, and E. Suffredini. 2020. "First Detection of Sars-Cov-2 in Untreated Wastewaters in Italy." *MedRxiv*.

Lanciotti, R. S., and A. J. Kerst. 2001. "Nucleic acid sequence-based amplification assays for rapid detection of west nile and st. louis encephalitis viruses." *Journal of Clinical Microbiology* 39 (12). American Society for Microbiology Journals: 4506–13. doi:10.1128/JCM.39.12.4506-4513.2001.

Leblanc, D., M. J. Gagné, É. Poitras, and J. Brassard. 2019. "Persistence of murine norovirus, bovine rotavirus, and hepatitis A virus on stainless steel surfaces, in spring water, and on blueberries." *Food Microbiology* 84 (December). Academic Press. doi:10.1016/j.fm.2019.103257.

Leclerc, H., and A. Moreau. 2020. "Microbiological Safety of Natural Mineral Water." Accessed October 1. doi:10.1111/j.1574-6976.2002.tb00611.x.

Le Guyader, F., L. Haugarreau, L. Miossec, E. Dubois, and M. Pommepuy. 2000. "Three-Year study to assess human enteric viruses in shellfish." *Applied and Environmental Microbiology* 66 (8): 3241–48. doi:10.1128/AEM.66.8.3241-3248.2000.

Lee, S. H., and S. J. Kim. 2002. "Detection of infectious enteroviruses and adenoviruses in tap water in Urban Areas in Korea." *Water Research* 36 (1). Elsevier Ltd: 248–56. doi:10.1016/S0043-1354(01)00199-3.

Lehtola, M. J., I. T. Miettinen, M. M. Keinänen, T. K. Kekki, O. Laine, A. Hirvonen, T. Vartiainen, and P. J. Martikainen. 2004. "Microbiology, chemistry and biofilm development in a pilot drinking water distribution system with copper and plastic pipes." *Water Research* 38 (17). Elsevier Ltd: 3769–79. doi:10.1016/j.watres.2004.06.024.

Liu, L., H. L. Johnson, J. Perin, M. Li, R.E. Black, L. Liu, et al. 2012. "Global, regional, and national causes of child mortality: An updated systematic analysis for 2010 with time trends since 2000." *The Lancet* 379: 2151–61. doi:10.1016/S0140.

Lodder, W. J., and A. M. de Roda Husman. 2005. "Presence of noroviruses and other enteric viruses in sewage and surface waters in The Netherlands." *Applied and Environmental Microbiology* 71 (3). American Society for Microbiology (ASM): 1453–61. doi:10.1128/AEM.71.3.1453-1461.2005.

Malik, Y. S., N. Kumar, K. Sharma, K. Dhama, M. Z. Shabbir, B. Ganesh, N. Kobayashi, and K. Banyai. 2014. "Epidemiology, phylogeny, and evolution of emerging enteric *Picobirnaviruses* of animal origin and their relationship to human strains." *BioMed Research International* 2014: 1–13. doi:10.1155/2014/780752.

Malik, Y. S., S. Sircar, S. Saurabh, J. J. Kattoor, R. Singh, B. Ganesh, S. Ghosh, K. Dhama, and R. K. Singh. 2018. "Epidemiologic status of picobirnavirus in India, A less explored viral disease." *The Open Virology Journal* 12 (1): 99–109. doi:10.2174/1874357901812010099.

Masachessi, G., M. B. Pisano, V. E. Prez, L. C. Martínez, J. F. Michelena, M. Martínez-Wassaf, M. O. Giordano, et al. 2018. "Enteric viruses in surface waters from Argentina: Molecular and viablevirus detection." *Applied and Environmental Microbiology* 84 (5). American Society for Microbiology. doi:10.1128/AEM.02327-17.

Mazumder, P., S. K. Sharma, K. Taki, A. S. Kalamdhad, and M. Kumar. 2020. "Microbes involved in arsenic mobilization and respiration: A review on isolation, identification, isolates and implications". *Environmental Geochemistry and Health* 42: 1–27.

Mukherjee, S., P. Mazumder, M. Joshi, C. Joshi, S. V. Dalvi, and M. Kumar. 2020. "Biomedical application, drug delivery and metabolic pathway of antiviral nanotherapeutics for combating viral pandemic: A review". *Environmental Research*, 191, 110119.

McCall, C., H. Wu, B. Miyani, and I. Xagoraraki. 2020. "Identification of multiple potential viral diseases in a large urban center using wastewater surveillance." *Water Research* 184 (October). Elsevier Ltd. doi:10.1016/j.watres.2020.116160.

McKinney, K., Y. Y. Gong, and T. Lewis. 2006. "Environmental Transmission of SARS at Amoy Gardens Background."

Medema, G., L. Heijnen, G. Elsinga, R. Italiaander, and A. Brouwer. 2020. "Presence of SARS-Coronavirus-2 in Sewage." *MedRxiv*. Cold Spring Harbor Laboratory Press. doi:10.1101/2020.03.29.20045880.

Mena, K. D. & Gerba, C. P. Waterborne Adenovirus. Reviews of Environmental Contamination and Toxicology 198, 133–167 (2008).

Mena, K. D., and C. P. Gerba. 2009 "Waterborne Adenovirus." In *Reviews of Environmental Contamination and Toxicology Volume 198*, pp. 1–35. doi:10.1007/978-0-387-09647-6_4.

Miettinen, I. T., O. Lepistö, T. Pitkänen, M. Kuusi, L. Maunula, J. Laine, J. Ikonen, and M.-L. Hänninen. 2012. "A waterborne outbreak caused by a severe faecal contamination of distribution network: Nokia case." In pp. 34–37. doi:10.1039/9781849735421-00034.

Mohapatra, S., N. G. Menon, G. Mohapatra, L. Pisharody, A. Pattnaik, N. G. Menon, P. L. Bhukya, M. Srivastava, M. Singh, M. K. Barman, K. Y. Gin, 2020. "The novel SARS-CoV-2 pandemic: Possible environmental transmission, detection, persistence and fate during wastewater and water treatment". *Science of the Total Environment* 6: 142746.

Mohapatra, S., A. Sargaonkar, and P. K. Labhasetwar, 2014. "Distribution network assessment using EPANET for intermittent and continuous water supply". *Water Resources Management*, 28 (11): 3745–3759.

Naik, S. R., R. Aggarwal, P. N. Salunke, and N. N. Mehrotra. 1992. "A large waterborne viral hepatitis e epidemic in Kanpur, India." *Bulletin of the World Health Organization* 70 (5). World Health Organization: 597–604. doi:10.1007/bf02821726.

Nims, R. W., and M. Plavsic. 2013. "Polyomavirus inactivation - A review." *Biologicals*. Academic Press. doi:10.1016/j.biologicals.2012.09.011.

O'Brien, E., J. Nakyazze, H. Wu, N. Kiwanuka, W. Cunningham, J. B. Kaneene, and I. Xagoraraki. 2017. "Viral diversity and abundance in polluted waters in Kampala, Uganda." *Water Research* 127. Elsevier Ltd: 41–49. doi:10.1016/j.watres.2017.09.063.

Opere, W. M., M. John, and O. Ombori. 2020. "Occurrence of enteric viruses in surface water and the relationship with changes in season and physical water quality dynamics." *Advances in Virology* 2020. Hindawi Limited. doi:10.1155/2020/9062041.

"Orthomyxoviridae - Negative Sense RNA Viruses - Negative Sense RNA Viruses (2011) - International Committee on Taxonomy of Viruses (ICTV)." 2020. Accessed October 6. https://talk.ictvonline.org/ictv-reports/ictv_9th_report/negative-sense-rna-viruses-2011/w/negrna_viruses/209/orthomyxoviridae.

Pandey, P. K., P. H. Kass, M. L. Soupir, S. Biswas, and V. P. Singh. 2014. "Contamination of water resources by pathogenic bacteria." *AMB Express* 4 (1). Springer Verlag: 1–16. doi:10.1186/s13568-014-0051-x.

Patel, M. M., A. J. Hall, J. Vinjé, and U. D. Parashar. 2009. "Noroviruses: A comprehensive review." *Journal of Clinical Virology*. doi:10.1016/j.jcv.2008.10.009.

Pinon, A., and M. Vialette. 2018. "Survival of viruses in water." *Intervirology* 61 (5): 214–22. doi:10.1159/000484899.

Prow, N. A., M. G. Mah, J. M. Deerain, D. Warrilow, A. M. G. Colmant, C. A. O'Brien, J. J. Harrison, et al. 2018. "New Genotypes of Liao Ning Virus (LNV) in Australia exhibit an insect-specific phenotype." *Journal of General Virology* 99 (4): 596–609. doi:10.1099/jgv.0.001038.

Radin, D. 2014. "New trends in food- and waterborne viral outbreaks." *Archives of Biological Sciences* 66 (1): 1–9. doi:10.2298/ABS1401001R.

Rames, A. 2020. "Reviewing seadornaviruses: The next dengue?" *Transactions on Science and Technology* 7. http://tost.unise.org/.

Reuter, G., P. Pankovics, and Á. Boros. 2017. "Saliviruses-the first knowledge about a newly discovered human picornavirus." *Reviews in Medical Virology* 27 (1). John Wiley and Sons Ltd: e1904. doi:10.1002/rmv.1904.

Risku, M., S. Lappalainen, S. Räsänen, and T. Vesikari. 2010. "Detection of human coronaviruses in children with acute gastroenteritis." *Journal of Clinical Virology* 48 (1): 27–30. doi:10.1016/j.jcv.2010.02.013.

Roden, R. B. S., D. R. Lowy, and J. T. Schiller. 1997. "Papillomavirus is resistant to desiccation." *Journal of Infectious Diseases* 176 (4). Oxford University Press: 1076–79. doi:10.1086/516515.

Rodríguez-Lázaro, D., N. Cook, F. M. Ruggeri, J. Sellwood, A. Nasser, M. S. J. Nascimento, M. D'Agostino, et al. 2012. "Virus hazards from food, water and other contaminated environments." *FEMS Microbiology Reviews* 36 (4): 786–814. doi:10.1111/j.1574-6976.2011.00306.x.

Rose, J. B., C. P. Gerba, S. N. Singh, G. A. Toranzos, and B. Keswick. 1986. "Isolating viruses from finished water." *Journal/American Water Works Association* 78 (1). American Water Works Association: 56–61. https://arizona.pure.elsevier.com/en/publications/isolating-viruses-from-finished-water.

Ruggeri, F. M., and L. Fiore. 2012. "Vaccine preventable viral diseases and risks associated with waterborne transmission." *Annali Dell'Istituto Superiore Di Sanita* 48 (4). Istituto Superiore di Sanita: 460–72. doi:10.4415/ANN_12_04_12.

Rusiñol, M., and R. Girones. 2019. "Summary of excreted and waterborne viruses." In *Global Water Pathogen Project*. Michigan State University. doi:10.14321/waterpathogens.19.

Rutjes, S. A., H. H. J. L. van den Berg, W. J. Lodder, and A. M. de Roda Husman. 2006. "Real-time detection of noroviruses in surface water by use of a broadly reactive nucleic acid sequence-based amplification assay." *Applied and Environmental Microbiology* 72 (8). American Society for Microbiology (ASM): 5349–58. doi:10.1128/AEM.00751-06.

Seitz, S. R., J. S. Leon, K. J. Schwab, G. Marshall Lyon, M. Dowd, M. McDaniels, G. Abdulhafid, et al. 2011. "Norovirus infectivity in humans and persistence in water." *Applied and Environmental Microbiology* 77 (19). American Society for Microbiology (ASM): 6884–88. doi:10.1128/AEM.05806-11.

Sharma, N., S. Mohapatra, L. P. Padhye, and S. Mukherji, 2021. "Role of precursors in the formation of trihalomethanes during chlorination of drinking water and wastewater effluents from a metropolitan region in western India." *Journal of Water Process Engineering* 40, 101928.

Shuval, H. I., B. F. S. Cymbalista, and N. Goldblum. 1969. "The phase-separation method for the concentration and detection of viruses in water." *Water Research* 3 (4): 225–40. doi:10.1016/0043-1354(69)90019-0.

Silva, H. D., M. T. A. García-Zapata, and C. E. Anunciação. 2011. "Why the use of adenoviruses as water quality virologic marker?" *Food and Environmental Virology* 3 (3–4): 138–40. doi:10.1007/s12560-011-9069-2.

Singh, A., A. K. Patel, J. P. Deka, A. Das, A. Kumar, and M. Kumar. 2019. "Prediction of arsenic vulnerable zones in the groundwater environment of a rapidly urbanizing setup, Guwahati, India". *Geochemistry*, 80: 125590.
Sinclair, R. G., E. L. Jones, and C. P. Gerba. 2009. "Viruses in recreational water-borne disease outbreaks: A Review." *Journal of Applied Microbiology* 107 (6): 1769–80. doi:10.1111/j.1365-2672.2009.04367.x.
Skraber, S., J. Schijven, C. Gantzer, and A. M. de Roda Husman. 2005. "Pathogenic viruses in drinking-water biofilms: A public health risk?" *Biofilms*. doi:10.1017/S1479050505001833.
Smits, S. L., E. E. Zijlstra, J. J. van Hellemond, C. M. E. Schapendonk, R. Bodewes, A. C. Schürch, B. L. Haagmans, and A. D. M. E. Osterhaus. 2013. "Novel cyclovirus in human cerebrospinal fluid, Malawi, 2010–2011." *Emerging Infectious Diseases* 19 (9). Centers for Disease Control and Prevention: 1511–13. doi:10.3201/eid1909.130404.
Soule, H., O. Genoulaz, B. Gratacap-Cavallier, P. Chevallier, J. X. Liu, and J. M. Seigneurin. 2000. "Ultrafiltration and reverse transcription-polymerase chain reaction: An efficient process for poliovirus, rotavirus and hepatitis A virus detection in water." *Water Research* 34 (3). Elsevier Science Ltd: 1063–67. doi:10.1016/S0043-1354(99)00197-9.
Steininger, C., M. Kundi, S. W. Aberle, J. H. Aberle, and T. Popow-Kraupp. 2002. "Effectiveness of reverse transcription-PCR, Virus isolation, and enzyme-linked immunosorbent assay for diagnosis of influenza a virus infection in different age groups." *Journal of Clinical Microbiology* 40 (6). American Society for Microbiology (ASM): 2051–56. doi:10.1128/JCM.40.6.2051-2056.2002.
Suttle, C. A. 2005. "Viruses in the sea." *Nature* 437 (7057): 356–61. doi:10.1038/nature04160.
Symonds, E. 2008. "Viruses found in raw sewage and their potential to indicate fecal pollution in coastal environments." *Graduate Theses and Dissertations*, June. https://scholarcommons.usf.edu/etd/522.
Tabraiz, S., S. Nasreen, L. A. Qureshi, M. Zeeshan, I. Ahmad, S. Ahmed, and Z. Hassan. 2016. "Sewage land disposal and unpaved drains: Threat to groundwater quality." *Desalination and Water Treatment* 57 (43). Taylor and Francis Inc.: 20464–69. doi:10.1080/19443994.2015.1110723.
Tiwari, S., and T. N. Dhole. 2018. "Assessment of enteroviruses from sewage water and clinical samples during eradication phase of polio in North India." *Virology Journal* 15 (1). BioMed Central Ltd.: 157. doi:10.1186/s12985-018-1075-7.
Todd, D. 2000. "Circoviruses: Immunosuppressive threats to avian species: A review." *Avian Pathology*. Carfax Publishing Company. doi:10.1080/030794500750047126.
Vabret, A., J. Dina, S. Gouarin, J. Petitjean, S. Corbet, and F. Freymuth. 2006. "Detection of the new human coronavirus HKU1: A report of 6 cases." *Clinical Infectious Diseases* 42 (5): 634–39. doi:10.1086/500136.
Vaidya, S. R., S. D. Chitambar, and V. A. Arankalle. 2002. "Polymerase chain reaction-based prevalence of Hepatitis A, Hepatitis E and TT viruses in sewage from an endemic area." *Journal of Hepatology* 37 (1): 131–36. doi:10.1016/S0168-8278(02)00106-X.
van der Poel, W.H.M. 2014. "Food and environmental routes of hepatitis e virus transmission." *Current Opinion in Virology*. Elsevier B.V. doi:10.1016/j.coviro.2014.01.006.
van Leeuwen, M., M. M. W. Williams, P. Koraka, J. H. Simon, S. L. Smits, and A. D. M. E. Osterhaus. 2010. "Human picobirnaviruses identified by molecular screening of diarrhea samples." *Journal of Clinical Microbiology* 48 (5): 1787–94. doi:10.1128/JCM.02452-09.
Vong, S., S. Ly, M. D. van Kerkhove, J. Achenbach, D. Holl, P. Buchy, S. Sorn, et al. 2009. "Risk factors associated with subclinical human infection with avian influenza A (H5N1) Virus—Cambodia, 2006." *The Journal of Infectious Diseases* 199 (12). Oxford Academic: 1744–52. doi:10.1086/599208.
Vong, S., S. Ly, S. Mardy, D. Holl, and P. Buchy. 2008. "Environmental contamination during influenza a virus (H5N1) Outbreaks, Cambodia, 2006." *Emerging Infectious Diseases* 14 (8): 1303–5. doi:10.3201/eid1408.070912.

Vu, D. L., A. Bosch, R. M. Pintó, and S. Guix. 2017. "Epidemiology of classic and novel human astrovirus: Gastroenteritis and beyond." *Viruses*. MDPI AG. doi:10.3390/v9020033.

Wallis, C., and J. L. Melnick. 1962. "Magnesium chloride enhancement of cell susceptibility to poliovirus." *Virology* 16 (2): 122–32. doi:10.1016/0042-6822(62)90287-8.

Wang, X. W., J. S. Li, M. Jin, B. Zhen, Q. X. Kong, N. Song, W. J. Xiao, et al. 2005a. "Study on the resistance of severe acute respiratory syndrome-associated coronavirus." *Journal of Virological Methods* 126 (1–2). Elsevier: 171–77. doi:10.1016/j.jviromet.2005.02.005.

Wang, X.-W., J.-S. Li, T.-K. Guo, B. Zhen, Q.-X. Kong, B. Yi, Z. Li, et al. 2005b. "Excretion and detection of SARS coronavirus and its nucleic acid from digestive system." *World Journal of Gastroenterology* 11 (28): 4390. doi:10.3748/wjg.v11.i28.4390.

"WHO | Guidelines for the Safe Use of Wastewater, Excreta and Greywater - Volume 2." 2016. *WHO*. World Health Organization. http://www.who.int/water_sanitation_health/publications/gsuweg2/en/.

"WHO | Water Quality and Health Strategy 2013–2020." 2016. *WHO*. World Health Organization. http://www.who.int/water_sanitation_health/publications/water_quality_strategy/en/.

Wigginton, K. R., Y. Ye, and R. M. Ellenberg. 2015. "Emerging investigators series: The source and fate of pandemic viruses in the urban water cycle." *Environmental Science: Water Research and Technology*. Royal Society of Chemistry. doi:10.1039/c5ew00125k.

Wu, F., A. Xiao, J. Zhang, X. Gu, W. L. Lee, K. Kauffman, W. Hanage, et al. 2020. "SARS-CoV-2 titers in wastewater are higher than expected from clinically confirmed cases." *MedRxiv*, April. Cold Spring Harbor Laboratory Press. doi:10.1101/2020.04.05.20051540.

Xagoraraki, I., Z. Yin, and Z. Svambayev. 2014. "Fate of viruses in water systems." *Journal of Environmental Engineering (United States)*. American Society of Civil Engineers (ASCE). doi:10.1061/(ASCE)EE.1943-7870.0000827.

Yang, B. Y., X. L. Liu, Y. M. Wei, J. Q. Wang, X. Q. He, Y. Jin, and Z. J. Wang. 2014. "Rapid and sensitive detection of human astrovirus in water samples by loop-mediated isothermal amplification with hydroxynaphthol blue dye." *BMC Microbiology* 14 (1). doi:10.1186/1471-2180-14-38.

Yates, M. V. 2013. "Astroviruses." In *Microbiology of Waterborne Diseases: Microbiological Aspects and Risks: Second Edition*, pp. 479–91. Elsevier Ltd. doi:10.1016/B978-0-12-415846-7.00024–X.

Yates, M. V., C. P. Gerba, and L. M. Kelley. 1985. "Virus persistence in groundwater." *Applied and Environmental Microbiology*. http://aem.asm.org/.

Ye, X. Y., X. Ming, Y. L. Zhang, W. Q. Xiao, X. N. Huang, Y. G. Cao, and K. D. Gu. 2012. "Real-time PCR detection of enteric viruses in source water and treated drinking water in Wuhan, China." *Current Microbiology* 65 (3): 244–53. doi:10.1007/s00284-012-0152-1.

Zhai, S. L., S. S. Lu, W. K. Wei, D. H. Lv, X. H. Wen, Q. Zhai, Q. L. Chen, Y. W. Sun, and Y. Xi. 2019. "Reservoirs of porcine circoviruses: A mini review." *Frontiers in Veterinary Science*. Frontiers Media S.A. doi:10.3389/fvets.2019.00319.

Chapter 8

Microbial indicators and methods for source tracking faecal contamination of groundwater

Salonee Martins
National Institute of Virology
ACTREC, Tata Memorial Centre

Aishwarya Pathare, Purva Salvi, Unnati Bhalerao, and Mahalaxmi U Bhat
National Institute of Virology

Meenakshi Singh
National Institute of Virology
ACTREC, Tata Memorial Centre

Muneesh Kumar Barman and Kailash Chand
National Centre for Cell Sciences

Sai Tejaswi Lavuri
Chalmeda Ananda Rao Institute of Medical Sciences

Nitish Venkateswarlu Mogili
National Institute of Technology Andhra Pradesh

Sanjeeb Mohapatra
National University of Singapore

Umesh Kumar
IMS Ghaziabad University Courses Campus

Ujjawal Sharma
Postgraduate Institute of Medical Education and Research

Manjita Srivastava and Prudhvilal Bhukya
National Institute of Virology
ACTREC, Tata Memorial Centre

DOI: 10.1201/9781003157465-10

CONTENTS

8.1 INTRODUCTION

The contamination of water bodies utilized for drinking, domestic uses, recreation and industrial activities with faeces of human or animal origin poses a great risk to public health and the economy (Mohapatra et al., 2014, 2020; Pantha et al., 2021). Potential zoonotic enteric pathogens as contaminants of water bodies include *E. coli*, *Salmonella spp.* and *Cryptosporidium spp.* The occurrence of enteric pathogens like *Shigella spp.*, *Salmonella spp.* and *Serovar typhi* and viruses such as the Norwalk-group of viruses and hepatitis A virus is highly associated with faecal pollution (Scott et al., 2002). Groundwater constitutes a major share of up to 95% of the world's usable freshwater (Guy et al., 2006) and is habitually used with no treatment (Pedley et al., 2006). Groundwater is essential for potable needs, and the presence of human pathogens in groundwater raises the great risk of human exposure to these harmful microorganisms and the spread of related waterborne diseases (Xagoraraki et al., 2014). Groundwater contamination with viruses, particularly human enteric viruses, is a critical issue in various parts of the world (Gotkowitz et al., 2016; Hynds et al., 2014; Kumar et al., 2020a,b) as consumption of such water can result in enteric illnesses and outbreaks (Beer et al., 2015; Borchardt et al., 2011; CHO et al., 2014; Fout et al., 2017; Kumar et al., 2021). To improve the water quality, the need of the hour is to upgrade the wastewater treatment plant (WWTP) technologies and urban sustainable drainage technologies (SuDs). Sustainability can be achieved by enforcing best management practices

(BMPs), minimizing the pollution caused by livestock farming and enhancing facilities to retain increased rainwater amounts (Rusiñol et al., 2014).

8.2 IMPORTANCE OF PROTECTING GROUNDWATER RESOURCES

Point sources like the discharge from sewage waste treatment plants and slaughterhouses are potentially increasing the levels of faecal pollution in surface water bodies (Collins et al., 2005). Other sources of faecal pollution can be traced back to urban surface water run-off from roofs or roads contaminated by avian, wild or domestic animal faeces or from cross-connecting urban sewage systems, voiding of animals in open waters or manure spreading, all of which have shown to contain high loads of microbes (Brownell et al., 2007; Rusiñol et al., 2014; Sercu et al., 2011, Thakur et al., 2020).

Despite the public health risks that can potentially impact human life, regulations controlling the quality of groundwater used as potable water vary significantly in different countries. Most of them abide by the WHO Guidelines for Drinking Water Quality (WHO | Guidelines for drinking-water quality, fourth edition, 2015). This mentions all potable water sources and suggests that *E. coli* per 100 ml or any kind of thermostable coliform bacteria should be completely absent from any drinking water resource. Canada has issued similar guidelines mentioning the water should not contain any detectable amounts of coliforms or *E. coli* per 100 ml (Guidelines for Canadian Drinking Water Quality - Summary Table - Canada.ca, n.d.). The Australian regulative authority for water has decreed to include coliphages and intestinal enterococci in addition to commonly tested faecal coliform indicators that should not be detected per 100 ml of water to preserve the potable quality of water (Australian Drinking Water Guidelines | NHMRC, n.d.). In some countries, monitoring of the bacteriological quality of drinking water resources is mandatory. For example, Japan has mandated screening for *E. coli* (Wakayama and 厚生労働省, n.d.). In Korea, a periodic check is carried out for enteric viruses, and they have specific legislation called the Groundwater Act that ensures mandatory testing and monitoring of *E. coli* in groundwater (Lee and Kwon, 2016; Kumar et al., 2020a,b). The Ground Water Rule (GWR) issued by the US regulatory body ensures that groundwater sources with faecal contamination are identified to implement corrective actions. Furthermore, source water monitoring and compliance monitoring are important components of GWR. Source monitoring is essential when the source of water and its quality are uncertain. Triggered monitoring is required to ascertain that none of the following faecal indicators *E. coli*, coliphages and enterococci are present in the groundwater source. The GWR also suggests that aquifers in gravel, limestone, igneous and metamorphic rock settings are potentially susceptible to contamination. In the case of gravel aquifers, the potential for contamination is higher in gravel/cobble aquifers than gravel/sand aquifers because of the grain size. Gravel/sand setting has grain sizes of 4 mm or larger, while the grains in gravel/cobble have sizes up to 256 mm (A Scale of Grade and Class Terms for Clastic Sediments on JSTOR, n.d.; Fout et al., 2017). The viral agents that are often reported in connection with disease outbreaks via groundwater contamination are Norwalk-like viruses and hepatitis A virus (de Serres et al., 1999; Lawson et al., 1991).

Monitoring groundwater for enteric viruses among public water systems has recently become the focus of many studies (Dixon, 2003). However, the real risk lies with the use of private wells, which are neither well-maintained nor tested routinely. Even though the distance between a domestic well and the nearest septic tank is regulated by most states, it may still lead to contamination of the well due to overloading of waste in the proximity of the well (Staley et al., 2018). In most countries, a major amount of potable water is sourced from underground water sources, and in the event of contamination, there would be increased health and economic burden on the community (Diston et al., 2015). With respect to water management, karst aquifers are of immense significance, since these are mostly found in areas with limestone and are vulnerable to contamination by faecal pollution carried with rainfall (Diston et al., 2015; Pochon et al., 2008). High permeability of the limestone helps in the rapid percolation of rainwater (Metcalfe et al., 2011), as a result of which there is easy transport of contaminants throughout the system. Groundwater derived from these aquifers is an essential resource of potable water in various parts of the world (Ford and Williams, 2007). Unfortunately, in contrast to other types of aquifers, karst springs display a strong time-based vulnerability with an amplified weakness to microbiological and chemical contamination (Heinz et al., 2009). Sparsely covered soil layer and the highly permeable local routes such as open fractures, sinkhole drainages and sinking streams with inconsequential particle removal of incoming water make it easy for the contaminants to percolate into groundwater. The conduit networks present inside the aquifers help in the rapid transport of contaminants, where attenuation is almost ineffective than in other aquifers (Karst Hydrogeology and Human Activities: Impacts, Consequences and Imp, n.d.).

Research studies have shown that viruses can travel through soil layers and contaminate groundwater during land applications (Keswick and Gerba, 1980). Viruses have been known to migrate up to 64 m into the subsurface and horizontally traverse up to 408 m from contamination sources (Gerba, 1999). The factors that influence virus removal include the nature of virus, characteristics of the soil and availability of cations and soluble organic matter. In general, the removal of viruses in sandy soil setting is low because of the absence of clay (Moore et al., 1982). The removal of viruses from the soil is believed to be possible by adsorption on the soil surface (Lance and Gerba, 1984). However, the viruses cannot be held immobilized for long and may eventually elute out from the soil, along with rainwater, and enter groundwater sources nearby (Gerba and Lance, 1978). The introduction of viruses into the subsurface can occur by various methods, like landfills, artificial recharge of groundwater, crop irrigation with treated wastewater and septic tank seepage. Burying disposable diapers in sanitary landfills may introduce untreated sewage into groundwater, and this is particularly hazardous in the case of infected faeces, which may contain up to 10^6 plaque-forming units (PFUs) of enteroviruses/g and 10^{10} rotaviruses/g. The enteric viruses of human importance are enteroviruses, rotaviruses, adenoviruses, reoviruses and Norwalk-like viruses (Abbaszadegan et al., 1999). Factors controlling the migration and survival of viruses in groundwater are varied, viz. the climatic conditions (e.g. heat and rainfall), the nature and composition of the soil (e.g. clay content and moisture-holding capacity) and the type of virus. Migration of viruses is possible for considerable distances in the subsurface areas, as mentioned before (Keswick and Gerba, 1980). This ability of viruses enables them to enter the resources used for drinking purposes and cause waterborne diseases outbreaks (Gerba, 1999). The practice of land application of wastewater, subsequent contamination with viruses and their prolonged persistence raised the concern of viruses as the leading cause of groundwater-related morbidity in the USA. The rules in place for the construction of a domestic well, regarding septic tanks and other potential contamination sources, do not consider the presence of treated effluents (Yates et al., 1985).

Contamination of groundwater by microorganisms is a serious threat to society and can lead to the occurrence of epidemics caused by waterborne diseases. For example, it was reported that between 1946 and 1977 a lack of or the improper treatment of groundwater led to the occurrence of 264 disease epidemics with a total of 62,273 cases (Colford et al., 2007), indicating that contaminated water is a serious public health concern. Besides the threat posed to human health by waterborne pathogens, other types of contaminants like pollutants discharged by industries, organic matter and nutrients have adverse impact on the surrounding environment and could lead to eutrophication, depletion of oxygen content, disruption of functioning of biological ecosystems and biomagnification that tend to have a huge economic impact on shellfish harvesting (Colford et al., 2007; Gourmelon et al., n.d.). An estimate of 663 million of the world's total population does not have access to potable water resources (Water and Sanitation – United Nations Sustainable Development, n.d.). In developing and low-income countries, improper sanitation facilities and improper waste management practices result in water resources contaminated with human or animal faeces (Schriewer et al., 2015). According to WHO, reliable approaches are needed for environmental monitoring, which is crucial for the evaluation of the effects of various components on the microbiological quality of drinking water (WHO | Guidelines for drinking-water quality, fourth edition, 2015).

8.3 MICROBIAL INDICATORS OF FAECAL POLLUTION

The age-old practice of using indicator microorganisms to detect faecal contamination of water by humans and animals is well-known. The efficiency of using conventional microbiological indicators to deduce the effect of faecal wastes of animal and human origin, and prediction of subsequent health risks upon consumption of contaminated products is limited as these microbes are ubiquitous (Scott et al., 2002). Testing for the presence of certain microorganisms, known as indicator microorganisms, is helpful in reducing the risk posed by pathogenic microorganisms. The detection of pathogenic microorganisms circumvents the need to detect individual pathogenic agents (Scott et al., 2002). Indicator organisms are mostly non-pathogenic, easily enumerated, quickly detected, and whose survival rates and environmental conditions required for the persistence are similar to pathogens. It is due to these reasons that the presence of these indicator organisms is strongly correlated with the presence of waterborne pathogens (Scott et al., 2002). Table 8.1 enlists the characteristics displayed by an ideal microbiological indicator organism for faecal contamination of water.

Table 8.1 Characteristics of an ideal microbial indicator of contamination (Fewtrell et al., 2001)

1. Must be a member of the intestinal microbiota of warm-blooded animals.
2. Must be present in the presence of its pathogenic counterparts and absent in uncontaminated samples.
3. Must be present in higher concentrations than the pathogen.
4. Must be at least equally resistant as the pathogen to environmental conditions and disinfection protocols of wastewater treatment plants.
5. Must not be able to multiply in the environment.
6. Must be easily detectable by relatively rapid and inexpensive detection techniques.
7. Must be non-pathogenic.

The detection of various microorganisms that are part of the normal human intestinal microbiota and that of other animals, which are mostly non-pathogenic, is possible using simple and inexpensive techniques (Fout et al., 2017). Owing to irregular occurrence and difficulty in detecting most pathogenic organisms, especially viruses, indicator microorganisms help in risk assessment and detecting faecal pollution (Fout et al., 2017). The most popular indicators of faecal contamination are coliforms and *E. coli*, with *E. coli* as a more effective indicator (Edberg et al., 2000). Bacterial indicators are proven to be effective testers for monitoring the bacteriological water quality; they are, however, regarded inadequate as indicators of viral contamination (Fout et al., 2017). This is because the characteristics such as survival, size, structure, composition and morphology of a virus are very different from those of a bacterium (Payment and Locas, 2011; WHO | Guidelines for drinking-water quality, fourth edition, 2015). The persistence of viruses in the environment is longer than vegetative bacteria (Nasser et al., 1993) and has shown to possess different transport properties. Aerobic and anaerobic endospores also have major distinguishing properties of survival and transport (Fout et al., 2017; Nasser et al., 1993; Payment and Locas, 2011). The traditional method of using faecal indicator bacteria (FIB), i.e. faecal coliforms and enterococci, for detecting poor microbiological water quality, however, offers neither any concrete proof of pollutant source nor that of the presence of pathogen (Harwood et al., 2014).

In order to carry out direct detection of pathogenic viruses, bacteria, and protozoan, cyst-forming parasites, a necessity for expensive, time-consuming techniques and well-skilled personnel arises. It is a huge and daunting task that would entail the monitoring of thousands of pathogenic and parasitic microorganisms routinely in water resources, wastewater plants, effluent receiving systems, soils and other sample matrices obtained from the environment. Hence, the need for faecal indicators increased, and around 1914, the US Public Health Service (USPHS) decided to adopt a group of coliforms as indicators for the detection of microbiological contamination of drinking water. As time passed, more organisms were added to the roster. In the last decades, intensive research in the field has shed light on the environmental fate of indicator microbes and their ability to represent the hardier contaminants such as protozoan cysts and viruses.

8.3.1 Coliform bacteria

The commonly used indicator microbes are coliform bacteria (total and faecal coliforms) (Leclerc et al., 2001; Standard Methods for the Examination of Water and Wastewater, 1999).

8.3.1.1 Characteristics

The family Enterobacteriaceae represents the total coliforms group that includes Gram-negative aerobes and facultative anaerobes which are non-spore-forming, rod-shaped bacteria. These bacteria are thermotolerant lactose fermenters at 44°C which evolve gas at 35°C within 48 hours (Standard Methods for the Examination of Water and Wastewater, 1999). These total coliforms include *Escherichia coli*, *Enterobacter sp.*, *Klebsiella pneumoniae.* and *Citrobacter sp.* which are discharged in large quantities (2×10^9 coliforms per day per capita) in human and animal wastes. However, not all of them are faecal in origin. These organisms are instrumental in the determination of potable water quality, and that of recreational waters, and shellfish harvesting waters. Agricultural and industrial wastes harbour bacteria of this group like *Klebsiella*. The efficiency of a water treatment plant is usually determined using the total coliform indicators used in the plant. The sole use of *E. coli* has been suggested by some investigators as it can be easily singled out from other members of this group by properties like presence of β-glucuronidase and absence of urease. However, as *E. coli* can be detected in environmental settings such as tropical rainforests and contaminated soil, it is suggested that it could prove to be an erroneous indicator of faecal pollution in such settings (Bermudez and Hazen, 1988).

8.3.1.2 Detection of total and faecal coliforms by standard methods

The lactose fermenting characteristic of coliforms and their ability to generate gas as a by-product at 35°C within 48 hours have paved the way to detection using either most probable number (MPN) or membrane filtration. Both these techniques are detailed in standard methods for the examination of water and wastewater (Standard Methods for the Examination of Water and Wastewater, 1999). The true mark of faecal coliform is the production of gas in *Escherichia coli* (EC) broth at 44.5°C in MPN procedure or in membrane filtration method; the colonies formed are blue in colour when grown on m-FC agar at the same temperature.

Factors influencing the recovery of coliforms from the environment are varied and include pore size of membrane filter, sample turbidity, the presence of non-coliforms, diluting solution and type of growth medium used. The occurrence of injured coliform bacteria is a crucial factor affecting detection. These injured bacteria cannot grow well in the selective growth medium (bile salts and deoxycholate are used) at higher temperatures than that of environmental conditions (Bermudez and Hazen, 1988). The low rate of recovery of debilitating bacteria in environmental samples may lead to false results providing underestimated numbers. Research has shown that even injured bacterial pathogens may retain their pathogenicity (Singh and McFeters, 1987). The m-T7 agar, a growth medium, was proposed to enable the recovery of injured microorganisms (LeChevallier et al., 1983).

8.3.1.3 Rapid detection methods for coliforms

An alternative approach for sensitive yet rapid detection of E. coli and total coliforms in environmental specimens is provided in the form of enzymatic assays. The detection is based on β-galactosidase activity. Chromogenic substrates such as 5-bromo-4-chloro-3-indolyl-β-D-galactopyranoside (X-GAL), o-nitrophenyl-β-D-galactopyranoside (ONPG), chlorophenol red-β-D-galactopyranoside (CPRG) or cyclohexenoesculetin-β-D-galactoside; and fluorogenic substrates such as fluorescein-di-β-D-galactopyranoside (FDG) or 4-methylumbelliferone-D-D-glucuronide (MUG) are commonly used (Bitton et al., 1995; James et al., 1996).

The rapid detection of E. coli is based on the hydrolysis of fluorogenic substrates like 4-methylumbelliferone glucuronide (MUG) by the enzyme β-glucuronidase, an enzyme found in E. coli and some Shigella spp. The assay comprises of sample incubation, in lauryl-tryptose broth supplemented with 100 mg/L of MUG, at 35°C for 24 hrs. The development of fluorescence is observed within the 24-hour incubation time. This method is useful for E. coli detection in clinical and environmental samples (Trepeta and Edberg, 1984). Furthermore, this enzymatic method can be adapted in membrane filter procedure; the colonies positive for β-glucuronidase develop a product, which exhibits fluorescence that can be detected using a long-wave ultraviolet lamp.

Colilert, a commercially developed assay, is used to detect E. coli as well as total coliforms simultaneously, within the 24-hour time period, from environmental samples (Edberg et al., 1990). The test constitutes of addition of the sample to test tubes containing a powdered mixture of salts and enzymatic substrates (carbon sources) specific to total coliforms and E. coli. In general, the chromogenic enzymatic substrate, ONPG, is used for the detection of total coliforms and fluorogenic substrate, MUG, for the detection of E. coli. The samples containing total coliforms turn yellow upon 24-hour incubation, while the fluorescence of the samples positive for E. coli is detected with the help of a UV lamp. Research surveys on drinking water sources have revealed that the sensitivity of the Colilert method is on a par with the typical methods of multiple tube fermentation or membrane filtration (Edberg et al., 1988). Another rapid detection test is ColiPADTM, which involves the hydrolysis of MUG and CPRG on assay pads. E. coli produces purple coloured spots, while total coliforms develop fluorescent spots. Similar to the Colilert method, data of wastewater effluents and lake samples have shown high correlation between the results by ColiPADTM and conventional methods (Bitton, 2005; Bitton et al., 1995). The Environment Protection Act (EPA) proposed, modified membrane thermotolerant E. coli (mTEC) method, uses a special medium containing the chromogen 5-bromo-6-chloro-3-indolyl-β-D-glucuronide. In this method, firstly, samples are passed through membrane filters. Sample filtration is followed by the placement of membrane filters on mTEC medium and incubation at 35°C for a period of 2 hours, and then incubation at 44.5°C for 20–22 hours. E. coli is detected as distinct magenta colonies (Bitton, 2005). Research studies have shown that coliforms (proxy indicators) are different from the pathogenic organisms in the aspects of their ecology, prevalence and resistance to stress (Desmarais et al., 2002; Sobsey et al., 1990). These stark differences have limited the use of coliforms as faecal pollution indicators. In place of them, some additional microorganisms such as Clostridium perfringens are suggested as faecal indicators (Griffin et al., 2001; Scott et al., 2002).

E. coli is a well-known and frequently used indicator of faecal pollution (Geldreich and Taft, 1966). The characteristics that make it a good indicator of faecal pollution are as follows: (i) generally, it is not pathogenic to humans and (ii) the concentration load of E. coli is high in comparison with other pathogens (Desmarais et al., 2002). The major Enterococcus bacteria include Enterococcus faecium, E. faecalis, E. gallinarum, E. avium and E. durans. Differentiation of these species from other streptococci is done on the basis of their ability to grow in the presence of 6.5% NaCl, alkaline pH 9.6 and incubation temperature of 45°C. The species of enterococci most prominently found in humans are E. faecium and E. faecalis. Additionally, enterococci are also considered to be reliable indicators of risk to human health in marine ecosystems and recreational waters (Cabelli, 1983). However, a well-known fact is that the environmental reservoirs of these organisms are present, and hence, their regrowth in the environment is possible (Desmarais et al., 2002).

8.3.2 Faecal streptococci

The members of this group, which commonly inhabit the intestinal cavity of humans and other warm-blooded animals, comprise of microorganisms like Streptococcus faecalis, S. equinus, S. avium and S. bovis. Their presence in the intestinal tract of warm-blooded creatures allows them to be used for the detection of faecal contamination of potable water. Although their multiplication rates in the environment are considerably low, they are known to survive longer than other indicator organisms. A subgroup exhibited the ability to grow in a medium with 6.5% NaCl, alkaline pH (9.6) and high temperature (45°C). These properties make them useful indicators of viral presence in biosolids and marine waters (Bitton, 2005). Detection of faecal streptococci is done using selective growth medium via methods such as MPN or membrane filtration. Rapid detection of this group of indicators is possible with the use of chromogenic or fluorogenic enzymatic assays. Enzyme-based assays for the detection of faecal streptococci are performed by the incorporation of fluorogenic substrates like 4-methylumbe lliferone-β-D-glucoside (MUD) or chromogenic substrates like indoxyl-β-D-glucoside into the selective media being used. These assays rely on the enzymatic activity of two enzymes, namely β-D-glucosidase and pyroglutamyl aminopeptidase (Manafi and Sommer, 1993). Microassays, using microplates and fluorogenic substrate MUD, have proven to be successful in selective detection of these microbes in environmental specimens (Hernandez et al., 1991). Based on the same principle as the aforementioned assay, Colilert, a 24-hour MPN test, has been approved and marketed for the detection of enterococci and is called Enterolert (Bitton, 2005; Budnick et al., 1996).

8.3.3 Anaerobic bacteria

Few anaerobic bacteria are considered important components of the faecal flora. Three main anaerobic microorganisms proposed as faecal contamination indicators are covered in this heading (Bitton, 2005).

8.3.3.1 Clostridium perfringens

Clostridia are mainly opportunistic pathogens and have been known to cause human diseases such as acute colitis (*C. difficile*), gas gangrene (*C. perfringens*), botulism (*C. botulinum*) and tetanus (*C. tetani*) (Bitton, n.d.). *Clostridium perfringens* is a Gram-positive, endospore-forming, sulphite-reducing, rod-shaped anaerobe which is found in the colon. It is estimated that around 0.5% of the faecal microflora of an average human body is *Clostridium perfringens*. It is commonly shed via human and animal faeces and in wastewater and is part of the sulphite-reducing Clostridia (SRC) group (Bitton, 2005). In European countries, SRCs are considered as indicators of water quality, but recent additions to the regulations of the European Union (EU) consider

C. perfringens as the indicator of choice. The standard set was 0/100 mL of drinking water (Council Directive 98/83/EC on the quality of water intended for human consumption, n.d.). Due to its nature to form highly resistant endospores, resistant to oxidizing agents and to UV, it could be a suitable indicator. They are mostly recommended in instances such as presence of waterborne viruses is predicted or remote faecal pollution (Fujioka and Shizumura, 1985; Payment and Franco, 1993). However, its use in microbial source tracking (MST) is still debatable (Bitton, 2005).

8.3.3.2 *Bacteroides spp.*

Bacteroides are obligate anaerobes as well as non-spore-forming bacteria that occur in warm-blooded intestines at a concentration of 10^{10} cells per gram of faeces. This microbial group comprises about one-third of the human faecal flora and, hence, outnumbers the faecal coliform bacteria. The growth of obligate anaerobes is restricted to warm-blooded animals as their survival in environmental waters is difficult, making them more desirable as indicators (Meays et al., 2004). However, as compared to enterococci, they are more difficult to grow in laboratories (Bitton, 2005).

8.3.3.3 *Bifidobacteria*

These are Gram-positive, non-spore-forming anaerobes that inhabit gut region of humans and animals. This group of microbes have been considered as good faecal indicators. *Bifidobacteria* are the third most commonly found bacterial genus in the intestinal tract of humans. Few examples are *B. adolescentis*, *B. dentium*, *B. infantis* and *B. bifidum*. These microorganisms also serve as a distinguishing tool for human faecal contamination from sources of animal faecal contamination (Bitton, 2005). Human isolates are sorbitol fermenters, making it possible for these bacteria to be selectively grown and identified on sorbitol agar (Mara and Oragui, 1983). rRNA probes can also be used to detect *Bifidobacteria* (Bonjoch et al., 2004). The only limiting factor in the consideration of *Bifidobacteria* as indicators is their sensitivity to environmental stresses (Bitton, 2005).

8.3.4 Bacteriophages

Bacteriophages are viruses with a varied range of bacteria as their hosts. The basic structural components of the bacteriophage are similar to other animal viruses. Upon infection of a host cell, a lytic cycle is initiated which leads to the generation of progeny phages and the eventual lysis of the host cell (Bitton, 2005). Phage detection in environmental samples comprises of three main steps: (i) sample concentration, (ii) decontamination of the concentrate and (iii) single- or double-layer phage assay methods. A wide range of bacterial host cells can be used. Although somatic coliphages are assayed using E. coli C strain as host cells, the assay for male-specific phages needs Salmonella typhimurium WG49 strain or the E. coli HS[pFamp]R strain as host cells. Two methods (1601 and 1602) have been proposed by the US EPA for the detection of somatic coliphages (host E. coli CN-13) and F-specific coliphages (host E. coli F-amp) in aquatic environments. The first method, 1601, includes the overnight enrichment step where water is supplemented with tryptic soy broth, $MgCl_2$ and the host cells, subsequent spotting of this suspension on a bacterial lawn comprising of the host bacteria. Method 1602 comprises of a 100 ml water sample supplemented with double strength molten agar, host bacteria and $MgCl_2$; and the mixture is then poured into Petri dishes. The plaques are enumerated at the end of overnight incubation (United States Environmental Protection Agency, 2001). Reverse transcription polymerase chain reaction (RT-PCR) technique was also used to detect F+ RNA coliphage in faecally contaminated marine waters (Rose et al., 1997). Of the many bacteriophage species, male-specific RNA coliphages (FRNA phages), somatic coliphages and phages infecting Bacteroides fragilis (McFeters et al., 1982) are considered as potential indicators.

8.3.4.1 Somatic coliphages

Somatic coliphages infect *E. coli,* but some strains can also infect other bacterial species of the Enterobacteriaceae family. They are used as indicators in monitoring water quality of biosolids, wastewater, potable waters, estuaries and seawater (Mocé-Llivina et al., 2003). They are also used in tracking of pollution sources in aquifers and surface waters, and to study the efficiency of treatment of wastewater in treatment plants (Paul et al., 1995). Genetically modified phages can be used to avoid the hindrance displayed by indigenous phages in environmental specimens. Using polymerase chain reaction or plaque hybridization methods, the unique DNA marker sequence added to phage can be used for the detection (Daniell et al., 2000).

8.3.4.2 F+ coliphages

F+ coliphages are mainly associated with the families *Inoviridae* and *Leviviridae*. They possess single-stranded DNA or RNA genomes and mainly infect *E. coli* that contain F plasmid (fertility factor) or sex pilus which is associated with phage adherence. The factors that influence the selection of these phages as indicators include higher concentration in wastewater and high resistance to chlorination (Harwood et al., 2014; Havelaar et al., 1990). With regard to the contamination of shellfish harvesting water, male-specific phages are suitable as model organism to understand the fate of viruses in shellfish (Doré and Lees, 1995). Hence, F+ coliphages could also indicate viral contamination in aquatic/marine environments (Bitton, 2005).

Although it is true that monitoring of FIB has helped to make improvements in public health protection regulations (Fewtrell et al., 2001), few epidemiological studies have failed to correlate the presence of FIB and human pathogens when the origin of the contamination is not a point source (Yates et al., 1985). The widespread distribution of FIB in animal and human wastes has made it difficult to track the exact source of the pollution (Harwood et al., 2014; Kelty et al., 2012). In order to address these issues more thoroughly, various MST markers have been identified (Bernhard and Field, 2000; Mieszkin et al., 2009; Reischer et al., 2006) and successfully tested in different parts of the world (Boehm et al., 2013; Reischer et al., 2013). This tool has the potential to help us achieve an accurate estimate of the quantitative microbial risk and manage the inflicted areas in an efficient manner (Gourmelon et al., n.d.; McQuaig et al., 2012).

The limitations of using FIB alone for monitoring water quality may lead to erroneous conclusions, and thereby, timely and effective corrective actions, improved assessment of health risk and accurate TMDL estimations cannot be enforced (Soller et al., 2010). Though the indicators like total coliforms (TC), faecal coliforms (FC), *Escherichia coli* and enterococci have been used for a long time, their mere presence does not indicate the source of the contamination (Duran et al., 2006). In order to trace the sources of contamination in water reserved for recreational use and combat the deficiencies associated with using FIB concentrations to monitor water quality, MST methods have been proposed for determining the source of faecal contamination (Field and Samadpour, 2007).

8.4 MICROBIAL SOURCE TRACKING: AN INTRODUCTION

Understanding the origin of pollutants in groundwater is a necessity to reduce the health-related problems associated with it. MST is defined as tracing the origin of faecal contamination using different approaches such as microbiological, genotypic, phenotypic and chemical methods (Scott et al., 2002). MST is a comparatively new and fast-developing technology which allows the discrimination between different sources of groundwater contaminating agents present in the environment. By using MST methods, one can trace back the origin of the causative agent of microbial contamination by the detection of host-specific gastrointestinal microflora that has been released into the water along with the excreta (Hagedorn et al., 2011).

The source of pollutants needs to be correctly identified in order to minimize the reduction of water quality. Pollution may occur as a result of either point or non-point sources. Point sources of pollution can be easily identified and include combined sewer overflow and wastewater discharge points. Non-point sources cannot be identified directly. Few examples of non-point sources are leaking sewers, run-off from urban rainfall and agricultural farmlands (Epa et al., n.d.). However, due to these potential contamination sources, water frequently contains traces of faeces (Roslev and Bukh, 2011).

The group of microorganisms, which are used as indicators, serve the purpose of recognizing the subtle differences between different microorganisms to identify the origin of microbial agent. MST is an essential tool to detect the faecal contamination of aquatic systems and also allows informed risk assessment of water usage. There is no universal standard available as of now for MST marker to identify either human or non-human faecal pollution of water bodies. The criteria of an ideal MST tool are enlisted in Table 8.2.

Table 8.2 Criteria of an ideal MST tool for faecal contamination source tracking (Field et al., 2003)

1. It must only exist in host faeces.
2. It should be unable to replicate in natural aquatic environments.
3. It should have a simple, reliable, rapid and inexpensive method for faecal contamination source tracking.
4. It should not require culturing of isolates.
5. It should not require a large reference strain library.
6. It should be easy to collect and store samples.

Methodology for tracking the human versus animal sources of faecal pollutant is essential because traditional indicator microorganisms are present in warm-blooded animals in general (Bitton, 2005). Microbial species used as MST markers are *Bacteroides dorei*, *Bacteroides thetaiotaomicron*, *Bifidobacterium adolescentis*, and *Clostridium perfringens*, *Faecalibacterium prausnitzii,* gut-associated bacteria (*Eubacterium rectale* and *Ruminococcus bromii)* and skin-associated bacteria (*Staphylococcus epidermidis* and *Propionibacterium acnes)* (Geldreich and Kenner, 1969; Staley et al., 2016).

8.5 MICROBIAL METHODS FOR SOURCE TRACKING

8.5.1 Ratio of faecal coliform/faecal streptococcus

A ratio of >4.0 faecal coliforms to faecal streptococci indicates faecal pollution of human origin, whereas a ratio of ≤0.7 would specify that faecal pollution is of non-human origin. This method is rapid and non-labour-intensive with minimal requirement for skills. The success rate for the method depends on various factors such as rate of survival of the varied species of faecal streptococci, choice of detection method and fluctuations in sensitivity to water treatments. Hence, the use of this method was abandoned for faecal source tracking (Standard Methods for the Examination of Water and Wastewater, 1999).

8.5.2 Bifidobacterium spp.

Bifidobacteria are found as part of gut microbiota. They are non-spore-forming and obligate anaerobes. As these are rarely found in animals, they are considered as potential indicator of faecal pollution of specifically human origin (Mara and Oragui, 1983). Additionally, Bifidobacteria are unable to reproduce once deposited in the environment, but the only limitation is that the survival rate of these organisms decreases by 3 to 4 orders of magnitude if examined within a span of 2 weeks. Therefore, the detection of these organisms provides evidence of recent faecal contamination. However, further research is needed to develop techniques with better specificity and sensitivity for the detection of these organisms (Scott et al., 2002).

8.5.3 Bacteroides fragilis phages

B. fragilis belongs to the Bacteroides group which are present in animal and human intestines. These are Gram-negative, pleomorphic rod-shaped obligate anaerobes. The phages that mainly infect Bacteroides fragilis HSP40 are detected in human faecal matter and aquatic environments (Cornax et al., 1990; Tartera and Jofre, 1987). In contrast to other coliform bacteria, these do not have the potential to survive in the environment. Its high resistance to chlorination, as compared to bacterial indicators, in sewage has been confirmed (Durán et al., 2002). Their resistance to water treatment processes is proposed to be higher than that of traditional microbiological indicators, phages (male-specific phages and somatic phages) and enteric viruses

(Jofre et al., 1995). They are also highly resistant to environmental inactivation in freshwater systems compared to faecal coliforms and other phages (Durán et al., 2002). The detection of bacteriophage that infects B. fragilis is a highly specific method for source tracking of human faecal pollution. The advantages are that these phages are unable to replicate in the environment, and their presence in the environment is correlated with the presence of human enteric viruses (Jofre et al., 1989). One of the major limitations is the inherent difficulty in performing the assay to detect these phages (Scott et al., 2002).

8.6 VIRUS AS MICROBIAL SOURCE TRACKING TOOL

Tracking viruses that contaminate groundwater can be an alternative way for MST because bacterial indicators are widely spread in the environment. Another major drawback is that bacteria like *E. coli* and enterococci lack host specificity for source tracking. Human enteroviruses (HEV), human adenoviruses (HAdV) and human polyomaviruses (HPyV) can serve as a MST tool for indication contamination of human origin (Fong et al., 2005; Noble et al., 2003). Viruses can be detected using cell culture. The methodology involves the infection of cell culture with specific virus, and then, the change in morphology (cytopathic effect) is visualized under a microscope. However, this technique is labour-intensive and certain viruses are not cytopathic. As compared to this traditional method, PCR is a better alternative. The advantages of PCR over cell culture include rapid results, higher specificity and sensitivity, and the ability to detect non-culturable viruses. Integrated cell culture-PCR technique is useful for the study of infectivity of viruses (Xagoraraki et al., 2014).

8.7 CONCLUSIONS

In general, no sole MST tool is adept at unequivocally ascertaining the precise origin of faecal contamination in the environment. There is no universal indicator, and many areas of yet, in the trial phase. Since no method is entirely foolproof, false results could lead to incorrect conclusions. Most methods entail a considerable expenditure of time, labour and capital for costly laboratory apparatus. Also, culture bias is an issue when culturing bacterial isolates. This bias may be removed by using real-time technology, which is also capable of massively contributing to the resolution of current pollution problems and similar occurrences in the future. No MST method is recommended as a standard method to date. Additionally, these methods require standardization and thorough comparison studies. The association between survival features of pathogens and that of the indicator organisms chosen as their predictors need to be studied further. Additional characterization and, by extension, the recognition of novel groups of host-specific indicators like *Bifidobacteria*, *B. fragilis* bacteriophages and F+ RNA coliphages would expand upon their existing benefits of MST. Nevertheless, the validation of these organisms is essential for their use as source trackers ahead. With increasing industrialization and urbanization, especially in developing nations, MST technologies perform a vital part in tracing both point and diffuse origins of faecal pollution.

REFERENCES

A Scale of Grade and Class Terms for Clastic Sediments on JSTOR [WWW Document], n.d. URL https://www.jstor.org/stable/30063207?seq=1#metadata_info_tab_contents (Accessed 2.14.21).

Abbaszadegan, M., Stewart, P., LeChevallier, M., 1999. A strategy for detection of viruses in groundwater by PCR. *Applied and Environmental Microbiology* 65, 444–449. https://doi.org/10.1128/aem.65.2.444-449.1999.

Australian Drinking Water Guidelines | NHMRC [WWW Document], n.d. URL https://www.nhmrc.gov.au/about-us/publications/australian-drinking-water-guidelines (Accessed 2.14.21).

Beer, K.D., Gargano, J.W., Roberts, V.A., Hill, V.R., Garrison, L.E., Kutty, P.K., Hilborn, E.D., Wade, T.J., Fullerton, K.E., Yoder, J.S., 2015. Surveillance for Waterborne Disease Outbreaks Associated with Drinking Water — United States, 2011–2012. MMWR. Morbidity and Mortality Weekly Report 64. https://doi.org/10.15585/mmwr.mm6431a2.

Bermudez, M., Hazen, T.C., 1988. Phenotypic and genotypic comparison of Escherichia coli from pristine tropical waters. *Applied and Environmental Microbiology* 54, 979–983. https://doi.org/10.1128/aem.54.4.979-983.1988.

Bernhard, A.E., Field, K.G., 2000. A PCR assay to discriminate human and ruminant feces on the basis of host differences in Bacteroides-Prevotella genes encoding 16S rRNA. *Applied and Environmental Microbiology* 66, 4571–4574. https://doi.org/10.1128/AEM.66.10.4571-4574.2000.

Bitton, G., 2005. *Wastewater Microbiology*, 3rd Edition. Wiley-Liss, Hoboken, NJ.

Bitton, G., n.d. *Encyclopedia of Environmental Microbiology, 6 Volume Set*. Wiley, Hoboken, NJ.

Bitton, G., Koopman, B., Jung, K., 1995. An assay for the enumeration of total coliforms and *Escherichia coli* in water and wastewater. *Water Environment Research* 67, 906–909. https://doi.org/10.2175/106143095X133121.

Boehm, A.B., van de Werfhorst, L.C., Griffith, J.F., Holden, P.A., Jay, J.A., Shanks, O.C., Wang, D., Weisberg, S.B., 2013. Performance of forty-one microbial source tracking methods: A twenty-seven lab evaluation study. *Water Research* 47, 6812–6828. https://doi.org/10.1016/j.watres.2012.12.046.

Bonjoch, X., Ballesté, E., Blanch, A.R., 2004. Multiplex PCR with 16S rRNA gene-targeted primers of Bifidobacterium spp. to identify sources of fecal pollution. *Applied and Environmental Microbiology* 70, 3171–3175. https://doi.org/10.1128/AEM.70.5.3171-3175.2004.

Borchardt, M.A., Bradbury, K.R., Alexander, E.C., Kolberg, R.J., Alexander, S.C., Archer, J.R., Braatz, L.A., Forest, B.M., Green, J.A., Spencer, S.K., 2011. Norovirus outbreak caused by a new septic system in a dolomite aquifer. *Ground Water* 49, 85–97. https://doi.org/10.1111/j.1745-6584.2010.00686.x.

Brownell, M.J., Harwood, V.J., Kurz, R.C., McQuaig, S.M., Lukasik, J., Scott, T.M., 2007. Confirmation of putative stormwater impact on water quality at a Florida beach by microbial source tracking methods and structure of indicator organism populations. *Water Research* 41, 3747–3757. https://doi.org/10.1016/j.watres.2007.04.001.

Budnick, G.E., Howard, R.T., Mayo, D.R., 1996. Evaluation of enterolert for enumeration of enterococci in recreational waters. *Applied and Environmental Microbiology* 62, 3881–3884. https://doi.org/10.1128/aem.62.10.3881-3884.1996.

Cabelli, V.J., 1983. Health Effects Criteria for Marine Recreational Waters. USEPA 600/1-80-031. Health Effects Research Laboratory, Research Triangle Park, N.C.

Cho, H.G., Lee, S.G., Kim, W.H., Lee, J.S., Park, P.H., Cheon, D.S., Jheong, W.H., Jho, E.H., Lee, J.B., Paik, S.Y., 2014. Acute gastroenteritis outbreaks associated with ground-waterborne norovirus in South Korea during 2008–2012. *Epidemiology and Infection* 142. https://doi.org/10.1017/S0950268814000247.

Colford, J.M., Wade, T.J., Schiff, K.C., Wright, C.C., Griffith, J.F., Sandhu, S.K., Burns, S., Sobsey, M., Lovelace, G., Weisberg, S.B., 2007. Water quality indicators and the risk of illness at beaches with nonpoint sources of fecal contamination. *Epidemiology* 18, 27–35. https://doi.org/10.1097/01.ede.0000249425.32990.b9.

Collins, R., Elliott, S., Adams, R., 2005. Overland flow delivery of faecal bacteria to a headwater pastoral stream. *Journal of Applied Microbiology* 99, 126–132. https://doi.org/10.1111/j.1365-2672.2005.02580.x.

Cornax, R., Morinigo, M.A., Paez, I.G., Munoz, M.A., Borrego, J.J., 1990. Application of direct plaque assay for detection and enumeration of bacteriophages of Bacteroides fragilis from contaminated-water samples. *Applied and Environmental Microbiology* 56, 3170–3173. https://doi.org/10.1128/aem.56.10.3170-3173.1990.

Council Directive 98/83/EC on the quality of water intended for human consumption. [WWW Document], n.d. URL https://www.ecolex.org/details/legislation/council-directive-9883ec-on-the-quality-of-water-intended-for-human-consumption-lex-faoc018700/ (Accessed 2.14.21).

Daniell, T.J., Davy, M.L., Smith, R.J., 2000. Development of a genetically modified bacteriophage for use in tracing sources of pollution. *Journal of Applied Microbiology* 88, 860–869. https://doi.org/10.1046/j.1365–2672.2000.01028.x.

de Serres, G., Cromeans, T.L., Levesque, B., Brassard, N., Barthe, C., Dionne, M., Prud'Homme, H., Paradis, D., Shapiro, C.N., Nainan, O. V., Margolis, H.S., 1999. Molecular confirmation of hepatitis A virus from well water: Epidemiology and public health implications. *Journal of Infectious Diseases* 179, 37–43. https://doi.org/10.1086/314565.

Desmarais, T.R., Solo-Gabriele, H.M., Palmer, C.J., 2002. Influence of soil on fecal indicator organisms in a tidally influenced subtropical environment. *Applied and Environmental Microbiology* 68, 1165–1172. https://doi.org/10.1128/AEM.68.3.1165-1172.2002.

Diston, D., Sinreich, M., Zimmermann, S., Baumgartner, A., Felleisen, R., 2015. Evaluation of molecular- and culture-dependent MST markers to detect fecal contamination and indicate viral presence in good quality groundwater. *Environmental Science & Technology* 49. https://doi.org/10.1021/acs.est.5b00515.

Dixon, P., 2003. VEGAN, a package of R functions for community ecology. *Journal of Vegetation Science* 14, 927–930. https://doi.org/10.1111/j.1654-1103.2003.tb02228.x.

Doré, W.J., Lees, D.N., 1995. Behavior of Escherichia coli and Male-Specific Bacteriophage in Environmentally Contaminated Bivalve Molluscs before and after Depuration, Applied and Environmental Microbiology.

Durán, A.E., Muniesa, M., Méndez, X., Valero, F., Lucena, F., Jofre, J., 2002. Removal and inactivation of indicator bacteriophages in fresh waters. *Journal of Applied Microbiology* 92, 338–347. https://doi.org/10.1046/j.1365–2672.2002.01536.x.

Duran, M., Haznedaroğlu, B.Z., Zitomer, D.H., 2006. Microbial source tracking using host specific FAME profiles of fecal coliforms. *Water Research* 40. https://doi.org/10.1016/j.watres.2005.10.019.

Edberg, S.C., Allen, M.J., Smith, D.B., Kriz, N.J., 1990. Enumeration of total coliforms and Escherichia coli from source water by the defined substrate technology. *Applied and Environmental Microbiology* 56, 366–369. https://doi.org/10.1128/aem.56.2.366-369.1990.

Edberg, S.C., Allen, M.J., Smith, D.B., LeChevallier, M., Kriz, N., Callan, D., Ward, R., Calver, D., Jackson, W., Uryc, M., Storms, C., Loriner, J., Trok, T., Burns, M., 1988. National field evaluation of a defined substrate method for the simultaneous enumeration of total coliforms and Escherichia coli from drinking water: Comparison with the standard multiple tube fermentation method. *Applied and Environmental Microbiology* 54, 1595–1601. https://doi.org/10.1128/aem.54.6.1595-1601.1988.

Edberg, S.C., Rice, E.W., Karlin, R.J., Allen, M.J., 2000. Escherichia coli: The best biological drinking water indicator for public health protection. *Journal of Applied Microbiology Symposium Supplement*. https://doi.org/10.1111/j.1365-2672.2000.tb05338.x.

Epa, Ow, Owow, Awpd, NPSCB, n.d. National Nonpoint Source Program-a catalyst for water quality improvements.

Fewtrell, L., Bartram, J., Ashbolt, N.J., Grabow, W.O.K., Snozzi, M., 2001. *Indicators of microbial water quality.*

Field, K.G., Chern, E.C., Dick, L.K., Fuhrman, J., Griffith, J., Holden, P.A., LaMontagne, M.G., Le, J., Olson, B., Simonich, M.T., 2003. A comparative study of culture-independent, library-independent genotypic methods of fecal source tracking. *Journal of Water and Health* 1, 181–194. https://doi.org/10.2166/wh.2003.0020.

Field, K.G., Samadpour, M., 2007. Fecal source tracking, the indicator paradigm, and managing water quality. *Water Research.* https://doi.org/10.1016/j.watres.2007.06.056.

Fong, T.T., Griffin, D.W., Lipp, E.K., 2005. Molecular assays for targeting human and bovine enteric viruses in coastal waters and their application for library-independent source tracking. *Applied and Environmental Microbiology* 71, 2070–2078. https://doi.org/10.1128/AEM.71.4.2070-2078.2005.

Ford, D., Williams, P., 2007. *Karst Hydrogeology and Geomorphology, Karst Hydrogeology and Geomorphology.* John Wiley & Sons Ltd., West Sussex, England. https://doi.org/10.1002/9781118684986.

Fout, G.S., Borchardt, M.A., Kieke, B.A., Karim, M.R., 2017. Human virus and microbial indicator occurrence in public-supply groundwater systems: meta-analysis of 12 international studies. *Hydrogeology Journal* 25. https://doi.org/10.1007/s10040-017-1581-5.

Fujioka, R.S., Shizumura, L.K., 1985. Clostridium perfringens, a reliable indicator of stream water quality on JSTOR. *Journal (Water Pollution Control Federation)* 5, 986–992.

Geldreich, E.E., Kenner, B.A., 1969. Concepts of fecal streptococci in stream pollution - PubMed. *Journal of Water Pollution Control Federation* 41, R336.

Geldreich, E.E., Taft, R.A., 1966. Sanitary significance of fecal coliforms in the environment. [Cincinnati]: U.S. Dept. of the Interior, Federal Water Pollution Control Administration.

Gerba, C.P., 1999. Virus survival and transport in groundwater. *Journal of Industrial Microbiology and Biotechnology* 22, 535–539. https://doi.org/10.1038/sj.jim.2900663.

Gerba, C.P., Lance, J.C., 1978. Poliovirus removal from primary and secondary sewage effluent by soil filtration. *Applied and Environmental Microbiology* 36, 247–251. https://doi.org/10.1128/aem.36.2.247-251.1978.

Gotkowitz, M.B., Bradbury, K.R., Borchardt, M.A., Zhu, J., Spencer, S.K., 2016. Effects of climate and sewer condition on virus transport to groundwater. *Environmental Science & Technology* 50. https://doi.org/10.1021/acs.est.6b01422.

Gourmelon, M., Lazure, P., Hervio-Heath, D., le Saux, J., Caprais, M., le Guyader, F., Catherine, M., Pommepuy, M., Rees, G., Pond, K., Kay, D., Bartram, J., Santo Domingo, J., n.d. Microbial modelling in coastal environments and early warning systems: Useful tools to limit shellfish microbial contamination.

Griffin, D.W., Lipp, E.K., Mclaughlin, M.R., Rose, J.B., 2001. Marine Recreation and Public Health Microbiology: Quest for the Ideal Indicator: This article addresses the historic, recent, and future directions in microbiological water quality indicator research. *BioScience.* Oxford Academic. https://doi.org/10.1641/0006-3568(2001)051[0817:MRAPHM]2.0.CO;2.

Guidelines for Canadian Drinking Water Quality - Summary Table - Canada.ca [WWW Document], n.d. URL https://www.canada.ca/en/health-canada/services/environmental-workplace-health/reports-publications/water-quality/guidelines-canadian-drinking-water-quality-summary-table.html (Accessed 2.14.21).

Guy, H., Bartram, J., Pedley, S., Schmoll, O., Chorus, I., Berger, P., Chilton, J., 2006. Groundwater and public health. In: O. Schmoll, G. Howard, J. Chilton, I. Chorus (eds.) *Protecting Groundwater for Health: Managing the Quality of Drinking-Water Sources.* World Health Organization, IWA Publishing, London.

Hagedorn, C., Harwood, V.J., Blanch, A.R. (Eds.), 2011. *Microbial Source Tracking: Methods, Applications, and Case Studies, Microbial Source Tracking: Methods, Applications, and Case Studies*. Springer, New York. https://doi.org/10.1007/978-1-4419-9386-1_1.

Harwood, V.J., Staley, C., Badgley, B.D., Borges, K., Korajkic, A., 2014. Microbial source tracking markers for detection of fecal contamination in environmental waters: relationships between pathogens and human health outcomes. *FEMS Microbiology Reviews* 38. https://doi.org/10.1111/1574-6976.12031.

Havelaar, A.H., Pot-Hogeboom, W.M., Furuse, K., Pot, R., Hormann, M.P., 1990. F-specific RNA bacteriophages and sensitive host strains in faeces and wastewater of human and animal origin. *Journal of Applied Bacteriology* 69, 30–37. https://doi.org/10.1111/j.1365-2672.1990.tb02908.x.

Heinz, B., Birk, S., Liedl, R., Geyer, T., Straub, K.L., Andresen, J., Bester, K., Kappler, A., 2009. Water quality deterioration at a karst spring (Gallusquelle, Germany) due to combined sewer overflow: Evidence of bacterial and micro-pollutant contamination. *Environmental Geology* 57. https://doi.org/10.1007/s00254-008-1359-0.

Hernandez, J.F., Guibert, J.M., Delattre, J.M., Oger, C., Charriere, C., Hughes, B., Serceau, R., Sinegre, F., 1991. Miniaturized fluorogenic assays for enumeration of E. coli and enterococci in marine water. *Water Science and Technology* 24, 137–141. https://doi.org/10.2166/wst.1991.0045.

Hynds, P.D., Thomas, M.K., Pintar, K.D.M., 2014. Contamination of groundwater systems in the US and Canada by Enteric Pathogens, 1990–2013: A review and pooled-analysis. *PLoS ONE* 9. https://doi.org/10.1371/journal.pone.0093301.

James, A.L., Perry, J.D., Ford, M., Armstrong, L., Gould, F.K., 1996. Evaluation of cyclohexenoesculetin-β-D-galactoside and 8- hydroxyquinoline-β-D-galactoside as substrates for the detection of β- galactosidase. *Applied and Environmental Microbiology* 62, 3868–3870. https://doi.org/10.1128/aem.62.10.3868-3870.1996.

Jofre, J., Blasi, M., Bosch, A., Lucena, F., 1989. Occurrence of bacteriophages infecting Bacteroides fragilis and other viruses in polluted marine sediments. *Water Science and Technology* 21, 15–19. https://doi.org/10.2166/wst.1989.0072.

Jofre, J., Olle, E., Ribas, F., Vidal, A., Lucena, F., 1995. Potential usefulness of bacteriophages that infect Bacteroides fragilis as model organisms for monitoring virus removal in drinking water treatment plants. *Applied and Environmental Microbiology* 61, 3227–3231. https://doi.org/10.1128/aem.61.9.3227-3231.1995.

Karst Hydrogeology and Human Activities: Impacts, Consequences and Imp [WWW Document], n.d. URL https://www.routledge.com/Karst-Hydrogeology-and-Human-Activities-Impacts-Consequences-and-Implications/Drew-Hotzl/p/book/9789054104643 (Accessed 2.14.21).

Kelty, C.A., Varma, M., Sivaganesan, M., Haugland, R.A., Shanks, O.C., 2012. Distribution of genetic marker concentrations for fecal indicator bacteria in sewage and animal feces. *Applied and Environmental Microbiology* 78, 4225–4232. https://doi.org/10.1128/AEM.07819-11.

Keswick, B.H., Gerba, C.P., 1980. Viruses in groundwater. *Environmental Science and Technology* 14, 1290–1297. https://doi.org/10.1021/es60171a602.

Kumar, M., Thakur, A. K., Mazumder, P., Kuroda, K., Mohapatra, S., Rinklebe, J., ... & Bhattacharya, P. 2020a. Frontier review on the propensity and repercussion of SARS-CoV-2 migration to aquatic environment. *Journal of Hazardous Materials Letters* 1, 100001.

Kumar, M., Chaminda, G. T., & Honda, R. 2020b. Seasonality impels the antibiotic resistance in Kelani River of the emerging economy of Sri Lanka. *NPJ Clean Water* 3(1), 1–8.

Kumar, M., Kuroda, K., A. K., Patel, N., Bhattacharya, P., Joshi, M., & Joshi, C. G. 2021. Decay of SARS-CoV-2 RNA along the wastewater treatment outfitted with Upflow Anaerobic Sludge Blanket (UASB) system evaluated through two sample concentration techniques. *Science of the Total Environment* 754, 142329.

Lance, J.C., Gerba, C.P., 1984. Virus movement in soil during saturated and unsaturated flow. *Applied and Environmental Microbiology* 47, 335–337. https://doi.org/10.1128/aem.47.2.335-337.1984.

Lawson, H., Braun, M., Glass, R., Stine, S., Monroe, S., Atrash, H., Englender, S., 1991. Waterborne outbreak of Norwalk virus gastroenteritis at a southwest US resort: role of geological formations in contamination of well water. *The Lancet* 337, 1200–1204. https://doi.org/10.1016/0140-6736(91)92868-3.

LeChevallier, M.W., Cameron, S.C., McFeters, G.A., 1983. New medium for improved recovery of coliform bacteria from drinking water. *Applied and Environmental Microbiology* 45, 484–492. https://doi.org/10.1128/aem.45.2.484-492.1983.

Leclerc, H., Edberg, S., Pierzo, V., Delattre, J.M., 2001. Bacteriophages as indicators of enteric viruses and public health risk in groundwaters. *Journal of Applied Microbiology* 88, 5–21. https://doi.org/10.1046/j.1365-2672.2000.00949.x.

Lee, J.-Y., Kwon, K., 2016. Current status of groundwater monitoring networks in Korea. *Water* 8. https://doi.org/10.3390/w8040168.

Manafi, M., Sommer, R., 1993. Rapid identification of enterococci with a new fluorogenic-chromogenic assay. *Water Science and Technology* 27, 271–274. https://doi.org/10.2166/wst.1993.0358.

Mara, D.D., Oragui, J.I., 1983. Sorbitol-fermenting bifidobacteria as specific indicators of human faecal pollution. *Journal of Applied Bacteriology* 55, 349–357. https://doi.org/10.1111/j.1365-2672.1983.tb01331.x.

McFeters, G.A., Cameron, S.C., LeChevallier, M.W., 1982. Influence of diluents, media, and membrane filters on detection of injured waterborne coliform bacteria. *Applied and Environmental Microbiology* 43, 97–103. https://doi.org/10.1128/aem.43.1.97-103.1982.

McQuaig, S., Griffith, J., Harwood, V.J., 2012. Association of fecal indicator bacteria with human viruses and microbial source tracking markers at coastal beaches impacted by non-point source pollution. *Applied and Environmental Microbiology* 78, 6423–6432. https://doi.org/10.1128/AEM.00024-12.

Meays, C.L., Broersma, K., Nordin, R., Mazumder, A., 2004. Source tracking fecal bacteria in water: A critical review of current methods. *Journal of Environmental Management*. https://doi.org/10.1016/j.jenvman.2004.06.001.

Metcalfe, C.D., Beddows, P.A., Bouchot, G.G., Metcalfe, T.L., Li, H., van Lavieren, H., 2011. Contaminants in the coastal karst aquifer system along the Caribbean coast of the Yucatan Peninsula, Mexico. *Environmental Pollution* 159, 991–997. https://doi.org/10.1016/j.envpol.2010.11.031.

Mieszkin, S., Furet, J.P., Corthier, G., Gourmelon, M., 2009. Estimation of pig fecal contamination in a river catchment by real-time PCR using two Pig-Specific Bacteroidales 16S rRNA genetic markers. *Applied and Environmental Microbiology* 75, 3045–3054. https://doi.org/10.1128/AEM.02343-08.

Mocé-Llivina, L., Muniesa, M., Pimenta-Vale, H., Lucena, F., Jofre, J., 2003. Survival of bacterial indicator species and bacteriophages after thermal treatment of sludge and sewage. *Applied and Environmental Microbiology* 69, 1452–1456. https://doi.org/10.1128/AEM.69.3.1452-1456.2003.

Mohapatra, S., Sargaonkar, A., Labhasetwar, P.K., 2014. Distribution network assessment using EPANET for intermittent and continuous water supply. *Water Resources Management* 28(11), 3745–3759.

Mohapatra, S., Menon, N.G., Mohapatra, G., Pisharody, L., Pattnaik, A., Menon, N.G., Bhukya, P.L., Srivastava, M., Singh, M., Barman, M.K., Gin, K.Y. 2020. The novel SARS-CoV-2 pandemic: Possible environmental transmission, detection, persistence and fate during wastewater and water treatment. *Science of the Total Environment* 765, 142746.

Moore, R.S., Taylor, D.H., Reddy, M.M.M., Sturman, L.S., 1982. Adsorption of reovirus by minerals and soils. *Applied and Environmental Microbiology* 44, 852–859. https://doi.org/10.1128/aem.44.4.852-859.1982.

Nasser, A.M., Tchorch, Y., Fattal, B., 1993. Comparative survival of E. coli, F+bacteriophages, HAV and poliovirus 1 in wastewater and groundwater. *Water Science and Technology* 27, 401–407. https://doi.org/10.2166/wst.1993.0381.

Noble, R.T., Allen, S.M., Blackwood, A.D., Chu Weiping, Jiang, S.C., Lovelace, G.L., Sobsey, M.D., Stewart, J.R., Wait, D.A., 2003. Use of viral pathogens and indicators to differentiate between human and non-human fecal contamination in a microbial source tracking comparison study - PubMed. *Journal of Water and Health* 1, 195–207.

Pantha, K., Acharya, K., Mohapatra, S., Khanal, S., Amatya, N., Ospina-Betancourth, C., Butte, G., Shrestha, S.D., Rajbhandari, P., Werner, D., 2021. Faecal pollution source tracking in the holy Bagmati River by portable 16S rRNA gene sequencing. *NPJ Clean Water* 4(1), 1–10.

Paul, J.H., Rose, J.B., Brown, J., Shinn, E.A., Miller, S., Farrah, S.R., 1995. Viral tracer studies indicate contamination of marine waters by sewage disposal practices in key largo, Florida. *Applied and Environmental Microbiology* 61, 2230–2234.

Payment, P., Franco, E., 1993. Clostridium perfringens and somatic coliphages as indicators of the efficiency of drinking water treatment for viruses and protozoan cysts. *Applied and Environmental Microbiology* 59, 2418–2424. https://doi.org/10.1128/aem.59.8.2418-2424.1993.

Payment, P., Locas, A., 2011. Pathogens in Water: Value and limits of correlation with microbial indicators. *Ground Water* 49, 4–11. https://doi.org/10.1111/j.1745-6584.2010.00710.x.

Pedley, S., Yates, M., Schijven, J.F., West, J., Howard, G., Barrett, M., 2006. Pathogens: Health relevance, transport and attenuation. In: O. Schmoll, G. Howard, J. Chilton, I. Chorus (eds.) *Protecting Groundwater for Health: Managing the Quality of Drinking-Water Sources*. WHO-IWA Publishing, London.

Pochon, A., Tripet, J.-P., Kozel, R., Meylan, B., Sinreich, M., Zwahlen, F., 2008. Groundwater protection in fractured media: A vulnerability-based approach for delineating protection zones in Switzerland. *Hydrogeology Journal* 16(7), 1267–1281. https://doi.org/10.1007/s10040-008-0323-0.

Reischer, G.H., Ebdon, J.E., Bauer, J.M., Schuster, N., Ahmed, W., Åström, J., Blanch, A.R., Blöschl, G., Byamukama, D., Coakley, T., Ferguson, C., Goshu, G., Ko, G., de Roda Husman, A.M., Mushi, D., Poma, R., Pradhan, B., Rajal, V., Schade, M.A., Sommer, R., Taylor, H., Toth, E.M., Vrajmasu, V., Wuertz, S., Mach, R.L., Farnleitner, A.H., 2013. Performance characteristics of qPCR assays targeting human- and ruminant-associated *Bacteroidetes* for microbial source tracking across sixteen countries on six continents. *Environmental Science & Technology* 47. https://doi.org/10.1021/es304367t.

Reischer, G.H., Kasper, D.C., Steinborn, R., Mach, R.L., Farnleitner, A.H., 2006. Quantitative PCR method for sensitive detection of ruminant fecal pollution in freshwater and evaluation of this method in alpine karstic regions. *Applied and Environmental Microbiology* 72, 5610–5614. https://doi.org/10.1128/AEM.00364-06.

Rose, J.B., Zhou, X., Griffin, D.W., Paul, J.H., 1997. Comparison of PCR and plaque assay for detection and enumeration of coliphage in polluted marine waters. *Applied and Environmental Microbiology* 63, 4564–4566. https://doi.org/10.1128/aem.63.11.4564-4566.1997.

Roslev, P., Bukh, A.S., 2011. State of the art molecular markers for fecal pollution source tracking in water. *Applied Microbiology and Biotechnology*. https://doi.org/10.1007/s00253-010-3080–7.

Rusiñol, M., Fernandez-Cassi, X., Hundesa, A., Vieira, C., Kern, A., Eriksson, I., Ziros, P., Kay, D., Miagostovich, M., Vargha, M., Allard, A., Vantarakis, A., Wyn-Jones, P., Bofill-Mas, S., Girones, R., 2014. Application of human and animal viral microbial source tracking tools in fresh and marine waters from five different geographical areas. *Water Research* 59. https://doi.org/10.1016/j.watres.2014.04.013.

Schriewer, A., Odagiri, M., Wuertz, S., Misra, P.R., Panigrahi, P., Clasen, T., Jenkins, M.W., 2015. Human and animal fecal contamination of community water sources, stored drinking water and hands in rural India measured with validated microbial source tracking assays. *The American Journal of Tropical Medicine and Hygiene* 93. https://doi.org/10.4269/ajtmh.14-0824.
Scott, T.M., Rose, J.B., Jenkins, T.M., Farrah, S.R., Lukasik, J., 2002. microbial source tracking: Current methodology and future directions. *Applied and Environmental Microbiology* 68. https://doi.org/10.1128/AEM.68.12.5796-5803.2002.
Sercu, B., van de Werfhorst, L.C., Murray, J.L.S., Holden, P.A., 2011. Sewage exfiltration as a source of storm drain contamination during dry weather in urban watersheds. *Environmental Science and Technology* 45, 7151–7157. https://doi.org/10.1021/es200981k.
Singh, A., McFeters, G.A., 1987. Survival and virulence of copper- and chlorine-stressed Yersinia enterocolitica in experimentally infected mice. *Applied and Environmental Microbiology* 53, 1768–1774. https://doi.org/10.1128/aem.53.8.1768-1774.1987.
Sobsey, M.D., Schwab, K.J., Handzel, T.R., 1990. Simple membrane filter method to concentrate and enumerate male-specific RNA coliphages. *Journal / American Water Works Association* 82, 52–59.
Soller, J.A., Schoen, M.E., Bartrand, T., Ravenscroft, J.E., Ashbolt, N.J., 2010. Estimated human health risks from exposure to recreational waters impacted by human and non-human sources of faecal contamination. *Water Research* 44. https://doi.org/10.1016/j.watres.2010.06.049.
Staley, Z.R., Chuong, J.D., Hill, S.J., Grabuski, J., Shokralla, S., Hajibabaei, M., Edge, T.A., 2018. Fecal source tracking and eDNA profiling in an urban creek following an extreme rain event. *Scientific Reports* 8. https://doi.org/10.1038/s41598-018-32680-z.
Staley, Z.R., Grabuski, J., Sverko, E., Edge, T.A., 2016. Comparison of microbial and chemical source tracking markers to identify fecal 1 contamination sources in the Humber River and associated stormwater outfalls. https://doi.org/10.1128/AEM.01675-16.
Standard Methods for the Examination of Water and Wastewater, 1999.
Tartera, C., Jofre, J., 1987. Bacteriophages active against Bacteroides fragilis in sewage-polluted waters. *Applied and Environmental Microbiology* 53, 1632–1637. https://doi.org/10.1128/aem.53.7.1632-1637.1987.
Thakur, A. K., Vithanage, M., Das, D., & Kumar, M., 2020. A review on design, material selection, mechanism, and modeling of permeable reactive materials for community-scale groundwater treatment. *Environmental Technology & Innovation* 19, 100917.
Trepeta, R.W., Edberg, S.C., 1984. Methylumbelliferyl-β-D-glucuronide-based medium for rapid isolation and identification of Escherichia coli. *Journal of Clinical Microbiology* 19, 172–174. https://doi.org/10.1128/jcm.19.2.172-174.1984.
United States Environmental Protection Agency, O. of W., 2001. Method 1601: Male-specific (F +) and Somatic Coliphage in Water by Two-step Enrichment Procedure.
Wakayama, H., 厚生労働省, J., n.d. Revision of Drinking Water Quality Standards in Japan 日本の水道水質基準の改正について.
Water and Sanitation – United Nations Sustainable Development [WWW Document], n.d. URL https://www.un.org/sustainabledevelopment/water-and-sanitation/ (Accessed 2.14.21).
WHO | Guidelines for drinking-water quality, fourth edition, 2015. WHO.
Xagoraraki, I., Yin, Z., Svambayev, Z., 2014. Fate of viruses in water systems. *Journal of Environmental Engineering* 140. https://doi.org/10.1061/(ASCE)EE.1943-7870.0000827.
Yates, M. V., Gerba, C.P., Kelley, L.M., 1985. Virus persistence in groundwater. *Applied and Environmental Microbiology* 49, 778–781.

Chapter 9

Pathways for migration of microbial and organic contaminants

Special emphasis on run-off from aquaculture industries

M. Divyagnaneswari
Gargi College

Parasuraman Aiya Subramani
Vels Institute of Science, Technology and Advanced Research

CONTENTS

9.1 INTRODUCTION

Globally fisheries and aquaculture are important sources of food, nutrition, income and livelihoods for hundreds of millions of people. Global food fish consumption increased at an average annual rate of 3.1% from 1961 to 2017, a rate almost twice that of annual world population growth (1.6%) for the same period, and higher than that of all other animal protein foods (meat, dairy, milk, etc.), which increased by 2.1% per year. World per capita fish supply reached to 19.7 kg (Figure 9.1) in 2016 from 13.5 kg in 1990 (FAO 2020a). This has great potential to contribute to the food security and economic benefits to the farmers as well as to the countries. Fish has been consumed as a source of protein by 3.2 billion people worldwide. Fish contain about 16%–20% protein and are excellent source of essential minerals (magnesium, phosphorus, iron, calcium and iodine) and vitamins (A&D) and also contain high level of polyunsaturated fatty acids,

DOI: 10.1201/9781003157465-11

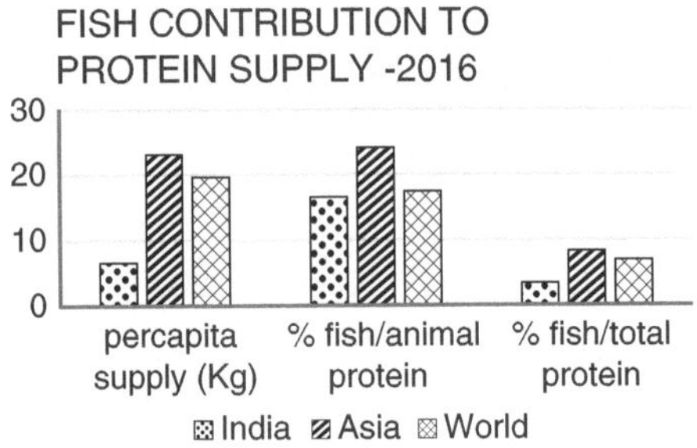

Figure 9.1 Contribution of fish to protein supply in 2016. (FAO 2020.)

especially omega-3 fatty acids. When compared with any other farm animal, aquaculture has high multiplication capacity, in terms of fecundity. The energy requirement for protein production is 22–400 KJ/g, whereas other livestock animal needs 550–3,400 KJ/g. Total aquaculture production from inland and marine has been increased from 61.8 million tonnes to 80 million tonnes in 2017. The drastic increase in world fisheries and aquaculture production was observed from 154 to 179 million tonnes from 2011 to 2018 (Table 9.1). Correspondingly, market value of fish produced is also increasing (Table 9.2) (FAO 2020b). India is facing a similar trend in terms of production (Table 9.3) and value (Figure 9.2).

Increase in aquaculture production directly relates to the increase in number of farms and its increased spatial distribution of the industry. As the industry expands geographically, it poses threat to the environment due to the usage of several chemicals and the waste generated in aquaculture. According to (FAO 2006), the expanding aquaculture has several negative impacts on the ecosystem as well as on the other animals. Some of them are change in water quality, destruction of natural habitats, competition for the use of freshwater, improper use of chemicals and transmission of aquatic animal diseases. There is an interlink between water quality, chemical usage and disease. Decrease in water quality is mainly due to overpopulation of fish and accumulation of faeces and excess feed (Bert 2007). This leads to poor immune system of fish and makes them susceptible to many viral, bacterial, fungal and protozoan diseases. This scenario forces the farmers to use several antibiotics and pharmaceuticals in order to protect them from diseases and also for the treatment of diseases (Halling-Sørensen et al. 1998; Richardson and Bowron 1985). The antibiotics reported to be used in aquaculture are chloramphenicol, avermectins, oxytetracyclines. Though there are several harmless antibiotics like erythromycin, some of them are responsible for potential risk like development of resistant bacteria, especially in aquatic environment, and also, their presence in fish and fish products is detrimental to the consumers. There are numerous reports on the presence of antimicrobials in the sediments as well as the presence of antibiotic-resistant bacteria in the aquaculture field/fish farms. Moreover, the development of antibiotic-resistant bacteria is responsible for more incidences of diseases. Several chemicals and hormones are also used to promote the growth of the fish, and also to promote the monoculture simply for the economic benefit. The most commonly used chemical for sex reversal in fish to promote monoculture is 17 α-methyltestosterone; other chemical compounds used for the same purpose are 17 α-methyldihydrotestosterone,

Table 9.1 World fisheries production by capture + aquaculture (1000t)

Year	*Freshwater fishes*	*Diadromous fishes*	*Marine fishes*	*Crustaceans*	*Molluscs*	*Other aquatic animals*	*Aquatic plants*
1997	5976 +15369	1645+1831	71963 + 695	5601 + 1177	7372 + 8150	640 + 99	1353 + 8271
2007	8787 + 26640	1803+3241	66099 +1725	5871 + 4798	7507 + 13031	415 + 507	1103 + 16338
2017	10560 + 44659	1926 + 5646	66413 + 3098	6812 + 8443	6326 + 17394	471 + 894	1111+ 31811

Table 9.2 World aquaculture production by inland and marine waters (USD 1000)

Year	*Inland*	*Marine waters*
2008	61 929 252	44 836 721
2017	138 863 598	98 686 181

Table 9.3 Fisheries production by capture and aquaculture in India (t)

Year	*Capture*	*Aquaculture*	*Total*
2017	5 427 678	6 180 000	11 607 678
2000	3 666 427	1 942 531	5 608 958
1990	2 782 586	1 017 136	3 799 722

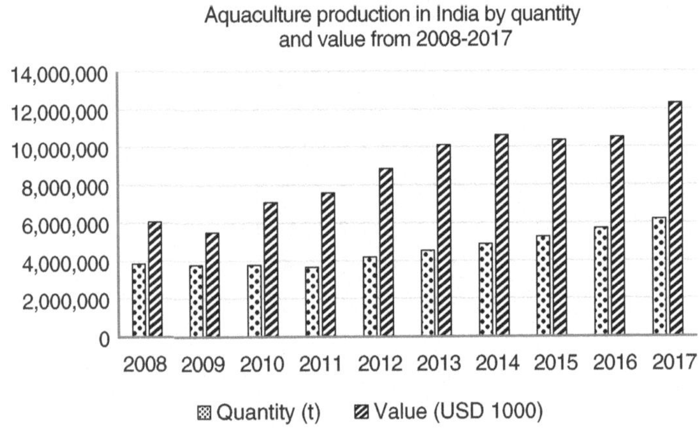

Figure 9.2 Aquaculture production in India. (FAO 2020.)

norethisterone, trenbolone and nortestosterone. These biochemical compounds have huge impact on the ecosystem and have several ecological consequences due to the reason that these hormones are neither metabolized nor degraded properly (Hamscher 2006; Rigos et al. 2004). Other chemicals like malachite green and crystal violet are used to treat fungal and protozoal diseases in fish. Despite the fact that these chemicals are being toxic, it is widely used in many fish farms. In Japan, 26 different compounds were permitted to be used for the treatment in aquaculture in the 1990s (Alderman and Hastings 1998). In another study, the use of at least 13 different antibiotics in shrimp pond in Thailand was reported by Holmström et al. (2003).

The chemicals and the hormones used in the aquaculture pond either settle down at the bottom or enter into the nearby water bodies and agricultural land if it is used for irrigation or into the seas or sea animals (Kennedy et al. 2000; Kolodziej et al. 2004; Vethaak et al. 2002). This leads to contamination of nearby water bodies which in turn results in complex and irreversible effect on the natural ecosystem. The enrichment of water bodies with organics and nutrients (nitrogen and phosphorus) leads to change in its physical (turbidity, temperature and pH) and chemical (dissolved oxygen, COD and BOD) properties, and one of the major consequences is eutrophication and harmful algal blooms (Chua 1993). These changes will definitely hamper the biotic component of the ecosystem in terms of species richness or diversity, density and abundance. The leaching of nutrients is mainly because of inappropriate amount of feed, method of feeding or overfeeding of farm animals and discharging the water directly into the nearby water bodies. Without proper water exchange facilities, the fish farms do not have any other option than to discharge it to nearby water bodies as holding the water may negatively impact the fish farm, leading to disease outbreak and decreased production. The dilution of organic matter and aerobic mineralization observed in seawater minimizes the impact of organic wastes on huge ocean bodies which is not the case in inland fisheries.

The chances of contamination increase wherever the integrated aquaculture is practised for the optimal use of resources. The release of harmful pollutants from the industry is also reported while processing the seafood and also during post-harvest handling of several fishes. The strengthening of methods to monitor the pollution caused by the aquaculture sector is very much needed to identify the aquaculture-related problems. This will also help us to prevent potentially irreversible damage to the ecosystem. Inadequate understanding of the aquatic environment and several misconceptions by the farmers and policymakers is another main reason for such malpractices that result in water pollution. This ultimately will have huge impact on the aquaculture sector itself where water quality is one of the prime concerns. It is our responsibility to make this economically viable sector to cause only minimum destruction to the ecosystem for long-term benefits and sustainability. The development of long-term policies and their enforcement is essential to create awareness among the farmers and in aquaculture-associated industries to accomplish environment-compatible aquaculture and for a sustainable expenditure of natural resources. The concentration of the pollutant can be reduced by using biofilters where bacteria, microalgae and macroalgae can be used. One of the successful methods implemented to reduce pollutants in marine water bodies is the culturing of seaweed. It helps to minimize the effect of eutrophication and also adds economic value to the aquaculture industry.

Integrating aquaculture with other activities to minimize the ecological footprint is a major challenge to the scientific community and farmers due to the non-availability of universal technology for its implementation to reduce the ecological disturbances. In-depth research could be encouraged to devise site-specific tools and technology to explore the capacity of fish farm and to identify the different organisms which can be co-cultured. This will also help us to identify the possibility of collecting, processing and recycling wastes/nutrients, especially organic wastes from fish farms. Traditional aquaculture practices should be explored and promoted if it is ecologically beneficial, less complicated and cost-effective. Traditional methods of aquaculture followed in India like bheries of West Bengal and pokkali fields of Kerala were shown to yield 500–750 kg/ha/year of shrimp without any manuring and feeding (Ayyappan 2006). We need to focus on conducting basic ecological research in fish farms to document the biotic and abiotic components of the system for better understanding and to devise the periodic monitoring protocols. This is possible only when we can have coordination between institutions, farmers and government. Government regulations to evaluate the risk associated with any aquaculture practices in farms as well as in associated industries should be emphasized for an ecologically sustainable aquaculture.

9.2 ORIGIN OF CONTAMINANTS

After knowing the different types of contaminants produced by the aquaculture industry, it is better to know how these contaminants originate and how they eventually end up in polluting the water bodies. The aquaculture industry became a noticeable source of pollution only after it has been industrialized, i.e. when the cultural practices have become extremely intensive and synthetic additives replaced the natural ones. Based on feed/fertilizer input, aquaculture production systems can be classified into:

a. Extensive systems are one of the traditional aquaculture practices in which fish are cultured in large natural water bodies such as ponds and lakes. Seedlings of a single/variety of fish species are introduced into the water body and are cultivated after a certain period. Virtually, the pond is devoid of farmers' care.
b. Semi-intensive systems are slightly improvised systems in which fish are fed intermittently and fertilizers are regularly added to the water body in order to develop supplementary feed (plankton, algae, etc.). However, water quality will not be maintained.
c. Intensive systems are industrial-scale systems in which fish are fed only with formulated feed and water quality is monitored regularly. With the advent of biotechnology, intensive systems use feed formulations containing growth promoters, antimicrobials, immunostimulants, etc. Fish farmers also use genetic and regular husbandry techniques to improve the stock used for culture.
d. As with many of the agricultural sectors, aquaculture suddenly underwent industrialization (intensive systems) in the late 20th century. The industrialization of aquaculture is inevitable because of the stagnant capture fishery and increasing demand for fish meat. On the other hand, the growth of intensive aquaculture is not without detrimental side effects. Fishes grown in intensive culture are more susceptible to infectious diseases which eventually lead to large-scale indiscriminate use of chemical substances (Quesada et al. 2013). One of the most threatening detrimental side effects discussed in detail in this chapter is the unabated release of contaminants into the biosphere. Of the many agents that trigger pollution, the foremost is the fierce competition for the use of limited available land and water resources.

9.2.1 Geographical constraints

It has been estimated that 40% of earth's land mass is used for agriculture and its allied occupations (Froehlich et al. 2018). Aquaculture farmers were found to prefer the use of surface water which has unpredictable water quality parameters over groundwater which has predictable water quality parameters (Larsson 1994). Hence, growing aquaculture must compete for the available, meagre land resources. Indirect pollution caused by the exploitation of land resources like deforestation, biodiversity disruption and increase in greenhouse gas release though important is not discussed in this chapter. Farmers of the leading aquaculture production countries like India have <2 hectares of land in their possession. Improper planning, unprecedented disease outbreak and calamities like drought will drive these small farmers out of aquaculture. Not all soil types and water sources can be used for aquaculture. Aquaculture production will be the highest in clay loamy soil and well-aerated water sources. There are many other geographical parameters that further reduce the suitability of land area available for aquaculture. In India, the abandonment of the aquaculture industry by small-time farmers is well studied. For instance, recent data by Jayanthi et al. (2019) showed that 30.6% of shrimp farms were abandoned in the year 2015 alone. The researchers also projected that in the year 2021, about 11,500 hectares of aquaculture land will be abandoned. This is analogous to the infamous slash-and-burn agriculture. On the other hand, due to the lack of proper knowledge and facilities, the farmers who abandon their farms are reluctant to restart them again but are poised to use the fresh agricultural lands available. If not properly regulated, the abandonment of old farms and opening farms in fresh land phenomenon will start affecting the nearby habitats and eventually human settlements. The abandonment of aquaculture farms severely affects the soil and water quality which is discussed in the next section. While inland aquaculture suffers these serious limitations, not much encouragement is seen in the marine counterparts. Very few sites are used by the farmers for cage and net-pen culture in the open seas. Of the many factors that influence the use of coastal areas for aquaculture, resistance by the coastal communities and accessibility to markets were found to be very important. This implies the fact that even if unlimited space is available, fish farmers tend to set up farms only in certain areas which is the main reason for competition (Hofherr et al. 2015). The factors that cause geographical constraints are given in Figure 9.3.

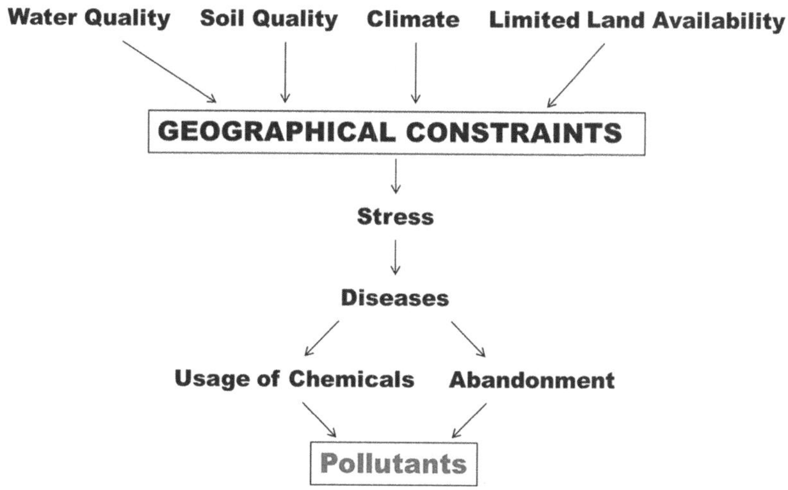

Figure 9.3 Various factors that cause geographical constraints for aquaculture farmers and eventually culminating in the use of chemicals/abandonment of farms.

Due to the limited land resources, fish farmers tend to stock more fish than the permissible level. It is a well-known fact that excessive crowding of fish causes stress by modulating the neuro-endocrine system. The stress caused by crowding may be temporary or chronic, and the higher the duration of stress, the lower the reversibility of damage to the host's immune system (Binuramesh et al. 2005). In the warming world, almost all agricultural sectors are affected. If we take into consideration the amount of plant protein used in the production of aquaculture feed, calamities like floods and droughts may drastically affect aquaculture production indirectly. The direct effects of climate change on aquaculture include raising water temperature causing inhospitable environment for fish, hypoxia due to increased atmospheric CO_2 levels, acidification of water bodies leading to sudden bloom of planktons and increased incidence of emergent pathogens (Barange et al. 2018). Each of these geographical constraints directly or indirectly leads to fish diseases. Most of the known aquatic pathogens are opportunistic, and stressed fish are easy targets for them. In order to treat diseases, farmers will start the indiscriminate use of chemicals including antimicrobials and disinfectants (hypochlorite, potassium permanganate, etc.). Farmers also use a variety of other chemicals like fertilizers, chelating agents (to reduce minerals), flocculent (to remove suspended particles), copper sulphate (herbicide and anti-algal agent) and agents such as lime to increase the water pH. These are harmful not only to the fishes but also to the environment.

9.2.2 Antimicrobials and other chemicals

Of the different pollutants effused by the aquaculture industries, the apparent effects of chemicals used for the control of diseases including antimicrobials and disinfectants are well studied. In the year 2010, farmed animals (excluding aquaculture) were estimated to be fed with >60 thousand tonnes of antimicrobials. It is also projected that by 2030, livestock would be fed with more than 100 thousand tonnes of antimicrobials which is greater than the amount of antimicrobials used by human beings (Van Boeckel et al. 2015). As mentioned elsewhere, antimicrobials are used indiscriminately in aquaculture and the amount of antimicrobials used does not always correlate with the reduction in diseases. Salmon farms alone accounted for the use of >5 thousand tonnes of antimicrobials (Ibrahim et al. 2020). On the other hand, *Piscirickettsia salmonis* an intracellular pathogenic bacterium of salmonid fishes was found to be uncontrollable by antimicrobials and killed nearly 90% of farmed salmons during the peak epizootic in countries like Chile (Makrinos and Bowden 2017). With the increase of earth's average temperature due to global warming, it has been predicted that aquaculture will be the worst affected among the meat industries by zoonotic diseases, particularly in the tropics. Correspondingly, the use of antimicrobials will increase in the tropical countries (Reverter et al. 2020). Tetracycline is the most widely used antimicrobial in aquaculture, and China tops the list of antibiotic users (by weight) in aquaculture. Though many countries claimed to have imposed stringent regulations and approved few (3–5) antimicrobials for commercial use, results from various studies show that the regulations are seldom followed by the fish farmers. In certain Asian countries like Thailand where aquaculture is practised in integrated farms (often along with poultry), antimicrobial usage is very high. Also, in these countries, the farmers are poorly informed about the regulations and directions of use of these chemicals. The farmers tend to add the antimicrobials directly into the water instead of injection or feed additive (which is scientifically preferred). Such acts will have a tremendous impact on the environment (Preena et al. 2020). Antimicrobials used in aquaculture have numerous detrimental effects, of which conferring antimicrobial resistance to pathogens, destruction of beneficial bacteria (in both host and environment) and toxicity to host are described in the next section.

Apart from the antimicrobials, fish farmers routinely use disinfectant chemicals such as chloramine, iodoform, hydrogen peroxide, potassium permanganate, formalin and malachite green. These chemicals are routinely used by the industry even when there is no imminent threat to the crop. The precise use of these chemicals is given in Table 9.4. Fish farmers use most of these chemicals indiscriminately in large quantities and sometimes ritually without understanding the harmful consequences.

Table 9.4 List of disinfectants used in aquaculture (Newman 2006)

Compound	*Description*	*Comments*
Ozone	Toxic derivative of oxygen	Widely used. Can form toxic residues and be very toxic to larval animals.
Quaternary ammonium compounds	Broad-spectrum chemical compounds	Widely used in water on farms in Southeast Asia. Kill bacteria and viruses.
Iodophors	Iodine-based compounds, complexed with PVP	Egg disinfectant, footbath, hand sanitizer. Corrosive.
Chlorinated compounds	Contain chlorine as the active ingredient	Footbath, water sterilization. Most are very toxic to fish and shrimp. Can also be hazardous to workers.
Copper compounds	Contain copper in an elemental or chelated form	Algaecide, although also used as molluscicide. Can be toxic.
Calcium hydroxide, calcium oxides	Lime, burnt lime, etc.	Algaecide, used to kill bacteria and viruses in water, as well. Caustic and potentially hazardous.
Peroxides	Use killing action of oxygen	Algaecide, also used to kill bacteria and viruses in water. Caustic and potentially hazardous.
Pesticides	Varied modes of action	Used to kill viral vectors. At least one is routinely used in shrimp hatcheries as a fungicide.
Aldehydes	Highly reactive chemical groups against proteins	Used to kill parasites and protozoa. Broad-spectrum activity. Potentially hazardous to workers.
Dyes (malachite green, etc.)	Varied modes of action	Used to kill fungi and other organisms. Some are banned for use in aquaculture.

www.aquaculturealliance.org/advocate/disinfectant-use-in-aquaculture/.

9.2.3 Hormones

Amidst controversies, farmed animals are deliberately exposed to various hormone-based additives for the stimulation of certain metabolic processes (Nachman and Smith 2015). It has been reported that fish in aquaculture farms in the USA having 1 tonne of live fish excrete around 1 and 1.3 g/day androstenedione and oestrone, respectively, under normal conditions into open waters (Kolodziej et al. 2004). Sex reversal by hormones is found to be widely used in aquaculture farms to increase production. This is because certain sex of fish attains maturity faster and grows quicker than their counterparts. For instance, all-male tilapia is widely used practice in tilapia farms worldwide (Abo-Al-Ela 2018). Apart from sex reversal, hormones are routinely used in aquaculture to induce reproduction in broodstock especially out of season. Some of the most widely used hormones in aquaculture are crude extracts of pituitary gland, androgens (testosterone) and oestrogens (progesterone and oestradiol). These natural hormones are considered to be safe for use in aquaculture and the carcasses are fit for human consumption as these natural compounds act locally and are degraded rapidly within few days after treatment. However, with the advent of bioengineering and biotechnology products, various synthetic hormones like Ovaprim® (salmon gonadotropin-releasing hormone analogue) are approved for use in various countries. To make the situation worse, hormones used as human medication such as tamoxifen and human chorionic gonadotropin (hCG) are being tested for their use in aquaculture. These synthetic hormones have long half-life and are not readily degraded like their natural counterparts. Hence, the use of these synthetic hormones in aquaculture may have a devastating effect on the health of human beings and other animals if not properly regulated (Celia A Hoga et al. 2018).

9.2.4 Exploitation of Mother Nature

Farmers tend to reuse the waste and leftover of livestock as fodder/manure. Wastes such as blood, feather, scales, skin, intestines and hooves are now considered to be by-products of the meat industry and are widely used in making products like gelatine and meat-and-bone meal-based animal feeds (Hicks and Verbeek 2016). However, in the animal meat industry, the use of animal by-product-based feed is highly regulated and not used in certain countries. In contrast, aquaculture uses fish meal as one of the main feed ingredients. Fish meal is becoming one of the important traded goods. For instance, countries like Thailand which were exporters of inexpensive fish meal are now relying almost entirely on imports from distant North European countries for its use in the shrimp industry. It has been estimated that 50% of fish meal is fed to fish that corresponds to only 10% (by weight) of total aquaculture production. However, this 10% aquaculture production accounts for >50% value in terms of US$. This is nothing but reckless exploitation of nature. The use of wild fish and bycatch fish meal for producing aquaculture feed is also found to have direct effects on the poor and undernourished people who could otherwise derive protein nutrition from the fish meal (Wijkström 2009). Despite the fact that this exploitation of Mother Nature has profound impacts on life quality of the poor, the socio-economic impacts of this exploitation remain widely unreported. Though scientists are trying to substitute fish meal with that of plant-based diets, the results are not encouraging (Deutsch et al. 2007). Hence, high-quality fish meal is an indispensable commodity in aquaculture without which the industry may become prone to diseases or compromise in the quality of fish meat. At that point, fish farmers will turn towards the increased use of chemical disinfectants and synthetic growth promoters. The only possible way out is to culture the fish species like tilapia and carp that require less fish meal in their diet for aquaculture. Even though these fish may be a little lower in terms of value in US$, production costs and exploitation of Mother Nature are comparatively lesser.

9.2.5 Plastics used in aquaculture

Global water bodies are being increasingly filled with plastic wastes, and it has been estimated that about 4–8 million tonnes of plastics reach oceans every year. Majority of this waste (about ¾) comes from land-based sources (Haward 2018). Almost all the gears including tanks, pond lining, ropes, nets, seining elements, pipes and transporting bags are made up of plastics (predominantly of low quality and single-use plastics). Aquaculture is found to be the largest source of microplastic contamination in Qin River which flows into the Maowei Sea (Zhang et al. 2020). Coastal aquaculture systems producing bivalves and shrimps were found to be the major contributors of plastic contamination, followed by cage and net-pen culture of finfish. These fishing gears are abandoned or lost or discarded and eventually find their way to open waters and end up in littering the oceans (Lusher et al. 2017). Unlike most of the other pollutants described in this chapter, plastic wastes degrade into smaller microplastics and eventually accumulate in aquatic animals culminating in the human intake of plastics via aquatic animals (Lyu et al. 2020). The mechanisms of degradation and migration pathways of aquaculture pollutants are described in the next sections.

9.3 EFFECT OF AQUACULTURE ON SURFACE WATER AND GROUNDWATER

The main sources of contamination in aquaculture are the uneaten feed, faeces and ammonia. Uneaten feed and faecal material are usually siphoned out and drained without any treatment. Fishes being ammonotelic organisms excrete ammonia directly into the surrounding water. In intensive culture system, ammonia accumulation is highly dangerous and is constantly removed from the culture systems by various methods. Modern techniques like Biofloc remove the excreted ammonia in situ by employing nitrifying bacteria. However, farmers who are unable to afford the ammonia removal systems clandestinely release the water into open water systems. Unregulated release of all the three contaminants wreaks havoc to surface water bodies including ponds, lakes, rivers, estuaries and seas. Feed of certain aquaculture species like shrimps contains 25%–30% protein which is very rich source of nitrogen. Uneaten feed forms turbid suspension and reduces light permeation and contains organic contaminants that may consume the dissolved oxygen during their mineralization process causing severe deterioration of water quality (Tacon and Forster 2003). Eutrophication and reduction of water qualities are the end results of aquaculture discharge into the open waters. Under controlled conditions, nitrogen-rich aquaculture fish feed stimulated the growth of algae, *Microcystis aeruginosa* (Kong et al. 2020). Enrichment of water bodies as large as the Gulf of Mexico with nitrogen and phosphorus contaminants is shown to create "dead zone" wherein dissolved oxygen levels are reduced to <2 mg/L (hypoxia) leading to mass death of aquatic organisms (Dodds 2006). Sediments rich in nitrogen and other nutrients cause secondary eutrophication of water bodies. Apart from this, various trace metals are added to feed as mineral mix. This mineral mix includes heavy metals such as zinc, copper, selenium and cadmium. Incidence of heavy metal toxicants is prevalent in waters near aquaculture farms (Dean et al. 2007). On the other hand, fish itself is susceptible to bioaccumulation of heavy metals (Karikari et al. 2020).

Groundwater is one of the precious renewable resources that are being unwantedly depleted by negligence. More than 50% of the world's agriculture is dependent upon groundwater. Some of the developing economies like India actively encourage the groundwater depletion by subsidizing the farmers with the electricity and digging charges (Famiglietti 2014). Hence, large-scale aquaculture will become a tough contender that competes with other forms of agriculture. Already, groundwater is becoming a costly affair and had caused several conflicts in some countries, the phenomenon which scientists term hydropolitics (De Stefano et al. 2010). Aquaculture not only depletes the groundwater but also deteriorates the quality of groundwater. While in countries like Armenia which depends almost entirely upon the groundwater for drinking, developmental and security purposes, conventional aquaculture methods are challenging and are considered as a burden. Hence, water usage for aquaculture is heavily regulated and practices like raceway and recirculating aquaculture systems are encouraged (Mirzoyan et al. 2018). Penmetsa et al. (2013) clearly correlated the poor groundwater quality of sampling points with the adjacency to aquaculture farms in East Godavari District of Andhra Pradesh, India. Of the many parameters tested by the investigators, nitrate, ammonia and BOD levels were above the permissible limits in areas closer to aquaculture farms. The investigators also concluded that the presence of both aquaculture and agriculture farms together has synergistic effect in deteriorating the groundwater quality. On the other hand, shrimp aquaculture did not affect the quality of groundwater in the coastal areas (Rekha et al. 2015). This is due to the fact that marine aquifers/seawater has better buffering capabilities compared to freshwater.

Antibiotic pollutants from aquaculture that get discharged into open water bodies will affect the aquatic animals by modulating their metabolism and immune system. At the same time, the presence of excess antibiotics will induce resistance in aquatic microbiota which gets transmitted horizontally and gives rise to multidrug-resistant pathogens. Numerous studies show that the presence of antibiotic resistance genes in aquatic bacteria is correlated with the amount of antibiotic used in the farm (Bojarski et al. 2020). Though the exact amount of residues present in the final products is not evaluated for most countries, certain products like fish fillet from Nigeria were found to have oxytetracycline residues greater than 500 μg/kg in it. Many studies show that consumers of such fish products have increased cancer risk apart from other toxic effects (Okocha et al. 2018). Antibiotics induce free radical reactions and suppress many immune-related signalling mechanisms. Antibiotics were also shown to induce mortality of embryo and larva of different aquatic animals. Antibiotics inhibit protein synthesis, teratogenesis and genotoxicity in algae and fish (Bilal et al. 2020).

A number of bacteria (Gauthier 2015) and protozoa (Lima dos Santos and Howgate 2011) from the aquaculture industry were reported as the causative factors of zoonosis. The outbreaks of *Aeromonas dhakensis* were well-studied clinical cases of aquatic ecosystems related to zoonosis (Chen et al. 2016). Such bacteria are considered to be causing new and emerging human infections which will aggravate in future due to global warming and increased incidence of multidrug resistance in aquatic environments. Hormones released from aquaculture industries affect the developmental stages of various aquatic animals and amphibians. In higher concentrations (>100 mg/L), certain hormone pollutants inhibited the growth of phytoplankton (Sacdal et al. 2020). Exposure to low doses of xenoestrogens causes the development of female characteristics in males, reproductive dysfunction in females and higher doses may become fatal.

9.4 FATE OF CONTAMINANTS FROM AQUACULTURE

The total reactive nitrogen (Nr) emission from aquaculture is increasing at a rate of ~5% per annum. In China, the topmost aquaculture producer, input of nitrogenous compounds into the aquaculture industry soared >100 times over a period of 37 years (1978–2015). To be precise, 1.8 million tons of nitrogen (in the form of feed, drugs, fish fry) was pumped into aquaculture farms in China. Due to various factors, nearly 50% of this input is converted to Nr and is released back into the environment in various (solid, liquid and gaseous) forms. At the present conditions, most of the Nr (~50%) released from aquaculture tend to remain solubilized in the water that holds the fish called water accumulation. It has been found that water containing 2.90 million tons of Nr is being discharged into the ocean by the Chinese aquaculture industry. Increased denitrification in aquaculture farms results in the release of N_2O which is a potent greenhouse gas. Depending on the water quality and animal behaviour, there are chances of Nr being released as NH_3 into the atmosphere by a process called ammonia volatilization. Ammonia volatilization is increasing in aquaculture recently and severely depletes the feed nutrients which in turn lead to increased usage of feed. Increased feed use will in turn increase the Nr burden of the environment. Another small portion of the Nr gets deposited as sediments in water bodies and river beds where the aquaculture effluents are discharged (Luo et al. 2018). Another principal nutrient in aquaculture, phosphorus also shows a similar fate as that of Nr. However, both nitrogen and phosphorus contaminants do not readily move/diffuse in water bodies unless deliberately discharged into water currents such as rivers or oceans (Guo et al. 2009).

Even before the explosion of aquaculture in the 21st century, reports about the prevalence of antibiotic contaminants in aquatic environment are widely available. Antibiotics expelled from aquaculture farms can either stay solubilized in surface water or settle as sediments along with faeces and other debris. According to Awad et al. (2014), >30% of antibiotics administered to farmed animals are not absorbed or metabolized and are directly excreted via faeces and urine into the surrounding environment. Since in aquaculture antibiotics are administered mainly via feed as in their terrestrial counterparts, fish excrete a major portion of it through faeces. The minuscule amount present in the gut may influence the gut microbiota to express their antibiotic resistance genes (ARGs) which in turn leads to horizontal transfer of ARGs to other bacteria in the environment. High amount of antibiotics excreted via faeces gets sedimented and gives rise to multidrug-resistant bacteria which include pathogens. Since bacteria can acquire antibiotic resistance both inside and outside of fish, the use of antibiotics is harmful to it (Bojarski et al. 2020). On the brighter side of this, many studies indicate that these induced ARGs are short-lived in the environment and are quickly degraded if the resistant bacteria are not transmitted to a suitable host (Kraemer et al. 2019).

Natural hormones used in aquaculture industries are usually metabolized well by the fishes and are excreted as non-toxic substances through faeces. Oestrogens that are administered to fish are rapidly metabolized by the liver to inactive conjugates and are excreted. However, these excreted hormone conjugates are converted back to their active form by microbial action when they are released into open waters. Once the compounds become activated, the hormones alter the physiology of various organisms and eventually human beings (Celia A. Hoga et al. 2018). Microplastic contaminants become persistent in their local dump site and gradually move to different places dependent upon water current and animal activity. The presence of microplastics in surface water and benthic sediments is widely studied. Eventually, microplastics are ingested by filter-feeding organisms such as molluscs (oysters) and shrimps. Biomagnification of microplastics across the higher trophic levels (carnivorous fish) was also reported. Finally, the microplastic contamination reaches human beings who consume those fish and shrimp (Lusher et al. 2017).

9.5 MIGRATION OF CONTAMINANTS

Due to overfeeding, leaching, etc., sizeable amount of the nutrients pumped into aquaculture via feed is lost. A significant fraction of nutrients is lost due to low feed conversion ratio and is excreted in the form of urine or faeces as mentioned in the previous sections. It has been estimated that 90% of nitrogen and 50% of phosphorus present in fish excreta dissolve and remaining portion sediments (Fernandes et al. 2007). About 70% of feed carbon input is lost for similar reasons. In addition, intensive aquaculture is also a source of dissolved CO_2 which is the end product of fish respiration. Nutrients and minerals from aquaculture effluents are either recycled naturally (follow carbon, nitrogen, phosphorus cycle, etc.) or can cause sudden, unwanted bloom of algae called eutrophication (Wang et al. 2012). Phytoplanktons fiercely compete for these nutrient resources and assimilate them before it gets sedimented. Even when sedimented, the benthic and detritus organisms utilize them. However, there are studies reporting the migration of inorganic nitrogen and other nutrients to groundwater (Giammarino and Quatto 2015).

Voitsitskiy et al. (2019) studied in detail the kinetics of migration of antibiotics along the various strata and trophic levels of water bodies. It has been found that the migration is entirely dependent upon the concentration of the antibiotics and hence follows the first-order kinetics. Antibiotics tend to aggregate and form colloidal suspension in stagnant water bodies. These colloidal particles later agglomerate, often with other contaminants, and are deposited as silt. Antibiotics also enter different trophic levels as benthic forms and plants accumulate the antibiotics from the deposited silt. This will transport the antibiotics vertically upwards when organisms of higher trophic level feed them. Studying the migration of antibiotics in flowing water bodies is rather a complex process as it involves several factors like water current, topography of the benthic zone and chemical composition. Though it takes a longer time, if the antibiotics are not taken up by any organism, it tends to settle to the bottom similar to the mechanisms described above. Sedimented antibiotics can percolate into the soil by diffusion of colloidal particles and eventually reach the groundwater aquifer. Antibiotics that percolated into the groundwater eventually brought back to the surface for human consumption or irrigation. The antibiotic resistance genes acquired by bacteria upon exposure to low dose of antibiotics were not found to be advantageous to the survival of the bacteria in natural aquatic environments. As mentioned above, resistance genes are usually discarded by the host under unlimited nutrient resources (Na et al. 2018). However, when the burden of antibiotics is heavy, bacteria with antibiotic resistance tend to thrive better. The pathway by which the chemicals migrate to the surface is given in Figure 9.4. Natural steroidal oestrogens are extremely photolabile and are hydrophobic. Hence, they are rapidly destroyed when present in surface. However, when sedimented, these hormones are relatively stable and have a half-life of up to 6 days. The stability of hormones increases under anaerobic conditions. Synthetic hormones are relatively stable and readily migrate to the subsurface, i.e. into the groundwater. Hormones are also readily assimilated by plants and animals (Adeel et al. 2017).

The migration of microplastics along various strata of water bodies is studied using mathematical methods. Hydrodynamics, temperature and water current are the important factors that play a key role in the transport of microplastics. Sinking (sedimentation) of microplastics is found to be the final resting place of microplastics as reported in most of the studies. Microplastic sedimentation is dependent upon its density and fouling (colonization on the surface by microbes). The more dense and higher fouling microplastics will settle faster. There are also numerous reports on the adsorption of various substances on sinking microplastics. These include heavy metals and antibiotics among others that critically affect the fouling part and consumption of these microplastics by aquatic organisms. Microplastics sedimented at the bottom of the water bodies tend to remain there for prolonged time period unless consumed by the organisms dwelling in the bottom. These ingested microplastics then migrate vertically and reach the surface again as particles suspended in faeces (Akdogan and Guven 2019).

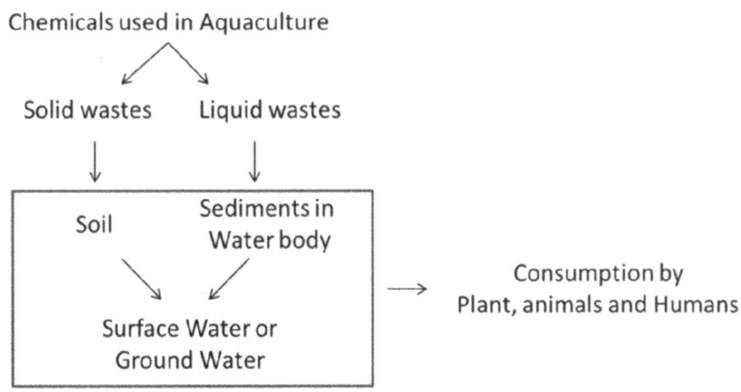

Figure 9.4 Migration of chemicals used in aquaculture.

9.6 CONCLUSION AND FUTURE REMARKS

Aquaculture is a growing meat industry which should quickly learn from the mishaps that happened in other industries. At present, most of the estimations conclude that aquaculture is underutilizing the natural resources. However, the negative impacts of aquaculture on water bodies are already evident. This implies that the worst is yet to come. Hence, fish farmers and policymakers should work together and should prevent any catastrophe that may hinder the growth rate of aquaculture. Effluents from aquaculture should not be considered as contaminants anymore and could be used as fertilizer for agriculture after minor treatments. Fish farmers should undertake more advanced and productive techniques like aquaponics which will benefit them economically. Feed production should be greatly regulated; especially, the protein and mineral additives should be minimal. Best sustainable alternative should replace fish meal in feed formulations. Antibiotics present in aquaculture effluents, if any, should be inactivated before being discarded. Land and water usage for aquaculture should be licensed and rationed. Governments should encourage the culture of traditional fish species which are more resistant to endemic pathogens. The traditional aquaculture practices in India include bheries of West Bengal and pokkali fields of Kerala. Bheries are man-made impoundments in coastal wetlands, and pokkali is salt-resistant deep-water paddy field in the coastal areas of Kerala. The production levels achieved here are between 500 and 750 kg/ha/year of shrimp without any manuring, feeding or any other additional input.

REFERENCES

Abo-Al-Ela, H.G., 2018. Hormones and fish monosex farming: A spotlight on immunity. *Fish and Shellfish Immunology* 72, 23–30.

Adeel, M., Song, X., Wang, Y., Francis, D., Yang, Y., 2017. Environmental impact of estrogens on human, animal and plant life: A critical review. *Environment International* 99, 107–119.

Akdogan, Z., Guven, B., 2019. Microplastics in the environment: A critical review of current understanding and identification of future research needs. *Environmental Pollution* 25, 113011.

Alderman, D., Hastings, T., 1998. Antibiotic use in aquaculture: development of antibiotic resistance–potential for consumer health risks. *International Journal of Food Science and Technology* 33, 139–155.

Awad, Y.M., Kim, S.-C., Abd El-Azeem, S.A., Kim, K.-H., Kim, K.-R., Kim, K., Jeon, C., Lee, S.S., Ok, Y.S., 2014. Veterinary antibiotics contamination in water, sediment, and soil near a swine manure composting facility. *Environmental Earth Sciences* 71, 1433–1440.

Ayyappan, S., 2006. National aquaculture sector overview: India.

Barange, M., Bahri, T., Beveridge, M.C., Cochrane, K.L., Funge-Smith, S., Poulain, F., 2018. Impacts of climate change on fisheries and aquaculture.

Bert, T.M. 2007. *Ecological and Genetic Implications of Aquaculture Activities* (Vol. 6): Springer Science & Business Media, Berlin.

Bilal, M., Mehmood, S., Rasheed, T., Iqbal, H.M.N., 2020. Antibiotics traces in the aquatic environment: persistence and adverse environmental impact. *Current Opinion in Environmental Science & Health* 13, 68–74.

Binuramesh, C., Prabakaran, M., Steinhagen, D., Michael, R.D., 2005. Effect of chronic confinement stress on the immune responses in different sex ratio groups of *Oreochromis mossambicus (Peters)*. *Aquaculture* 250, 47–59.

Bojarski, B., Kot, B., Witeska, M., 2020. Antibacterials in Aquatic Environment and Their Toxicity to Fish.

Chen, P.-L., Lamy, B., Ko, W.-C., 2016. Aeromonas dhakensis, an Increasingly Recognized Human Pathogen.

Chua, T.-E. 1993. *Environmental management of coastal aquaculture development.* Paper presented at the Environment and Aquaculture in Developing Countries. *ICLARM Conf. Proc.*

De Stefano, L., Edwards, P., De Silva, L., Wolf, A.T., 2010. Tracking cooperation and conflict in international basins: historic and recent trends. *Water Policy* 12, 871–884.

Dean, R.J., Shimmield, T.M., Black, K.D., 2007. Copper, zinc and cadmium in marine cage fish farm sediments: An extensive survey. *Environmental Pollution* 145, 84–95.

Deutsch, L., Gräslund, S., Folke, C., Troell, M., Huitric, M., Kautsky, N., Lebel, L., 2007. Feeding aquaculture growth through globalization: Exploitation of marine ecosystems for fishmeal. *Global Environmental Change* 17, 238–249.

Dodds, W.K., 2006. Nutrients and the "dead zone": The link between nutrient ratios and dissolved oxygen in the northern Gulf of Mexico. *Frontiers in Ecology and the Environment* 4, 211–217

Famiglietti, J.S., 2014. The global groundwater crisis. *Nature Climate Change* 4, 945–948.

FAO, 2006. State of World Aquaculture: 2006.

FAO, 2020a. FAO yearbook. Fishery and Aquaculture Statistics. 2018/FAO annuaire. In: FAO Rome.

FAO, 2020b. The state of world fisheries and aquaculture 2020. Sustainability in action.

Fernandes, M., Angove, M., Sedawie, T., Cheshire, A., 2007. Dissolved nutrient release from solid wastes of southern bluefin tuna (Thunnus maccoyii, Castelnau) aquaculture. *Aquaculture Research* 38, 388–397.

Froehlich, H.E., Runge, C.A., Gentry, R.R., Gaines, S.D., Halpern, B.S., 2018. Comparative terrestrial feed and land use of an aquaculture-dominant world. *Proceedings of the National Academy of Sciences* 115, 5295–5300.

Gauthier, D.T., 2015. Bacterial zoonoses of fishes: A review and appraisal of evidence for linkages between fish and human infections. *The Veterinary Journal* 203, 27–35.

Giammarino, M., Quatto, P., 2015. Nitrates in drinking water: Relation with intensive livestock production. *Journal of Preventive Medicine and Hygiene* 56, E187–E189.

Guo, L., Li, Z., Xie, P., Ni, L., 2009. Assessment effects of cage culture on nitrogen and phosphorus dynamics in relation to fallowing in a shallow lake in China. *Aquaculture International* 17, 229–241.

Halling-Sørensen, B., Nielsen, S.N., Lanzky, P., Ingerslev, F., Lützhøft, H.H., Jørgensen, S., 1998. Occurrence, fate and effects of pharmaceutical substances in the environment-A review. *Chemosphere* 36, 357–393.

Hamscher, G., 2006. Veterinary pharmaceuticals. In: Reemtsma T., Jekel M. (eds.) *Organic Pollutants in the Water Cycle* (pp. 99–120). Wiley-VCH, Weinheim.

Haward, M., 2018. Plastic pollution of the world's seas and oceans as a contemporary challenge in ocean governance. *Nature Communications* 9, 667.

Hicks, T.M., Verbeek, C.J.R. 2016. Chapter 3- Meat industry protein by-products: Sources and characteristics. In G. Singh Dhillon (ed.) *Protein Byproducts* (pp. 37–61). Academic Press, Cambridge, MA.

Hofherr, J., Natale, F., Trujillo, P., 2015. Is lack of space a limiting factor for the development of aquaculture in EU coastal areas? *Ocean & Coastal Management* 116, 27–36.

Hoga, C.A., Almeida, F.L., Reyes, F.G., 2018. A review on the use of hormones in fish farming: Analytical methods to determine their residues. *CyTA - Journal of Food* 16, 679–691.

Holmström, K., Gräslund, S., Wahlström, A., Poungshompoo, S., Bengtsson, B.E., Kautsky, N., 2003. Antibiotic use in shrimp farming and implications for environmental impacts and human health. *International Journal of Food Science & Technology* 38, 255–266.

Ibrahim, M., Ahmad, F., Yaqub, B., Ramzan, A., Imran, A., Afzaal, M., Mirza, S.A., Mazhar, I., Younus, M., Akram, Q., Ali Taseer, M.S., Ahmad, A., Ahmed, S., 2020. Current trends of antimicrobials used in food animals and aquaculture, pp. 39–69.doi:10.1016/B978-0-12-818882-8.00004-8.

Jayanthi, M., Ravisankar, T., Nagaraj, G., Thirumurthy, S., Muralidhar, M., Saraswathy, R., 2019. Is aquaculture abandonment a threat to sustainable coastal resource use?–A case study of Andhra Pradesh, India, with options for reuse. *Land Use Policy* 86, 54–66.

Karikari, A.Y., Asmah, R., Anku, W.W., Amisah, S., Agbo, N.W., Telfer, T.C., Ross, L.G., 2020. Heavy metal concentrations and sediment quality of a cage farm on Lake Volta, Ghana. *Aquaculture Research* 51, 2041–2051.

Kennedy, D.G., Cannavan, A., McCracken, R.J., 2000. Regulatory problems caused by contamination, a frequently overlooked cause of veterinary drug residues. *Journal of Chromatography A* 882, 37–52.

Kolodziej, E.P., Harter, T., Sedlak, D.L., 2004. Dairy wastewater, aquaculture, and spawning fish as sources of steroid hormones in the aquatic environment. *Environmental Science and Technology* 38, 6377–6384.

Kong, W., Huang, S., Yang, Z., Shi, F., Feng, Y., Khatoon, Z., 2020. Fish feed quality is a key factor in impacting aquaculture water environment: Evidence from incubator experiments. *Scientific Reports* 10, 1–15.

Kraemer, S.A., Ramachandran, A., Perron, G.G., 2019. Antibiotic Pollution in the Environment: From Microbial Ecology to Public Policy.

Larsson, B., 1994. Three overviews on environment and aquaculture in the tropics and sub-tropics.

Lima dos Santos, C.A.M., Howgate, P., 2011. Fishborne zoonotic parasites and aquaculture: A review. *Aquaculture* 318, 253–261.

Luo, Z., Hu, S., Chen, D., 2018. The trends of aquacultural nitrogen budget and its environmental implications in China. *Scientific Reports* 8, 10877.

Lusher, A., Hollman, P., Mendoza-Hill, J. 2017. Microplastics in fisheries and aquaculture: status of knowledge on their occurrence and implications for aquatic organisms and food safety: FAO.

Lyu, W., Chen, Q., Cheng, L., Zhou, W., 2020. Microplastics in aquaculture systems and their transfer in the food chain. doi:10.1007/698_2020_455.

Makrinos, D.L., Bowden, T.J., 2017. Growth characteristics of the intracellular pathogen, Piscirickettsia salmonis, in tissue culture and cell-free media. *Journal of Fish Diseases* 40, 1115–1127.

Mirzoyan, N., Avetisyan, N., Mnatsakanyan, H., Tadevosyan, L., 2018. Groundwater use and efficiency in small- and medium-sized aquaculture farms in Ararat Valley, Armenia. *Groundwater for Sustainable Development* 6, 1–5.

Na, G., Lu, Z., Gao, H., Zhang, L., Li, Q., Li, R., Yang, F., Huo, C., Yao, Z., 2018. The effect of environmental factors and migration dynamics on the prevalence of antibiotic-resistant Escherichia coli in estuary environments. *Scientific Reports* 8, 1663.

Nachman, K.E., Smith, T.J., 2015. Hormone use in food animal production: Assessing potential dietary exposures and breast cancer risk. *Current Environmental Health Reports* 2, 1–14.

Okocha, R.C., Olatoye, I.O., Adedeji, O.B., 2018. Food safety impacts of antimicrobial use and their residues in aquaculture. *Public Health Reviews Volume* 39, 21.

Penmetsa, A., Muppidi, S.R., Popuri, R., Golla, S.B., Tenneti, R., 2013. Aquaculture and its impact on ground water in East Godavari District Andhra Pradesh, India–A case study.

Preena, P.G., Swaminathan, T.R., Kumar, V.J.R., Singh, I.S.B., 2020. Antimicrobial resistance in aquaculture: a crisis for concern. *Biologia* 75, 1497–1517.

Quesada, S.P., Paschoal, J.A.R., Reyes, F.G.R., 2013. Considerations on the aquaculture development and on the use of veterinary drugs: Special issue for fluoroquinolones—A review. *Journal of Food Science* 78, R1321–R1333.

Rekha, P.N., Gangadharan, R., Ravichandran, P., Mahalakshmi, P., Panigrahi, A., Pillai, S., 2015. Assessment of impact of shrimp farming on coastal groundwater using Geographical Information System based Analytical Hierarchy Process. *Aquaculture* 448, 491–506.

Reverter, M., Sarter, S., Caruso, D., Avarre, J.-C., Combe, M., Pepey, E., Pouyaud, L., Vega-Heredía, S., de Verdal, H., Gozlan, R.E., 2020. Aquaculture at the crossroads of global warming and antimicrobial resistance. *Nature Communications* 11, 1870.

Richardson, M.L., Bowron, J.M., 1985. The fate of pharmaceutical chemicals in the aquatic environment. *Journal of Pharmacy and Pharmacology* 37, 1–12.

Rigos, G., Nengas, I., Alexis, M., Troisi, G.M., 2004. Potential drug (oxytetracycline and oxolinic acid) pollution from Mediterranean sparid fish farms. *Aquatic Toxicology* 69, 281–288.

Sacdal, R., Madriaga, J., Espino, M.P., 2020. Overview of the analysis, occurrence and ecological effects of hormones in lake waters in Asia. *Environmental Research* 18, 109091.

Tacon, A.G.J., Forster, I.P., 2003. Aquafeeds and the environment: Policy implications. *Aquaculture* 226, 181–189.

Van Boeckel, T.P., Brower, C., Gilbert, M., Grenfell, B.T., Levin, S.A., Robinson, T.P., Teillant, A., Laxminarayan, R., 2015. Global trends in antimicrobial use in food animals. *Proceedings of the National Academy of Sciences* 112, 5649.

Vethaak, A., Rijs, G., Schrap, S., Ruiter, H., Gerritsen, A., Lahr, J., 2002. Estrogens and xeno-estrogens in the aquatic environment of The Netherlands: Occurrence, potency and biological effects.

Voitsitskiy, V., Danchuk, V., Ushkalov, V., Midyk, S., Kepple, O.Y., Danchuk, O., Shevchenko, L., 2019. Migration of antibiotics residual quantities in aquatic ecosystems.

Wang, X., Olsen, L.M., Reitan, K.I., Olsen, Y., 2012. Discharge of nutrient wastes from salmon farms: Environmental effects, and potential for integrated multi-trophic aquaculture. *Aquaculture Environment Interactions* 2, 267–283.

Wijkström, U.N. 2009. The use of wild fish as aquaculture feed and its effects on income and food for the poor and the undernourished. In: M.R. Hasan and M. Halwart (eds.) *Fish as Feed Inputs for Aquaculture: Practices, Sustainability and Implications* (Vol. 518, pp. 371–407): FAO, Rome.

Zhang, L., Liu, J., Xie, Y., Zhong, S., Yang, B., Lu, D., Zhong, Q., 2020. Distribution of microplastics in surface water and sediments of Qin river in Beibu Gulf, China. *Science of the Total Environment* 708, 135176.

Chapter 10

Analysis of emerging microbial contaminants through next-generation sequencing

Shubha Rani Sharma
Birla Institute of Technology

Avinash Singh, Neha Singh, and Amit Prasad
IIT Mandi

Vinod Kumar Nigam
Birla Institute of Technology

CONTENTS

DOI: 10.1201/9781003157465-12

10.1 INTRODUCTION

The genetic material in almost all organisms found on earth is either deoxyribonucleic acid (DNA) or ribonucleic acid (RNA). Approximately, all contagious agents carry DNA or RNA as their genomes whose sequencing can prove to be a marvelous approach in the detection of pathogenic agents. When the sample is contaminated with microbes, it is difficult to correctly analyze the data in the biological research. Microbial contamination is a concern among all cereals, but the greatest problem is with malt barley where moderate temperatures and high humidity during malting provide ideal conditions for microbial growth. Microbial contamination of fresh meat occurs during slaughtering, carcass dressing, and packaging. Microbial contamination of dairy products is often through the release of bacteria from biofilms developing on almost any surface of food manufacturing unit. Although there are certain conventional methods for the detection of microbial contamination like various sequencing methods including Sanger's method of sequencing as well as Maxim and Gilbert's method of sequencing, now there is a need for a smarter approach for their detection. The problem with the previous sequencing methods is that they need considerable use of hazardous chemicals. They also have a complex setup and cannot be utilized to investigate more than 500 base pairs. With the advent of the various computational techniques which are now using new strategies like next-generation sequencing (NGS) which happens to be a very conducive method for detection of the microbial contaminants (Gu et al., 2019; Mohapatra et al., 2020). The microbial contamination of our samples can be due to various types of microorganisms within the same sample. Thus, there is a need for microbial typing, which is the process of identification of this type of microbes. Even though the identification of the microbial contaminants is still being done by the conventional techniques like phenotyping as well as molecular methods like sequencing, an urgent need for computational methods like NGS has been invented and is being used on a wide scale to identify the microbial contaminants. NGS is a resourceful tool that can be applied for the identification of a wide variety of microbes like viruses, bacteria, and fungi. The choice of the identification of technique depends upon the purpose of sequencing as well as the considerations like accuracy, efficiency, and the cost involved. Short-read sequencing that generates about 1,000 base pairs long sequences has been applied for the regular sequencing of the microbial contaminants in the clinical and medical laboratories of the USA (Gwinn et al., 2019).

NGS can be defined as an assemblage of highly developed technologies that establish the precise array of nucleotides in the DNA molecules. The NGS technology can identify the microbial contaminants down to the genus as well as species level based on the sequence of genes. Rapid and accurate identification of pathogens through NGS technologies has been a big achievement that can provide novel vision in scientific research. An NGS technique for microbial identification commences with the whole genome sequencing (WGS) with the complete DNA extracted from the sample for metagenomic studies (Mayo et al., 2014). From the extracted DNA, a genomic library is prepared with the help of polymerase chain reaction (PCR) and a bioinformatics tool is used for data analysis (Figure 10.1). Though the technique was a very costly affair previously price of high-throughput sequencing (HTS) or NGS has been reduced very abruptly since 2004 when this technique was introduced (Reuter et al., 2015). This review aims at encompassing a whole lot of applications of NGS in the diagnosis of microbial contamination in various areas of scientific research as well as technologies. In recent times, NGS technologies have become a versatile tool for the classification of foodborne pathogens, microbial contaminants, economically important microbes, etc. NGS is also being used for studying the functionality of the genome by extrapolating at the level of RNA with the metatranscriptomic technique.

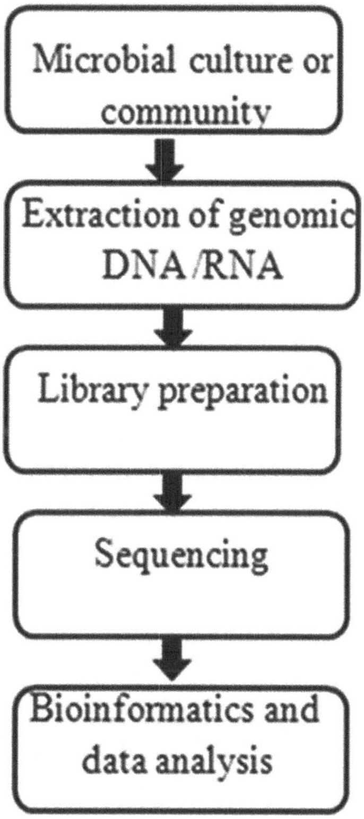

Figure 10.1 The workflow of next-generation sequencing (NGS). The steps include obtaining microbial community from the sample, then extraction of DNA, followed by the DNA library preparation. Now, sequencing is done and the data analyzed by various bioinformatics tools.

10.2 DIFFERENT MICROBIAL CONTAMINATIONS

The term microbiological contamination means the unintentional entry of the microbes like bacteria and viruses along with the toxins secreted by them. It has been observed that the genus *Bacillus* is the most common aerobic bacteria to present as a contaminant in laboratory and medical samples (Farrar & Reboli, 2006). Anaerobic bacteria like *Bifidobacterium* and *Clostridium* are the most common proponents in indoor environments (Novakova et al., 2014). Among the fungi, *Penicillium* and *Aspergillus* were found with the highest incidence (Khan & Karuppayil, 2012). These microorganisms, because of their enzymatic or toxigenic activities, have posed grave problems in pharmaceutical, food, agricultural, and analytical areas. Microbial contamination of food has been due to various microbes like bacteria and viruses. To quote a few examples the undercooking of chicken can result in contamination of bacteria *Campylobacter*, bacteria like *Salmonella* can be transferred in their products during the process of rearing and slaughtering of animals (Facciolà et al., 2017). The humans who eat fish and shellfish are prone to get infected as the fishes may be involved in consuming toxin-producing microbes. Prions, viruses, bacteria, and other prokaryotes are also known to contaminate the various samples. A human who ingests beef contaminated with bovine spongiform encephalopathy (BSE) is found to suffer from human spongiform encephalopathies. So it can be said that the food, as well as water contaminated with prions, may be a major problem for human beings (Saunders et al., 2008, Mazumder et al., 2020, Kumar et al., 2020a,b). Additionally, these microbes are extremely small as compared to other microbes that are difficult to eliminate (Bartelt-Hunt & Bartz, 2013). Also, their resistance to various physical and chemical agents endows them with the property to stay in the environment for a longer time.

10.3 CONVENTIONAL METHODS FOR DETECTION OF MICROBIAL CONTAMINATION

Several conventional methods are used for the detection of contamination of microbes.

a. **Detection of viral contaminants by culture method**: Various enteric viruses can be cultured in susceptible hosts of animal models or mammalian cell lines and thus can be detected further.
b. **Combined method of cell culture and nucleic acid detection as well as amplification**: There are some protozoan parasites and enteric viruses whose detection and quantitation can be enhanced by a combination of two different techniques like cell culture and amplification by PCR or RT-PCR. The sample to be analyzed is inoculated into the vulnerable host cell cultures and further incubated to permit the viruses or parasites to contaminate the cells and propagate.
c. **Detection of the pathogens and contaminants by nucleic acid method**: The different steps followed in the detection of pathogens and contaminants are cloning of nucleic acids, synthesis, hybridization, sequencing, and various bioinformatic analysis tools. But due to several shortcomings like requirement of large sample volumes, ambiguities, and failures in the hybridization reaction, uncertainties of positive reactions indicating infection are the problems associated with this method.

d. **Detection of microbes in foods by test kits**: The toxins produced by the various contaminants in food can be examined with the help of test kits, which are based on assays by ELISA, real-time PCR tests, etc. RIDASCREEN® assay is the immunological method (ELISA), which is based on the sandwich technology for the detection of the contaminants. In this method, the bacterial antigen is arrested in the microplate well and detected by adding a second antibody, which is marked with a special detection enzyme.

e. **Detection of microbial contaminants in the production conditions: In the process of production of various substances, the inadequately cleaned surfaces may carry microbes, which can be detected in the food samples.** Handy Compact Dry plates are used for colony counting in foods as well as different surfaces (Kodaka et al., 2005). The microbes are cultivated on standard nutrient media with chromogenic substrates, which are then converted to colored products by the microbial metabolism.

f. **Rapid methods**: The qualitative detection of microbes can be used for checking the presence of microbes by just measuring the variation in the pressure, impedance, etc. The detection is based on the measurement of variation in conductance due to bacterial growth. The direct impedance technique uses the variation in the conductivity of the medium as a parameter for measuring growth, while in the indirect impedentiometry, the amount of gas absorbed from the bacterial culture is correlated with the variation in the electrical conductivity of a reaction suspension.

10.4 ADVENT OF NEXT-GENERATION SEQUENCING AND VARIOUS NGS TECHNOLOGIES

a. **Solexa sequencing technology or Roche 454 sequencing**: The method of sequencing is based on pyrosequencing that identifies the release of pyrophosphate utilizing fluorescence of the nucleotides that are added to the new strand by the enzyme polymerase (Heather & Chain, 2016).

b. **Ion torrent**: This technique also known as Proton or Personal Genome Machine (PGM) sequencing measures the direct liberation of H+ ions from the integration of bases by DNA polymerase (Lahens et al., 2017).

c. **Illumina technology**: The Illumina technology involves the terminator technology in sequencing like in Sanger's sequencing method. In this method, a fluorescently labeled base is incorporated in the growing strand of DNA, and just after the incorporation of the fluorescent base, the elongation of the strand is stopped. The only difference of this method with Sanger's sequencing is that the terminated strand is one base longer as incorporated in fluorescently labeled base.

   When the sequencing is taking place, four different fluorescently labeled nucleotides are offered to the growing strand, and then, the blocked bases are added to terminate the reaction. During the reaction, bases complementary to the template strand are added and a fluorescent signal is detected. The cycle will be repeated and will only be terminated after the addition of 3'OH (Figure 10.2).

d. **Nanopore technology**: Nanopore technology is also known as third-generation sequencing (TGS). This is considered a single-molecule sequencing (SMS) technology (Maitra et al., 2012; Pantha et al., 2021). Nanopore sequencing has an option of either with or without PCR amplification, so here any replication enzymatic

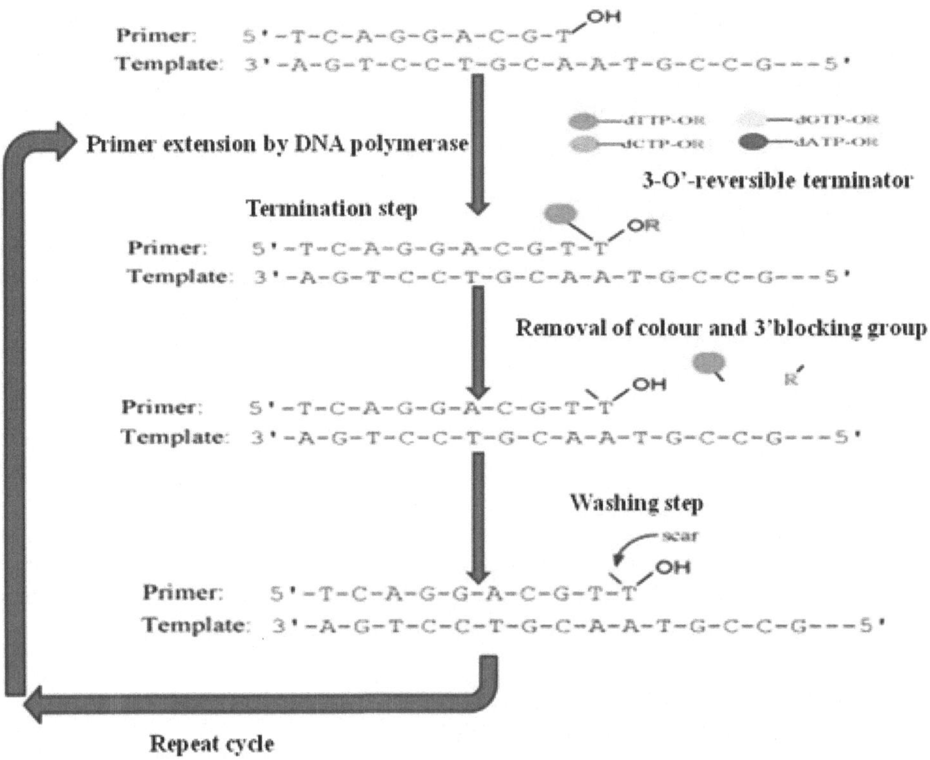

Figure 10.2 Reversible terminator technologies during sequencing by synthesis.

system may not be required and also this is a real-time sequencing method. This method of sequencing provides sequencing of log reads up to 1Mb, which is advantageous in the assembly of the genome. In this technique, there is no need for the PCR amplification step, so after high-molecular-weight DNA extraction, only a genomic library is created. Adapters are attached to the ends of DNA along with a processive enzyme. As the sequence–enzyme complex interacts with the nanopore, the sequencing starts.

10.5 ADVANTAGES OF METAGENOMIC NEXT-GENERATION SEQUENCING FOR MICROBIAL DETECTION

Previously, the traditional culturing methods were used to detect the presence of microbes in various clinical or environmental samples. However, few microbes are very hard to culture and hence hard to detect by using traditional methods. Metagenomic sequencing arises as an alternate of traditionally used microbial diagnostic methods for its detection; this includes the characterization of microbial nucleic acid from both viable and nonviable microbes (Greninger & Naccache, 2019). In clinical microbiology laboratories, metagenomic next-generation sequencing (mNGS) can be used for the identification of the pathogen, mapping of resistance genes, and investigation of outbreaks, with reasonably high accuracy (Miller et al., 2013, Kumar et al., 2020a,b). Thousands to billions of reads can be generated in a single run by sequencing of metagenome of microbes including viruses, bacteria, fungi, and parasites. The mNGS efficiently decreases the time required for the identification of unknown microbes in any clinical sample. Two different approaches are used in mNGS: (i) approaches based on shotgun method, this provides uncountable accesses to genomic features of microbes, which have some clinical relevance like profiling of resistance genes, detection of virulence factors, and characterization of all microbial strains, and (ii) targeted metagenomic sequencing based on PCR amplification, which includes ITS-based approaches or 16S rRNA gene, which provides microbial broad range taxonomical classification (Chiu & Miller, 2019; Couto et al., 2018). 16S, ITS, and shotgun metagenomic sequencing enables peerless perception of the microbial assemblage, helping the researchers to explore various microbiomes such as soil and intestinal flora. The process of sequencing and analysis of 16S rRNA gene is done for bacterial communities, whereas the internal transcribed spacer (ITS) is analyzed for fungal communities. Features of mNGS like extraction of high-quality nucleic acid, impartial sampling, and storage/transport of standard sample are used by diagnostic laboratories. Recently, mNGS has been exploited for the detection of the SARS-CoV-2 pandemic too.

In antibody neutralizing test, observation of cytopathic effects by a virus on the monolayer culture is the method, which is traditionally in use for virus identification, but in laboratory conditions, some viruses are uncultivable and the quality of antiserum also affects the antibody neutralization test. Real-time PCR requires nucleic acid sequence information, which may not be available for a novel virus. Nowadays, mNGS has become popular among researchers because it facilitates the identification of novel viruses and unbiased detection of complete taxonomic composition. It can also be used to identify those novel or uncultivable viruses that are responsible for uncommon etiologies (Nooij et al., 2018). First time in 2008 mNGS was used to identify novel arenavirus (Delgado et al., 2008). mNGS also had helped in developing therapeutic measures/strategies for various bacterial diseases (Fancello et al., 2012). The shotgun approach of mNGS helps in deciphering both the microbial population and its function. The mNGS allows low cost and HTS of multiple genes and genomes at the same time. Conventional culturing methods are not able to detect complete microbial profiles in any sample. The conventional method of bacterial identification depends upon different factors like existing organism's viability and different optimum growth requirements (Abayasekara et al., 2017). Unlike the conventional method for bacterial identification, mNGS does not depend upon the growth requirement of bacteria or viability of the organism, but can easily detect the entire genome of a microbe. Shotgun-based approaches can provide important data for functional capabilities of microbes, for example, the presence of mobile genetic elements, antibiotic resistance gene (ARG), and virulence factors. For fungal and bacterial detection, the targeted approach of mNGS, for example, using ITS sequences and primers from 16S rRNA, has been used in species-level identification. Evolutionary tracing and strain identification of any microbe required auxiliary genomic information, which is provided by mNGS. mNGS provides the semiquantitative and quantitative data by counting the sequenced reads that give information about the concentration of organisms in a given sample (Salipante et al., 2014). Thus, the treatment of microbial infection and its accurate diagnosis can be done by exploiting applications of mNGS.

10.6 LIMITATIONS OF METAGENOMIC NEXT-GENERATION SEQUENCING FOR MICROBIAL DETECTION

Despite multiple merits of mNGS for microbial detection, there are some limitations too, and that needs to be discussed. The major shortcomings of mNGS apart from cost and requirement of highly skilled manpower are as follows: (i) the overwhelming amount of human host DNA is present in a clinical specimen, (ii) clinical specimen contains a lower concentration of microbial genome, **and** (iii) for laboratory validation of mNGS assays, there is an absence of established protocol (Dulanto et al., 2019). The mNGS sensitivity is compromised as the relative concentration of human host background to microbial specimen increases, which ultimately reduces the proportion and number of microbial genome reads and reduces the sensitivity of mNGS. For discrepancy testing, the availability of specimens is limited and the costs of mNGS-based clinical diagnostic testing in laboratories are some of the shortcomings. Microbiologists and clinicians find a barrier between understanding the recent advances from the concept and their application. It generates a relative scarcity of microbial genome reads. However, this can be overcome to a certain extent by using either the host depletion method or targeted sequencing. Another limitation of mNGS is that reagents used for processing, laboratory environment, and contamination present in a sample can increase the complications during analysis and interpretation of results while detecting microbe from the sample. Contaminating DNA is ubiquitous in commonly used DNA extraction kits and other laboratory reagents, varies greatly in composition between different kits and kit batches, and that this contamination critically impacts results obtained from samples containing a low microbial biomass. During the collection of clinical samples routinely, it can contaminate the site of biopsies that are presumably sterile. Procedures like fine-needle aspiration may include contamination from skin flora. To avoid any false positives from any experiment, negative controls are used and reagent assessment is required. In mNGS, it is important to distinguish between colonization of true pathogen from contaminant to avoid problems from findings. For minimal database error, strict quality control of bioinformatics is needed to get high-quality and validated genome information, and for interpretation of recorded data or findings, bioinformatics personnel is required, but it is generally absent from most of the clinical microbiology laboratories. Another shortcoming of NGS is that it cannot differentiate between viable and non-viable bacteria, i.e., genetic materials of all microbes whether alive or dead are detected. So new molecular viability analyses have been developed that can differentiate the living from the dead in natural samples, without relying on microbiological culture with its many limitations. Although each strategy has limitations, molecular viability testing (MVT) and viability polymerase chain reaction (viability PCR) have complementary advantages that make them useful additions to the molecular microbiological toolbox.

10.7 APPLICATION OF NGS IN MICROBIAL DETECTION IN VARIOUS FIELDS

Numerous applications of NGS are witnessed in the field of detection of microbes as shown in Figure 10.3.

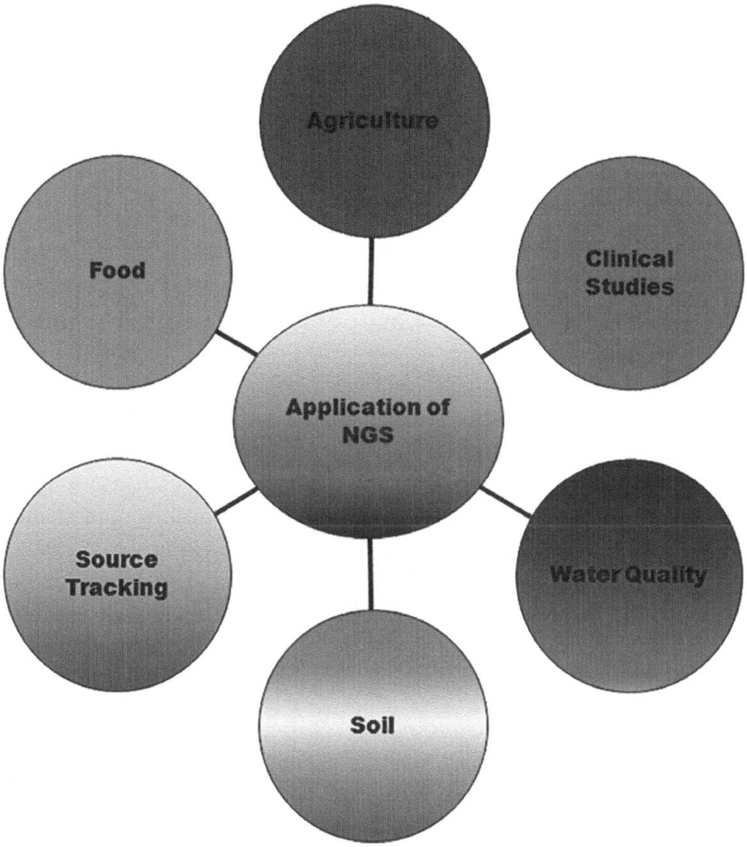

Figure 10.3 Various applications of NGS.

10.7.1 NGS in plant pathogen detection and their application in agriculture

NGS can be defined as a technology with the help of which genome sequencing at an affordable price. NGS can be used for the identification of microbial contaminants in agriculture, which may prove to be pathogenic for the plants (Jagadeesan et al., 2019). This technique can be used for developing portable equipment for detecting pathogens as well as contaminants. The emergence of novel detection techniques for speedy as well as the precise discovery of various contaminants and pathogens is the need of the hour. Nanopore technology or single DNA sequencing technology is one of the recent promising technologies. The detection of viral contaminants is an essential component of the thriving produce of crops and has enormous implications in provisions of world food safety. Molecular methods like RT-PCR have amplified the pace and precision of viral diagnostics, but they only permit the discovery of known viruses; thus, the unknown viruses may remain undetected making the testing slow as well as costly. With the advent of the NGS method, the detection of all types of viruses is possible and is proving to be key progress for crops where the symptoms of a disease are absent or the presence of multiple viruses cumulatively causing the disease. RNA sequence analysis has been used for the detection of viral content of garlic, pepper plants, etc. (Li et al., 2016). Two dissimilar pepper cultivars were sequenced with the help of HiSeq 2000 technology. RNA sequencing was also performed in the case of citrus sudden death (CSD) disease in Brazil in which more than 4 million orange trees died and a variant of citrus tristeza virus (CTV) was found to be the causative agent (Bové and Ayres, 2007). Genome Analyzer II technology was used to sequence the sRNAs from tomato plants that had viral disease symptoms (Jones et al., 2017).

10.7.2 NGS in the detection of pathogenic microbes in clinical samples from patients

Integration of diagnostic microbiology and public health microbiology had resulted in clinical microbiology. Diagnostic microbiology comprises the detection and identification of the pathogen from clinical samples to develop treatment strategies for the patients having an infection. The conventional method of pathogen identification from the sample of a patient includes isolation and growth of microbes in culture media, detecting antigen or antibody specific to pathogen. It also includes identification of pathogen by molecular methods, mainly PCR and detecting pathogen's DNA or RNA (nucleic acids), detection of virulence gene, and determining resistome. Only limited numbers of the pathogen were identified using the conventional method of pathogen detection in a single reaction. The mNGS approaches help in the characterization of all the nucleic acids (DNA/RNA) present in the patient's sample, thus increasing its throughput. It not only facilitates the analysis of the entire microbiome and detection of culture-shy microbes but also helps in the analysis of the transcriptome. Nowadays, multiple novel viral pathogens are being identified using mNGS. It also helped in comparing virome in both diseased and normal conditions (Moustafa et al., 2017). Recently, mNGS had been used as a diagnostic tool for the discovery of the SARS-CoV-2, too (Carbo et al., 2020). For the gene profiling of cancer cells consisting of mutations, for the very first time, mNGS was used for the diagnostic purpose (Voelkerding et al., 2009). Clinical samples containing microbial genome can be analyzed in a single run generating thousand to billions of reads. In 2004, the cost of NGS decreased by several orders of magnitude, and the development of several available sequencing platforms and less cumbersome kits started attracting microbiologists to use mNGS as a diagnostic tool for the identification and characterization of the pathogen in clinical samples on a routine basis. One of the most general approaches for increasing the efficiency is the multiplexing of samples, and multiple samples are collected together for sequencing in the same run. After sequencing the library pool including the indexes, each read would then be reassigned to its corresponding sample according to the unique index sequence. In the very first report of the use of mNGS for clinical diagnosis, mNGS was used for the diagnosis of neuroleptospirosis in a 14-year-old boy suffering from meningoencephalitis (Wilson et al., 2014). This was the very first case in which mNGS was used for successful diagnosis, followed by developing personalized treatment.

The ability of mNGS to characterize nucleic acids of all potential pathogens like viruses, bacteria, fungi, and parasites including uncultivable microbes present in a clinical sample made it one of the best methods for microbial diagnosis. In addition to pathogen identification, it can carry out resistance gene mapping and outbreak investigation with high accuracy. It has also reduced the time of microbial identification in some cases as sequencing occurs directly from the clinical specimen. The shotgun approach of mNGS provides access to several genomic features that are clinically relevant like virulence factor detection, resistance gene profiling, and strain-level typing of all microbial strains present in a clinical sample. And another approach is targeted metagenomic sequencing, which is based on PCR amplification integrates ITS or 16S rRNA-based approach which helps in deciphering only the pathogen's taxonomical classification. For implementing clinical metagenomics, several workflows are available (Balloux et al., 2018). The workflow of mNGS mainly includes four steps: (i) collection of specimen/sample, their handling, and storage, (ii) extraction of nucleic acid, (iii) preparation of library and sequencing, and (iv) analysis based on bioinformatics. The overall output of the experiment depends on the technical factors involved in selecting the sequencing and extraction protocol.

It is not easy to detect a complete microbial profile present in a clinical sample using conventional culturing methods. There are multiple limitations of the traditional culturing method, which can hinder the detection of the pathogen from the sample, like mostly culturing techniques were designed for aerobic pathogens. Various factors like the viability of existing organisms, different growth requirements, or inhibition of growth of pathogenic bacteria because of bacteriocin production affect the characterization and identification of bacterial isolates. Phenotypical characteristics and requirement of culture do not affect the metagenomic sequencing and can detect microbial genome entirely. mNGS can provide complete information about bacterial pathogen functional capabilities like virulence factors and ARGs. mNGS facilitates WGS of all microorganisms present in the sample, which helps in the treatment of infection and accurate diagnosis. For novel treatment, a comprehensive diagnosis of bacterial infection, untargeted mNGS was exploited by the researchers. It serves as the most unbiased approach. It provides rapid determination of antimicrobial susceptibility pattern, etiology of disease, and potential pathogenicity, which is required to initiate a proper therapy. During therapy, metagenomic sequencing of the clinical sample after regular time interval has important implications (Wu et al., 2018).

mNGS helps in reducing the time required for pathogen detection from clinical samples, which promotes pathogen's rapid identification and can improve patients' outcomes, and can reduce the use of empiric, broad-spectrum antibiotics. DNA NGS can reveal, for example, the presence of antimicrobial genes, while RNA NGS (or RNA sequencing) detects global gene expression, including the expression of antimicrobial resistance genes. A number of NGS studies have been reported of identification of pathogen genomes with at least 70 genes that might be involved in the drug resistance of *Mycobacterium tuberculosis*, which were most probably unidentified by the other molecular techniques (Warner & Mizrahi, 2013). The workflow of mNGS for infection of bacteria like sepsis, pneumonia, and meningitis was developed and is now available for testing of patients (Blauwkamp et al., 2019). Multiple pathogenic bacteria were detected using mNGS such as *Haemophilus influenza & Klebsiella pneumonia* (Yan et al., 2016), *Streptococcus pneumonia* (Bogaert et al., 2011), in the patients suffering from pneumonia. Patients suffering from respiratory failure were also infected with *Mycobacterium tuberculosis*. Patients infected with sepsis (transfusion-related) were also infected with *K. pneumoniae*, *Pseudomonas aeruginosa*, and *Acinetobacter baumannii* (Crawford et al., 2019) and were also detected using mNGS.

Viral pathogens had been conventionally detected by using an antibody neutralizing test or by detecting a cytopathic effect on monolayer on cell culture. Prerequisite knowledge of sequence is required for molecular methods like real-time PCR. Multiple novel or uncultivable viruses that were responsible for uncommon symptoms can be identified. In 2008, a novel arenavirus was detected using metagenomics from a patient suffering from a transplant-associated disease. Using Roche-based/454 viral mNGS study, many previously unknown (novel) and potential pathogenic viruses were also identified using a similar workflow. In clinical virology, the major contribution of mNGS includes the diagnosis of some fatal human cases of Ebola, Zika fever, encephalitis, etc. Additionally, fevers of unknown origin caused by dengue virus, chikungunya virus, West Nile virus, and hepatitis C virus were also identified using mNGS.

10.7.3 Analysis of microbial water quality with next-generation sequencing (NGS)

The quality of water is an evolving estate that comprises a complicated system that includes interacting microbial populations and introduced microbial and chemical contaminants. The use of NGS for water quality testing had provided a new perception into the biology of microbial processes that may affect freshwater quality, like pathogen dissemination, algal growth, and biodegradation of chemical contaminants. Nucleic acid sequencing methods targeting small subunit (SSU) like rRNA hypervariable regions have permitted the identification of unique microbial species that can serve as biological indicators for sewage material contamination in a specific water body. Apart from this, the amplicon sequencing and metagenomic and metatranscriptomic analyses of microbial populations/microbiome in an aquatic environment uncover the genetic potentials and interplay of waterborne microorganisms, which help in understanding the mechanisms for production and biodegradation of toxins and other contaminants (Jiang et al., 2016). So far, integrated multi-omics approaches have been used to decipher the effect of microbial composition and communities present in a variety of aquatic environments (fresh or marine) (Fondi & Liò, 2015). The multi-omics analysis techniques have been used for microbial water quality assessment too by probing the microbial components and their associated ecological tasks (e.g., propagation of ARGs and pollutant degradation) in the urban water cycle, such as ponds, lakes, and rivers, engineered waste tailings dam/ponds, sewage treatment units, and water circulation systems. These findings had generated required understandings into the activities facilitated by the microbiome in its aquatic environment and possible biological risk elements in the form of presence or absence of specific microbes or microbial populations composition and their relevant genes for virulence, toxin degradation, and biosynthesis or ARGs, which can be crucial in maintaining microbial water quality for its intended use. The mNGS can be used specifically to (i) identify indicators of human sewage and human fecal contamination for water quality assessment, (ii) examine the fate and transport of human pathogens in water and wastewater systems, (iii) understand the ecological drivers of harmful algal bloom persistence and toxin production/degradation, (iv) observe the spread and distribution of ARG in freshwater environments, and (v) investigate mechanisms of natural and stimulated or facilitated biodegradation of detrimental pollutants that can interfere with water quality. While the use of mNGS-based findings is providing exceptional insights into the composition and function of microbial communities, the open challenge remains to convert the observed trends from various studies into actionable data for water quality managers. The MinION method of NGS is considered as a low-cost, portable NGS tool, which generates long sequencing reads. This is portable and is of the size of a memory stick. It furnishes a reasonable substitute to the Illumina platform and is recently the most commonly used technique for monitoring the microbial quality of water and environmental inspection based on NGS (Acharya et al., 2019; Acharya et al., 2020). There are reports of the value of a multi-pronged approach that integrates culture-based quantification of fecal indicator and multidrug-resistant bacteria with gene-specific quantitative PCR and MinION NGS for the identification of microbiome-induced fragmentation and AMR sources in the river (Ho et al., 2021).

10.7.4 Use of metagenomics or NGS in profiling soil microbial communities

The determination of the microbial community in the samples obtained from the soil of different environments is done by amplicon NGS. The microbial communities in the soil are indicative of various interactions between organisms as well as the microbes concerning the mobilization of nutrients and their breakdown. Any microbial contamination other than the normal microflora can be studied with the help of NGS to provide information about environmental pollution due to microbial contamination. Direct DNA extraction from soils where the DNA is extracted directly without isolating the microbes is rapid and more precise than indirect extraction, so this technique is always applied (Soliman et al., 2017). The application of sequencing methods to soil microbial communities will aid in unraveling the unrevealed resources such as peculiar genes, biomolecules, and other valuable products, which are yet to be unearthed. Soil metagenomics is an auspicious technique in narrating the functional potential of the soil microbial community, which might provide greater insight into the health of soil than taxonomy-based metrics.

10.7.5 Monitoring of contamination in the microbiological laboratory with the help of NGS

High-throughput technology like NGS offers numerous advantages and a promising approach in the risk assessment, management, and monitoring of potential laboratory contaminants. It is a prerequisite of safety for any biological laboratory to be free from contaminants like pathogenic/non-pathogenic microorganisms. Therefore, it becomes elementary to develop models and methods for efficient and unbiased monitoring systems for such contaminants. A very important aspect of biosafety management is proper disposal, prevention as well as control of contaminants (cultured microorganisms) and infectious materials. The surrounding environment is so enriched with microorganisms, and the laboratory settings that undertake frequent microbiology work report contamination in instruments, operators, and rooms due to leaking/easy spread of contaminants. These contaminations arise majorly from isolation and culture of microbes, plasmid extraction, gene cloning, or handling of clinical samples. Under these settings in laboratories with frequent exposure to an array of microbes, tracing down the source of contamination, surveillance of common laboratory, lack of contaminants detectors, etc., are many challenges faced in keeping up with the sanitized environment. Hence, as a safety measure, it becomes of utmost importance to have a robust system for monitoring contaminants. The isolation of microbial species from the sample is difficult due to the small sample size and associated protocol. Molecular technology like PCR is a sensitive tool, but it is limited to detecting the gene of known pathogen only. Consequently, it is empirical to have one-point highly sensitive technology capable of detecting the potential contaminates of unknown resources. NGS does exactly that it provides a culture-independent microorganism identification resource with high sensitivity and finds wide application in the detection of pathogenic microorganisms (Ren et al., 2017). Experiments were carried on, and reports were shared that NGS method was capable of monitoring the nucleic acids contaminations from different sources in the laboratory, demonstrating its promising utility in monitoring and assessing the risk of potential laboratory contamination. The risk of contamination from reagents, remnant DNA, and environment should be considered in data analysis and result interpretation (Xiao et al., 2019).

10.7.6 NGS as a tool for investigation of contamination in the environment due to the disastrous incidents

Catastrophic incidents are the incidents that involve a chief hysterical emission, fire, or blast that causes major damage, injuries, etc., and produces various contaminants that create a polluted environment. The environmental samples that are contaminated may include soil, water, and air. The contamination of the samples with microbes is a vital feature of the environment. NGS can prove to be a wonderful tool for the analysis of various samples after the catastrophic events have taken place. The microbial contamination can be very well detected and identified by the NGS method, and a comparative chart can be mapped and prepared, which will include the microbial population that previously existed in that region, and the variation in the microbial population can be analyzed detecting the contaminants, which have inhabited the region after the catastrophe. Now, the environmental variations can be correlated with the presence of these contaminants. One of the best examples is the study of the diversity of microbes during the oil spill in the marine ecosystem in Kaohsiung Port, Taiwan, China. The analysis of the dominant bacteria from the soil was found to be instrumental in oil degradation as well as the weathering of UCMs (unresolved complex mixtures). The relative abundance (RA) of different microbes by the NGS method was found to be elevated the boundary area of spilled-oily-water in comparison with the control and center of the oil spill area (Maity et al., 2020). So after the oil spill catastrophe, the environment promotes the growth of the bacteria, such as *Staphylococcus saprophyticus* and *Bacillus megaterium* (CYCTW3), which are involved in the degradation of the C10–C30 including UCM of oil. Consequently, the diversity in the marine microbial community to degrade the oil and normal conditions that were present before the catastrophe took place was observed. This is achieved by changing the diversity of the microbes by the process of ecological selection as well as the adaptation process.

10.7.7 NGS in the detection of pathogenic microbes in food samples

NGS finds growing application as a primary detection tool and supportive detection tool in foodborne microbiology. It is widely used in detecting common bacterial foodborne pathogens *Salmonella*, *Shigella*, *Escherichia*, *Clostridium*, etc. Foodborne illness arising from 31 pathogens resulted in ~9.4 million occurrences of illness in the USA with reports of 1,351 deaths per year out of 55,961 hospitalizations; over the same period of the time, acute gastroenteritis with unspecified agents reported 38.4 million with 5,072 deaths from 473,832 hospitalizations (Scallan et al., 2011). The popularly studied foodborne pathogens are well-defined and often a major focus of public health programs and hence follow routine regulatory procedures for their detection and verification. First, retrospective analyses of foodborne illness outbreaks were done using pulsed-field gel electrophoresis (PFGE). In recent years, countries like the UK, USA, France, and Denmark were introduced to WGS of microbial genomes for a potential approach for the surveillance of foodborne pathogens (Allard et al., 2016). WGS implementation turned out to be revolutionary in terms of outbreak surveillance. With the robust strengths that WGS offers, it is being presented as a replacement technique in health considerations (Kwong et al., 2015). It is aimed to replace old school microbiological laboratory practices in identification and characterization like virulence profiling, serotyping antimicrobial resistance, and other molecular methods. The WGS introduction for probable listeriosis surveillance in the USA has led to further and even minor outbreaks detectable. These outbreaks were observed much earlier, the cause of outbreaks was acknowledged more frequently, and the total number of outbreak-associated cases identified increased by the use of WGS. WGS can help replace the plethora of traditional practices in microbiology with a single efficient analytical WGS workflow. Not only the application implementation of WGS is cost-effective but also its introduction will provide the public health sector with more reliable and actionable data than previously followed traditionally. Looking at the promising aspects of WGS in public health, its bright prospects draw attention toward its use in the food safety requirement in the food industry. The positive impact of the technology on the food industry can be assessed as enhanced root cause analysis in a spoilage contamination incident or pathogen. For instance, WGS can help discriminate among new and persistent occurrence of an old pathogen into the production environment. It can also be employed for risk assessment by forecasting traits like antimicrobial resistance or virulence of a pathogen or the capacity of an organism that causes spoilage to break the preservation barriers of a product. Though industry food safety checking does not require the comprehensive microbial characterization required by standard reference laboratories, WGS technology is being progressively explored for tracing the sources/root causes of microbial contamination (Rantsiou et al., 2017). With time and advances in technology, the drop in the cost of sequencing will make it more economical and viable for the industry to incorporate it and use it.

10.7.8 Detection of viral contaminants in clinical samples with the help of metagenomics

Viral infections cause several dreadful diseases that pose a great risk to public health. Thus, a quick and precise detection of the viral contaminants through the NGS technique can be vital in the management of the outbreak and analyzing the disease prevalent in a particular region. The metagenomic approaches that have been developed in recent times have been confirmed to be potent tools for real-time detection of the pathogen. A systematic study conducted on the samples obtained from the patients suffering from acute lower respiratory tract infections by metagenomics revealed seven species of known respiratory viral agents. They used ultradeep sequencing and an interpretation method, which was found to be consistent with the results of conventional PCR techniques. The metagenomic approach based on short reads proves to be adequate for the accurate and similar screening of viral contaminants. Also, paired-end sequencing strategies can be employed to ease *de novo* assembly or produce longer sequences *in silico*. The release of protocols to detect a minimum of 0.5 μg of initial DNA for sequencing on the 454 sequencers by Roche makes the Roche 454 platform competitive in clinical metagenomic studies (Yang et al., 2011). With these advances, both long-read-based like Roche 454 as well as short-read-based like Illumina Solexa, metagenomics may prove to be a versatile method for the analysis of small volumes of clinical samples. There are large numbers of viral contaminants and pathogens that cannot be isolated and cultured so the molecular tests cannot prove to be a complete solution for the identification of the contaminants. So there is a need for an extremely efficient method for the detection of microbial contaminants that cannot be cultured and the answer is NGS technology. The application of NGS may be very well observed in the WGS of purified bacterial isolates for recognition, detection of antibiotic resistance, profiling of pathogenicity, etc. (Gilchrist et al., 2015). The method of NGS can be used for the discovery of infection of cell cultures with bovine viral diarrhea virus (BVDV) (Uphoff et al., 2019). As the viruses are very infective, so the safety of the laboratory may be at stake and may be very difficult to analyze them in the cell cultures. Conventional techniques like PCR, immunological assays, and enzyme activity tests have been used from time immemorial, but with the contagious nature of the viruses, an alternative method was needed that did not involve the process of culturing. More than 300 cancer cell lines have been analyzed by Encyclopedia RNA sequencing and 60 whole exome sequencing human cell lines (Arora et al., 2019). IDbyDNA has developed a Taxonomer bioinformatics tool that analyzes the data sets for specific viral sequences and general viral nucleotide and protein sequence assessment. The results obtained from both the direct alignment method and the Taxonomer alignment method have exposed an absolute similarity with the results obtained by PCR (Flygare et al., 2016). Bovine polyomavirus infection in the breast cancer cell line SK-BR-3 has been detected by Taxonomer, which was introduced by contaminated fetal bovine serum (FBS). A promising novel technique for identifying microbes in clinical specimens has been recognized—unbiased HTS of whole metagenome shotgun DNA libraries. This is highly sensitive to the contaminations introduced by the processing of the human clinical samples. The evaluation of the degree of contamination by different DNA sequences was done by aligning 57 sequencing runs against the NCBI BLAST databases. It was found that the most common contaminant was of the genera *Bradyrhizobium* (Laurence *et al.*, 2014).

10.7.9 NGS in the detection of unbiased analysis by high-throughput sequencing of the viral diversity in fetal bovine serum and trypsin used in cell culture

The viral contaminants can be introduced in the pharmaceutical products by the addition of reagents such as FBS and trypsin, which are essential reagents in cell culture. HTS of two sets of marketable batches of FBS and three marketable batches of trypsin was performed (Gagnieur et al., 2014). In comparison of the sequences of contigs and singleton with the nucleic acid and protein databases, different taxonomies were attributed. Same viral species were detected in the batches of a specific reagent, but the fraction of viral reads among total reads differed noticeably between samples, which were observed between 0.002% and 22.7%. The sequences mainly from bovine viral diarrhea virus (BVDV) 1 to 3 and *bovine parvovirus* 3 (BPV3) were detected in the FBS. The sequences of BVDV resulting from FBS exhibited only slight inconsistencies with primers usually used for the screening of BVDV. In contrary to the viral sequences found in FBS, we found that the viral sequences chiefly from porcine circovirus type 2 were found in trypsin. Other known viral sequences at lower read counts and potential new viral species (*bovine parvovirus* and *bovine pegivirus*) were observed.

10.7.10 Detection of bacterial phage-like DNA sequences in reagents of commercially available Taq DNA polymerase

NGS or metagenomics significantly transformed our outlook about the diversity in microbial communities in various ecosystems. Significant mapping of the entire microbial component on large scale was only possible due to this novel technique. But there are certain shortcomings in this technique that prove to be blemish in its character and need special care for rectification. The problem associated with it is the genomic sequences that may have been unintentionally added into the sample during the handling of the samples. These are the contaminations that without any tantrums are sequenced with the targeted sequences, and this further adds to the background noise masking the signals obtained from the target sequences. The various microbial contaminations may have been introduced into the system during different steps of execution of the sequencing that may give rise to the genomic background noise. Just, for instance, the unintentional addition of bacteriophage ΦX174 genomic DNA as a positive control and even after filtering the sequences may be found luxuriously in the resultant metagenomic studies (Santiago-Rodriguez & Hollister, 2020). It is frequently observed that the cross-species contamination from bacterial and mammalian DNA takes place, thus resulting in background noise that furnishes faulty metagenomic mapping. These may be the result of the introduction of the contaminations during human handling of the samples or through the contaminations introduced by the enzymes required in the DNA extraction as well as the commercially available kits. Anything can add to the contamination by foreign DNA, even the plain buffers. Even though we use the far-reaching and strict filtering of the data by the bioinformatics approach, yet we are not able to avoid these background noises. It was found that the enzyme, *Taq* polymerase, approximately has between 10^2 and 10^5 genomes of bacteria-derived DNA per unit of the enzyme, also holds murine, human retroviruses, as well as bacteriophage-derived sequences. Contamination by the independently replicating plasmids has also been found along with the partial sequences of the expression vectors. While examining samples for the presence of ARGs such as $Bla_{Tem\text{-}1}$ deliberately, scientists used native *Taq* polymerase to circumvent the results of the tests, which may be false positive. The faulty identification of ARGs may result in a very wrong diagnosis of the patient's disease and also an erroneous treatment of the patient. In the current metagenomic study of the virus genome, several contaminations like plasmids and sequences of a horse retroviral pathogen in human samples were found. The special sequence of pol gene present in the virus, *Equine Infectious Anemia Virus* (EIAV), contaminated the samples of the urine as well as pharyngeal lavage of healthy humans.

10.7.11 Detection and identification of common contaminants in next-generation sequencing hampering the unearthing of microbes that exhibit low abundance

Several microbes have also been found as contaminants in the ultrapure water systems. These kinds of studies and conclusions can detect the novel microbes, and the information can be exploited to develop preventive procedures like UV irradiation of the water as well as cross-checking by different individual techniques. It is very complicated to differentiate the concerned microbes from new contagious sequences by just examining the individual sequences. The problem of detecting and identifying the new microbes with low prevalence has found no place in the NCBI BLAST databases. Now, this problem can be solved by just removing the cause of contagion in the laboratory or at the reagent level. Also, the solution to the problem can be sorted by completely sequencing the contaminant genome that is present in the sequencing so that they may be eliminated while the process of data analysis. A dose of 8,000 J/m^2 of ultraviolet C radiation, which is about four times more than the radiation fixed by the United States Environmental Protection Agency (US EPA), is adequate for purifying the surface water sources for safe drinking. Millions of short sequences are produced by NGS experiments to present biological perception at the genomic, transcriptomic, and epigenomic levels. A small number of sequences are unmapped each time in the sequencing, which may be of the microbial contamination present at a very low level. This problem may be due to low-quality bases or sequence differences between the sample reads and the reference genome.

10.8 FUTURE PERSPECTIVES

Even though the sequencing of the microbial contaminants through NGS has been commercialized recently and the data are still being extensively studied, it has become of the finest and accurate techniques for microbial detection and identification. In the future, several cost-effective sequencing techniques will be available for sequencing the microbial genomes. A sharp decrease in the price of NGS has been witnessed since it was invented. Previously, it could be afforded by only those laboratories that were funded heavily as well as the medical research institutions that were having a sound financial background, but as of now, the common laboratories are also able to go for

NGS owing to its low price. Numerous modifications and developments in the field of NGS will offer more versatile and more cost-effective techniques in the identification of the microbial contaminants. The robustness and speed of the technique can offer a very efficient method for the detection of contaminants in the future and can aid in the development of preventative measures against microbial contaminants. Additionally, this will permit for an additional and precise accurate explanation for the functions of the genomes of the microbes. The development of the third wave of NGS tools offers the promising application of genome sequencing and utility-based sequencing. The amplification biases can be eliminated with the advent of the SMS method, which is a great advancement in the NGS. Increased precision and profiling of the difficult transcriptomes can be achieved by the longer reads sequenced by the novel methods, which will provide greater coverage and depth of data. Modern and prospective NGS technologies pledge to provide new insights into individual microbial genomes, the structure of the communities they inhabit, and their impact on human health and disease.

10.9 CONCLUSION

With the initiation of the technology of NGS, the zeal for enlightening the microbial contaminants has soared high. The NGS proves to be a tremendously sensitive technology that is reasonably unbiased and can be used in the complete assessment of diverse contaminations in a variety of samples. In this chapter, the various NGS methods for the assessment and depiction of the microbial contaminants in the biological as well as medical research areas have been covered. The spotlight of the chapter is on the versatile technique of NGS, its types as well as its application in the field of detection of microbial contaminants. The prospect of the various modern developments in the field of NGS as well as the advancements in the technology for the detection of microbial contaminants has also been discussed. Profiling microbial populations in water samples taken from deep ocean vents or in soil samples from human-made environments like active mineral mines. Environmental metagenomic study data are used for agricultural microbiome analysis, ecological remediation, or other biological investigations.

REFERENCES

Abayasekara, L.M., Perera, J., Chandrasekharan, V., et al., 2017. Detection of bacterial pathogens from clinical specimens using conventional microbial culture and 16S metagenomics: A comparative study. *BMC Infect Dis*; 17:631. doi:10.1186/s12879-017-2727-8.

Acharya, K., Khanal, S., Pantha, K., et al., 2019. A comparative assessment of conventional and molecular methods, including MinION nanopore sequencing, for surveying water quality. *Sci Rep*; 9:15726. doi:10.1038/s41598-019-51997-x.

Acharya, K., Blackburn, A., Mohammed, J., et al., 2020. Metagenomic water quality monitoring with a portable laboratory. *Water Res*; 184:116112. doi:10.1016/j.watres.2020.116112.

Allard, M.W., Strain, E., Melka, D., et al., 2016. Practical value of food pathogen traceability through building a whole-genome sequencing network and database. *J Clin Microbiol*; 54(8):1975–1983. doi: 10.1128/JCM.00081-16.

Arora, K., Shah, M., Johnson, M., et al., 2019. Deep whole-genome sequencing of 3 cancer cell lines on 2 sequencing platforms. *Sci Rep*; 9:19123. doi:10.1038/s41598-019-55636-3.

Balloux, F., Brønstad Brynildsrud, O., van Dorp, L., et al., 2018. From theory to practice: Translating whole-genome sequencing (WGS) into the clinic. *Trends Microbiol*; 26:1035–1048. doi:10.1016/j.tim.2018.08.004.

Bartelt-Hunt, S.L., Bartz, J.C., 2013. Behavior of prions in the environment: Implications for prion biology. *PLoS Pathog*; 9(2):e1003113. doi:10.1371/journal.ppat.1003113.

Blauwkamp, T.A., Thair, S., Rosen, M.J., et al., 2019. Analytical and clinical validation of a microbial cell-free DNA sequencing test for infectious disease. *Nat Microbiol*; 4:663–674. doi:10.1038/s41564-018-0349-6.

Bogaert, D., Keijser, B, Huse, S., et al., 2011. Variability and diversity of nasopharyngeal microbiota in children: A metagenomic analysis. *PLoS One*; 6:e17035. doi:10.1371/journal.pone.0017035.

Bové, J.M., Ayres, A.J., 2007. Etiology of three recent diseases of citrus in São Paulo State: Sudden death, variegated chlorosis and huanglongbing. *IUBMB Life*; 59(4–5):346–354. doi:10.1080/15216540701299326.

Carbo, E.C., Sidorov, I.A., Zevenhoven-Dobbe, J.C., et al., 2020. Coronavirus discovery by metagenomic sequencing: A tool for pandemic preparedness. *J Clin Virol*; 131:104594. doi:10.1016/j.jcv.2020.104594.

Chiang, A.D., Dekker, J.P., 2019. From the pipeline to the bedside: Advances and challenges in clinical metagenomics. *J Infect Dis*. pii: jiz15. doi:10.1093/infdis/jiz151.

Chiu, C.Y., Miller, S.A., 2019. Clinical metagenomics. *Nat Rev Genet*; 20:341–355. doi:10.1038/s41576-019-0113-7.

Couto, N., Schuele, L., Raangs, T.F., et al., 2018. Critical steps in clinical shotgun metagenomics for the concomitant detection and typing of microbial pathogens. *Sci Rep*; 8:13767. doi:10.1038/s41598-018-31873-w.

Crawford, E., Kamm, J., Miller, S., et al., 2019. Investigating transfusion-related sepsis using culture-independent metagenomic sequencing. *Clin Infect Dis*. pii: ciz960. doi:10.1093/cid/ciz960.

Delgado, S., Erickson, B.R., Agudo, R., et al., 2008. Chapare virus, a newly discovered arenavirus isolated from a fatal hemorrhagic fever case in Bolivia. *PLoS Pathog*; 4:e1000047. doi:10.1371/journal.ppat.1000047.

Dulanto C.A., & Dekker, J. P. (2020). From the pipeline to the bedside: advances and challenges in clinical metagenomics. *The Journal of infectious diseases, 221*(Supplement_3), S331–S340.

Facciolà, A., Riso, R., Avventuroso, E., et al., 2017. Campylobacter: From microbiology to prevention. *J Prev Med Hyg*; 58(2):E79–E92. PMID: 28900347 PMCID: PMC5584092.

Fancello, L., Raoult, D., Desnues, C., 2012. Computational tools for viral metagenomics and their application in clinical research. *Virology*; 434:162–174. doi:10.1016/j.virol.2012.09.025.

Farrar, W.E., Reboli, A.C., 2006. The Genus Bacillus—Medical. In: Dworkin M., Falkow S., Rosenberg E., Schleifer K.H., Stackebrandt E. (eds) *The Prokaryotes*. Springer, New York.

Flygare, S., Simmon, K., Miller, C. et al., 2016. Taxonomer: An interactive metagenomics analysis portal for universal pathogen detection and host mRNA expression profiling. *Genome Biol*; 17:111. doi:10.1186/s13059-016-0969-1.

Fondi, M., Liò, P., 2015. Multi-omics and metabolic modelling pipelines: Challenges and tools for systems microbiology. *Microbiol Res*; 171:52–64. doi:10.1016/j.micres.2015.01.003.

Gagnieur, L., Cheval, J., Gratigny, M., et al., 2014. Unbiased analysis by high throughput sequencing of the viral diversity in fetal bovine serum and trypsin used in cell culture. *Biologicals*; 42(3):145–152. doi:10.1016/j.biologicals.2014.02.002.

Gilchrist, C.A., Turner, S.D., Riley, M.F., et al., 2015. Whole-genome sequencing in outbreak analysis. *Clin Microbiol Rev*; 28(3):541–563. doi:10.1128/CMR.00075-13.

Greninger, A.L., Naccache, S.N., 2019. Metagenomics to assist in the diagnosis of bloodstream infection. *J Appl Lab Med*; 3:643–653. doi:10.1373/jalm.2018.026120.

Gu, W., Miller, S., Chiu, C.Y., 2019. Clinical metagenomic next-generation sequencing for pathogen detection. *Annu Rev Pathol*; 14:319–338. doi:10.1146/annurev-pathmechdis-012418-012751.

Gwinn, M., MacCannell, D., Armstrong, G.L., 2019. Next-generation sequencing of infectious pathogens. *JAMA*; 321(9):893–894. doi:10.1001/jama.2018.21669.

Heather, J.M., Chain, B., 2016. The sequence of sequencers: The history of sequencing DNA. *Genomics*; 107(1):1–8. doi:10.1016/j.ygeno.2015.11.003.

Ho, J.Y., Jong, M.C., Acharya, K., et al., 2021. Multidrug-resistant bacteria and microbial communities in a river estuary with fragmented suburban waste management. *J Hazard Mater*; 405:124687. doi:10.1016/j.jhazmat.2020.124687.

Jagadeesan, B., Gerner-Smidt, P., Allard, M.W., et al., 2019. The use of next-generation sequencing for improving food safety: Translation into practice. *Food Microbiol*; 79:96–115. doi:10.1016/j.fm.2018.11.005.

Jiang, Y., Xiong, X., Danska, J. et al., 2016. Metatranscriptomic analysis of diverse microbial communities reveals core metabolic pathways and microbiome-specific functionality. *Microbiome*; 4:2. doi:10.1186/s40168-015-0146-x.

Jones, S., Baizan-Edge, A., MacFarlane, S., Torrance, L., 2017. Viral diagnostics in plants using next-generation sequencing: Computational analysis in practice. *Front Plant Sci*; 8:1770. doi:10.3389/fpls.2017.01770.

Khan, A.A.H., Karuppayil, S.M., 2012. Fungal pollution of indoor environments and its management. *Saudi J Biol Sci*; 19(4):405–426. doi:10.1016/j.sjbs.2012.06.002.

Kodaka, H., Mizuochi, S., Teramura, H., Nirazuka, T., 2005. Comparison of the compact dry TC method with the standard pour plate method (AOAC official method 966.23) for determining aerobic colony counts in food samples: Performance-tested method. *J AOAC Int*; 88(6):1702–1713. doi:10.1093/jaoac/88.6.1702.

Kumar, M., Chaminda, G.T., Honda, R., 2020a. Seasonality impels the antibiotic resistance in Kelani River of the emerging economy of Sri Lanka. *NPJ Clean Water*; 3(1):1–8.

Kumar, M., Thakur, A.K., Mazumder, P., Kuroda, K., Mohapatra, S., Rinklebe, J., et al., 2020b. Frontier review on the propensity and repercussion of SARS-CoV-2 migration to aquatic environment. *J Hazard Mater Letters*; 1:100001.

Kwong, J.C., McCallum, N., Sintchenko, V., Howden, B.P., 2015. Whole genome sequencing in clinical and public health microbiology. *Pathology*; 47(3):199–210. doi:10.1097/PAT.0000000000000235.

Lahens, N.F., Ricciotti, E., Smirnova, O., et al., 2017. A comparison of Illumina and Ion Torrent sequencing platforms in the context of differential gene expression. *BMC Genomics*; 18:602. doi:10.1186/s12864-017-4011-0.

Laurence, M., Hatzis, C., Brash, D.E., 2014. Common contaminants in next-generation sequencing that hinder discovery of low-abundance microbes. *PLoS One*; 9(5):e97876. doi:10.1371/journal.pone.0097876.

Li, J., Yang, P., Kang, J., et al., 2016. Transcriptome analysis of pepper (Capsicum annuum) revealed a role of 24-epibrassinolide in response to chilling. *Front Plant Sci*; 7:1281. doi:10.3389/fpls.2016.01281.

Maitra, R. D., Kim, J., Dunbar, W. B., 2012. Recent advances in nanopore sequencing. *Electrophoresis*; 33(23):3418–3428. doi:10.1002/elps.201200272.

Maity, J.P., Huang, Y.H., Lin, H.F., et al., 2020. Variation of microbial diversity in catastrophic oil spill area in marine ecosystem and hydrocarbon degradation of UCMs (unresolved complex mixtures) by marine indigenous bacteria. *Appl Biochem Biotechnol*. doi:10.1007/s12010-020-03335-5.

Mayo, B., Rachid, C.T., Alegría, A., et al., 2014. Impact of next generation sequencing techniques in food microbiology. *Curr Genomics*; 15(4):293–309. doi:10.2174/138920291566614061 6233211.

Mazumder, P., Sharma, S.K., Taki, K., Kalamdhad, A.S., Kumar, M., 2020. Microbes involved in arsenic mobilization and respiration: A review on isolation, identification, isolates and implications. *Environ Geochem Health*:1–27.

Miller, R.R., Montoya, V., Gardy, J.L., et al., 2013. Metagenomics for pathogen detection in public health. *Genome Med*; 5:81. doi:10.1186/gm485.

Mohapatra, S., Menon, N.G., Mohapatra, G., Pisharody, L., Pattnaik, A., Menon, N.G., Bhukya, P.L., Srivastava, M., Singh, M., Barman, M.K., Gin, K.Y.H., 2020. The novel SARS-CoV-2 pandemic: Possible environmental transmission, detection, persistence and fate during wastewater and water treatment. *Sci Total Environ*:142746.

Moustafa, A., Xie, C., Kirkness, E., et al., 2017. The blood DNA virome in 8,000 humans. *PLoS Pathog*; 13(3):e1006292. doi:10.1371/journal.ppat.1006292.

Nooij, S., Schmitz, D., Vennema, H., et al., 2018. Overview of virus metagenomic classification methods and their biological applications. *Front Microbiol*; 9:749. doi:10.3389/fmicb.2018.00749.

Novakova, J., Dzunkova, M., Musilova, S., et al., 2014. Selective growth-inhibitory effect of 8-hydroxyquinoline towards Clostridium difficile and Bifidobacterium longum subsp. longum in co-culture analysed by flow cytometry. *J Med Microbiol*; 63:1663–1669. doi:10.1099/jmm.0.080796-0.

Pantha, K., Acharya, K., Mohapatra, S., Khanal, S., Amatya, N., Ospina-Betancourth, C., Butte, G., Shrestha, S.D., Rajbhandari, P., Werner, D., 2021. Faecal pollution source tracking in the holy Bagmati River by portable 16S rRNA gene sequencing. *NPJ Clean Water*; 4(1):1–10.

Rantsiou, K., Kathariou, S., Winkler, A., et al., 2017. Next-generation microbiological risk assessment: Opportunities of whole-genome sequencing (WGS) for foodborne pathogen surveillance, source tracking, and risk assessment. *Int J Food Microbiol*; 287:3–9. doi:10.1016/j.ijfoodmicro.2017.11.007.

Ren, L., Yang, D., Ren, X., et al., 2017. Genotyping of human rhinovirus in adult patients with acute respiratory virus infections identified predominant infections of genotype A21. *Sci Rep*; 7:41601. doi:10.1038/srep41601.

Reuter, J.A., Spacek, D.V., Snyder, M.P., 2015. High-throughput sequencing technologies. *Mol Cell*; 58(4):586–597. doi:10.1016/j.molcel.2015.05.004.

Salipante, S.J., Hoogestraat, D.R., Abbott, A.N., et al., 2014. Coinfection of Fusobacterium nucleatum and Actinomyces israelii in mastoiditis diagnosed by next-generation DNA sequencing. *Clin Microbiol*; 52(5):1789–1792. doi:10.1128/JCM.03133-13.

Santiago-Rodriguez, T.M., Hollister, E.B., 2020. Potential applications of human viral metagenomics and reference materials: Considerations for current and future viruses. *Appl Environ Microbiol*; 86(22):e01794–20. doi:10.1128/AEM.01794-20.

Saunders, S.E., Bartelt-Hunt, S.L., Bartz, J.C., 2008. Prions in the environment: Occurrence, fate and mitigation. *Prion*; 2(4):162–169. doi:10.4161/pri.2.4.7951.

Scallan, E., Griffin, P.M., Angulo, F. J., Tauxe, R.V., Hoekstra, R.M., 2011. Foodborne illness acquired in the United States—unspecified agents. *Emerg Infect Dis*; 17:16–22. doi:10.3201/eid1701.091101p2.

Soliman, T., Yang, S.Y., Yamazaki, T., Jenke-Kodama, H., 2017. Profiling soil microbial communities with next-generation sequencing: The influence of DNA kit selection and technician technical expertise. *Peer J*; 5:e4178. doi:10.7717/peerj.4178.

Uphoff, C.C., Pommerenke, C., Denkmann, S.A., Drexler, H.G., 2019. Screening human cell lines for viral infections applying RNA-Seq data analysis. *PLoS One*; 14(1):e0210404. doi:10.1371/journal.pone.0210404.

Voelkerding, K.V., Dames, S.A., Durtschi, J.D., 2009. Next-generation sequencing: From basic research to diagnostics. *Clin Chem*; 55, 641–658. doi:10.1373/clinchem.2008.112789.

Warner, D.F., Mizrahi, V. 2013, Complex genetics of drug resistance in Mycobacterium tuberculosis. *Nat Genet*; 45(10):1107–1108. doi:10.1038/ng.2769.

Wilson, M.R., Naccache, S.N., Samayoa, E., et al., 2014. Actionable diagnosis of neuroleptospirosis by next-generation sequencing. *N Engl J Med*; 370(25):2408–2417. doi:10.1056/NEJMoa1401268.

Wu, S.C., Rau, C.S., Liu, H.T., Kuo, P.J., Chien, P.C., Hsieh, T.M., et al., 2018. Metagenome analysis as a tool to study bacterial infection associated with acute surgical abdomen. *J Clin Med*; 7. pii: E346. doi:10.3390/jcm7100346.

Xiao, Y., Zhang, L., Yang, B., et al., 2019. Application of next generation sequencing technology on contamination monitoring in microbiology laboratory. *Biosaf Health*; 1(1):25–31.

Yan, Q., Cui, S., Chen, C., et al., 2016. Metagenomic analysis of sputum microbiome as a tool toward culture-independent pathogen detection of patients with ventilator-associated pneumonia. *Am J Respir Crit Care Med*; 194:636–639. doi:10.1164/rccm.201601-0034LE.

Yang, J., Yang, F., Ren, L., et al., 2011. Unbiased parallel detection of viral pathogens in clinical samples by use of a metagenomic approach. *J Clin Microbiol*; 49(10):3463–3469. doi: 10.1128/JCM.00273-11.

Section 3

Environmental and health risk of anthropogenic and geogenic contaminants

Chapter 11

Emerging contaminants of potential concern in aquatic Ecosystem

Fate and transport in microalgae

Sarat Chandra T.
Trinity College Dublin

Madhu Balaji C. K.
CSIR- Central Food Technological Research Institute

Ramesh Babu P.
Trinity College Dublin
University College Dublin

CONTENTS

11.1 INTRODUCTION: BACKGROUND AND IMPORTANCE

India is home to 17.5% of the world's population, but it has only 4% of the world's water resources. The availability of water person has significantly decreased due to the increase in the population. Further, India's per capita water availability is estimated to decline from 1,508 cubic meters (as of 2014) to 1,465 cubic meters by 2025. In this context, the Government of India has launched ambitious projects like Swachh Bharat and Namami Gange, with a budget of more than Rs 20, 000 crores for the recovery of polluted water resources. One of the key objectives of these flagship programs is to constantly monitor grossly polluting industries for their water treatment and management capabilities. The Government of India's mission of developing

DOI: 10.1201/9781003157465-14

smart cities to promote clean, sustainable urban cities also contributes to increase in the demand for water within the next decade. In this regard, the depletion of groundwater levels due to irrigation in major river basins is expected to be a reality by 2050. On the other side, the surface waters have been contaminated by recalcitrant toxic compounds considered as emerging contaminants posing a threat to the availability of safe and clean water for human consumption. Emerging contaminants are a group of micropollutants such as antibiotics, insecticides, pesticides, steroids, and organic molecules from personal care products. These are released to the surface waters/groundwaters by domestic human activities as well as from industrial emissions (Jennifer et al. 2017, Kumar et al. 2017). Given the toxicity of these chemicals, the remediation of these contaminants by conventional wastewater treatment processes is a serious challenge (Tran et al., 2018). The presence of these compounds has been detected in rivers and lakes, and their presence demerits the quality of water, making it unsafe for consumption (Rivera-Jaimes et al., 2018, Wu et al., 2017, Kumar et al. 2019). Various kinds of pollutants categorized as emerging contaminants that have been detected in surface and ground waters are presented in Table 11.1. Increase in population, as well as emergence of new diseases and infections, has led to increase in the production and consumption of pharmaceuticals. After ingestion of the drug into the body, depending on the bioavailability of the drug, a certain concentration of the drug is excreted from the body through the urine. This would enter the domestic water streams and contaminate the aquatic ecosystems and the surface water resources (Hernández et al., 2014). Among the pharmaceutical products, antibiotics are at a higher rank, given their toxicity and resistance toward biodegradation by conventional aerobic degradation processes. In a recent case study, the presence of pharmaceuticals such as ibuprofen, diclofenac, naproxen, gemfibrozil, and ketoprofen was identified in the surface waters as well as in groundwaters of Mexico City (Félix-Cañedo et al., 2013). Further, 93 different types of pharmaceuticals that include antibiotics, antidepressants, analgesics, and anti-hypertensive have been detected in surface waters across the USA (Deo, 2014).

Table 11.1 List of pollutants of emerging concern and their concentrations reported in the literature

Compound	*Class*	*Concentration (ng/L)*	*Reference*
Acetaminophen	Analgesic drug	3,422	Rivera-Jaimes et al. (2018)
Carbamazepine	Anticonvulsant drug	29	Wanda et al. (2017)
Ibuprofen	Non-steroidal anti-inflammatory drug	1,830	Meierjohann et al. (2016)
Triclosan	Antibacterial and antifungal agent	105	Yang et al. (2018)
17β-estradiol	Steroid hormone	3,700	Olatunji et al. (2017)
Erythromycin	Antibiotic	183	Yang et al. (2018)
Benzophenone-3	Organic compound used as sunscreen agent	68	Fent et al. (2010)
Diclofenac	Non-steroidal anti-inflammatory drug	260	Gago-Ferrero et al. (2017)
Atrazine	Herbicide	1,650	Glinski et al. (2018)

Endocrine-disrupting compounds (EDCs) are another class of emerging contaminants commonly detected in the surface waters (Menon et al., 2020, Priyanka and Mohapatra, 2020). Steroids, estrogens, progesterone, bisphenol A, estriol, nonylphenol, octylphenol, and their corresponding ethoxylates are a few EDCs. These compounds can be found in different types of products used on a daily basis, such as plastic bottles, metal cans, toys, cosmetics, food, flame retardants, liners of the metals, and pesticides. According to the US Environmental Protection Agency report 1997, an EDC can be defined as "an exogenous agent that interferes with the synthesis, secretion, transport, binding, action, or elimination of natural hormones in the body that are responsible for the maintenance of homeostasis, reproduction, development, and/or behavior" (Crisp et al., 1997). As per the regulations, the concentrations of EDC in the surface water should be below 1 ng/L (Seibert et al., 2019). EDCs enter into the human body and affect human fertility and may pose the risk of cancer. Various types of EDCs have been reported viz. bisphenol A (BPA), phthalic acid esters (PAEs), phenol, alkanes, nicotine, polycyclic aromatic hydrocarbon, halogen, parabens, organotin compounds, and carboxylic acid (Dubey et al., 2021, Boonnorat et al., 2016, Fudala-Ksiazek et al., 2018, Kapelewska et al., 2016, Khalil et al., 2018, Li et al., 2016, Reis et al., 2017, and Vahčič et al., 2011, Thakur et al., 2020). Pesticides and personal care products are also considered to be potential contaminants posing a risk to human health. An increase in agricultural activities and the use of personal products has resulted in the detection of pesticides and toxic organic compounds in surface waters. Given the toxicity of these contaminants, the removal efficiency of these compounds is low through the conventional water treatment systems. Hence, new efficient methodologies need to be developed for the effective removal of these compounds from surface water or groundwater resources. Various physicochemical processes such as ozonation, wet air oxidation, flocculation, enzymatic degradation, membrane filtration, and adsorption have been explored for efficient removal of the emerging contaminants (Leng et al., 2020, Mohapatra et al., 2021, Furuichi et al., 2006, Mailler et al., 2016, Hai et al., 2011, Racz et al., 2012, Li et al., 2014, Samaras et al., 2013, Huerta-Fontela et al., 2011, Barrios et al., 2015, Gogoi et al., 2020). However, these technologies are characterized by higher energy requirements and are not economical. Hence, alternative processes such as biodegradation or adsorption by microalgae, a renewable resource, could be explored. Detailed insights into various physicochemical processes employed for the removal of emerging contaminants have been discussed in detail by Gogoi et al. (2018). The current chapter focused on the use of microalgae for the removal of emerging contaminants from water has been reviewed, and the possible mechanisms reported for the degradation of various emerging contaminants are discussed. The overview of microalgae-based remediation of emerging contaminants is summarized in Figure 11.1.

11.2 MICROALGAE AS A TOOL FOR REMOVAL OF EMERGING CONTAMINANTS

Microalgae are photosynthetic organisms that utilize sunlight and carbon dioxide from the atmosphere and convert it into biomass. Microalgae play an important role in the sustainable functioning of the aquatic ecosystem and account for the production of more than 75% of the oxygen in the atmosphere. Further, these are the sources of various value-added products such as lipids, pigments, carbohydrates, and proteins.

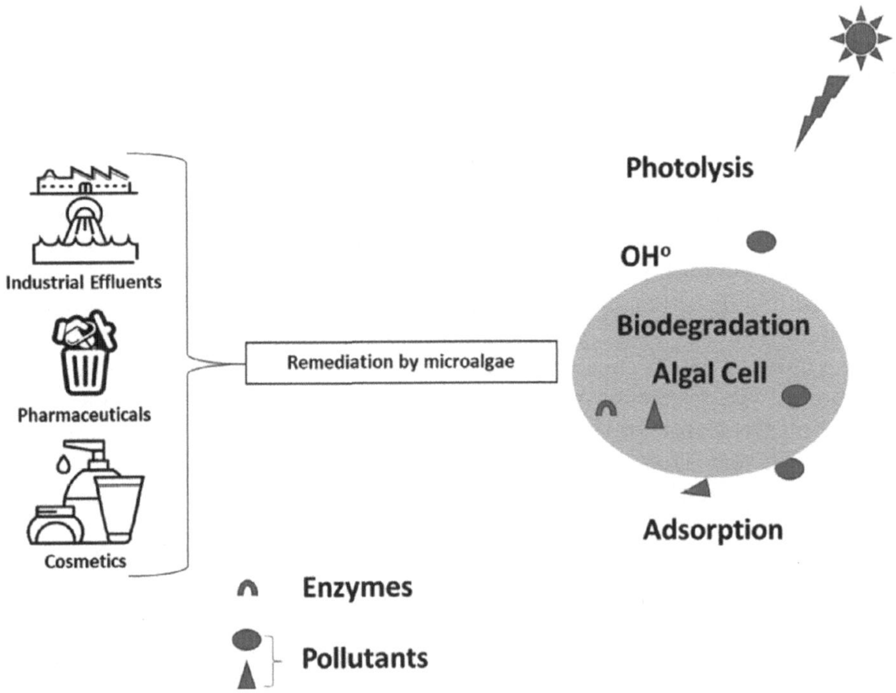

Figure 11.1 Bioremediation of emerging pollutants by microalgae.

As microalgae exist widely across the aquatic ecosystems and are exposed to a wide variety of pollutants, exploring microalgae as a tool for remediation of emerging contaminants is of great interest. The Organization for Economic Co-operation and Development (OECD) has demonstrated the use of microalgae as a reliable tool for assessing the toxicity of chemical contaminants as microalgae are the key components of freshwater/marine ecosystem. The antioxidant enzymes present in microalgae viz. superoxide dismutase (SOD) and catalase (CAT) could act as biomarkers for prior detection of pollution in the water (Chen and Jiang, 2011, Huang et al., 2012, Torres et al., 2008, Kurade et al., 2016). The presence of toxic contaminants in the water will trigger the generation of reactive oxygen species (ROS) in the microalgae. The antioxidant enzymes viz. SOD and CAT play a crucial role in minimizing the ROS levels inside the cell. When the intracellular ROS levels increase, SOD catalyzes the conversion of superoxide anions (O2•−) to hydrogen peroxide (H_2O_2) and oxygen (O_2). Following this, the CAT enzyme acts upon H_2O_2 and hydrolyzes it to H_2O and O_2. Though microalgal metabolic activity is complex, tracking the activity of these antioxidant enzymes would facilitate identifying the physical presence of contaminants in the water.

11.2.1 Possible routes of remediation by microalgae

Microalgae have been widely explored for the removal of pollutants from water resources. There are three possible ways of remediation of contaminants by microalgae viz.

a. adsorption,
b. absorption (accumulation in biomass), and
c. biodegradation.

(a) Adsorption/absorption: The cell membranes of microalgae have a net negative charge and thus can remediate contaminants from water by adsorption (Sarat Chandra et al., 2015). Microalgae have been effectively used as adsorbents for the removal of heavy metals from wastewaters (Yu et al., 2020, Gunasundaria et al., 2020, Chin et al., 2020, Bordoloi et al., 2020). Also, various researchers have explored the use of the microalgal biomass for remediation of antibiotics, dyes, etc., via the process of adsorption. Choi et al. (2020) have studied the removal of tetracycline using biochar derived from pyrolysis of *Spirulina* biomass. The yield and properties of biochar vary with the pyrolytic temperatures, and hence, the impact of pyrolysis temperature on the overall removal efficiencies of tetracycline was assessed. The authors observed that increase in pyrolytic temperatures has led to increase in aromaticity, the specific surface area of the biochar. A maximum tetracycline removal efficiency of 132.8 mg of tetracycline per g of biochar has been reported. The authors have reported that the mechanism of removal is majorly due to the surface adsorption, and the process of adsorption could be well-explained by pseudo-first-order kinetic models. For the last decade, microalgae have been explored as a potential candidate for the production of biodiesel. During the process of biodiesel production, 70% of the biomass is left as waste and has the potential to be used as adsorbent. Lipid-extracted microalga residues have also been used as a tool for the removal of trace contaminants from water. Sarat Chandra et al. (2015) have studied the removal of dyes from wastewaters. The authors have explored the use of lipid-extracted *Scenedesmus* biomass for the removal of methylene blue (MB) from water resources. The authors have reported the removal efficiency of >95% with a maximum adsorption capacity of 7.8 mg/g of biomass. Further, the removal of Cephalexin, a commonly used broad-spectrum antibiotic from wastewater by lipid-extracted microalgal biomass residues, has been studied (Angulo et al., 2018). The potential of raw and lipid-extracted biomass residues of freshwater microalga, *Chlorella vulgaris,* as a low-cost adsorbent has been evaluated for the removal of cephalexin. A maximum of 82.77% removal of cephalexin was observed with full fat biomass, whereas with lipid-extracted residues, it was observed to be ~70%, indicating the potential of microalgal biomass residues for the removal of contaminants from wastewater.

(b) Degradation: As microalgae are photosynthetic aquatic organisms, photolysis of water takes place during the process of photosynthesis and leads to the generation of hydroxyl radicals. Research has been conducted to understand the possibility of degradation or adsorption of emerging contaminants by microalgae. Bai and Acharya (2017) have evaluated the removal of five common contaminants viz. trimethoprim, sulfamethoxazole, carbamazepine, ciprofloxacin, and triclosan originated from pharmaceutical and personal care products via green microalga, *Nannochloropsis* sp. The microalgae are inoculated in the wastewater with known concentrations of the above-mentioned contaminants and were incubated for 14 days. Among these contaminants, the concentrations of trimethoprim and carbamazepine were remained >90% after 14 days of incubation, whereas the concentrations of sulfamethoxazole were observed to be >60%. The decrease in the concentration of the sulfamethoxazole is induced by algae-mediated photolysis of water. Further, the concentrations of ciprofloxacin and triclosan were observed to be less than 5% after 7 days of the incubation period. Algae-mediated photolysis was observed to be dominant mechanism for the removal of ciprofloxacin and triclosan, whereas adsorption was observed to predominant mechanism for the removal of sulfamethoxazole, carbamazepine, and trimethoprim with removal efficiencies of 11%, 13%, and 11%, respectively. This indicates that in the live microalgal cells degradation could be the major cause for the removal of contaminants. The use of *Chlorella* for biodegradation of florfenicol, a broad-spectrum antibiotic, has also been investigated (Song et al., 2019). The effect of initial concentration (40–160 mg/L) on the mechanism of removal has been investigated. With the initial concentrations of 40–50 mg/L, the mechanism of removal was major via biodegradation and was observed to be the only pathway. However, with an increase in the concentration to ~160 mg/L, the removal of florfenicol was observed to be via bioaccumulation and biosorption. Further, with the increase in the concentration of florfenicol from ~50 to 160 mg/L, the removal efficiency has decreased from 97% to 76%.

11.2.2 Assessing the toxicity of emerging contaminants on microalgal growth

In order to assess the toxicity of the emerging contaminant on microalgal growth, parameters such as total chlorophyll, total carotenoids, malondialdehyde content, and activity of antioxidant enzymes such as SOD and CAT need to be investigated. Further, the concentration of malondialdehyde can be correlated with the activity of various intracellular enzymes such as NADPH-dependent peroxidases and oxidases. NAD(P)H-oxidase would result in the generation of superoxide and hydrogen peroxide, whereas peroxidases can result in the generation of hydroxyl radicals. This will further induce the production of malondialdehyde (Apel and Hirt, 2004). Peroxidation of lipids results in an increase in dielectric constant and further increase in permeability of membranes. This would result in the transport of the contaminants across the cell membrane (Tang et al., 2016; Wong-Ekkabut et al., 2007). In microalgae, the presence of growth-inhibiting factors in the medium will induce the generation of ROS such as superoxide ions and hydroxyl radicals. Generation of ROS is a primary defense mechanism of microalgae, and increase in intracellular concentration of ROS will lead to an increase in activity of SOD and CAT, which form the antioxidative mechanisms. Hence, the activity of SOD or CAT can be used as markers for assessing the toxicity of an emerging contaminant to microalgae. Xiong et al. (2016) have observed an increase in CAT activity with the increase in the concentration of carbamazepine in *Chlamydomonas* and *Scenedesmus* sp. Enrofloxacin, a fluoroquinolone antibiotic, was studied for its ecotoxicity across five different microalgal species viz. *Scenedesmus obliquus*, *Chlamydomonas mexicana*, *Chlorella vulgaris*, *Ourococcus multisporus*, and *Micractinium resseri* (Xiong et al., 2017). The microalgae are exposed to different concentrations of enrofloxacin ranging from 1 to 100 mg/L individually and in the consortium. Inhibition of growth was negligible at a concentration of <1 mg/L, whereas growth inhibition of ~70% was observed at a concentration of 100 mg/L when cultivated individually. However, the growth rate inhibition was decreased with an increase in the cultivation time, indicating the adaptability of microalgae toward the contaminant. The toxicity of enrofloxacin on microalgae was measured by analyzing the malondialdehyde content in the cells. The malondialdehyde content has increased across five different microalgal species with an increase in the concentration of enrofloxacin from 20 to 100 mg/L, whereas increase in concentration from 0 to 20 mg/L has not resulted in a significant increase in malondialdehyde content. Various studies have been reported on evaluating the toxicity of emerging contaminants and understanding the biodegradation potential of these contaminants by microalgae. The assessment of toxicity of various pollutants of emerging concern (PECs) by microalgae is reported below.

a. **Assessing toxicity of methyl parabens**: Toxicity tests of methyl parabens, a commonly used preservative, on microalgae have been studied by Puerta et al. (2020). Three microalgal species viz. *Pseudopediastrum boryanum, Desmodesmus communis, and Raphidocelis subcapitata* have been cultivated in the presence of methyl parabens to understand the toxicity of the MB on aquatic ecosystems. EC50 values were estimated and were observed to be in the range of 18–125 g/L, indicating that the presence of these contaminants in trace concentrations is also lethal to the aquatic ecosystems. Hence, efficient removal of these trace contaminants is thus essential.
b. **Assessing toxicity of sulfonamides**: Sulfonamides, class of antibiotics, used for humans to treat infectious diseases are also recognized as emerging contaminants. Drugs such as sulfamethazine and sulfamethoxazole are examples of sulfonamides used commonly and were widely reported as common groundwater and wastewater contaminants. The toxicity of these drugs and their subsequent removal from wastewater by a freshwater microalga, *Scenedesmus obliquus,* have been investigated (Xiong et al., 2019). The toxicity of these drugs in the concentration range of 0.05–0.5 mg/L on the growth of *Scenedesmus obliquus* was studied. The authors have observed that the increase in the concentration of the drug has resulted in increased growth inhibition. However, the inhibition has decreased with increase in cultivation time, indicating that the microalgal species have adapted to the contaminant stress via the degradation of the compounds. The rate of degradation of sulfonamides has increased with an increase in the cultivation time, which was evident from the increase in sulfur content of the biomass with respect to the time. The results indicated biodegradation as a predominant mechanism for microalgae toward the removal of emerging contaminants from wastewater.
c. **Assessing toxicity of carbamazepine**: Carbamazepine is another emerging contaminant commonly used as antiepileptic agent (Zhang et al., 2008). It is discharged to the ecosystem through the urine and feces. Due to the toxicity of carbamazepine, the removal efficiency of this drug is low and is hence widely detected as trace contaminant in aquatic ecosystems. The toxicity and biodegradation of carbamazepine by freshwater microalgal species viz. *Scenedesmus obliquus* and *Chlamydomonas mexicana* have been evaluated (Xiong et al., 2016). The toxicity of carbamazepine was tested at different initial concentrations ranging from 1 to 200 mg/mL. Among the two different microalgal species, *Chlamydomonas* has shown higher tolerance toward the carbamazepine compared to the *Scenedesmus* sp. The effective concentration (EC50) defined as the concentration at which 50% of the growth rate was inhibited has been found and estimated to be 797 and

149 mg/L for *Chlamydomonas* and *Scenedesmus* sp, respectively. The toxicity of carbamazepine on microalgae was assessed on other freshwater microalgal species such as *Dunaliella tertiolecta* and *Chlorella pyrenoidosa*. Matamoros et al. (2016) have studied the degradation of carbamazepine using microalgal consortium consisting of *Chlorella* and *Scenedesmus* sp. Zhang et al. (2012) have studied the toxicity of carbamazepine on *Scenedesmus obliquus* and *Chlorella pyrenoidosa*. The authors have observed that the growth rate of *Chlorella* and *Scenedesmus* was greatly affected by increase in concentration of the drug. Tsiaka et al. (2013) have performed toxicity assessment of carbamazepine on green microalgae *Dunaliella tertiolecta*. The author has observed an increase in malondialdehyde content, a measure of intracellular lipid peroxidation, with the increase in concentration of carbamazepine, indicating the intracellular stress levels in the presence of the drug. The biodegradation studies indicated that the *Chlamydomonas* and *Scenedesmus* have resulted in 37% and 30% removal of carbamazepine, respectively, at an initial concentration of 1 mg/L. The highest overall rate of carbamazepine degradation has been observed with *Chlamydomonas* (~35%) compared to *Scenedesmus* sp (28%).

11.2.3 Mechanism of degradation by microalgae

Various researchers have attempted to develop insights into the mechanism of degradation and toxicity of emerging contaminants such as antibiotics, pesticides, and contaminants from personal care products. Few case studies regarding remediation of emerging contaminants by microalgae reported in the literature are discussed in the following subsections:

11.2.3.1 Case study 1: biodegradation of benzophenone

Benzophenone-3, one of the emerging contaminants used widely in personal care products used as UV filters, is frequently detected in the domestic wastewaters in the concentration of ~4 mg/L. Ecotoxicity and removal of benzophenone-3 in the concentration range of 0.1–3 mg/L have been studied using *Scenedesmus obliquus* (Lee et al., 2020). The removal efficiency of 23%–29% has been observed and was induced by bioaccumulation, biosorption, and biodegradation. The biodegradation mechanism has been assessed by conducting metabolic fate analysis, and it was observed that the degradation of benzophenone by *Scenedesmus obliquus* was mediated via hydroxylation and methylation. The main enzyme responsible for the degradation of emerging contaminants such as pharmaceuticals and endocrine disruptors is the oxidoreductase, a cytochrome P450 enzyme. Three major degradation products have been identified, and the author has demonstrated the possible mechanism of degradation. Based on the hypothesis, primarily, the cytochrome p450 enzyme acts on the contaminants through ring cleavage reaction, followed by subsequent action of enzymes such as glutathione S-transferase and monooxygenase, leading to the degradation of the organic contaminants. The possible mechanism of benzophenone degradation by *Scenedesmus obliquus* is presented in Figure 11.2. Similar observations regarding the degradation mechanism of pharmaceutical compounds by microalgae have been reported by various researchers. Peng et al. (2014) have reported degradation of steroids by *Scenedesmus obliquus*

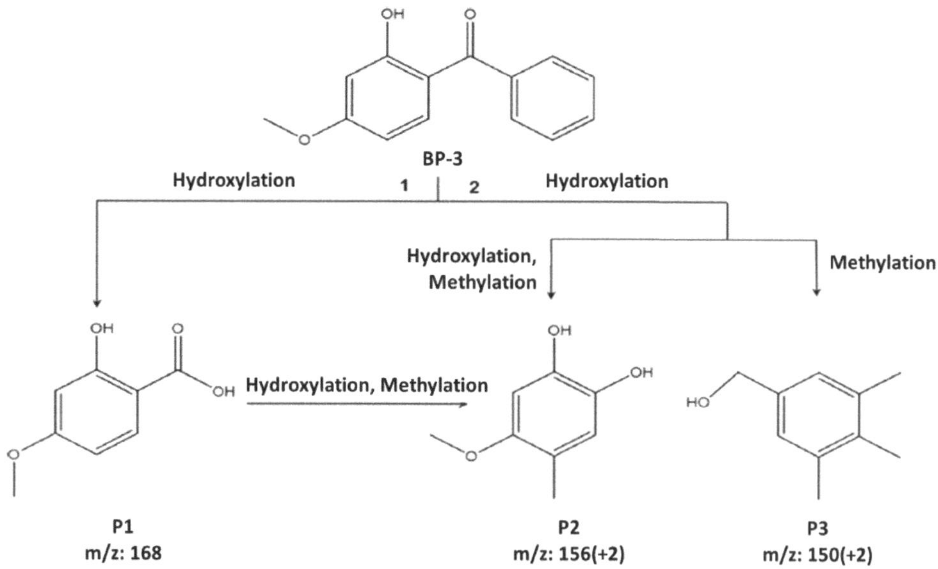

Figure 11.2 Mechanism of benzophenone degradation by freshwater microalga, Scenedesmus obliquus. (Adapted from Lee et al., 2020.)

and *Chlorella pyrenoidosa* via ring cleavage, hydroxylation, and dehydrogenation reactions. Ding et al. (2017) have also reported the degradation of ibuprofen by *Navicula* sp via hydroxylation, acylation, demethylation, and glucuronidation. Wang et al. (2017) have reported the degradation of bisphenol A by *Desmodesmus* sp. via a series of steps involving hydroxylation oxidation and methylation. Further, the toxicity of intermediate products has also been assessed. Lee et al. (2020) have observed that the toxicity of the intermediate products resulted from the degradation of benzophenone by *Scenedesmus obliquus* is less toxic compared to benzophenone. These results indicate that microalgae could be employed for the degradation and toxicity reduction of emerging contaminants.

11.2.3.2 Case study 2: biodegradation of diazinon

Agricultural runoff waters with pesticides are one of the major sources of emerging contaminants. Diazinon is one of the most commonly used organophosphorus insecticides and is widely detected as a contaminant in agricultural runoff waters. Diazinon is highly toxic, and the removal efficiencies of water treatment plants are reported to be low. The toxicity and biodegradation of diazinon by four freshwater green algae viz. *Scenedesmus obliquus*, *Chlorella vulgaris*, *Chlamydomonas Mexicana*, and *Chlamydomonas pitschmanni* have been investigated. Among the four different microalgae, *Chlorella vulgaris* has shown a maximum removal efficiency of >90% at an initial diazinon concentration of 20 mg/L. However, the increase in concentration of diazinon to 40 mg/L has resulted in growth rate inhibition of >30%. Further, enzyme activity studies revealed that increase in concentration of diazinon has resulted in an increase

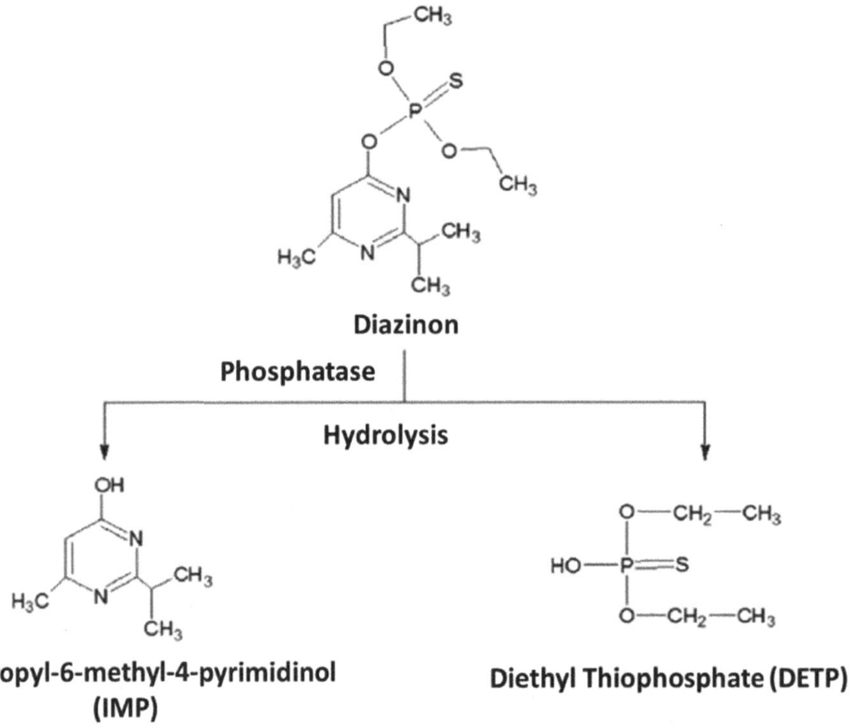

Figure 11.3 Mechanism of diazinon degradation by freshwater microalga, *Chlorella vulgaris*. (Adapted from Kurade et al., 2016.)

in the activity of SOD and CAT. Overall, the results indicated that growth inhibition of *Chlorella vulgaris* was not significant and is due to the formation of degradation products with lower toxicity. Biodegradation has been observed as a major removal mechanism of diazinon by *Chlorella vulgaris*. The degradation products of diazinon have been identified by mass spectroscopic techniques and observed that 2-isopropyl-6-methyl-4-pyrimidinol (IMP) and diethyl thiophosphate as major intermediates formed during the degradation of diazinon by *Chlorella vulgaris*. Kurade et al. (2016) have hypothesized that hydrolysis of diazinon to IMP and DETP would have been activated by phosphor-hydrolytic enzymes such as alkaline phosphatases. The proposed pathway for the degradation of diazinon by *Chlorella vulgaris* is presented in Figure 11.3. Further, IMP was also observed as a major degradation product of diazinon during its degradation via other physicochemical treatment methodologies such as photocatalysis and ultrasonication.

11.2.3.3 Case study 3: biodegradation of ibuprofen

Ibuprofen is a commonly used drug and is being detected as a common trace contaminant in surface waters. The toxicity and mechanism of ibuprofen degradation by *Navicula* sp. a diatom are explored (Ding et al., 2017). The toxicity of ibuprofen at a different initial concentration ranging from 0.1 to 100 mg/L on *Navicula* sp. was assessed. The author has observed that ibuprofen at lower concentrations in the range of 0.1–1 mg/L has stimulated the growth of algae, whereas higher concentrations in the range of 100 mg/L have shown significant inhibition on the growth rate of *Navicula* sp. Geiger et al. (2016) have also studied the toxicity of ibuprofen on freshwater microalgae *Chlorella vulgaris*. The authors observed a similar trend of growth rate inhibition when exposed to higher ibuprofen concentrations (~80 mg/L). With the increase in exposure time, the growth rate inhibition of *Navicula* sp was decreased, indicating the adaptability of the species to the contaminant. A maximum of 60% degradation was achieved at an initial concentration of 1 mg/L, whereas 20% degradation of ibuprofen was observed at higher concentrations of 50 mg/L. The degradation products of ibuprofen were identified, and the subcellular distribution of ibuprofen and the degradation products are assessed. The metabolic fate of ibuprofen in *Navicula* sp. was hypothesized based on the eight different intermediary products obtained. The intermediate products are characterized using ultrahigh-performance liquid chromatography-mass spectrometry (UPLC-MS/MS). The possible mechanism of degradation is presented in Figure 11.4.

As microalgae capture CO_2 from the atmosphere, the use of microalgae for the remediation of trace contaminants serves the dual purpose of carbon dioxide capture and the removal of contaminants. Du et al. (2018) have studied the use of microalgae for simultaneous capture of CO_2 from the atmosphere and removal of toxic antibiotics viz. amoxicillin and cefradine from wastewater. The effect of light intensity and the CO_2 concentration on microalgal growth and antibiotic removal efficiencies were studied. Two different microalgal species viz. *Chlorella pyrenoidosa* and *Microcystis Aeruginosa* have been considered. It was observed that the light intensity and the CO_2

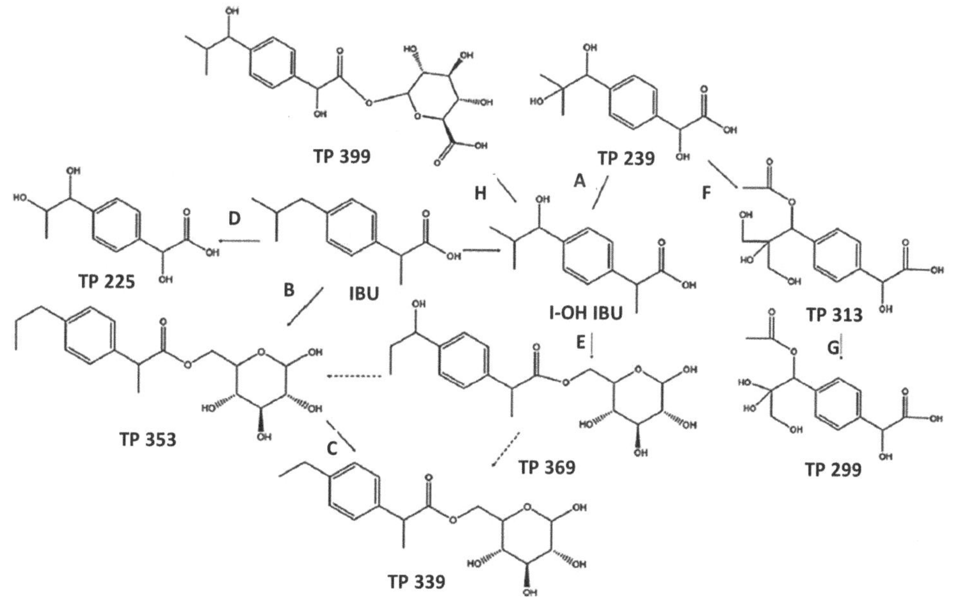

Figure 11.4 Mechanism of ibuprofen degradation by *Navicula* sp. a diatom. (Adapted from Ding et al., 2017.)

concentration had a significant effect on the uptake of antibiotics from the growth medium and are species-dependent. Maximum antibiotic removal was observed at higher CO_2 concentrations (20%) for *Microcystis*, whereas for *Chlorella*, maximum removal efficiencies were noticed at 10% CO_2. The mechanism of degradation has been studied by analyzing the intracellular metabolites, and a possible mechanism of degradation has also been elucidated. The microalgae-mediated removal of emerging contaminants reported by various researchers is summarized in Table 11.2.

Table 11.2 Summary of studies on microalgae-mediated remediation of emerging contaminants

S. no	*Microalgae*	*Emerging contaminant*	*Removal efficiency and mechanism*	*References*
1.	*Chlorella sp.*	Florfenicol	97%, removal efficiency	Song et al. (2019)
2.	*Nannochloropsis sp.*	Sulfamethoxazole Triclosan	32%, photodegradation 100%, photodegradation	Bai and Acharya (2017)
3.	*Scenedesmus sp.*	Anionic surfactants' removal	90%–97%	Serejo et al. (2020)
4.	*Scenedesmus obliquus*	Benzophenone-3	29%, biodegradation	Lee et al. (2020)
5.	*Spirogyra sp.*	Sulfathiazole Oxytetracycline Sulfapyridine Tetracycline Sulfamethazine Sulfamethoxazole	36%, photodegradation 93%, photodegradation 37%, photodegradation 89%, photodegradation 15%, photodegradation 14%, photodegradation	Garcia-Rodríguez et al. (2013)
6.	*Scenedesmus obliquus*	Sulfamethazine and Sulfamethoxazole	62.3% and 46.8%, respectively	Xiong et al. (2019)
7.	*Scenedesmus obliquus*	Carbamazepine	35%, biodegradation	Xiong et al. (2016)
8.	*Chlamydomonas Mexicana*	Carbamazepine	28%, biodegradation	Xiong et al. (2016)
9.	*Spirulina sp.*	Tetracycline adsorption	132.8 mg of tetracycline/g of biochar, adsorption	Choi et al. (2020)
10.	*Arthrospira platensis*	^{137}Cs, ^{233}U, ^{239}Pu, ^{241}Am, ^{90}Sr, and ^{237}Np	Bioaccumulation/ biosorption	Zinicovscaia et al. (2020)
11.	*Chlorella pyrenoidosa*	Triclosan	55%–77%, photodegradation	Wang et al. (2013)
12.	*Chlorella vulgaris*	Norfloxacin	36.9%, photodegradation	Zhang et al. (2012)
13.	*Chlorella pyrenoidosa*	Cefradine	76%, photodegradation	Chen et al. (2015)
14.	*Chlorella sorokiniana*	Salicylic acid and paracetamol	70%	Escapa et al. (2015)
15.	*Scenedesmus obliquus* and *Chlorella pyrenoidosa*	Progesterone and norgestrel	~95%	Peng et al. (2014)

(Continued)

Table 11.2 (Continued) Summary of studies on microalgae-mediated remediation of emerging contaminants

S. no	*Microalgae*	*Emerging contaminant*	*Removal efficiency and mechanism*	*References*
16.	*C. vulgaris*	Metronidazole	100%, adsorption	Hena et al. (2018)
17.	*Desmodesmus subspicatus*	17α-ethinylestradiol	68%	Maes et al. (2014)
18.	*Selenastrum capricornutum* and *Chlamydomonas Reinhardtii*	ß-estradiol and 17α-ethynylestradiol	47%–100%, photodegradation	Hom-Diaz et al. (2015)
19.	*Fischerella* sp.	Methyl parathion	>80%, biodegradation	Tiwari et al. (2017)
20.	*Phaeodactylum tricornutum*	Oxytetracycline	97%, biodegradation	Santaeufemia et al. (2016)
21.	*Chlorella sorokiniana*	5-Methyl-1H-benzotriazole	97%, photodegradation	Gatidou et al. (2019)
1.		5-Chlorobenzotriazole	52%, photodegradation	
1.		Xylyltriazole	42%, photodegradation	
1.		1H-Benzotriazole	79%, photodegradation	
22.	*Scenedesmus obliquus*	Enrofloxacin	23%, biodegradation	Xiong et al. (2017)
1.	*Chlorella vulgaris*		26%, biodegradation	
1.	*Micractinium resseri*		20%, biodegradation	
1.	*Ourococcus multisporus*		18%, biodegradation	
1.	*Chlamydomonas mexicana*		25%, biodegradation	
23.	*Chlorella sorokiniana*	Diclofenac	40%–60%	de Wilt et al. (2016)
		Trimethoprim	40%	
		Carbamazepine	30%	
		Metoprolol	100%	
		Paracetamol	100%	
		Ibuprofen	100%	

11.3 CONCLUSION

The present chapter summarizes the potential of microalgae for the remediation of PECs from water resources. The literature suggests that microalgae could be a promising candidate for the remediation of PECs; however, currently, the removal efficiencies are noticed to be low. Developing detailed insights into the mechanism of remediation would help to achieve higher removal efficiencies. Adsorption and degradation are the major pathways observed for the remediation of PECs by microalgae. As biodegradation is a slow process, integrated approaches for the removal of EC via microalgae could be explored. Strategies such as use of microalgal biomass as an adsorbent and subsequent processing of the biomass for the generation of energy could be studied. Further, only few species such as *Chlorella*, *Scenedesmus obliquus*, and *Chlamydomonas* have been investigated for the removal of PECs and there is a need to conduct large-scale screening of various microalgal strains for the removal of PECs.

ACKNOWLEDGMENTS

The author Sarat Chandra Togarcheti is thankful to Irish Research Council (IRC) for CAROLINE Research Fellowship. This project has received funding from IRC and from European Union's Horizon 2020 research and innovation program under Marie Skłodowska-Curie grant agreement No 713279. The author Ramesh Babu duly acknowledges the funding from Science Foundation Ireland (SFI), AMBER (Grant No: 12/RC/2278_P2), and BiOrbic (Grant No: 16/RC/3889).

REFERENCES

Angulo, E., Bula, L., Mercado, I., Montaño, A., & Cubillán, N., 2018. Bioremediation of Cephalexin with non-living *Chlorella* sp., biomass after lipid extraction. *Bioresour. Technol.*, 257, 17–22.

Apel, K., & Hirt, H., 2004. Reactive oxygen species: Metabolism, oxidative stress, and signal transduction. *Annu. Rev. Plant Biol.*, 55, 373–399.

Bai, X., & Acharya, K., 2017. Algae-mediated removal of selected pharmaceutical and personal care products (PPCPs) from Lake Mead water. *Sci. Total Environ.*, 581, 734–740.

Barrios, J. A., Becerril, E., De León, C., Barrera-Díaz, C., & Jiménez, B., 2015. Electrooxidation treatment for removal of emerging pollutants in wastewater sludge. *Fuel*, 149, 26–33.

Boonnorat, J., Chiemchaisri, C., Chiemchaisri, W., & Yamamoto, K., 2016. Kinetics of phenolic and phthalic acid esters biodegradation in membrane bioreactor (MBR) treating municipal landfill leachate. *Chemosphere*, 150, 639–649.

Bordoloi, N., Tiwari, J., Kumar, S., Korstad, J., & Bauddh, K., 2020. Efficiency of Algae for Heavy Metal Removal, Bioenergy Production, and Carbon Sequestration. In *Emerging Eco-Friendly Green Technologies for Wastewater Treatment* (pp. 77–101). Springer, Singapore.

Chen, H., & Jiang, J. G., 2011. Toxic effects of chemical pesticides (trichlorfon and dimehypo) on *Dunaliella salina*. *Chemosphere*, 84(5), 664–670.

Chen, J., Zheng, F., & Guo, R., 2015. Algal feedback and removal efficiency in a sequencing batch reactor algae process (SBAR) to treat the antibiotic cefradine. *PLoS One*, 10(7), e0133273.

Chin, J. Y., Chng, L. M., Leong, S. S., Yeap, S. P., Yasin, N. H. M., & Toh, P. Y., 2020. Removal of synthetic dye by *Chlorella vulgaris* microalgae as natural adsorbent. *Arabian J. Sci. Eng.*, 45(9), 7385–7395.

Choi, Y. K., Choi, T. R., Gurav, R., Bhatia, S. K., Park, Y. L., Kim, H. J., … & Yang, Y. H., 2020. Adsorption behavior of tetracycline onto *Spirulina* sp. (microalgae)-derived biochars produced at different temperatures. *Sci. Total Environ.*, 710, 136282.

Crisp, T. M., Clegg, E. D., Cooper, R. L., Anderson, D. G., Baetcke, K. P., Hoffmann, J. L., … & Touart, L. W., 1997. Special report on environmental endocrine disruption: An effects assessment and analysis. Washington DC: US Environmental Protection Agency.

de Wilt, A., Butkovskyi, A., Tuantet, K., Leal, L. H., Fernandes, T. V., Langenhoff, A., & Zeeman, G., 2016. Micropollutant removal in an algal treatment system fed with source separated wastewater streams. *J. Hazard. Mater.*, 304, 84–92.

Deo, R. P., 2014. Pharmaceuticals in the surface water of the USA: A review. *Curr. Environ. Health Rep.*, 1(2), 113–122.

Ding, T., Yang, M., Zhang, J., Yang, B., Lin, K., Li, J., & Gan, J., 2017. Toxicity, degradation and metabolic fate of ibuprofen on freshwater diatom *Navicula* sp. *J. Hazard. Mater.*, 330, 127–134.

Du, Y., Wang, J., Li, H., Mao, S., Wang, D., Xiang, Z., … & Chen, J., 2018. The dual function of the algal treatment: Antibiotic elimination combined with CO_2 fixation. *Chemosphere*, 211, 192–201.

Dubey, M., Mohapatra, S., Tyagi, V. K., Suthar, S., & Kazmi, A. A., 2021. Occurrence, fate, and persistence of emerging micropollutants in sewage sludge treatment. *Environ. Pollut.*, 273, 116515.

Escapa, C., Coimbra, R. N., Paniagua, S., García, A. I., & Otero, M., 2015. Nutrients and pharmaceuticals removal from wastewater by culture and harvesting of *Chlorella sorokiniana. Bioresour. Technol.*, 185, 276–284.

Félix–Cañedo, T. E., Durán–Álvarez, J. C., & Jiménez–Cisneros, B., 2013. The occurrence and distribution of a group of organic micropollutants in Mexico City's water sources. *Sci. Total Environ.*, 454, 109–118.

Fent, K., Zenker, A., & Rapp, M., 2010. Widespread occurrence of estrogenic UV-filters in aquatic ecosystems in Switzerland. *Environ. Pollut.*, 158(5), 1817–1824.

Fudala-Ksiazek, S., Pierpaoli, M., & Luczkiewicz, A., 2018. Efficiency of landfill leachate treatment in a MBR/UF system combined with NF, with a special focus on phthalates and bisphenol A removal. *Waste Manage.*, 78, 94–103.

Furuichi, T., Kannan, K., Suzuki, K., Tanaka, S., Giesy, J. P., & Masunaga, S., 2006. Occurrence of estrogenic compounds in and removal by a swine farm waste treatment plant. *Environ. Sci. Technol.*, 40(24), 7896–7902.

Gago-Ferrero, P., Gros, M., Ahrens, L., & Wiberg, K., 2017. Impact of on-site, small and large scale wastewater treatment facilities on levels and fate of pharmaceuticals, personal care products, artificial sweeteners, pesticides, and perfluoroalkyl substances in recipient waters. *Sci. Total Environ.*, 601, 1289–1297.

Garcia-Rodríguez, A., Matamoros, V., Fontàs, C., & Salvadó, V., 2013. The influence of light exposure, water quality and vegetation on the removal of sulfonamides and tetracyclines: A laboratory-scale study. *Chemosphere*, 90(8), 2297–2302.

Gatidou, G., Anastopoulou, P., Aloupi, M., & Stasinakis, A. S., 2019. Growth inhibition and fate of benzotriazoles in *Chlorella sorokiniana* cultures. *Sci. Total Environ.*, 663, 580–586.

Geiger, E., Hornek-Gausterer, R., & Saçan, M. T., 2016. Single and mixture toxicity of pharmaceuticals and chlorophenols to freshwater algae *Chlorella vulgaris. Ecotoxicol. Environ. Saf.*, 129, 189–198.

Glinski, D. A., Purucker, S. T., Van Meter, R. J., Black, M. C., & Henderson, W. M., 2018. Analysis of pesticides in surface water, stemflow, and throughfall in an agricultural area in South Georgia, USA. *Chemosphere*, 209, 496–507.

Gogoi, A., Mazumder, P., Tyagi, V. K., Chaminda, G. T., An, A. K., & Kumar, M., 2018. Occurrence and fate of emerging contaminants in water environment: A review. *Groundw. Sustain. Dev.*, 6, 169–180.

Gogoi, A., Taki, K., & Kumar, M. 2020. Seasonal dynamics of metal phase distributions in the perennial tropical (Brahmaputra) river: Environmental fate and transport perspective. *Environ. Res.*, 183, 109265.

Gunasundaria, E., Kumara, P. S., Rajamohanb, N., & Vellaichamyc, P., 2020. Feasibility of naphthol green-B dye adsorption using microalgae: Thermodynamic and kinetic analysis. *Desalin. Water Treat.*, 192, 358–370.

Hai, F. I., Li, X., Price, W. E., & Nghiem, L. D., 2011. Removal of carbamazepine and sulfamethoxazole by MBR under anoxic and aerobic conditions. *Bioresour. Technol.*, 102(22), 10386–10390.

Hena, S., Znad, H., Heong, K. T., & Judd, S., 2018. Dairy farm wastewater treatment and lipid accumulation by *Arthrospira platensis. Water Res.*, 128, 267–277.

Hernández, F., Ibáñez, M., Bade, R., Bijlsma, L., & Sancho, J. V., 2014. Investigation of pharmaceuticals and illicit drugs in waters by liquid chromatography-high-resolution mass spectrometry. *Trends Analyt. Chem.*, 63, 140–157.

Hom-Diaz, A., Llorca, M., Rodríguez-Mozaz, S., Vicent, T., Barceló, D., & Blánquez, P., 2015. Microalgae cultivation on wastewater digestate: β-estradiol and 17α-ethynylestradiol degradation and transformation products identification. *J. Environ. Manage.*, 155, 106–113.

Huang, L. D., Lu, D. H., Diao, J. L., & Zhou, Z. Q., 2012. Enantioselective toxic effects and biodegradation of benalaxyl in *Scenedesmus obliquus*, *Chemosphere*, 87, 7–11.

Huerta-Fontela, M., Galceran, M. T., & Ventura, F., 2011. Occurrence and removal of pharmaceuticals and hormones through drinking water treatment. *Water Res.*, 45(3), 1432–1442.

Jennifer, A., Ae, A. M., & Harrad, S., 2017. Pharmaceuticals and Personal Care Products (PPCPs) in the freshwater aquatic environment. *Emerg Contam.*, 3(1), 1–16.

Jin, Z. P., Luo, K., Zhang, S., Zheng, Q., & Yang, H., 2012. Bioaccumulation and catabolism of prometryne in green algae. *Chemosphere*, 87(3), 278–284.

Kapelewska, J., Kotowska, U., & Wiśniewska, K., 2016. Determination of personal care products and hormones in leachate and groundwater from Polish MSW landfills by ultrasound-assisted emulsification microextraction and GC-MS. *Environ. Sci. Pollut. Res.*, 23(2), 1642–1652.

Khalil, C., Al Hageh, C., Korfali, S., & Khnayzer, R. S., 2018. Municipal leachates health risks: Chemical and cytotoxicity assessment from regulated and unregulated municipal dumpsites in Lebanon. *Chemosphere*, 208, 1–13.

Kumar, M., Patel, A. K., Das, A., Kumar, P., Goswami, R., Deka, P., & Das, N. 2017. Hydrogeochemical controls on mobilization of arsenic and associated health risk in Nagaon district of the central Brahmaputra Plain, India. *Environ. Geochem. Health*, 39(1), 161–178.

Kumar, M., Snow, D. D., Li, Y., & Shea, P. J. 2019. Perchlorate behavior in the context of black carbon and metal cogeneration following fireworks emission at Oak Lake, Lincoln, Nebraska, USA. *Environ. Pollut., 253*, 930–938.

Kurade, M. B., Kim, J. R., Govindwar, S. P., & Jeon, B. H., 2016. Insights into microalgae mediated biodegradation of diazinon by *Chlorella vulgaris*: Microalgal tolerance to xenobiotic pollutants and metabolism. *Algal Res.*, 20, 126–134.

Lee, S. H., Xiong, J. Q., Ru, S., Patil, S. M., Kurade, M. B., Govindwar, S. P., ... & Jeon, B. H., 2020. Toxicity of benzophenone-3 and its biodegradation in a freshwater microalga *Scenedesmus obliquus*. *J. Hazard. Mater.*, 389, 122149.

Leng, Y., Bao, J., Xiao, H., Song, D., Du, J., Mohapatra, S., ... & Wang, J., 2020. Transformation mechanisms of tetracycline by horseradish peroxidase with/without redox mediator ABTS for variable water chemistry. *Chemosphere*, 258, 127306.

Li, J., Zhao, L., Qin, L., Tian, X., Wang, A., Zhou, Y., ... & Chen, Y., 2016. Removal of refractory organics in nanofiltration concentrates of municipal solid waste leachate treatment plants by combined Fenton oxidative-coagulation with photo–Fenton processes. *Chemosphere*, 146, 442–449.

Li, Y., Zhu, G., Ng, W. J., & Tan, S. K., 2014. A review on removing pharmaceutical contaminants from wastewater by constructed wetlands: Design, performance and mechanism. *Sci. Total Environ.*, 468, 908–932.

Maes, H. M., Maletz, S. X., Ratte, H. T., Hollender, J., & Schaeffer, A., 2014. Uptake, elimination, and biotransformation of 17α-ethinylestradiol by the freshwater alga *Desmodesmus subspicatus*. *Environ. Sci. Technol.*, 48(20), 12354–12361.

Mailler, R., Gasperi, J., Coquet, Y., Buleté, A., Vulliet, E., Deshayes, S., ... & Caupos, E., 2016. Removal of a wide range of emerging pollutants from wastewater treatment plant discharges by micro-grain activated carbon in fluidized bed as tertiary treatment at large pilot scale. *Sci. Total. Environ.*, 542, 983–996.

Matamoros, V., Uggetti, E., García, J., & Bayona, J. M., 2016. Assessment of the mechanisms involved in the removal of emerging contaminants by microalgae from wastewater: A laboratory scale study. *J. Hazard. Mater.*, 301, 197–205.

Meierjohann, A., Brozinski, J. M., & Kronberg, L., 2016. Seasonal variation of pharmaceutical concentrations in a river/lake system in Eastern Finland. *Environ. Sci. Processes & Impacts*, 18(3), 342–349.

Menon, N. G., Mohapatra, S., Padhye, L. P., Tatiparti, S. S. V., Mukherji, S., 2020. Review on Occurrence and Toxicity of Pharmaceutical Contamination in Southeast Asia. In: Kumar M., Snow D., Honda R. (eds) *Emerging Issues in the Water Environment during Anthropocene*. Springer Transactions in Civil and Environmental Engineering (pp. 63–91). Springer, Singapore.

Mohapatra, S., Menon, N. G., Padhye, L. P., Tatiparti, S. S. V., & Mukherji, S., 2021. Natural Attenuation of Pharmaceuticals in the Aquatic Environment and Role of Phototransformation. In *Contaminants in Drinking and Wastewater Sources* (pp. 65–94). Springer, Singapore.

Olatunji, O. S., Fatoki, O. S., Opeolu, B. O., Ximba, B. J., & Chitongo, R., 2017. Determination of selected steroid hormones in some surface water around animal farms in Cape Town using HPLC-DAD. *Environ. Monit. Assess.*, 189(7), 363.

Peng, F. Q., Ying, G. G., Yang, B., Liu, S., Lai, H. J., Liu, Y. S., ... & Zhou, G. J., 2014. Biotransformation of progesterone and norgestrel by two freshwater microalgae (*Scenedesmus obliquus* and *Chlorella pyrenoidosa*): Transformation kinetics and products identification. *Chemosphere*, 95, 581–588.

Priyanka, M.S., Mohapatra, S. 2020. Shifts and Trends in Analysis of Contaminants of Emerging Concern: Sulfonamides. In: Kumar M., Munoz-Arriola F., Furumai H., Chaminda T. (eds) *Resilience, Response, and Risk in Water Systems*. Springer Transactions in Civil and Environmental Engineering. Springer, Singapore.

Puerta, Y. T., Guimarães, P. S., Martins, S. E., & Martins, C. D. M. G., 2020. Toxicity of methylparaben to green microalgae species and derivation of a predicted no effect concentration (PNEC) in freshwater ecosystems. *Ecotoxicol. Environ. Saf.*, 188, 109916.

Racz, L., Muller, J. G., & Goel, R. K., 2012. Fate of selected estrogens in two laboratory scale sequencing batch reactors fed with different organic carbon sources under varying solids retention times. *Bioresour. Technol.*, 110, 35–42.

Reis, B. G., Silveira, A. L., Teixeira, L. P. T., Okuma, A. A., Lange, L. C., & Amaral, M. C. S., 2017. Organic compounds removal and toxicity reduction of landfill leachate by commercial bakers' yeast and conventional bacteria based membrane bioreactor integrated with nanofiltration. *Waste Manage.*, 70, 170–180.

Rivera-Jaimes, J. A., Postigo, C., Melgoza-Alemán, R. M., Aceña, J., Barceló, D., & de Alda, M. L., 2018. Study of pharmaceuticals in surface and wastewater from Cuernavaca, Morelos, Mexico: Occurrence and environmental risk assessment. *Sci. Total Environ.*, 613, 1263–1274.

Samaras, V. G., Stasinakis, A. S., Mamais, D., Thomaidis, N. S., & Lekkas, T. D., 2013. Fate of selected pharmaceuticals and synthetic endocrine disrupting compounds during wastewater treatment and sludge anaerobic digestion. *J. Hazard. Mater.*, 244, 259–267.

Santaeufemia, S., Torres, E., Mera, R., & Abalde, J., 2016. Bioremediation of oxytetracycline in seawater by living and dead biomass of the microalga *Phaeodactylum tricornutum*. *J. Hazard. Mater.*, 320, 315–325.

Sarat Chandra, T., Mudliar, S. N., Vidyashankar, S., Mukherji, S., Sarada, R., Krishnamurthi, K., & Chauhan, V. S., 2015. Defatted algal biomass as a non-conventional low-cost adsorbent: Surface characterization and methylene blue adsorption characteristics. *Bioresour. Technol.*, 184, 395–404.

Seibert, D., Quesada, H., Bergamasco, R., Borba, F. H., & Pellenz, L., 2019. Presence of endocrine disrupting chemicals in sanitary landfill leachate, its treatment and degradation by Fenton based processes: A review. *Process Saf. Environ. Prot.*, 131, 255–267.

Serejo, M. L., Farias, S. L., Ruas, G., Paulo, P. L., & Boncz, M. A., 2020. Surfactant removal and biomass production in a microalgal-bacterial process: Effect of feeding regime. *Water Sci. Technol.*, 82(6), 1176–1183.

Song, C., Wei, Y., Qiu, Y., Qi, Y., Li, Y., & Kitamura, Y., 2019. Biodegradability and mechanism of florfenicol via *Chlorella* sp. UTEX1602 and L38: Experimental study. *Bioresour. Technol.*, 272, 529–534.

Tang, S., Yin, H., Chen, S., Peng, H., Chang, J., Liu, Z., & Dang, Z., 2016. Aerobic degradation of BDE-209 by *Enterococcus casseliflavus*: Isolation, identification and cell changes during degradation process. *J. Hazard. Mater.*, 308, 335–342.

Thakur, A. K., Vithanage, M., Das, D., & Kumar, M., 2020. A review on design, material selection, mechanism, and modeling of permeable reactive materials for community-scale groundwater treatment. *Environ. Technol. Innov.*, 19, 100917.

Tiwari, B., Sellamuthu, B., Ouarda, Y., Drogui, P., Tyagi, R. D., & Buelna, G., 2017. Review on fate and mechanism of removal of pharmaceutical pollutants from wastewater using biological approach. *Bioresour. Technol.*, 224, 1–12.

Torres, M. A., Barros, M. P., Campos, S. C., Pinto, E., Rajamani, S., Sayre, R. T., & Colepicolo, P., 2008. Biochemical biomarkers in algae and marine pollution: A review. *Ecotoxicol. Environ. Saf.*, 71(1), 1–15.

Tran, N. H., Reinhard, M., & Gin, K. Y. H., 2018. Occurrence and fate of emerging contaminants in municipal wastewater treatment plants from different geographical regions-a review. *Water Res.*, 133, 182–207.

Tsiaka, P., Tsarpali, V., Ntaikou, I., Kostopoulou, M. N., Lyberatos, G., & Dailianis, S., 2013. Carbamazepine-mediated pro-oxidant effects on the unicellular marine algal species *Dunaliella tertiolecta* and the hemocytes of mussel Mytilus galloprovincialis. *Ecotoxicology*, 22(8), 1208–1220.

Vahčič, M., Milačič, R., & Ščančar, J., 2011. Development of analytical procedure for the determination of methyltin, butyltin, phenyltin and octyltin compounds in landfill leachates by gas chromatography–inductively coupled plasma mass spectrometry. *Anal. Chim. Acta*, 694(1–2), 21–30.

Wanda, E. M., Nyoni, H., Mamba, B. B., & Msagati, T. A., 2017. Occurrence of emerging micropollutants in water systems in Gauteng, Mpumalanga, and North West Provinces, South Africa. *Int. J. Environ. Res. Public Health*, 14(1), 79.

Wang, R., Diao, P., Chen, Q., Wu, H., Xu, N., & Duan, S., 2017. Identification of novel pathways for biodegradation of bisphenol A by the green alga *Desmodesmus* sp. WR1, combined with mechanistic analysis at the transcriptome level. *Chem. Eng. J.*, 321, 424–431.

Wang, S., Wang, X., Poon, K., Wang, Y., Li, S., Liu, H., ... & Cai, Z., 2013. Removal and reductive dechlorination of triclosan by *Chlorella pyrenoidosa*. *Chemosphere*, 92(11), 1498–1505.

Wong-Ekkabut, J., Xu, Z., Triampo, W., Tang, I. M., Tieleman, D. P., & Monticelli, L., 2007. Effect of lipid peroxidation on the properties of lipid bilayers: A molecular dynamics study. *Biophys. J.*, 93(12), 4225–4236.

Wu, Z., Zhang, D., Cai, Y., Wang, X., Zhang, L., & Chen, Y., 2017. Water quality assessment based on the water quality index method in Lake Poyang: The largest freshwater lake in China. *Sci. Rep.*, 7(1), 1–10.

Xiong, J. Q., Govindwar, S., Kurade, M. B., Paeng, K. J., Roh, H. S., Khan, M. A., & Jeon, B. H., 2019. Toxicity of sulfamethazine and sulfamethoxazole and their removal by a green microalga, *Scenedesmus obliquus*. *Chemosphere*, 218, 551–558.

Xiong, J. Q., Kurade, M. B., Abou-Shanab, R. A., Ji, M. K., Choi, J., Kim, J. O., & Jeon, B. H., 2016. Biodegradation of carbamazepine using freshwater microalgae *Chlamydomonas mexicana* and *Scenedesmus obliquus* and the determination of its metabolic fate. *Bioresour. Technol.*, 205, 183–190.

Xiong, J. Q., Kurade, M. B., & Jeon, B. H., 2017. Ecotoxicological effects of enrofloxacin and its removal by monoculture of microalgal species and their consortium. *Environ. Pollut.*, 226, 486–493.

Yang, Y. Y., Zhao, J. L., Liu, Y. S., Liu, W. R., Zhang, Q. Q., Yao, L., ... & Ying, G. G., 2018. Pharmaceuticals and personal care products (PPCPs) and artificial sweeteners (ASs) in surface and ground waters and their application as indication of wastewater contamination. *Sci. Total Environ.*, 616, 816–823.

Yu, K. L., Lee, X. J., Ong, H. C., Chen, W. H., Chang, J. S., Lin, C. S., ... & Ling, T. C., 2020. Adsorptive removal of cationic methylene blue and anionic Congo red dyes using wet-torrefied microalgal biochar: Equilibrium, kinetic and mechanism modeling. *Environ. Pollut.*, 272, 115986.

Zhang, J., Fu, D., & Wu, J., 2012a. Photodegradation of Norfloxacin in aqueous solution containing algae. *J. Environ. Sci.*, 24(4), 743–749.

Zhang, W., Zhang, M., Lin, K., Sun, W., Xiong, B., Guo, M., ... & Fu, R., 2012b. Ecotoxicological effect of Carbamazepine on *Scenedesmus obliquus* and *Chlorella pyrenoidosa*. *Environ. Toxicol. Pharmacol.*, 33(2), 344–352.

Zhang, Y., Geißen, S. U., & Gal, C., 2008. Carbamazepine and diclofenac: Removal in wastewater treatment plants and occurrence in water bodies. *Chemosphere*, 73(8), 1151–1161.

Zinicovscaia, I., Safonov, A., Zelenina, D., Ershova, Y., & Boldyrev, K., 2020. Evaluation of biosorption and bioaccumulation capacity of cyanobacteria *Arthrospira (spirulina) platensis* for radionuclides. *Algal Res.*, 51, 102075.

Chapter 12

Contaminants of emerging concerns and advanced treatment technologies

Neha Garg and Akshay Jakhete
Indian Institute of Technology Bombay

CONTENTS

12.1 INTRODUCTION

Due to the industrial revolution and change in the human lifestyle, there has been an exponential increase in contaminants in the environment. Contaminants of emerging concerns (CECs) are either the chemicals that have been present in the environment for a long time and their effects have been recently established, or which have been newly found and most of its implications and properties are not known (Fagan et al., 2016; Sauvé and Desrosiers, 2014). The concern emerging for these chemicals started probably from the 1962 book *Silent Spring* by Rachel Carson, who talked about potential concerns of using DDT pesticides (Sauvé and Desrosiers, 2014). In 1996, it became more relevant when Theo Colborn, Dianne Dumanoski, and John Myers published the book *Our Stolen Future: Are We Threatening Our Fertility, Intelligence, and Survival? A Scientific Detective Story*, which highlighted the problems caused by various chemicals (Vieira et al., 2020).

The CECs include pharmaceuticals, pesticides, personal care products (PCPs), disinfection by-products (DBPs), endocrine-disrupting chemicals (EDCs), metals,

DOI: 10.1201/9781003157465-15

and microplastics (Eregowda and Mohapatra, 2020; Kroon et al., 2020; Nawaz and Sengupta, 2019, Patel et al., 2020, Kumar et al., 2019, Kumar et al., 2020). These have been found in treated wastewater, surface water, groundwater, and also drinking water, generally in the range of 0.1 to 20 μg/L (Miranda-García et al., 2011, Singh et al., 2019). Most of the CECs are unregulated; however, some of the regulatory authorities, such as the United States Environmental Protection Agency (USEPA), the World Health Organization (WHO), and the United States Geological Survey (USGS) had raised concern and thereby started EDSP and REACH programs in the USA and European countries, respectively, to understand the toxicity of CECs (Appa et al., 2018).

The CECs are any natural or synthetic chemicals that are not typically monitored in the waste stream due to their low content (typically in trace amounts); however, they can harm the environment and can have toxic effects on aquatic organisms, humans, and animals. It has been observed that most of the wastewater and water treatment plants are unsuccessful in removing these contaminants (Fagan et al., 2016; Kroon et al., 2020; Nawaz and Sengupta, 2019, Kumar et al., 2021). Treatment of CECs is also dependent on the properties after they are released in water, such as solubility, distribution in the solid and liquid surface, polarity, vapor pressure, pH, temperature, and organic content (Nilsen et al., 2019, Kumar et al., 2017). Hence, it is important to understand these CECs, their fate in the environment and possible health hazards, and their removal mechanism.

One of the major problems associated with CECs is their detection in waste streams. Normally they are present in ng/L to mg/L, which limits its detection. These require highly sensitive equipment for their detection such as high-resolution liquid chromatography-mass spectrometry (HR-LCMS), time-of-flight mass spectrometry with liquid chromatography (LC-TOF MS), or gas chromatography-mass spectrometry (GC-MS) (Nawaz and Sengupta, 2019). As per the Environmental Protection Agency (EPA) regulation, for the detection of pharmaceuticals and PCPs, high-pressure liquid chromatography (HPLC/MS/MS), and the detection of steroids and hormones, high-resolution gas chromatography (HRGC)/high-resolution mass spectrometry (HRMS) is advised (Nawaz and Sengupta, 2019). For the detection of EDCs, solid-phase extraction (SPE) and biosensors can also be used (Vieira et al., 2020). Appa et al. (2018) had used SPE/LC-(ESI)MS-MS for the detection of pharmaceuticals as well as hormones. Along with detection, it is important to control CECs in the environment. Some basic approaches which can control the CECs are: ban on its production and development of remedial technologies using bioremediation methods (Alharbi et al., 2018, Shim et al., 2018) and incorporation of efficient treatment methods to existing wastewater treatment plants (WWTPs), which is not present in all the plants as of now.

The second problem is the perseverance and mobility of CEC in the waste stream. The sources of CECs can be very wide, from point sources (urban and industrial WWTPs) to diffuse sources (agricultural and crop production) (Kroon et al., 2020, Kumar, 2012). CEC can originate from households (pharmaceuticals, PCPs, and metals), agricultural areas (pesticides), and hospitals (hormones and pharmaceuticals) along industrial waste. These all find their way into either aquatic bodies or WWTPs (Nawaz and Sengupta, 2019; Nilsen et al., 2019). CECs also enter the aquatic system from leachates originating from landfills, erosion and deposition, surface runoff, and application of contaminated manure to land (Nilsen et al., 2019).

Another problem is that it is not just individual CEC that affects the population when released in water; however, it is a mixture of potentially toxic CECs that can affect and kill the inhabitants. The complex mixture of various CECs is present in all types of environments, and it was observed that they can cause serious effects on organisms. Some organisms accumulate higher chemical doses as compared to others. For example, the concentration of PFAs was higher in piscivorous fish species compared with non-piscivorous species in a lake in Ethiopia (Nilsen et al., 2019).

This chapter deals with the introduction to different CECs present in water and wastewater and methods which are used for the removal of these contaminants and challenges involved with these processes.

12.2 TYPES OF CECs

There are various types of CECs present in the environment; however, the focus of this chapter is mainly on five important ones, such as EDCs, DBPs, persistent organic pollutants (POPs), heavy metals, and microplastics.

12.2.1 Endocrine-disrupting chemicals

Endocrine-disrupting chemicals are "an agent that interferes with the synthesis, secretion, transport, binding, or elimination of natural hormones in the body which are responsible for the maintenance of homeostasis, reproduction, development, and/or behavior" as per EPA. In other words, these are natural or synthetic chemicals that interfere with the body's hormone system (Kabir et al., 2015). It is not always the actual chemical that is EDC; however, the by-product of the chemical is more hazardous, which causes harm to the body. Natural chemicals are those which are found in plant and animal food, such as phytoestrogen, while synthetic chemicals include industrial used solvents, PCPs, plastics, pesticides, fungicide, heavy metals, pharmaceuticals, and metals (Kabir et al., 2015).

Some of the sources of EDCs in the environment include the combustion of waste, vehicular exhaust, pesticide and herbicide sprayings, volatilization of chemicals, air fresheners, cosmetics, aerosols, dust, and particulates (Annamalai and Namasivayam, 2015). EDCs are found to be present everywhere around us. It is said to disrupt or alter the hormonal system of animals, including humans. They are thereby dysregulating various important functions, such as reproduction, metabolism, growth, and development of the body (Vieira et al., 2020) and fertility biomarkers, such as sperm counts (Sifakis et al., 2017). The major mechanism involved in disruption is either agonist or antagonist toward the receptors, which either enhances or inhibits the synthesis of hormones in the body (Annamalai and Namasivayam, 2015). EDCs can transfer into the body through three sources: water, soil, and fat. In water, some EDCs are found to be quite soluble in the levels of parts per trillion to parts per billion and hence harmful to aquatic organisms. They are also found to be accumulated in the soil, thereby ending up in the food chain through biomagnification. Whenever they are present inside the body, they have a high affinity toward fat tissue and thereby also transfer into offspring (Kabir et al., 2015; Vieira et al., 2020).

The EDCs are also present significantly in sewage and thereby into nearby surface water. It was found that even trace levels of steroid hormones can cause feminization and masculinization in aquatic species (Appa et al., 2018). Some of the pesticides which act as EDCs include organochlorines and organophosphorus. These can cause infertility, cancer, hypospadias, and many other problems related to the reproductive system (Sifakis et al., 2017). Organochlorines, such as DDT and dioxins, are not metabolized by humans and thereby get accumulated in the body. Some of the pesticides, such as endosulfan, DDT, endrin, and epoxide, were said to possess a high toxic risk to aquatic organisms, such as *Clarias gariepinus* and *Tilapia zilli,* in Owan River (Sifakis et al., 2017). DDT, an example of pesticide and polychlorinated biphenyls (PCBs), is a classic example of EDCs that were used extensively; however, due to their toxic effects, they were banned. Phthalates, which are used in children's products, packaging, and cosmetics, are also an example of EDCs (Vieira et al., 2020).

In 1988, USEPA had made the Endocrine Disruptor Screening Program for the evaluation of humans and wildlife (Kim et al., 2018; Vieira et al., 2020). However, only a few EDCs are listed in USEPA's drinking water contaminant list, which includes erythromycin (ETM), estrone (E1), 17β-estradiol (E2), 17α-ethinyl estradiol (EE2), and estriol (E3) (Kim et al., 2018). Bisphenol A (BPA) is a synthetic EDC, which is used in plastic production primarily. It is said to have estrogenic and toxic risks to organisms (Fagan et al., 2016; Subagio et al., 2010; Vieira et al., 2020). It is highly recalcitrant and found commonly in treated wastewater due to the ineffectiveness of treatment units (Subagio et al., 2010). USEPA reference dose for BPA is 50 μg/kg/day; however, actual daily consumption is quite higher due to its wide use in plastic and epoxy resins (Sifakis et al., 2017). Another important synthetic EDC is EE2, which is a steroid. It is most commonly found in oral contraceptive pills and thereby enters into the environment. It is said to affect the fishes mostly (Fagan et al., 2016). Anti-epileptic drug carbamazepine (CBZ) and anti-inflammatory drug diclofenac (DCF) are EDCs that are found at high concentrations in secondary wastewater effluent (Naddeo et al., 2020). Steroid hormones can cause secretion-related disorders, hormonal-based cancers, and a decrease in sperm count in mammals (Xu et al., 2020a).

12.2.2 Disinfection by-products

The disinfection step is an integral part of any water and WWTPs. Disinfection plays a key role in maintaining the drinking water quality by eliminating pathogenic organisms (Gilca et al., 2020). However, a significant unintended consequence of the water disinfection process has been the formation of DBPs (Wang et al., 2015). When natural organic matter (NOM) or inorganic matter or any anthropogenic contaminant present in water reacts with the disinfection chemicals, it can result in the formation of DBPs (Alexandrou et al., 2018). The presence of DBPs in water can be detrimental to human beings. In a study by Stalter et al. (2016), cellular toxicity was tested for 50 DBPs present in drinking water. It was found in the study that 49 DBPs activated oxidative stress; however, 34 were responsible for having genotoxic properties. These 34 were also responsible for showing carcinogenic effects.

The formation of different types of DBPs can be attributed to the precursors that exist in the water and the type of disinfection process or disinfectants used. The majority of DBP compounds are highly halogenated, mostly chlorine-based (Alexandrou et al., 2018). With advances in analytical techniques, more DBP compounds are being identified. Some common types of DBP groups include trihalomethanes

Table 12.1 DBP limits in drinking water

DBP class	*Limit in drinking water (mg/L)*	*Regulatory authority*
Total trihalomethanes (TTHMs)	0.08	US Regulations
HAA	0.06	US Regulations
Chlorate	0.7	WHO guidelines
Chlorite	0.7	
N-Nitrosodimethylamine (NDMA)	0.0001	WHO guidelines
Dichloroacetonitrile (DCAN)	0.02	WHO guidelines
Dibromoacetonitrile (DBAN)	0.07	

Source: Adapted from Wang et al. (2015); (WHO, 2017)

(THMs), oxyhalides, haloacetic acids (HAAs), haloacetonitriles (HANs), haloketones (HKs), nitrosamines, halofuranones, halonitromethanes (HNMs), halobenzoquinones, and aldehydes (Alexandrou et al., 2018; Tang et al., 2020). All the aforementioned DBP groups have been known to be harmful to human health with most having carcinogenic effects on the human body (Gilca et al., 2020). The different DBPs and their limits in drinking water as prescribed by various institutions are shown in Table 12.1.

The DBP exposure may occur in three pathways—inhalation, dermal absorption, and ingestion (Yang et al., 2018). With the ubiquitous presence of DBPs in our waters and their adverse effects on human health, it is important to understand the efficacy of our treatment systems in DBP remediation. Common DBP removal approaches include removing precursor compounds (both organic and inorganic) and the use of alternative techniques to carry out disinfection (Tang et al., 2020; Zazouli and Kalankesh, 2017). NOMs can be categorized into particulate organic matters and dissolved organic matters (DOMs). With different precursors resulting in the formation of a wide array of DBPs, there is not one single solution for DBP remediation. Different technologies have been studied and employed for effective DBP and/or its precursor removals, such as adsorption, coagulation, membrane filtration (Me), advanced oxidation processes (AOPs), ion exchange, and integrated technologies (Chaukura et al., 2020; Tang et al., 2020). The selection of the treatment process along with design and operating conditions is greatly influenced by the types and characteristics of NOMs (Bhatnagar and Sillanpää, 2017). The combination of two or more treatment technologies ensures better water quality while preventing DBP formation (Gilca et al., 2020).

12.2.3 Microplastics

Microplastics are synthetic polymers having a maximum size of 5mm. It is categorized into primary and secondary microplastics. Primary ones are those which are manufactured to be in size less than 5mm; they are usually used in textiles, cosmetics, lifestyle products, and medicines. However, secondary are those which get fragmented or are a part of large plastic debris through mechanical or photooxidation (Li et al., 2018). It has been observed that most of the microplastics present in aquatic environments belong to secondary. A matter of concern is that these microplastics can also be fragmented into nanoplastic, which can have severe environmental implications.

It can range from 10^{-5} to 10^{5} pieces/m^{3} in surface water to 40–400 pieces/L in sediments. It was also observed recently that microplastics were found in Antarctica water (Li et al., 2018).

A study was done by Talvitie et al. (2017) for the removal analysis of microlitter, particularly microplastics, during various stages of WWTPs. It was found that WWTPs can reduce around 97% of the microplastics in the pretreatment step only and 7%–20% in the activated sludge process. The tertiary treatment using biologically activated filters did not enhance the removal. However, after the removal, a considerable amount of microplastics make their way to the treated water (Li et al., 2018). It was also observed that microplastic sources are not only WWTPs, surface water runoff, and waste disposal; however, the fallout from the atmosphere also contributes to microplastics (Dris et al., 2016).

The major effect is the ingestion of these microplastics or nanoplastics by literally all types of organisms. It can cause toxic effects by translocation in the organism (Dris et al., 2016). It was found that microplastics can be bioaccumulated in the body, which can be transmitted to other organisms through the food chain and biomagnification. Microplastics not only cause direct toxicity but also adsorb various toxic chemicals, such as polyaromatic carbons, pesticides, and biphenyls, and thereby increasing toxicity. They cause oxidative stress, ulcers, cancer, decrease growth rate, decrease reproductive ability, organ damage, weakening, and even sometimes death in humans. In plants, it can be toxic to soil microbial growth, which can indirectly affect plant growth and maintenance (Sana et al., 2020). Most importantly, microplastics affect aquatic organisms, such as fishes, turtles, and seabirds, by entanglement, suffocation, starvation, and ingestion of particles. They can accumulate in the stomach and cause reproductive disruption. Even in drinking water, huge amounts of microplastics are present, which can make their way to the human and animal body and can cause long-term toxic responses (Li et al., 2018). Tunali et al. (2020) had studied different concentrations of microplastics on the growth and chlorophyll-a concentration of algae *Chlorella vulgaris*. It was found that at a low concentration (1–5 mg/L of microplastic), no change was observed. However, at a higher concentration (50–1,000 mg/L), maximum inhibition of 28.86% and 21.3% on growth and chlorophyll, respectively, was observed. It was found that microplastics get absorbed in the algal cells.

12.2.4 Metals

Metals are a challenging part of the environment due to their ubiquitous nature and toxic effects. Initial work related to metals' toxic effect on aquatic water bodies was started in the early 1930s. It is found in groundwater, surface water, soil and sediments, wastewater, and other sources due to natural and anthropogenic activities (Singh et al., 2020). Due to their toxic effects, it is present in most of the regulatory reports under water quality/environmental guidelines (Adams et al., 2020). However, the toxicity of metals is also dependent on the characteristics of surface water (Adams et al., 2020), the oxidation state of metals, solubility in water, complexity with organic matter (Fatima et al., 2020), and others. Metals can enter into the environment through smelting, fertilizers, leachates, or directly through industrial effluents (Adams et al., 2020; Fatima et al., 2020).

Metals and metalloids are present naturally in rocks and soils, which easily make their way into the aquatic environment and groundwater. Aquatic organisms can accumulate the metals in their body that causes toxicity, thereby affecting other organisms through biomagnification. It was observed that ionic species are more toxic and bioavailable than the ones which are bound to particles or other compounds. Even phosphate-based fertilizers and fungicides are a good source of metals, such as As, Cd, and Hg (Kroon et al., 2020). In a review by Kroon et al. (2020), it was shown that various marine organisms were growing and surviving in the Great Barrier Reef, Australia, and it had high concentrations of various metals and metalloids. Tropical microalgae *Isochrysis galbana*, tropical dog whelk *Nassarius dorsatus*, and marine diatom *Ceratoneis closterium* had high concentrations of Cu, Cu and Al, and Al, respectively. Out of all the metals, Al, Cd, Co, Cu, Ni, Pb, and Zn are mostly studied due to their toxic nature in the environment (Adams et al., 2020).

Metals cause mutagenesis, carcinogenic effect, low reproductive health, low fitness level, reduced fetal growth, and can be lethal to organisms (Fatima et al., 2020). It can cause tissue mortality in tropical sponges, photosynthesis reduction in corals, change in larval and embryo development in oysters, liver dysfunction in green turtles, and other development-related issues. They are also said to be potential EDCs. Metals, such as Cu, Zn, and Mn, along with microplastics can cause a severe negative impact on the growth rate and chlorophyll-a content of algae. There was about 49.57% and 70.43% growth reduction and 50.25% and 64.09% chlorophyll reduction when single-metal– and triple-metal–microplastic combinations were present (Tunali et al., 2020). In a study by Lee and Lee (2020), the effect of three metals Cd, Pb, and As was observed on the digestive enzyme coding genes in the brackish water flea *Diaphanosoma celebensis*. It was found that Cd and a high concentration of lead were found to be most toxic for the genes. There are numerous such studies, which demonstrate the toxic effect of various metal ions on the plant, aquatic organisms, soil microbes, animals, and humans.

12.2.5 Persistent organic pollutants

Persistent organic pollutants are carbon-based compounds that tend to persist in the environment for an extended period (Ashraf, 2017). These compounds are highly toxic even at low concentration, accumulate in the fats due to their high lipophilicity (bioaccumulative), and can transport over long distances. POPs can volatilize from soils and/or land, move into the air and travel long distances (Ashraf, 2017; Gaur et al., 2018). Due to these abilities, these are detected across the globe, even in the polar regions of the planet (Ashraf, 2017; Guo et al., 2019; Kumari et al., 2020). Halogenated POPs have a strong bond between the carbon and the halogen groups, while non-halogenated POPs have a stable structure, which makes both these resistant toward degradation, thus persisting in the environment for a long time (Guo et al., 2019). The presence of POPs in the environment means it inadvertently makes its way into the food chain, thus affecting the environment and human health (Gaur et al., 2018).

Alharbi et al. (2018) reviewed that the impact of POPs has on the environment and human health. The study reviewed and explored conditions, such as endocrine disruption and reproductive problems, cardiovascular problems, cancer, obesity, diabetes, chloracne, porphyria, transient hepatotoxicity, peripheral atherosclerosis, and neurophysiological impairment, which have been linked to different POPs. All life forms, i.e., insects, birds, and animals, have been exposed to POPs in one way or another, causing a shift in their sex

Table 12.2 POP list based on annexures and their significance (Stockholm Convention, 2001)

Annexure	*POPs*	*Annexure significance*
A	Aldrin Mirex Chlordane Dieldrin Heptachlor HCB PCB Endrin Toxaphene	Substances listed proposed to be eliminated from production and use
B	DDT	Restricted use
C	PCB and HCB PCDD and PCDF	Unintentionally produced POPs, for which the emissions need to be reduced

ratios, impaired fertility, physical abnormalities, and among other pathological changes. All these reasons have compelled the POPs to be regulated and technologies to be developed for their remediation. In 2001, The Stockholm Convention listed 12 POPs also called "dirty dozen" or "legacy POP" (as given in Table 12.2) and the countries agreed to lower the usage and/or eliminate the listed chemicals (Ashraf, 2017; Gaur et al., 2018; Guo et al., 2019). By 2019, 17 more POPs were added to the list (Guo et al., 2019).

Bioaccumulation of POPs in humans can occur through oral, inhalation, or dermal route (Alharbi et al., 2018). POPs being recalcitrant move into the atmosphere from various industrial processes, household furnaces, pesticide/insecticide sprayings, landfill transport, incineration, and combustions, such as forest fires (Ashraf, 2017). Natural sources of POPs (dioxins and dibenzofurans) are volcanic activities and vegetation fires (Gaur et al., 2018).

POPs can be classified into two groups: intentionally produced POPs and unintentionally produced POPs (Ashraf, 2017; Gaur et al., 2018). Industrial chemicals and organochlorine pesticides (OCP) are classified as intentionally produced chemicals while combustion by-products, such as polyaromatic hydrocarbons (PAHs), furans, and dioxins are unintentionally produced POPs (Ashraf, 2017; Gaur et al., 2018). Industries, such as textiles and dyes, pharmaceuticals, glassworks, iron foundries, petroleum and petrochemical, paper mills, and thermal power plants, contribute to POP generation by not properly treating the wastewaters (Kumari et al., 2020). Owing to the toxic effects of OCP, many countries have stopped or restricted its usage; however, these are still used in low-income countries (Gaur et al., 2018; Guo et al., 2019). PCBs were used in industries as coolants and lubricants in many pieces of electrical equipment while polybrominated diphenyl ether (PBDE) has been used as flame retardants in electrical appliances, such as TV, computers, and others (Guo et al., 2019). The US Congress had passed legislation banning the manufacture and use of PCBs while in 2004, the USA and EU banned PBDE (Alharbi et al., 2018; Guo et al., 2019). Dioxins and furans are generated as a by-product during the incineration process. Human exposure to dioxins and furans occurs through food consumption. Many countries have reported accidental poisoning cases resulting in exposures to dioxins and other POPs, and this has caused public attention to be focused on the safety of food that has led to increased

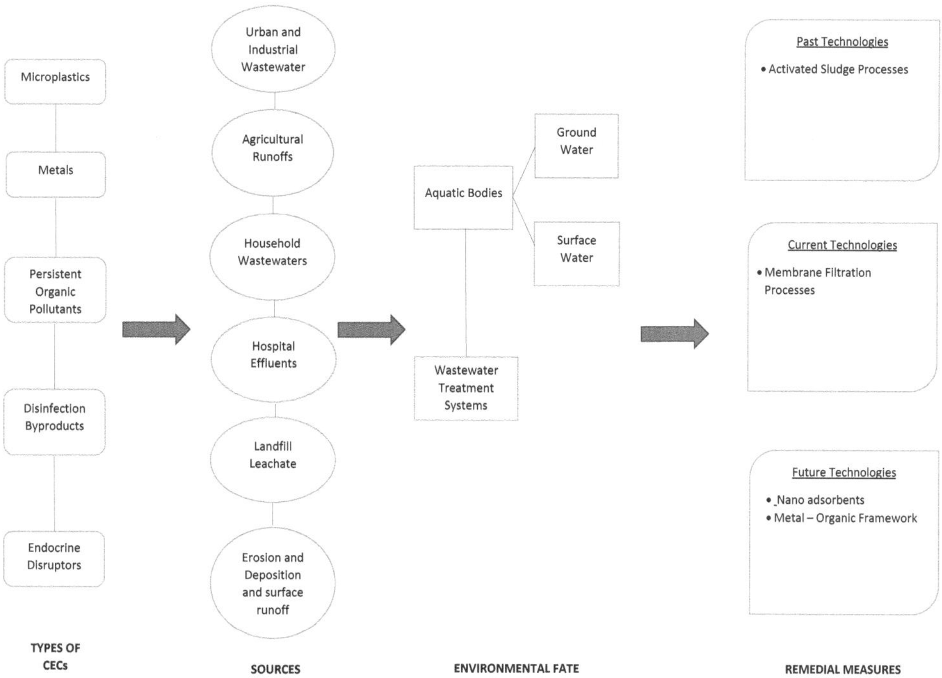

Figure 12.1 Flow sheet showing origin of contaminants, their environmental fate, and remedial approaches.

surveillance and setting up of limits to regulate POPs in food (Guo et al., 2019). In 1987, the EPA stressed on reducing the furans/dioxins, and as a result, an 85% decrease in furan and dioxin levels has been observed (Alharbi et al., 2018) (Figure 12.1).

12.3 TREATMENT METHODS

There are various methods for the removal of CECs from waste streams. WWTPs use biological processes (activated sludge process) for the removal of organics; however, they are not efficient in the removal of CECs (Kovačič et al., 2020). Some of the advanced methods used for pollutants removal are adsorption on activated carbon (AC), ozonation, membrane-based treatments, ultraviolet (UV) irradiation, and others (Kim et al., 2018). In this review, different treatment methods are discussed in detail concerning the majority of CECs present in the environment.

12.3.1 Coagulation

The coagulation process is one of the most common steps employed in the water and wastewater treatment process. It is a physicochemical process that uses certain coagulants to overcome the repulsive charge of colloids resulting in agglomeration into larger particles. The process has conventionally been used for turbidity reduction and total suspended solids (TSS) removal (Sillanpää et al., 2018b). However, few works of literature use a coagulation process for the removal of CECs.

The efficiency of the coagulation process is dependent on pH and type of coagulant used and its dosage. Excess coagulant addition results in sludge generation with the decrease in pH, while lesser coagulant addition would result in the presence of metals in water (Sillanpää et al., 2018b). Metal salt coagulants are the most commonly used coagulants, which include salts like aluminum salts and ferric salts (Chaukura et al., 2020), due to their low cost and high availability. Apart from using conventional coagulants, natural coagulants (biocoagulants) are also being utilized nowadays. These have been derived from natural sources, with chitosan- and tannin-based coagulants being available commercially. Advantages offered by biocoagulants are that it is renewable, non-toxic, biodegradable, and relatively cost-effective. While selecting the desired coagulation method, it is important to know the properties of targeted pollutants and the characteristics of targeted water sources (Sillanpää et al., 2018b).

The coagulation process is one of the most suitable processes for the NOM precursor removal to control the DBP formation. Sillanpää et al. (2018b), in their review, posited that integrating two or more technologies can help in the enhancement of NOM removal. The integrated coagulation–Me process requires a pretreatment step as the NOMs can result in fouling of membranes, which reduces the efficiency of the Me process; hence, various types of coagulants and membranes are being studied in different combinations. The integrated coagulation–adsorption process is another alternative that is being studied for varying adsorbents and coagulants. One of the most widely used adsorbents in water treatment is AC. One study has shown that using integrated AC along with enhanced coagulation can significantly reduce the NOM precursor hence reducing the chances of DBP formation, however, it also highlighted an increase in the brominated DBP formation. In the case of POPs, the coagulation–flocculation–sedimentation process has a poor efficiency in the range of 10% to 40% for the removal of pesticides or pharmaceutical compounds. Along with poor removal efficiency, sludge generation is also a major problem with the coagulation–flocculation process (Gaur et al., 2018; Kumari et al., 2020; Pi et al., 2018).

In a study done by Bogunović et al. (2020), based on the WWTP in Serbia, 48 micropollutants (including BPs, pharmaceuticals, and parabens) were removed using a natural coagulant (isolated from beans) and a conventional coagulant (ferric chloride). It was observed that natural coagulant was more effective in removing the CEC (BP F and BPA, and pharmaceuticals) as compared to the ferric chloride. The removal efficiency using natural coagulants was around 50% for BPs. This higher removal can be due to either higher content of colloidal effluent organic matter, which can be subsequently removed via coagulation, or by complexing and hydrogen bond formation. They had also studied the simultaneous process of powdered activated carbon (PAC) adsorption as well as coagulation. It was found that maximum removal efficiency was found in the case when natural coagulant was used along with PAC. In another study, iron-based coagulants were used with the UV/H_2O_2 process for enhancing the removal of sulfamethazine from real wastewater. It was found that coagulation enhanced the removal by 40%. The possible reason was the formation of reactive oxygen species (ROS) on the surface of iron coagulant, which enhances the removal. However, it also competes with photons, thereby weakening the process (Li et al., 2020).

Electrocoagulation (EC) has been in use for quite some time now and is being used for the removal of contaminants in drinking water supplies, municipal and industrial wastewaters (Sillanpää et al., 2018b). The EC process follows three steps—formation of coagulants by electrolytic oxidation of sacrificial anodes, destabilization of pollutants in colloidal suspensions, and aggregation to form flocs (Sillanpää et al., 2018b). As coagulants are formed in situ, there is no need to add other chemicals. Other advantages include fast reaction, low toxicity, and easy availability (El-Ashtoukhy et al., 2020). The efficiency of EC depends on water chemistry—pH, conductivity, along with the concentration and properties of pollutants (Sillanpää et al., 2018b). EC was used for the removal of heavy metals from the wastewater using a self-gas stirred batch system. This process was found to consume low energy as well as less electrolysis time. Nickel and copper were efficiently removed from the system; however, removal was faster for Cu, due to cathodic deposition. The removal efficiency was in the range of 36% to 100% based on the operating parameters used (El-Ashtoukhy et al., 2020). EC offers various advantages over conventional coagulation processes; however, it was also found that this process alone is not sufficient for the removal of estrogenic compounds. It is highly beneficial for the removal of organic matter (Maher et al., 2020). However, one of the major disadvantages of EC is high energy consumption mostly during the scale-up process (El-Ashtoukhy et al., 2020). Removal in the case of EC is mostly due to the oxidation process. Hence, in a study by Maher et al. (2020), EC was used along with the electrooxidation (EO) process to remove estrogen. This was said to drastically reduce energy consumption (more than half compared to EO alone). EC was able to reduce 70%–75% of the DOC concentration. Due to the reduction in DOC, the removal of estrogenic compounds was increased. It was found that 80%–90% of removal of these compounds was found during the EO process. However, there are possible chances of the formation of DBPs (such as chloroform, dichloroacetic acid, and trichloroacetic acid) when the EC–EO process is used for water having organic matter and *in situ* generations of chlorine species. In a study using EC-treated landfill leachate samples, there was a formation of a variety of DBPs. It was later on treated with granular activated carbon (GAC) for the reduction in DBPs and overall toxicity (Xu et al., 2020a).

12.3.2 Adsorption

Like coagulation, adsorption is a physicochemical process and is one of the most commonly used in the wastewater treatment industries. It offers several advantages, such as ease of operation, high CEC removal and removal of hydrophobic organics, simple design, and convenience, that make it the preferred mode of technology for water and wastewater treatment (Bhatnagar and Sillanpää, 2017). There are various types of adsorbents, such as GAC, modified carbon, single- and multiwalled carbon nanotubes (CNTs), iron oxide particles, polypyrrole-coated glass beads, animated polyacrylonitrile fibers, chitosan/zeolite composite, rice husk ash, surfactant-modified bentonite, and ion-exchange resin for removal of various pollutants (Bhatnagar and Sillanpää, 2017).

Adsorption has been reported to be very efficient in the removal of CECs of high concentrations. It also does not produce any new compounds. However, it lacks selectivity for very low concentrations while its large-scale application is hindered by high costs of adsorbent regeneration and high carbon/waste disposal (Bogunović et al., 2020; Gaur et al., 2018; Kumari et al., 2020). Adsorption is widely used for the removal of DBPs by removing NOM. In a review by Bhatnagar and Sillanpää (2017), it was found that adsorbent modification aided in increasing the NOM removal efficiency. However, modified adsorbents pose a higher risk of chemical leaching into the environment that could be detrimental to aquatic organisms. Similar is the case of using nanomaterials as adsorbents. Another drawback was that the disposal of saturated adsorbents would result in secondary pollution. With current studies being limited to laboratory scale, to make NOM removal via adsorption of a more viable process for full-scale application, it is important to investigate regeneration and reuse of adsorbents as it would be economically feasible. Another area that needs investigation is the combination of adsorption with other technologies that could enhance NOM removal efficiency (Bhatnagar and Sillanpää, 2017).

Flores et al. (2020) used clay minerals montmorillonites (Mts) and organo-Mts for the adsorption of fungicide thiophanate-methyl. Organo-Mt was activated using a surfactant, and it was the surfactant's functional group, which plays an important role in adsorption. Another type of clay that is commonly used for adsorption is bentonites. In a study by Gallouze et al. (2021), modified organic (Trp-Na-bent) and inorganic (Fe-Na-bent) bentonites were used for the adsorption of EDC EE2. Various operational parameters were also optimized for increasing the adsorption capability. It was found that organic bentonite was successfully able to adsorb the EE2 (initial concentration of 10 mg/L) with maximum adsorption of 4.2 mg/g, achieving equilibrium in 2 hours. It was also found that with the increase in EE2 concentration, adsorption capacity also increases. Zeolites are also a low-cost adsorbent used for the adsorption of various pollutants from wastewater. Zeolites prepared from waste provide an efficient way of the circular economy. In a study by Bogunović et al. (2020), PAC was used for the removal of CECs from the wastewater treatment effluent in Serbia. It was found that around 83%–99% removal efficiency was found for various micropollutants. However, a higher adsorbent dose of 20 mg/L was required. In another study by Noufel et al. (2020), three different types of adsorbents (organo-activated bentonite (OAB), AC, and AC–OAB) were cross-linked with alginate (A) beads for the efficient removal of BPA and trichlorophenol (TCP) from water. It was found that maximum removal was shown by the A–AC adsorbent, and when the binary system was used, removal efficiency decreased.

Electrosorption is a newly used method that uses an electric field along with adsorption. Xiong et al. (2020) had used electrosorption using AC for the removal of nanoplastics. These nanoplastics adsorb a high amount of BPA under influence of UV radiation, which later on electrosorbed having a capacity of 0.707 g nanoplastic/g AC.

12.3.3 Advanced oxidation processes (AOPs)

Advanced oxidation processes utilize the production of hydroxyl radicals (oxidation potential of 2.80 V), which can degrade most of the organic pollutants. They usually either mineralize compounds into less toxic products or carbon dioxide (Miranda-García et al., 2011). Various AOP technologies, such as ozone- and UV light-based, Fenton and photo-/electro-Fenton processes, heterogeneous catalytic and photocatalytic oxidation,

AOPs integrated with other technologies, and non-conventional AOPs, such as the use of ultrasound, microwaves, electron beam irradiation, gamma irradiation, and plasma technologies, were used for the removal of pollutants. The complete oxidation of organic matter may not be possible as the large-sized breaks down into smaller organic compounds that either can be easily treated or could be toxic, which hinders its large-scale application (Sillanpää et al., 2018a). Sgroi et al. (2020) had compared various AOPs for the treatment of several CECs. UV/H_2O_2, UV/Cl_2, O_3, O_3/UV, $H_2O_2/O_3/UV$, and $Cl_2/O_3/UV$ were used for tertiary wastewater treatment. Ozonation was found to be a highly efficient method for pollutant removal; however, it was least effective for microbial deactivation. This process, however, forms various DBPs.

In AOPs, heterogeneous photocatalysis using UV/TiO_2 is gaining huge importance (Miranda-García et al., 2011; Vieira et al., 2020). It has high photoactivity even without the addition of any chemicals, is reasonably inexpensive, is chemically and biologically stable, and has low toxicity. Hence, it is found to be a highly suitable semiconductor photocatalyst in water treatment systems. Due to the large bandgap of TiO_2, UV irradiation is required for the excitation as well as hydroxyl radical production (Subagio et al., 2010). However, it is commonly mixed in the slurry, which makes its removal from water difficult. Hence, the modified method is the immobilization of TiO_2 into borosilicate glass or glass beads, ceramics, or polymer, which makes the removal of contaminants easy and TiO_2 reusable (Fagan et al., 2016; Miranda-García et al., 2011). For ideal support, it is important to have the following properties: should be transparent to irradiation, should be resistant toward oxidizing conditions, and should enhance contact between contaminant and photocatalyst. Usually, glass substrates are used as they are resistant to high temperatures and help in catalyst crystallinity (Miranda-García et al., 2011). Miranda-García et al. (2011) had used 15 different CECs in simulated as well as real municipal WWTPs. They had observed around 85% degradation of 100 μg/L CECs using UV/TiO_2 after 120 minutes of illumination period.

If using sunlight, then it is required to dope TiO_2 with anions. Nitrogen acts as one of the most promising ones in recent times. Also, the addition of light-emitting diodes (LEDs) as a light source is used as a replacement for solar irradiation. Due to their various properties, such as long lifetime, energy efficiency, robustness, smaller size, and high reusability, it can enhance the effectiveness of the process (Subagio et al., 2010). In a study by Subagio et al. (2010), BPA was 90% degraded using nitrogen-doped TiO_2 hollow sphere at blue light (visible LED at 465 nm) after 2 hours. After 6 hours, 66% of TOC reduction was also observed. It was seen that when hollow spheres were incorporated into the $N–TiO_2$ system, surface area, and porosity were increased, which increased its phototonic activity (Fagan et al., 2016; Subagio et al., 2010). Wang and Lim (2010) showed BPA degradation using carbon and nitrogen codoped with TiO_2. They observed >99% degradation of BPA in white and blue lights; however, it was only 84% and 24% for a green and yellow light, respectively, after 5 hours of irradiation. Mineralization was 70%, 60%, 45%, and 9% using white, blue, green, and yellow lights, respectively. In the case of white light, more valence bands get excited to the conduction band, which enhances the degradation activity. Significant other challenges with AOP are catalyst poisoning and inactivation and costs associated due to high energy consumption (Kumari et al., 2020).

Another newer method is the use of metal-free nanocarbon materials due to the problem of metal leaching. Kang et al. (2019) used Mn-based magnetic N-doped nanocarbon springs along with hydrolysis for the removal of microplastics. It was found that they can reduce 50% of MPs from the solution. Also, the intermediates formed during the process were found to be non-toxic. Another type of AOPs is hydrodynamic cavitation (HC), which uses a sudden drop in pressure to form cavities. In a study by Kovačič et al. (2020), laboratory- and pilot-scale studies were done for the removal of various CECs. In laboratory and pilot scales, the removal of various CECs ranges from 15%–63% to 15%–90%, respectively. They also combined HC with UV and enhance the temperature to increase the removal; however, removal performance does not increase.

12.3.4 Membrane-based treatment

Membrane filtration processes include technologies, such as ultrafiltration (UF), microfiltration (MF), reverse osmosis (RO), nanofiltration (NF), and ceramic technologies. The aforementioned technologies are quite similar in their working mainly due to their structures and preparation. Membrane selection is dependent on the type of pollutant in water (Zazouli and Kalankesh, 2017). One of the major problems with membrane technologies is the fouling of membrane resulting in the flux and efficiency reduction, and hence, it is recommended to have the pretreatment of water before applying the membrane-based processes (Chaukura et al., 2020). Another drawback of using membrane-based processes is that they are expensive and energy-intensive (Bhatnagar and Sillanpää, 2017).

Chaukura et al. (2020) highlighted the role of ceramic membranes in NOM/DBP removal and stated that ceramic membranes were less prone to fouling compared to other polymeric membranes; however, their higher cost makes the large-scale process unsuitable. The study further states coating nanoparticles on the surface of ceramic membranes could significantly improve the performance. Zazouli and Kalankesh (2017) reviewed the removal of precursors and DBPs from water using Me processes and compared the MF, UF, RO, and NF technologies. The review highlighted the importance of pore size in the membrane-based processes, stating that NOM removal can be increased by decreasing the pore size. Negatively charged UF membranes can effectively remove larger hydrophobic NOM fractions while also helping in controlling AOC and bacterial regrowth in drinking water distribution systems (Zazouli and Kalankesh, 2017). UF is economically attractive but is susceptible to fouling and requires frequent cleaning (Chaukura et al., 2020). NF is very efficient in the removal of all kinds of compounds, but it is also very sensitive to membrane fouling and hence requires extensive pretreatment. RO is capable of removing 99% of dissolved matter and can remove both organic and inorganic precursors; however, it is relatively expensive, requires pretreatment, has high energy consumption, and produces brine that needs disposal. All these factors have restricted the widespread application of RO and have been limited to drinking water treatment plants (Zazouli and Kalankesh, 2017).

Pramanik et al. (2017) studied the removal efficiency of PFOA and PFOS from lake water using NF, UF, and GAC treatment techniques. The study looked at the trend of PFAS removal and the influence of organics (proteins, carbohydrates, and DOC) on the rejection of PFAS. The study concluded that NF had a greater removal when compared to UF and GAC. It was also seen that increased concentration of organics in water resulted in decreased removal efficiency.

Catalytic membrane reactors (CMRs) are a type of multifunctional system that has been developed by merging membrane-based separation and chemical reaction in a single system (Abdallah, 2017). CMRs tend to be cost-effective as they reduce the footprint of the system and also decrease the downstream treatment costs (Abdallah, 2017; Kumari et al., 2020). The type of CMR depends on the choice of catalytic membranes used, i.e., organic (polymeric) and inorganic (metallic and ceramic), and the process conditions (Abdallah, 2017). The catalytic membranes can also be classified based on the roles of the membrane, i.e., the membranes could be of the following: (i) extractor type: it withdraws one of the products in the system to enhance the conversion and is generally selective for one of the products; (ii) distributor type: here this membrane allows one of the reactants (especially gases) to pass through in a controlled manner that results in enhancing the selectivity; and (iii) contactor type: it helps in contacting between reactants and catalysts. Here the membrane can itself act as a catalyst or have a catalyst supported on its surfaces (Abdallah, 2017).

CMR technology can be applied on an industrial scale in various sectors such as pharma, biotech, energy, and the environment. Some of the reactions where CMRs have been applied are hydrogenation, dehydrogenation, esterification, and photocatalytic reactions (Abdallah, 2017). The use of photocatalytic membrane reactors (PCMRs) for the treatment of POPs was reviewed and discussed by Kumari et al. (2020). PCMR can be the slurry type or immobilized photocatalyst. Slurry-based PCMR systems for POPs offer an alternative solution by enhancing mass transfer and helping in degradation. However, they require lots of accessories, such as additional piping, pumps, and tanks, thus increasing costs and also causing membrane fouling. Similarly, immobilized photocatalysts have certain advantages, such as greater hydrophilicity, better antifouling properties, and protection against UV-sensitive membranes. Non-uniform coating and leaching of photocatalyst are some of the drawbacks of the immobilized photocatalyst PCMR. One of the key advantages of PCMR is that POPs could be mineralized to form non-toxic by-products, with both degradation and separation occurring simultaneously.

12.4 FUTURE TREATMENT TECHNOLOGIES AND CHALLENGES

Various hybrid methods are used for the removal of CECs from the wastewater systems. One such treatment method is the use of UF, AC adsorption, and ultrasound irradiation (USAMe®—UltraSound, Adsorption, and Membrane ultrafiltration) hybrid technology. It was observed that this hybrid method was able to completely remove DCF, CBZ, and amoxicillin (potential EDCs) from the wastewater. The main mechanism found was adsorption on the PAC as well as sonolytic degradation, which enhances the removal (Secondes et al., 2014). Since CECs are not alone in the wastewater, they are usually present with the organic pollutants. It was found that when only Me was used, then fouling in the membrane was caused. In adsorption (A), both NOM and CECs compete for the adsorption sites, thereby increasing adsorption doses. Hence, in another study by Naddeo et al. (2020), USAMe® was used as post-treatment of a biological method, as a tertiary method. It was found that more than 99% removal was obtained when USAMe was used, as compared to 90% when AMe treatment was used. It was also found that treated effluent did not cause any toxicity to *D. magna* seedlings.

Another newer method is the use of photobioreactors (PBRs). In a study by García-Galán et al. (2020), a semiclosed, tubular, and horizontal algal photobioreactor was used for the removal of CECs. The CEC treated using this method was two organophosphate flame retardants: tributyl phosphate (TBP) and tri-(2-chloroethyl) phosphate (TCEP), and one surfactant 2,4,7,9-tetramethyl-5-decyne-4,7-diol (TMDD). It was found that TBP was removed by 43%; however, no TCEP removal was found. It was because PBRs are made up of plastic, which releases TCEP in the solution. TMDD removal was 33%, majorly adsorbed on the algal biomass.

Metal-organic frameworks (MOFs) are a class of hybrid porous materials, having both organic and inorganic species (Bhadra et al., 2020). They have been reported to possess properties, such as high surface area, high pore volume, better stability, better topology, and larger mass transfer (Pi et al., 2018; Xu et al., 2020b). The potential of using MOFs for CECs degradation is very high with processes being energy and cost-efficient. In a study by Bhadra et al. (2020), Zn-doped MOFs were used for the removal of CECs. They have used the carbon derived from MOF for the purification of water. It was observed that they were able to remove mostly all the CECs present in water, and a major mechanism is said to be H-bonding during adsorption. Xu et al. (2020b) used Fe-based MOFs with different functional groups (MIL-101-R) for the removal of EDCs. Due to hydrogen and pi–pi interaction between MIL-101-NH_2 and EDC, adsorption efficiency was quite high. The pore size of MOF and EDC was an important factor for the adsorption. However, one of the major challenges highlighted was that in the presence of water, many MOFs were unstable, which resulted in lower recoveries and a possibility of secondary pollutants with the leaching of metals. Another critical issue was concerning its scale-up, i.e., on an industrial scale, the amount of MOFs required would be in kilograms. For energy-effective purposes, the study suggests combining the adsorption and photocatalysis processes and it could be advantageous for effective POP degradation (Pi et al., 2018).

One of the major challenges that have been highlighted across all aforementioned CEC removal studies and reviews is that a majority of the process development and applications have been carried out on synthetic waste. Fewer studies have been reported to use natural waters for testing the CEC removal technologies. Studying the effects of removal technologies using natural waters is crucial due to its complex nature and presence of all sorts of contaminants. Understanding this would be important when going for scaling up the technology. Similarly, carrying out studies at laboratory scale is insufficient to understand the complexities involved in CEC removal and/or its transformation and it would be prudent to carry out pilot-scale studies with natural waters to get the actual efficiency of the proposed solution. Apart from technical challenges, there are economic challenges involved with CEC removal technologies and have been discussed in the next section.

12.5 ECONOMICS

The large-scale application of any process is dependent on its economics. No matter how efficient any treatment system is, if it is not economically viable, it would not have a widespread application. Chaukura et al. (2020) reviewed the economics of various DBP precursor removal processes. The coagulation process is a low-cost and highly efficient technique for the removal of various contaminants; however, it is efficient in the removal of settleable particles and turbidity. Hence, it needs to be coupled with other processes. Both AOP and membrane processes are capable of removing NOM; however, they are energy-intensive and have high energy and maintenance costs. While proper membrane selection can help in removing said contaminants effectively, AOPs are recommended to be integrated with other processes to increase their effectiveness. With NF and RO being an expensive option, places, and countries with a high number of yearly solar irradiation days, solar-power-based AOP processes like the solar photo-Fenton process could be an attractive option for CEC removal from water and wastewaters (Rizzo et al., 2019). According to a study done by Sgroi et al. (2020), for removing CECs using AOPs, it was found that the ozone-based process was less energy-intensive as compared to UV-based. However, when a UV-based process is combined with ozone, then the process will be most efficient.

To better understand the long-term prospects and large-scale implementation of desired CEC removal technology, it would be advisable to carry out a techno-economic assessment of different processes. This would give a good idea regarding the economic viability of the process. Performing a life cycle assessment study is also recommended.

12.6 CONCLUSION

A variety of CECs present around us pollute our environment and cause harm to humans and other living organisms alike. Advances in analytical techniques have been crucial in the detection of CECs. This review aimed to focus on some specific CECs and the different challenges involved with their removal technologies. Different treatment methodologies were reviewed, and it could be seen that there have been tremendous strides in the development of technologies for the removal of various CECs. Unfortunately, due to the diverse nature of different CECs, there cannot be a single solution for CEC removal. It has been recommended that treatment technologies should be selected depending on the nature of the contaminant and its characteristics, environmental factors, and economics. A set of each of the treatment technologies have their pros and cons and it is important to select that based on the need. Several studies have suggested using combined or integrated technology instead of a single treatment that can negate each other's drawbacks and help in the efficient reduction of CECs.

REFERENCES

Abdallah, H. (2017). A review on catalytic membranes production and applications. *Bulletin of Chemical Reaction Engineering & Catalysis*, 12(2), 136–156.

Adams, W., Blust, R., Dwyer, R., Mount, D., Nordheim, E., Rodriguez, P. H. & Spry, D. (2020). Bioavailability assessment of metals in freshwater environments: A historical review. *Environmental Toxicology and Chemistry*, 39(1), 48–59.

Alexandrou, L., Meehan, B. J. & Jones, O. A. (2018). Regulated and emerging disinfection by-products in recycled waters. *Science of the Total Environment*, 637, 1607–1616.

Alharbi, O. M., Khattab, R. A. & Ali, I. (2018). Health and environmental effects of persistent organic pollutants. *Journal of Molecular Liquids*, 263, 442–453.

Annamalai, J. & Namasivayam, V. (2015). Endocrine disrupting chemicals in the atmosphere: their effects on humans and wildlife. *Environment International*, 76, 78–97.

Appa, R., Mhaisalkar, V. A., Bafana, A., Devi, S. S., Krishnamurthi, K., Chakrabarti, T. & Naoghare, P. K. (2018). Simultaneous quantitative monitoring of four indicator contaminants of emerging concern (CEC) in different water sources of Central India using SPE/LC-(ESI) MS-MS. *Environmental Monitoring and Assessment*, 190(8), 489.

Ashraf, M. A. (2017). Persistent organic pollutants (POPs): a global issue, a global challenge. *Environmental Science and Pollution Research*, 24, 4223–4227.

Bhadra, B. N., Yoo, D. K. & Jhung, S. H. (2020). Carbon-derived from metal-organic framework MOF-74: A remarkable adsorbent to remove a wide range of contaminants of emerging concern from water. *Applied Surface Science*, 504, 144348.

Bhatnagar, A. & Sillanpää, M. (2017). Removal of natural organic matter (NOM) and its constituents from water by adsorption—a review. *Chemosphere*, 166, 497–510.

Bogunović, M., Ivančev-Tumbas, I., Česen, M., Sekulić, T. D., Prodanović, J., Tubić, A., Heath, D. & Heath, E. (2020). Removal of selected emerging micropollutants from wastewater treatment plant effluent by advanced non-oxidative treatment-a lab-scale case study from Serbia. *Science of the Total Environment*, 765, 142764.

Chaukura, N., Marais, S. S., Moyo, W., Mbali, N., Thakalekoala, L. C., Ingwani, T., Mamba, B. B., Jarvis, P. & Nkambule, T. T. (2020). Contemporary issues on the occurrence and removal of disinfection byproducts in drinking water-A review. *Journal of Environmental Chemical Engineering*, 8(2), 103659.

Dris, R., Gasperi, J., Saad, M., Mirande, C. & Tassin, B. (2016). Synthetic fibers in atmospheric fallout: a source of microplastics in the environment? *Marine Pollution Bulletin*, 104(1–2), 290–293.

El-Ashtoukhy, E. Z., Amin, N. K., Fouad, Y. O. & Hamad, H. A. (2020). Intensification of a new electrocoagulation system characterized by minimum energy consumption and maximum removal efficiency of heavy metals from simulated wastewater. *Chemical Engineering and Processing-Process Intensification*, 154, 108026.

Eregowda, T., Mohapatra, S. (2020). Fate of Micropollutants in Engineered and Natural Environment. In: Kumar M., Munoz-Arriola F., Furumai H., Chaminda T. (eds) *Resilience, Response, and Risk in Water Systems.* Springer Transactions in Civil and Environmental Engineering (pp. 283–301). Springer, Singapore.

Fatima, S., Muzammal, M., Rehman, A., Rustam, S. A., Shehzadi, Z., Mehmood, A. & Waqar, M. (2020). Water pollution on heavy metals and its effects on fishes. *International Journal of Fisheries and Aquatic Studies*, 8(3), 6–14.

Fagan, R., McCormack, D. E., Dionysiou, D. D. & Pillai, S. C. (2016). A review of solar and visible light active TiO_2 photocatalysis for treating bacteria, cyanotoxins and contaminants of emerging concern. *Materials Science in Semiconductor Processing*, 42, 2–14.

Flores, F. M., Undabeytia, T., Jaworski, M., Morillo, E. & Sánchez, R. M. T. (2020). Organo-montmorillonites as adsorbent materials for thiophanate-methyl removal: adsorption-desorption studies and technological applications. *Journal of Environmental Chemical Engineering*, 8, 103806.

Gallouze, H., Akretche, D. E., Daniel, C., Coelhoso, I. & Crespo, J. G. (2021). Removal of Synthetic Estrogen from water by adsorption on modified bentonites. *Environmental Engineering Science*, 38, 4–14.

García-Galán, M. J., Matamoros, V., Uggetti, E., Díez-Montero, R. & García, J. (2020). Removal and environmental risk assessment of contaminants of emerging concern from irrigation waters in a semi-closed microalgae photobioreactor. *Environmental Research*, 194, 110278.

Gaur, N., Narasimhulu, K. & PydiSetty, Y. (2018). Recent advances in the bio-remediation of persistent organic pollutants and its effect on environment. *Journal of Cleaner Production*, 198, 1602–1631.

Gilca, A. F., Teodosiu, C., Fiore, S. & Musteret, C. P. (2020). Emerging disinfection byproducts: a review on their occurrence and control in drinking water treatment processes. *Chemosphere*, 127476.

Guo, W., Pan, B., Sakkiah, S., Yavas, G., Ge, W., Zou, W., Tong, W. & Hong, H. (2019). Persistent organic pollutants in food: contamination sources, health effects and detection methods. *International Journal of Environmental Research and Public Health*, 16(22), 4361.

Kang, J., Zhou, L., Duan, X., Sun, H., Ao, Z. & Wang, S. (2019). Degradation of cosmetic microplastics via functionalized carbon nanosprings. *Matter*, 1(3), 745–758.

Kabir, E. R., Rahman, M. S. & Rahman, I. (2015). A review on endocrine disruptors and their possible impacts on human health. *Environmental Toxicology and Pharmacology*, 40(1), 241–258.

Kim, S., Chu, K. H., Al-Hamadani, Y. A., Park, C. M., Jang, M., Kim, D. H., Yu, M., Heo, J. & Yoon, Y. (2018). Removal of contaminants of emerging concern by membranes in water and wastewater: a review. *Chemical Engineering Journal*, 335, 896–914.

Kovačič, A., Škufca, D., Zupanc, M., Gostiša, J., Bizjan, B., Krištofelc, N., Dolenc, M. S. & Heath, E. (2020). The removal of bisphenols and other contaminants of emerging concern by hydrodynamic cavitation: From lab-scale to pilot-scale. *Science of the Total Environment*, 743, 140724.

Kroon, F. J., Berry, K. L., Brinkman, D. L., Kookana, R., Leusch, F. D., Melvin, S. D., Neale, P. A., Negri, A. P., Puotinen, M., Tsang, J. J., van de Merwe, J. P. & Williams, M. (2020). Sources, presence and potential effects of contaminants of emerging concern in the marine environments of the Great Barrier Reef and Torres Strait, Australia. *Science of the Total Environment*, 719, 135140.

Kumar, M. (Ed.). (2012). *Problems, Perspectives and Challenges of Agricultural Water Management.* In-Tech International Publication, ISBN 979-953-307-706-1. https://www.intechopen.com/books/problems-perspectives-and-challenges-of-agricultural-water-management.

Kumar, M., Patel, A. K., Das, A., Kumar, P., Goswami, R., Deka, P. & Das, N. (2017). Hydrogeochemical controls on mobilization of arsenic and associated health risk in Nagaon district of the central Brahmaputra Plain, India. *Environmental Geochemistry and Health*, 39(1), 161–178.

Kumar, M., Snow, D. D., Li, Y. & Shea, P. J. (2019). Perchlorate behavior in the context of black carbon and metal cogeneration following fireworks emission at Oak Lake, Lincoln, Nebraska, USA. *Environmental Pollution*, 253, 930–938.

Kumar, M., Chaminda, T., Patel, A. K., Sewwandi, H., Mazumder, P., Joshi, M. & Honda, R. (2020). Prevalence of antibiotic resistance in the tropical rivers of Sri Lanka and India. *Environmental research, 188*, 109765.

Kumar, M., Kuroda, K., A. K., Patel, N., Bhattacharya, P., Joshi, M. & Joshi, C. G. (2021). Decay of SARS-CoV-2 RNA along the wastewater treatment outfitted with Upflow Anaerobic Sludge Blanket (UASB) system evaluated through two sample concentration techniques. *Science of the Total Environment*, 754, 142329.

Kumari, P., Bahadur, N. & Dumée, L. F. (2020). Photo-catalytic membrane reactors for the remediation of persistent organic pollutants–A review. *Separation and Purification Technology*, 230, 115878.

Lee, J. & Lee, Y. M. (2020). Effects of heavy metals on the expression of digestive enzyme-coding genes in the brackish water flea *Diaphanosoma celebensis. Toxicology and Environmental Health Sciences*, 1–8.

Li, J., Liu, H. & Chen, J. P. (2018). Microplastics in freshwater systems: A review on occurrence, environmental effects, and methods for microplastics detection. *Water Research*, 137, 362–374.

Li, M., Sun, M., Dong, H., Zhang, J., Su, Y. & Qiang, Z. (2020). Enhancement of micropollutant degradation in UV/H_2O_2 process via iron-containing coagulants. *Water Research*, 172, 115497.

Maher, E. K., O'Malley, K. N., Dollhopf, M. E., Mayer, B. K. & McNamara, P. J. (2020). Removal of estrogenic compounds from water via energy efficient sequential electrocoagulation-electrooxidation. *Environmental Engineering Science*, 37(2), 99–108.

Miranda-García, N., Suárez, S., Sánchez, B., Coronado, J. M., Malato, S. & Maldonado, M. I. (2011). Photocatalytic degradation of emerging contaminants in municipal wastewater treatment plant effluents using immobilized TiO_2 in a solar pilot plant. *Applied Catalysis B: Environmental*, 103(3–4), 294–301.

Naddeo, V., Secondes, M. F. N., Borea, L., Hasan, S. W., Ballesteros Jr, F. & Belgiorno, V. (2020). Removal of contaminants of emerging concern from real wastewater by an innovative hybrid membrane process–UltraSound, Adsorption, and Membrane ultrafiltration (USAMe®). *Ultrasonics Sonochemistry*, 68, 105237.

Nawaz, T. & Sengupta, S. (2019). Contaminants of emerging concern: Occurrence, fate, and remediation. In *Advances in Water Purification Techniques* (pp. 67–114). Elsevier.

Nilsen, E., Smalling, K. L., Ahrens, L., Gros, M., Miglioranza, K. S., Picó, Y. & Schoenfuss, H. L. (2019). Critical review: grand challenges in assessing the adverse effects of contaminants of emerging concern on aquatic food webs. *Environmental Toxicology and Chemistry*, 38(1), 46–60.

Noufel, K., Djebri, N., Boukhalfa, N., Boutahala, M. & Dakhouche, A. (2020). Removal of bisphenol A and trichlorophenol from aqueous solutions by adsorption with organically modified bentonite, activated carbon composites: A comparative study in single and binary systems. *Groundwater for Sustainable Development*, 11, 100477.

Pi, Y., Li, X., Xia, Q., Wu, J., Li, Y., Xiao, J. & Li, Z. (2018). Adsorptive and photocatalytic removal of Persistent Organic Pollutants (POPs) in water by metal-organic frameworks (MOFs). *Chemical Engineering Journal*, 337, 351–371.

Pramanik, B. K., Pramanik, S. K., Sarker, D. C. & Suja, F. (2017). Removal of emerging perfluorooctanoic acid and perfluorooctane sulfonate contaminants from lake water. *Environmental Technology*, 38(15), 1937–1942.

Patel, A. K., Bhagat, C., Taki, K., & Kumar, M. (2020). Microplastic vulnerability in the sediments of the Sabarmati river of India. In *Resilience, Response, and Risk in Water Systems* (pp. 127–138). (ISBN 978-981-15-4668-6) Springer, Singapore.

Rizzo, L., Malato, S., Antakyali, D., Beretsou, V. G., Đolić, M. B., Gernjak, W., Heath, E., Ivančev-Tumbas, I., Karaolia, P., Ribeiro, A. R. L., Mascolo, G., McArdell, C. S., Schaar, H., Silva, A. M. T. & Fatta-Kassinos, D. (2019). Consolidated vs new advanced treatment methods for the removal of contaminants of emerging concern from urban wastewater. *Science of the Total Environment*, 655, 986–1008.

Sana, S. S., Dogiparthi, L. K., Gangadhar, L., Chakravorty, A. & Abhishek, N. (2020). Effects of microplastics and nanoplastics on marine environment and human health. *Environmental Science and Pollution Research*, 1–14.

Sauvé, S. & Desrosiers, M. (2014). A review of what is an emerging contaminant. *Chemistry Central Journal*, 8(1), 1–7.

Secondes, M. F. N., Naddeo, V., Belgiorno, V. & Ballesteros Jr, F. (2014). Removal of emerging contaminants by simultaneous application of membrane ultrafiltration, activated carbon adsorption, and ultrasound irradiation. *Journal of Hazardous Materials*, 264, 342–349.

Sgroi, M., Snyder, S. A. & Roccaro, P. (2020). Comparison of AOPs at pilot scale: Energy costs for micro-pollutants oxidation, disinfection byproducts formation and pathogens inactivation. *Chemosphere*, 273, 128527.

Sifakis, S., Androutsopoulos, V. P., Tsatsakis, A. M. & Spandidos, D. A. (2017). Human exposure to endocrine disrupting chemicals: effects on the male and female reproductive systems. *Environmental Toxicology and Pharmacology*, 51, 56–70.

Sillanpää, M., Ncibi, M. C. & Matilainen, A. (2018a). Advanced oxidation processes for the removal of natural organic matter from drinking water sources: A comprehensive review. *Journal of environmental management*, 208, 56–76.

Sillanpää, M., Ncibi, M. C., Matilainen, A. & Vepsäläinen, M. (2018b). Removal of natural organic matter in drinking water treatment by coagulation: A comprehensive review. *Chemosphere*, 190, 54–71.

Stalter, D., O'Malley, E., Von Gunten, U. & Escher, B. I. (2016). Fingerprinting the reactive toxicity pathways of 50 drinking water disinfection byproducts. *Water Research*, 91, 19–30.

Subagio, D. P., Srinivasan, M., Lim, M. & Lim, T. T. (2010). Photocatalytic degradation of bisphenol-A by nitrogen-doped TiO_2 hollow sphere in a vis-LED photoreactor. *Applied Catalysis B: Environmental*, 95(3–4), 414–422.

Singh, A., Patel, A. K., Deka, J. P. & Kumar, M. (2020). Natural recharge transcends anthropogenic forcing that influences arsenic vulnerability of the quaternary alluviums of the Mid-Gangetic Plain. *NPJ Clean Water*, 3(1), 1–12.

Shim, J., Mazumder, P. & Kumar, M. (2018). Corn cob silica as an antibacterial support for silver nanoparticles: efficacy on Escherichia coli and Listeria monocytogenes. *Environmental Monitoring and Assessment*, 190(10), 1–10.

Singh, A., Patel, A. K., Deka, J. P., Das, A., Kumar, A. & Kumar, M. (2019). Prediction of arsenic vulnerable zones in the groundwater environment of a rapidly urbanizing setup, Guwahati, India. *Geochemistry*, 125590.

Stockholm Convention, 2001. http://www.pops.int/TheConvention/ThePOPs/The12InitialPOPs/tabid/296/Default.aspx.

Talvitie, J., Mikola, A., Setälä, O., Heinonen, M. & Koistinen, A. (2017). How well is microlitter purified from wastewater?–A detailed study on the stepwise removal of microlitter in a tertiary level wastewater treatment plant. *Water Research*, 109, 164–172.

Tang, Y., Long, X., Wu, M., Yang, S., Gao, N., Xu, B. & Dutta, S. (2020). Bibliometric review of research trends on disinfection by-products in drinking water during 1975–2018. *Separation and Purification Technology*, 241, 116741.

Tunali, M., Uzoefuna, E. N., Tunali, M. M. & Yenigun, O. (2020). Effect of microplastics and microplastic-metal combinations on growth and chlorophyll a concentration of *Chlorella vulgaris*. *Science of the Total Environment*, 743, 140479.

Vieira, W. T., de Farias, M. B., Spaolonzi, M. P., da Silva, M. G. C. & Vieira, M. G. A. (2020). Endocrine-disrupting compounds: occurrence, detection methods, effects and promising treatment pathways–a critical review. *Journal of Environmental Chemical Engineering*, 104558.

Wang, X. & Lim, T. T. (2010). Solvothermal synthesis of C–N codoped TiO_2 and photocatalytic evaluation for bisphenol A degradation using a visible-light irradiated LED photoreactor. *Applied Catalysis B: Environmental*, 100(1–2), 355–364.

Wang, X., Mao, Y., Tang, S., Yang, H. & Xie, Y. F. (2015). Disinfection byproducts in drinking water and regulatory compliance: a critical review. *Frontiers of Environmental Science & Engineering*, 9(1), 3–15.

WHO (2017). *Guidelines for Drinking Water Quality*, Fourth Edition. World Health Organization (WHO).

Xu, B., Iskander, S. M. & He, Z. (2020a). Dominant formation of unregulated disinfection byproducts during electrocoagulation treatment of landfill leachate. *Environmental Research*, 182, 109006.

Xu, J., Li, K., Zhang, S., Chen, Y. & Ning, L. (2020b). Removal of endocrine-disrupting chemicals from environment using a robust platform based on metal–organic framework nanoparticles. *ACS Applied Nano Materials*, 3(4), 3646–3651.

Xiong, Y., Zhao, J., Li, L., Wang, Y., Dai, X., Yu, F. & Ma, J. (2020). Interfacial interaction between micro/nanoplastics and typical PPCPs and nanoplastics removal via electrosorption from an aqueous solution. *Water Research*, 184, 116100.

Yang, L., Chen, X., She, Q., Cao, G., Liu, Y., Chang, V. W. C. & Tang, C. Y. (2018). Regulation, formation, exposure, and treatment of disinfection byproducts (DBPs) in swimming pool waters: A critical review. *Environment International*, 121, 1039–1057.

Zazouli, M. A. & Kalankesh, L. R. (2017). Removal of precursors and disinfection byproducts (DBPs) by membrane filtration from water; a review. *Journal of Environmental Health Science and Engineering*, 15(1), 25.

Chapter 13

Occurrence and fate of phenolic compounds in groundwater and their associated risks

Vijay Laxmi Mohanta and Brijesh Kumar Mishra
Indian Institute of Technology (ISM) Dhanbad

CONTENTS

13.1 INTRODUCTION

Rapid population growth, combined with industrial and farming activities, may lead to increased demand for water and wastewater generation. Groundwater can be polluted from many sources, such as wastewater leaking into sewage systems, runoff from rivers, and the use of agrochemicals and fertilizers (Singh et al., 2020). In aquifers, a large variety of organic pollutants can therefore be found which pose a risk to the quality of groundwater. Phenolic compounds are among the major substantial organic pollutants because they persist for a longer time in the environment. The accumulation of organic pollutants causes toxic consequences, resulting in the deterioration of the health of humans and animals (Bruce et al., 1987; Wolf et al., 2004; Vázquez-Suñé

DOI: 10.1201/9781003157465-16

et al., 2010; Kumar et al., 2017). Phenol can be found in water and air after being released from the production and waste material containing phenol as it is used in making plastics and as a disinfectant in households (EPA, 2008).

Their existence is feasibly attributed to the deterioration or disintegration of natural organic matter in the water by dumping domestic and industrial waste in water bodies and by the flux of farmland. These chemicals are more toxic than the initial compounds when converted into other metabolites in the water. Their association with physical, chemical, and microbial influences is usually due to this change (Kulkarni and Kaware, 2013; Kumar et al., 2021; Kumar et al., 2020). The United States Environmental Protection Agency (USEPA) and the European Union (EU) have classified phenolic compounds as key contaminants of concern. The fact that these compounds are harmful and significantly have short-term as well as long-term consequences for living beings is responsible for this classification. The Environmental Protection Agency (EPA) recognizes the most severe dangerous sites in the country. Then, all these locations are included on the NPL and designated for federal cleanup for a long period. Phenol was reported on at least 595 of the 1,678 existing or former NPL sites. Therefore, the presence of phenolic compounds in the water bodies is not only troublesome and harmful but also causes a serious threat to public health and biodiversity (USEPA, 2008). Once exposed to phenolic compounds, several factors will decide whether a person can get affected. These considerations include dosage, duration, age, sex, diet, family characteristics, lifestyle, and health status (USEPA, 2008).

13.2 CLASSIFICATIONS OF PHENOLIC COMPOUNDS

Phenolic compounds are defined as an assembly of molecules distinguished by having a hydroxyl group(s) bonded directly to at least one aromatic ring on their structures. Phenol is also termed as carbolic acid, benzophenone, or hydroxybenzene (C_6H_5OH) which is the first compound to be in this organic group. They are synthesized in the industries as well as produced by plants and microorganisms. Based on different factors like carbon chain and the fundamental structure, phenolic compounds can be classified into several subgroups, such as lignans, flavonoids, phenolic acids, quinones, stilbenes, coumarins, curcuminoids, and tannins as shown in Figure 13.1 (Hättenschwiler and Vitousek, 2000; Anku et al., 2017; Gan et al., 2019).

13.3 REGULATIONS OF PHENOLIC COMPOUNDS IN GROUNDWATER

The phenolic micropollutants usually incorporate chloro-, bromo-, nitro-, and alkylphenol. Due to their toxicity and perseverance in the climate, these compounds are contemplated as priority pollutants, which are included in a record of hazardous substances of the USEPA. The release of these compounds without treatment may prompt genuine well-being dangers to humans, animals, and aquatic systems. For a sustainable environment, worldwide administrative bodies have established severe release limits for phenolic compounds. For instance, the EPA has set a standard limit for purity of water for phenol contamination in surface water as within 1 µg/L. Usually, the harmfulness levels are ranged from 9 to 25 mg/L for human as well as amphibian life (Kazemi

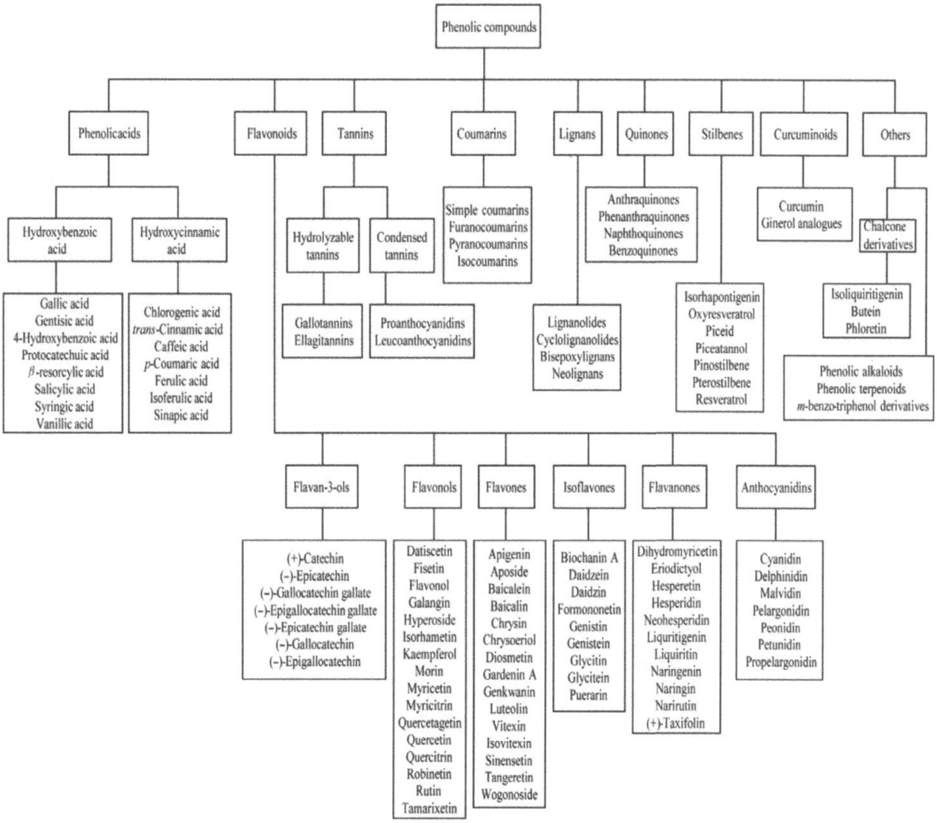

Figure 13.1 Classifications of phenolic compounds. (Reprinted from "Sprouted Grains, Ren-YouGan, Chak-Lun Chan, Qiong-Qiong Yang, Hua-Bin Li, Dan Zhang, Ying-Ying Ge, Anil Gunaratne, Jiao Ge, Harold Corke, Bioactive compounds and beneficial functions of sprouted grains, 191–246, Copyright (2019), with permission from Elsevier".)

et al., 2014; Sharma and Bhattacharya, 2017). An acute-duration oral Minimum Risk Levels (MRLs) of 1 mg/kg/day have been inferred by ATSDR, which is dependent on a 125 mg/kg/day of benchmark dose limit (BMDL) for a gain of body weight changes in pregnant rats that came in contact with phenol through isolated dosing in the course of 6–15 gestation days (GDs). The oral reference dosage (RfD) in the case of Sprague-Drawly rodents dosed with phenol in the pregnancy was 0.3 mg/kg/day on the basis of a BMDL of 93 mg/kg/day for the reduced pregnancy weight (York, 1997; IRIS, 2006).

The International Agency of Research on Cancer (IARC) has characterized phenol as Group 3, not classifiable concerning its cancer-causing nature to humans, while it is categorized as Group D by EPA, not classifiable as to carcinogenicity in humans. Also, phenol has not been listed as human carcinogenic by the National Toxicology Program. Phenol has been classified as an A4 cancer-causing agent (not identified as a human carcinogen) by ACGIH, i.e., the United States Conference of Governmental Industrial Hygienists (IARC, 2004; NTP, 2005; IRIS, 2006 and ACGIH, 2005). To minimize and keep exposed employees to or below the allowable exposure limits (PELs), OSHA has needed employees who are exposed to phenols for their work to develop engineering controls and work practices. Besides, for phenol, a 5 ppm time-sighted average (TWA) limit should be suggested for occupational exposure by the company for a 5 ppm average TWA of 8 hours (OSHA, 2005; ACGIH, 2005 and NIOSH, 2005).

Phenol is regulated by the subsequent industrial source's recommendations on clean water effluents: electrical plate, natural synthetic compounds, steam, asbestos, wood processing, metal processing, covering and roofing processing, painting formulation, gum and wood inks, dark carbon, casting and molding, shaped aluminum, and electric and electronic segments; see the guidelines for corresponding industrial sources (Gahlot et al., 2020). EPA controls phenol as a toxic material under the Clean Water Act (CWA) and a harmful air pollutant (HAP) under the Clean Air Act (CAA). Phenol has an Emergency Planning and Community Right to Know Act of 1986 listing of chemical items contained in "Toxic Substances Subject to Section 313," and a reporter quantity (RQ) of 1,000-pound cutoff is given. The RQ reflects the quantity of a hazardous material that must be reported to the relevant authority when delivered to the climate. Besides, phenol is considered an extremely dangerous pollutant (EPA, 2006a, b).

13.4 SOURCES AND PATHWAYS FOR GROUNDWATER POLLUTED BY PHENOLIC COMPOUNDS

Phenol has been recognized for recognition on the National Priorities List (NPL) of EPA at least 595 of the 1,678 hazardous waste hubs. Essentials for phenol have been given in effluents (maximum 53 ppm), groundwater (1.9–10 ppb), atmospheric water (1.5–100 ppb), and potable water and rainwater (0.075–1.2 ppb). Locational exposures occur with dermal and inhalation exposure; the regulated air concentration varies from 0.1 to 12.5 mg/m^3 (0.03–32 ppm) in various working environments. The dermal interaction of phenol or phenol-containing products may also lead to occupational as well as public exposure (USEPA, 2008). After discharge from the processing, usage, disposal of phenol-containing products, and natural decomposition of dead plants and animals, phenol can be present in the air and water. They are also synthesized in the aquatic environment by microorganisms and plants. Phenol would potentially pass into groundwater in the soil (USEPA, 2008). Groundwater pollution studies involve the identification of the pollution source as well as the passage of the pollutants in the soil. When pollutants are introduced, effective resource management and prevention measures are required to ensure the proper creation and remediation of contaminated sites. In groundwater, the main transport processes are associated with advection, dispersion, diffusion, adsorption, biodegradation, and chemical reactions (Sahajanand et al., 2020).

13.4.1 Natural sources of phenol pollution in groundwater

Animal waste and the decaying of agricultural waste are two natural causes of phenol in marine environments. Phenolic compounds represent components of several marines or terrestrial plant species. Some of them are produced by amino acids present with ultraviolet radiation in the hemicelluloses present in some plants. Human and animal bodies, without any external contact, contain phenol that is eventually excreted.

Thus, phenol is also present in metabolic wastes produced by living beings. They are also being contained in creosote and coal tar naturally. Phenol is also created in the course of the deterioration of benzene and natural fires in the atmosphere as a result of ultraviolet light emissions. Their introduction directly in the water systems or rivers and their indirect precipitations cause contamination of these compounds by water bodies (Sheets, 2007; Anku et al., 2017).

Chlorophyll synthesis of these compounds in plants is dependent on certain external stimuli, including solar radiation, chemical stressors, and microbial infections. Chlorophyll is often influenced by such external stimuli or factors. The precursor of phenolic plant synthesis is phenylalanine. Phenylalanine is typically deamined to cinnamate by catalysis of phenylalanine ammonia. A method of the hydroxylation of cinnamate-4 leads to the transformation into coumaric acid that, among other catalysts of phenolic groups, becomes the primary raw material for synthesis by separate enzymes of stilbenes, flavonosides, and furanocoumarins. In leaves, roots, and plant stem, the phenolic compounds are processed. These compounds are found in roots and leaving plant exudates. Runoffs from the field drain the substances into surrounding bodies of water (Daniel et al., 1999; Anku et al., 2017).

13.4.2 Anthropogenic sources of phenol pollution in groundwater

The most common sources of phenol from anthropogenic sources on water systems are coal tar and manufacturing wastewater, such as saps, plastics, fabrics, sticks, iron, aluminum, steel, leather, synthetic fuel, and rubber effluents (EPA, 1981a, b; Thurman, 1982; USEPA, 2008). Phenol will also be supplied from wood treatment plants and paper pulp mills (Keith, 1976). Additional phenol releases are the industrial by-products of phenol as well as phenol-containing products, such as basic disinfectants, slimicides, and pharmaceutical systems, such as throat capsules, antiseptic lotions, and mouthwashes (Goerlitz et al., 1985; Budavari et al., 1989; Darisimall, 2006). They are acquainted with the environment in an assortment of ways such as effluents from coal handling, paper manufacturing, farming, drugs, petrochemical industrial wastes, and municipal wastes (Sun et al., 2014). The phenolic micropollutants usually incorporate bromo-, chloro-, nitro-, and alkylphenols. Chlorine, bromine, and nitrous, and alkylphenols commonly are found in phenolic micropollutants. Owing to their toxicity and environmental consistency, these phenolic compounds are treated by the EPA as essential pollutants and are listed on the list of hazardous substances. Essentially, bisphenols and phenyl phenols are broadly utilized in clinics and families as a disinfectant. They are additionally used in makeup, paper, textile, and leather industries. The wide scope for the use of these phenolic compounds has affected environmental quality in numerous areas, for example, surface water, marine sediments, and sludge from sewage systems. It was accounted that their existence influences the endocrine system, demonstrating estrogenic movement (Olak et al., 2012).

13.4.3 Pathways for groundwater polluted by phenolic compounds

Phenolic compounds are discharged into air and water bodies from various sources. Based on its high water solubility and its presence in rainwater, certain phenol can be washed out from the atmosphere. Due to the short atmospheric half-life of phenol, it is possible that only small quantities would wash away. In daylight, when hydroxyl radical concentrations that are produced photochemically are most elevated in the air, a very small atmospheric movement of phenol is probably going to happen. Important transport frameworks are not necessary in water for either volatilization or sorption of sediments and suspended particulates. The biological treatment of phenol-contaminated wastewater has demonstrated the removal of <1% of phenol by the stripping process (Kincannon et al., 1983; Petrasek et al., 1983; Shim et al., 2018).

Phenol up to 608 ppm is reported in dry weight sediments. A hazardous waste dump is not known in the area of the site where this accumulation was represented, for example, whether or nearly a point cause of the release. Phenol sorption on aquifer materials is almost not available, indicating that the sedimentation of phenol may also be negligible. Phenol is used to philter to groundwater in the light of the soil adsorption coefficient (Ehrlich et al., 1982). Seepage of phenol from sediment from waste sites, from landfills, and harmful waste sites found phenol in groundwater. The phenol sorption coefficient increases with soil-based organic soil matter and can suggest that organic soil matter is the predominant soil-based phenol. The pKa value is 10 for phenol, which indicated that phenol essentially occurs as the protonated acid at pH levels. Phenol is partly present as an anion in alkaline soils and liquids, which may impair its destiny and transport processes (Xing et al., 1994; USEPA, 2008).

Due to hydrological parameters and complex geological strata, the process for determining the spread of contaminants like phenol into groundwater is complicated. Phenol is briskly deteriorated in the air by a gas-phase reaction with an estimated half-life of 14 hours but can continue for a longer period in the groundwater. Its concentration gets affected by the permeability of the aquifers. The predominant processes regulating phenol transportation in subsurface migration include sorption to clay minerals and organic matter, ion exchange in soil and groundwater, and microbial degradations (Lapworth et al., 2012; Sahajanand et al., 2020).

13.5 OCCURRENCES AND CONCENTRATIONS OF PHENOLIC COMPOUNDS IN GROUNDWATER

Phenolic compounds have been found in drinking water, rainwater, surface waters, sediments, sewage, urban runoff, industrial discharge, and hazardous waste dumping sites. The presence of high amounts of phenol in groundwater indicates that they are present in aquifers and surface water. Their presence in aquifers is likely to be the result of their discharge into the soil, possibly from industrial activity, release from waste dumps, and the consumer use of phenol-containing materials (USEPA, 2008).

The occurrence of phenol in drinking water is apparently because of such polluted groundwater or surface water as a source. Their persistence in groundwater is most likely due to soil release, mostly from industries or seeping from waste dumping sites, and successive percolation of phenol through the surface to aquifers. Table 13.1 shows some of the occurrences of phenolic compounds in groundwater worldwide. Italy has reported phenol concentrations of 0.58, 0.051, and 0.161 μg/L, respectively, in tap water, source water, and mineral water (Achilli et al., 1995). In Wisconsin, after a spill, it was found in nine wells up to 1.130 ppm (Delfino and Dube, 1976). Phenol was detected in groundwater (up to 10.4 ppm) of a waste pond in a wood-protection facility

Table 13.1 Occurrence of phenol in groundwater

Phenolic compounds	*Location*	*Concentration*	*Units*	*References*
Cresols and dimethylphenols	Denmark	29	μg/L	Reitzel and Ledin (2002)
Chlorophenols (average)	Finland	4,454	μg/L	Mikkonen et al. (2018)
Phenol, cresols, and xylenols	Not given	1,440	micromol	Hedbavna et al. (2016)
Phenol	China	0.686	mg/L	Han et al. (2016)
Total phenol	Taiwan	0.016–0.584	mg/L	Lin et al. (2007)
2.4.5-Trichlorophenol	Narangba	1.9–31	μg/L	GES (2014)
2.4.6-Trichlorophenol	Narangba	1.7–56.5	μg/L	GES (2014)
2.4-Dichlorophenol	Narangba	14.1–17,500	μg/L	GES (2014)
4-Chloro-3-methylphenol	Narangba	1.9–149	μg/L	GES (2014)
2.4-Dichlorophenyl acetic acid	Narangba	96.2–128	μg/L	GES (2014)
Tannins and lignans	Alberta	170	μg/L	Wallis et al. (1981)
4-hydroxybenzoic acid	England	21–104	μM	Hedbavna et al. (2016)

in Florida and found in two aquifers fifteen months after a coal gasification project was completed at 6,510,000 ppb in two aquifers (Stuermer et al., 1982; Goerlitz et al., 1985). Phenol in leachates from Illinois landfill sites was identified by a maximum concentration of 1.9 ppm. Phenol was nearly deposited in central Florida at approximately 17 and 15 ppb, respectively, in groundwater and surface water (Clark and Piskin, 1977; Chen and Zoltek, 1995). Several phenolic compounds such as 2-chlorophenol, 2,4,6-trichlorophenol, 2,6-dichlorophenol, 2,4-dichlorophenol, 2-methylphenol, and phenols were detected in 4 monitoring wells in Binary Industries Site, Narangba, Queensland. Among these compounds, 2,4-dichlorophenol was present in abundance, i.e., up to 17,500 μg/L (GEC 2014).

13.6 ANALYTICAL METHODS OF PHENOLIC COMPOUNDS IN WATER

Numerous research methods were previously proposed for the assurance in natural environmental waters and wastes of phenol and its derivatives. High-performance liquid chromatography (HPLC), gas chromatography (GC), LC-MS/MS, etc., were the most commonly used methods. Besides, electrochemical methods are the reliable means to test the phenolic compounds (phenolic) in the mixture easily and at a low cost. Techniques for phenolic compounds that rely on several enzymes, such as tyrosinase and laccaemia, have been used in literature. Researched significant organic pollutants (mainly industrially produced), such as phenol and o-cresol, are potentially dangerous to humans and the environment. Also, researchers proposed another electrochemical methodology that could concurrently detect phenol and o-cresol (Dobes et al., 2013; Monasterio et al., 2014; Liu et al., 2016).

For each phenolic compound, a regulatory threshold of 0.1 and 0.5 ppm for the amount of all substances in water for human use has been established by the European

Community (EC) (EC, 1980). As the maximum pollutant level (MCL) for several pollutants is low (sub ng/mL), there is a very strong need for highly sensitive analytical techniques. Several methods, including spectrophotometry, GC, electrochemical methods, and liquid chromatography (LC), for analytic methods of water-based phenolic compounds, were described in various studies (Lupetti et al., 2004; Kochana et al., 2008; Kovács et al., 2008; Huang et al., 2009).

The preparation of the samples represents one of the most critical factors for the measuring of any organic pollutant in the water. Solid-phase microextraction (SPME) had been adopted as a method of preparation for samples in the early 1990s and has been generally accepted for different types of analytes, including phenolic compounds. However, before sample preparation, the optimization of SPME techniques is essential. The following consideration has been shown by researchers to optimize SPME techniques.

a. **Extraction temperature**: Temperature is an important parameter influencing the efficiency of extraction. Analysts are moving from the solution to the front region and increasing the temperature extraction. Alternatively, the SPME's distribution constant is dependent on temperature and declines with the increase in temperature.
b. **Extraction time**: As SPME is a technique based on equilibrium, the equilibrium time provides the highest level of the analyte that the fibers can extract. Yazdi et al. (2012) have investigated the effect of extraction time (20–50 minutes) during the measurement of phenolic compounds and found that 40-minute extraction time has a more significant value.
c. **Effect of pH**: Adjusting the pH will increase extraction because the dissociation balance and the acid/basic aim analysis are affected.
d. **Effect of ionic strength**: The ionic force of the solution during the extraction process is affected by the extraction efficiency. Adding salt to the water decreases the solubility of organic compounds; this is referred to as the salting effect. The use of water molecules in hydration spheres around the ionic salt may explain this to reduce the water concentration required for the dissolution of organic compounds. It allows more of these analytes to migrate into the headspace stage.
e. **Stirring speed**: It is a significant parameter that is responsible for the extraction yield as mass transfer speed and extraction time are improved.
f. **Desorption time and temperature**: Temperature affects the amount of the analyte to which the GC injector may be released and varies depending on the type of analysis and fiber type. Increasing the temperature of desorption leads to a reduction in the analysis partition coefficient in between the fiber-coating and the gaseous phase.

Spectrophotometry is one of the popular techniques for the determination of pollutant concentration in water and wastewater, but the spectrophotometry method is not popular in the case of phenol measurement as it can measure only phenol, not phenolic compounds. The most popular techniques for the estimation of phenolic compounds in water as well as wastewater have been described by USEPA as methods 204 and 625.1. Some of the techniques and instrumental protocols have been described in the following subsection.

13.6.1 Phenolic compounds in water and wastewater by EPA 204 (USEPA, 2004)

This approach refers to the determination of phenol as well as some substitutes of phenols, i.e., 2-chlorophenol, 4-chloro-3-methylphenol, 4-nitrophenol, 2,4-dichlorophenol, pentachlorophenol, 2,4-dimethylphenol, 2,4,6-trichlorophenol, 2,4-dinitrophenol, 2-nitrophenol, 2-methyl-4,6-dinitrophenol, and phenol. The procedure for the determination of compounds described above in municipal and industrial discharges is flame ionization detector gas chromatography (FIDGC). This approach describes the analytical context for deriving, washing, and electron capture detector gas chromatography (ECDGC) detectors that can be used to validate FIDGC measurements. Method 625.1 offers the conditions for the gas chromatograph/mass spectrometer (GC/MS) suitable for the quantitative and qualitative authentication of results provided by this method for all the parameters mentioned earlier.

Summary of Method

a. A measured sample volume of approximately 1 L is acidified and collected employing a separatory funnel using methylene chloride. During concentration, the methylene chloride extract is dried and exchanged with 2-propanol to a volume of 10 mL or less. By gas chromatography, the extract is isolated and the phenols then were analyzed with an FID.
b. In samples with high general organic and organic-basis interferences, a preliminary wash can be used under basic conditions.
c. A derivative and chromatography cleanup protocol for columns to help remove interferences is also provided by this process. ECDGC analyzes the derivatives.

13.7 HUMAN HEALTH RISK ASSESSMENT DUE TO PHENOLIC COMPOUNDS IN DRINKING WATER

The use of some medicines (including throat lozenge and ointments) can cause exposure to phenol. After acute dermal or inhalation exposures, phenol can be severely irritated to skin, eyes, and mucous membranes in humans. Phenol from oral exposure is known to be very harmful to people. People, who have chronically (long-term) been exposed to the disease, have been reported anorexia, gradual loss of weight, diarrhea, vertigo, salivation, dark urine coloration as well as blood and liver symptoms. Studies on animals have demonstrated decreased fetal weights, slow growth as well as irregular development of phenol-exposed animals by the oral route in the offspring. Phenol was listed by EPA as Group D, not human carcinogenic (USEPA, 2008).

13.7.1 Acute and chronic effects

Phenol inhalation and dermal penetration to the eyes, skin, and mucous membranes of humans are highly irritating. The symptoms of acute human toxicity include erratic breathing, weakness of the muscle and tremors, loss of symptoms of acute toxicity include erratic breathing, weakness in the muscles and tremors, controlling loss, convulsions, coma, and lethal arrest. Animal testing of acute rat, mice, and rabbit has shown that phenol is extremely acutely toxic due to oral exposure (USEPA, 2008; Calabrese and Kenyon, 1991).

Chronically exposed humans have been reported for anorexia, gradual loss of weight, salivation, vertigo diarrhea, and dark coloration of the urine. There has also been reporting of stomach irritation, blood symptoms, and hepatic impact. One research reported muscle pain, faintness, expanded liver, and high levels of hepatic enzymes in a person following dermal and inhalation exposure to phenol and some other chemicals. Phenolic exposure to skin contributes to dermal inflammation called necrosis. In humans, elevated concentrations of phenol have also been recorded for cardiac arrhythmias. Central nervous system (CNS), kidney, liver, respiratory, and cardiovascular symptoms have been found to have persistent phenol inhalation toxicity in animals (USEPA, 2008).

The RfD is an estimation of uncertainty that can apply to a continuum of everyday oral exposure of humans including vulnerable subgroups that are expected to run no significant risk of nocivic consequences of non-cancer in the course of life. It is more a reference point to calculate the future consequences, rather than a precise risk estimator. The risk of adverse health effects improves at exposures more often than RfD. Exposure to life outside RfD does not inherently mean an adverse health effect (USEPA, 1999). Studies in animals reported decreased weight of the fetus body, delayed growth, and premature production of oral phenol-exhibited animals in the offspring. There has since been a reduction in maternal weight gain and a rise in maternal mortality (Calabrese and Kenyon, 1991; USEPA, 2008).

In occupationally exposed employees, minor, nonsignificant excesses in certain forms of cancer have been identified, but the results have not been specifically linked to exposure to phenol. Tumors from oral phenol exposures have not been seen by animal studies, although dermal studies have indicated that skin phenol can be a stimulating tumor and a low skin carcinogen in mice. Due to the lack of evidence concerning the effects of carcinogenicity in living beings, the EPA classed phenol as Group D, not classifiable to be humanly carcinogenic (USEPA, 2000). The majority of phenolic compounds are readily absorbed into the skin and absorbed by the human gastrointestinal tract. After metabolism, they are converted into different reactive proximate forms, especially quinone moieties, which are responsible for the formation of covalent protein bonds easily and thus have the potential to have toxic effects on humans. All of the adverse effects on humans have been characterized by chlorophenols, amino phenols, chlorocatechols, nitrophenols, methylphenols, and other phenols (Schweigert et al., 2001).

Bisphenol A and certain alkylphenols have been recognized to perturb the endocrine effect in human beings by modifying the mammary gland production in exposed species. Related work further shows the ability of bisphenol A to prolong the onset of girls' puberty. The ingestion of water that includes a high level of phenol, like drinking water, induces stomach issues and muscle shiver with walking difficulties. The administration of products that contain a high phenolic concentration to the skin leads to skin bubbles and burns; exposure to high levels of phenol can result in heart, kidney, and liver damage (ATSDR, 1999; Muñoz-de-Toro et al., 2005; Vom and Hughes, 2005). Due to their readily oxidizing ability to more reactive quinone radicals, catechols tend to cause DNA injury and arylation, degrade certain proteins present in the body, and interrupt electron transfer of energy transducing membranes. The presence of copper, coffee, and dihydrocaffeic acids often destroys DNA. (Schweigert et al., 2001; Li and Trush, 1994). Chlorophenol toxicity causes burning of the mouth, throat, and necrotic mouth, stomach, and esophagus. Chlorophenol also contributes to irregular temperature, spikes in pulses, weakness in muscles, and convulsions. Additional influences of poisoning due to chlorophenol include kidney, liver, respiratory, skin, and distress in the digestive tract (Gosselin et al., 1984; Anku et al., 2017).

13.7.2 Metabolism of phenols in the human body

Figure 13.2 demonstrates the metabolism that converts phenol into the urine before it is excreted. The reactions that process phenol are catalyzed by three distinct systems of enzymes. Cytosolic phenol sulfotransferases catalyzed transition to the phenol hydroxyl group of inorganic sulfur from the activated 3′-phosphoadenosine-5′-donor molecule. Molecules form an O-glucuronide conjugate in the hydroxyl mobility of phenol. The microsomal P4502E1 cytochrome catalyzes phenol hydroxylation into a much smaller volume of catechol and the enzymes in the phase II form. In exchange,

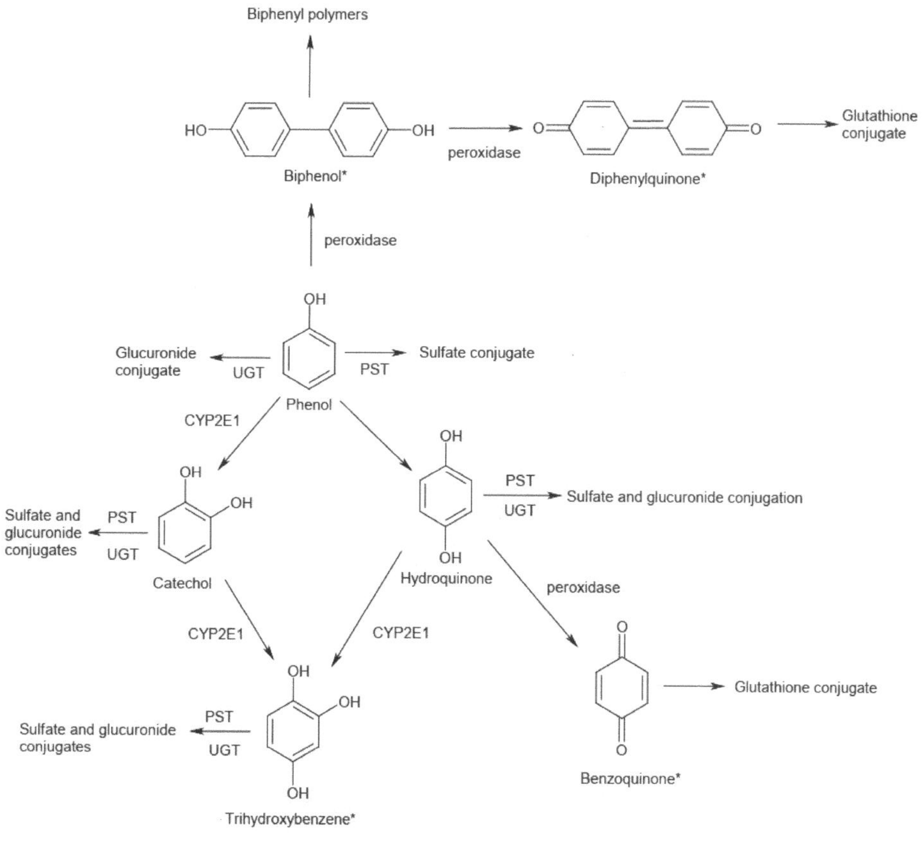

Figure 13.2 Metabolism of phenol in the human body. (Adapted from USEPA 2008.)

hydroquinone can form conjugates, become benzoquinone peroxidized, or be further oxidized into trihydroxybenzenes. There is rivalry not just for phenol but also for subsequent phenol among the three enzyme systems that is responsible for metabolizing the phenol in multiple tissues including hydroquinone and oxidative materials. As a result, the relative quantity of the products produced can differ according to the species, dosage, and path (USEPA, 2008).

13.8 HUMAN HEALTH RISK ASSESSMENT USING HAZARD QUOTIENT (HQ)

Risk analyses predict that a certain pollutant may cause adverse health effects over a certain period (Naz et al., 2016; Laxmi et al., 2019; Mohanta et al., 2020). The assessment of risks on human health due to any harmful material relies on a comprehensive risk assessment, and this model is described under the USEPA (2005) cancer and non-cancer risk. The USEPA 2005 system calculates the pollutant's risk quotient (HQ), the value of which is less protected than the unit when the occurrence of health hazards is warned.

For the assessment of health risk by principal exposure routes, i.e., direct intake, dermal as well as inhalations of phenol-contaminated groundwater, would be considered. Cancer and the non-cancer risks for the three groups, i.e., men, women, and children (≤10 years old), have been measured by an exposure assessment. Table 13.2 represents the input parameters based on Indian conditions such as time, exposure length, rate of absorption, body weight, and surface area. Chronic daily intake (CDI) for phenol was evaluated by all routes of exposure for all three categories as per IRIS (2009) (Mohanta et al., 2020; Mohanta and Mishra, 2020).

$$CDI_{Oral} = \frac{C_W \times EF \times IR \times ED}{B_W \times AT} \quad (13.1)$$

$$CDI_{Dermal} = \frac{C_W \times PC \times SA \times ET \times F \times EF \times ED}{B_W \times AT} \quad (13.2)$$

$$CDI_{inhalation} = \frac{C_{air} \times AE \times VR \times EF \times ET \times ED}{B_W \times AT} \quad (13.3)$$

C_{air} was evaluated by the equations with regard to phenol concentration in the shower room proposed by Little's theory (Mishra et al., 2014; Mohanta and Mishra, 2020);

$$C_{air} = \frac{Y_{s(t)} + Y_{s(i)}}{2} \quad (13.4)$$

where

$Y_{s(i)}$ = initial phenol concentration in the bathroom (assumed ≈ 0 mg/L) and
$Y_{s(t)}$ = phenol concentration in the bathroom at time t (mins).

$$Y_{s(t)} = \left[1 - e^{-bt}\right] \times \frac{a}{b} \quad (13.5)$$

Table 13.2 Input parameters for human health risk assessment

S. no	*Input parameters*		*Values*	*Unit*	*References*
1.	C_w	Phenol/phenolic compound concentration in water	As per study	mg/L	As per study
2.	C_{air}	Concentration of phenol in the air	Little's model	mg/L	Little (1992)
3.	IR	Rate of ingestion	Men:4; women:3; children:1	L/day	Chowdhury et al. (2001); Giri et al. (2005)
4.	EF	Exposure frequency	365	Events/ year	Chowdhury et al. (2001)
5.	ED	Exposure duration	Men:64; women:67; children:10	year	USEPA (2004)
6.	ET	Exposure time	35	min/day	RAIS (2012)
7.	F	Fraction	1	-	Mishra et al. (2014)
8.	AT	Average time	Men: 64*365; women: 67*365; children: 10*365	days	USEPA (2004)
9.	BW	Body weight	Men: 65; women: 55; children: 15	kg	Giri and Singh (2015)
10.	PC/Kp	Permeability coefficient	0.0011 (total phenol)	m/h	RAIS (2016)
11.	SA	Surface area	Men:1.66; women:1.53; children: 0.66	m^2	USEPA (2015); Mohanta and Mishra 2020
12.	IR	Inhalation rate	Men:0.84 women:0.66; Children: 0.38	m^3/h	USEPA (1997); Kawahara et al. (2011)
13.	VR	Ventilation rate	0.83	m^3/h	USEPA (1997)
14.	Vs	Bathroom volume	5	m^3	Little (1992)
15.	Q_L	Water flow rate	5	L/min	Little (1992)
16.	Q_G	Airflow rate	50	L/min	Little (1992)
17.	H	Henry's law constant	0.022	-	RAIS (2016)
18.	K_{OL} A	Mass transfer coefficient	$8.23*10^{-5}$	L/min	RAIS (2016)
19.	RfD	Reference dosage	0.3 (total phenol)	mg/kg-d	RAIS (2016)
20.	RfC	Reference concentration	0.2 (total phenol)	mg/m^3	RAIS (2016)

$$a = \frac{Q_L \times C_W\left(1-e^{-N}\right)}{V_s} \tag{13.6}$$

$$b = \frac{\left[\frac{Q_L}{H} \times C_W\left(1-e^{-N}\right)\right] + Q_G}{V_s} \tag{13.7}$$

$$N = \frac{K_{OL}A}{Q_L} \tag{13.8}$$

where

N = dimensionless coefficient;
Q_L = water flow rate (L/min);
a and b = factors;
t = time of contact (min).

The HQ implies that the risk of non-cancer can be determined by the division of a CDI by RfD concerning oral and dermal contacts of phenol in the groundwater, while the CDI was divided by RfC (reference concentration) in the formula mentioned below:

$$HQ_{Oral/Dermal} = \frac{CDI}{RfD_{Oral/Dermal}} \tag{13.9}$$

$$HQ_{inhalation} = \frac{CDI}{RfC} \tag{13.10}$$

$$HI_{total} = HQ_{oral} + HQ_{dermal} + HQ_{inhalation} \tag{13.11}$$

As per USEPA (2005), the threshold limit of 1 is considered for HI.

13.8.1 Multi-pathway of exposure of phenol and associated risks

Inhalation exposure to phenol is less likely than oral and dermal exposure. It is understood that phenol is found in small quantities in both cigars and tobacco, with respect to the detectable amount of phenol being generated indoors by consuming these items. Children can be exposed to low phenol levels when they are present in indoor conditions contaminated with tobacco smoke (Jenkins et al., 2000; IARC, 2004).

The RfD value of a pollutant is responsible for the extent of its associated non-cancer risk due to the oral and dermal routes of exposure. The lower the RfD value of a pollutant, the higher will be its associated risks. Similarly, the RfC value defines whether and the extent to which a pollutant is responsible for non-cancer risks due to the inhalation route of exposure. The RfD value of total phenol is 0.3 mg/kg-day and its RfC value is 0.2 mg/m^3 (RAIS, 2016). The total phenol which was reported in the literature in Table 13.2 was assessed for non-cancer risk using the method by USEPA and it found that hazard quotient due to oral route of exposure varied from 3.28×10^{-3} to 1.26×10^{-1}, 2.91×10^{-3} to 1.11×10^{-1}, and 3.55×10^{-3} to 1.36×10^{-1} for men, women, and children, respectively. Similarly, HQ because of the dermal exposure route was ranged from 9.46×10^{-6} to 3.64×10^{-4}, 9.52×10^{-6} to 3.66×10^{-4}, and 1.50×10^{-5} to 5.79×10^{-4} for men, women, and children, respectively. HQ because of inhalation exposure route was also assessed and was found to be varied from 9.96×10^{-5} to 3.83×10^{-3}, 1.92×10^{-4} to 3.5×10^{-3}, and 1.92×10^{-4} to 7.39×10^{-3} for men, women, and children, respectively. It was observed and validated from the literature that the non-cancer risk in children is highest, followed by men and then women due to a higher ingestion rate in children concerning their body weight and also higher body weight and surface area (Mohanta and Mishra, 2020). The non-cancer risk in total phenol is highest while ingested than that of the inhalation route of exposure. The lowest risk was observed via the dermal route of exposure due to the low permeability coefficient of the total phenol in human skin.

Although the total phenol does not pose a severe health hazard in terms of non-cancer risk based on the evaluated result, it may form various deadly carcinogenic by-products such as trichlorophenol and pentachlorophenol, whose cancer slope factors (CSFs) are 0.02 mg/kg-day and 0.4 mg/kg-day, respectively, when it reacts with other compounds such as chlorine. CSF value determines whether a pollutant is carcinogenic or not. It is calculated by the following expression.

$$\text{Cancer Risk}_{\text{oral/dermal/inhalation}} = \text{CDI}_{\text{oral/dermal/inhalation}} \times \text{CSF}$$

Another phenolic compound, i.e., phenyl phenol, is also carcinogenic in nature whose CSF value is 1.9×10^{-3} mg/kg-day (RAIS, 2016). The higher the CSF value, the higher will be the likelihood of cancer due to all the exposure routes, which makes pentachlorophenol to be the dangerous cancer-causing phenolic compound. Numerous phenolic compounds are present in nature due to natural as well as anthropogenic processes. The hazardous effects on human health can be identified similarly using the described method. This method of analyzing cancer and non-cancer risk is adopted by many researchers and is suggested to be effective in risk assessment.

13.9 SUMMARY AND FUTURE ASPECTS

In water systems, phenolic compounds are triggered by natural phenomena or manufacturing, domestic, and agricultural activity waste products. Phenolic compounds can enter the aquifer mostly by decomposing dead plants and animals or by releasing them from different industrial sectors. A broad variety of phenolic compounds as well as subterranean metabolites and by-products, even in limited quantities, can endanger groundwater systems under certain conditions because of their aggregation over a relatively long period. Phenol has a short atmospheric half-life, so it may wash out of the atmosphere. In water, phenol does not have to be transported either volatilized or sorted to sediments and suspended particles. While the highest concentrations are in the air photochemically produced hydroxyl radical, almost no air transport.

They may be transmitted to human beings by ingestion, inhalation, and dermal routes of exposure that may pose significant health hazards. Although the total phenol in groundwater and surface water does not post any significant threat itself on human health due to high RfD value and no CSF value, the other phenolic compounds or the by-products of phenols, some of which are carcinogenic in nature, pose a severe threat to human health due to lower RfD and high CSF value. Also, human beings are more prone to health risks when they ingest the phenol or phenolic compounds contaminated water when compared to the exposure due to inhalation and dermal routes of exposure. However, more and detailed research is needed to understand the transformation of total phenol to by-product formation of phenolic compounds in the aquatic environment.

ACKNOWLEDGMENT

The authors are grateful to the Department of Environmental Science and Engineering, Indian Institute of Technology (ISM) Dhanbad, India, for providing technical and financial support.

CONFLICT OF INTEREST

The authors declare that they have no conflict of interest.

REFERENCES

ACGIH. 2001. *Documentation of the Threshold Limit Values and Biological Exposure Indices. Phenol.* 7th ed. Cincinnati, OH: American Conference of Governmental Industrial Hygienists.

ACGIH. 2005. *Threshold Limit Values for Chemical Substances and Physical Agents and Biological Indices.* Cincinnati, OH: American Conference of Government Industrial Hygienists.

Achilli, G., Cellerino, G. P., d'Eril, G. M., Bird, S., 1995. Simultaneous determination of 27 phenols and herbicides in water by high-performance liquid chromatography with multielectrode electrochemical detection. *Journal of Chromatography A*, 697(1–2), 357–362.

Anku, W. W., Mamo, M. A., Govender, P. P., 2017. Phenolic compounds in water: Sources, reactivity, toxicity and treatment methods. *Phenolic Compounds-Natural Sources, Importance and Applications*, 420–443.

ATSDR. 1999. Phenol. Potential for human exposure. https://www.atsdr.cdc.gov/toxprofiles/tp115-c6.pdf.

Bruce, R. M., Santodonato, J., Neal, M. W., 1987. Summary review of the health effects associated with phenol. *Toxicology and Industrial Health*, 3(4), 535–568.

Budavari, S., O'Neil, M. J., Smith, A., et al., eds., 1989. *Phenol. The Merck Index.* 11th ed. Rahway, NJ: Merck & Co., Inc., p. 1150.

Calabrese, E. J., Kenyon, E., 1991. *Air Toxics and Risk Assessment.* Boca Rotan, FL: CRC Press.

Chen, C.S., Zoltek Jr, J., 1995. Organic priority pollutants in wetland-treated leachates at a landfill in central Florida. *Chemosphere*, 31(6), 3455–3464.

Chowdhury, U. K., Rahman, M. M., Mandal, B. K., Paul, K., Lodh, D., Biswas, B. K., et al., 2001. Groundwater arsenic contamination and human suffering in West Bengal, India and Bangladesh. *Environmental Science*, 8(5), 393–415.

Clark, T. P., Piskin, R., 1977. Chemical quality and indicator parameters for monitoring landfill leachate in Illinois. *Environmental Geology*, 1(6), 329–340.

Daniel, O., Meier, M. S., Schlatter, J., Frischknecht, P., 1999. Selected phenolic compounds in cultivated plants: Ecologic functions, health implications, and modulation by pesticides. *Environmental Health Perspectives*, 107(suppl 1), 109–114.

Darisimall. 2006. Sore throat lozenges and sprays. http://store.darisimall.com/sothlosp.html. July 27, 2006.

Delfino, J. J., Dube, D. J., 1976. Persistent contamination of ground water by phenol. *Journal of Environmental Science & Health Part A*, 11(6), 345–355.

Dobes, J., Sochor, J., Babula, P., Kynicky, J., Hubalek, J., Klejdus, B., Kizek, R., 2013. Electrochemical tools for determination of phenolic compounds in plants. A review.

Ehrlich, G. G., Goerlitz, D. F., Godsy, E. M., Hult, M., 1982. Degradation of phenolic contaminants in ground water by anaerobic bacteria: St. Louis Park, Minnesota. Groundwater, 20(6), 703–710.

EPA. 1981a. Exposure and risk assessment for phenol (revised). Washington, DC: U.S. Environmental Protection Agency. PB85221695.

EPA. 1981b. Treatability manual. Washington, DC: U.S. Environmental Protection Agency, Office of Research and Development. EPA600282001A, I.8. 1–1–1–5.

EPA. 2006a. Identification and listing of hazardous waste. Washington, DC: U.S. Environmental Protection Agency. Code of Federal Regulations. 40 CFR 261, Appendix VIII. http://www.epa.gov/epacfr40/chapt-1.info/chi-toc.htm. March 07, 2006.

EPA. 2006b. National recommended water quality criteria. Washington, DC: U.S. Environmental Protection Agency, Office of Water, Office of Science and Technology. http://www.epa.gov/waterscience/criteria/nrwqc-2006.pdf. January 8, 2008.

Eurpean Commission. 1980. Drinking Water Directive 80/778/EEC as amended by Directive 98/83.

Gaffney, J. S., Streit, G. E., Spall, W. D., Hall, J. H., 1987. Beyond acid rain. Do soluble oxidants and organic toxins interact with SO_2 and NO_x to increase ecosystem effects? *Environmental Science & Technology*, 21(6), 519–524.

Gahlot, R., Taki, K., Kumar, M., 2020. Efficacy of nanoclays as the potential adsorbent for dyes and metal removal from the wastewater: A review. *Environmental Nanotechnology, Monitoring & Management*, 14, 100339.

Gan, R. Y., Chan, C. L., Yang, Q. Q., Li, H. B., Zhang, D., Ge, Y. Y., Corke, H., 2019. Bioactive compounds and beneficial functions of sprouted grains. In *Sprouted Grains* (pp. 191–246). AACC International Press.

GEC (GeoEnvironmental Consultants). 2014. Groundwater monitoring- binary industries site magnesium court, Narangba, QLD. https://environment.des.qld.gov.au/__data/assets/pdf_file/0033/87657/groundwater-monitoring-may2014.pdf.

Giri, S., Singh, A. K., 2015. Human health risk assessment via drinking water pathway due to metal contamination in the groundwater of Subarnarekha River Basin, India. *Environmental Monitoring and Assessment*, 187(3), 63.

Goerlitz, D. F., Troutman, D. E., Godsy, E. M., Franks, B. J., 1985. Migration of wood-preserving chemicals in contaminated groundwater in a sand aquifer at Pensacola, Florida. *Environmental Science & Technology*, 19(10), 955–961.

Gosselin, R. E., Smith, R. P., Hodge, H. C., 1984. *Clinical Toxicology of Commercial Products.* Baltimore, MD: Williams and Wilkins.

Han, Z., Ma, H., Shi, G., He, L., Wei, L., Shi, Q., 2016. A review of groundwater contamination near municipal solid waste landfill sites in China. *Science of the Total Environment*, 569, 1255–1264.

Hättenschwiler, S., Vitousek, P. M., 2000. The role of polyphenols in terrestrial ecosystem nutrient cycling. *Trends in Ecology & Evolution*, 15(6), 238–243.

Hedbavna, P., Rolfe, S.A., Huang, W.E., Thornton, S.F., 2016. Biodegradation of phenolic compounds and their metabolites in contaminated groundwater using microbial fuel cells. *Bioresource Technology*, 200, 426–434.

Huang, X., Qiu, N., Yuan, D., Huang, B., 2009. A novel stir bar sorptive extraction coating based on monolithic material for a polar, polar organic compounds and heavy metal ions. *Talanta*, 78(1), 101–106.

IARC. 2004. Overall evaluations on carcinogenicity to humans: As evaluated in IARC Monographs.

IRIS. 2006. Phenol. Washington, DC: Integrated Risk Information System. U.S. Environmental Protection Agency. http://www.epa.gov/iris/subst/index.html. March 8, 2006.

IRIS. 2009. USEPA (electronic database), Integrated risk information system. Retrieved on 30 September 2017 from http://www.epa.gov/iris/.

Jenkins, R.A., Tomkins, B., Guerin, M.R., 2000. *The Chemistry of Environmental Tobacco Smoke: Composition and Measurement.* Boca Raton, FL: CRC Press.

Kawahara, J., et al., 2011. Estimation of daily inhalation rate in preschool children using a tri-axial accelerometer: A pilot study. *Science of the Total Environment*, 409(16), 3073–3077. (Retrieved on 29 December 2020).

Kazemi, P., Peydayesh, M., Bandegi, A., Mohammadi, T., Bakhtiari, O., 2014. Stability and extraction study of phenolic wastewater treatment by supported liquid membrane using tributyl phosphate and sesame oil as liquid membrane. *Chemical Engineering Research and Design*, 92(2), 375–383.

Kincannon, D. F., Stover, E. L., Nichols, V., Medley, D., 1983. Removal mechanisms for toxic priority pollutants. *Journal (Water Pollution Control Federation)*, 157–163.

Kochana, J., Gala, A., Parczewski, A., Adamski, J., 2008. Titania sol–gel-derived tyrosinase-based amperometric biosensor for determination of phenolic compounds in water samples. Examination of interference effects. *Analytical and Bioanalytical Chemistry*, 391(4), 1275–1281.

Kovács, Á., Kende, A., Mörtl, M., Volk, G., Rikker, T., Torkos, K., 2008. Determination of phenols and chlorophenols as trimethylsilyl derivatives using gas chromatography–mass spectrometry. *Journal of Chromatography A*, 1194(1), 139–142.

Kulkarni, S. J., Kaware, J. P., 2013. Review on research for removal of phenol from wastewater. *International Journal of Scientific and Research Publications*, 3(4), 1–5.

Kumar, M., Patel, A. K., Das, A., Kumar, P., Goswami, R., Deka, P., Das, N., 2017. Hydrogeochemical controls on mobilization of arsenic and associated health risk in Nagaon district of the central Brahmaputra Plain, India. *Environmental Geochemistry and Health*, 39(1), 161–178.

Kumar, M., Thakur, A. K., Mazumder, P., Kuroda, K., Mohapatra, S., Rinklebe, J., et al., 2020. Frontier review on the propensity and repercussion of SARS-CoV-2 migration to aquatic environment. *Journal of Hazardous Materials Letters*, 1, 100001.

Lapworth, D. J., Baran, N., Stuart, M. E., Ward, R. S., 2012. Emerging organic contaminants in groundwater: A review of sources, fate and occurrence. *Environmental Pollution*, 163, 287–303.

Laxmi Mohanta, V., Naz, A., Kumar Mishra, B., 2019. Distribution of heavy metals in the water, sediments, and fishes from Damodar river basin at steel city, India: A probabilistic risk assessment. *Human and Ecological Risk Assessment: An International Journal*, 26, 1–24.

Li, Y., Trush, M. A., 1994. Reactive oxygen-dependent DNA damage resulting from the oxidation of phenolic compounds by a copper-redox cycle mechanism. *Cancer Research*, 54(7 Supplement), 1895s–1898s.

Lin, C. W., Lai, C. Y., Chen, L. H., Chiang, W. F., 2007. Microbial community structure during oxygen-stimulated bioremediation in phenol-contaminated groundwater. *Journal of Hazardous Materials*, 140(1–2), 221–229.

Little, J. C. 1992. Applying the two-resistance theory to contaminant volatilization in showers. *Environmental Science and Technology*, 26(7), 1341–1349.

Liu, J., Huang, H., Zhong, S., She, X., Yin, D., 2016. Electrochemical simultaneously determination of phenol and o-cresol in water based on ZnO nanosheets. *International Journal of Electrochemical Science*, 11, 3921–3930.

Lupetti, K. O., Rocha, F. R., Fatibello-Filho, O., 2004. An improved flow system for phenols determination exploiting multicommutation and long pathlength spectrophotometry. *Talanta*, 62(3), 463–467.

Mikkonen, A., Yläranta, K., Tiirola, M., Dutra, L. A. L., Salmi, P., Romantschuk, M., Sinkkonen, A., 2018. Successful aerobic bioremediation of groundwater contaminated with higher chlorinated phenols by indigenous degrader bacteria. *Water Research*, 138, 118–128.

Mishra, B. K., Gupta, S. K., Sinha, A., 2014. Human health risk analysis from disinfection by-products (DBPs) in drinking and bathing water of some Indian cities. *Journal of Environmental Health Science and Engineering*, 12(1), 73.

Mohanta, V. L., Singh, S., Mishra, B. K., 2020. Human health risk assessment of fluoride-rich groundwater using fuzzy-analytical process over the conventional technique. *Groundwater for Sustainable Development*, 10, 100291.

Monasterio, R.P., Fontana, A.R., Silva, M.F., 2014. Matrix solid-phase dispersion: A simple and fast technique for the determination of phenolic compounds in olive oil by liquid chromatography. *Analytical Methods*, 6(22), 8986–8995.

Muñoz-de-Toro, M., Markey, C. M., Wadia, P. R., Luque, E. H., Rubin, B. S., Sonnenschein, C., Soto, A. M., 2005. Perinatal exposure to bisphenol-A alters peripubertal mammary gland development in mice. *Endocrinology*, 146(9), 4138–4147.

Naz, A., Chowdhury, A., Mishra, B. K., Gupta, S. K., 2016. Metal pollution in water environment and the associated human health risk from drinking water: A case study of Sukinda chromite mine, India. *Human and Ecological Risk Assessment: An International Journal*, 22(7), 1433–1455.

NIOSH. 2005. Phenol. NIOSH pocket guide to chemical hazards. Atlanta, GA: National Institute for Occupational Safety and Health, Centers for Disease Control and Prevention. http://www.cdc.gov/niosh/npg/. March 8, 2006.

NTP. 2005. Report on carcinogens. 11th ed. Research Triangle Park, NC: National Toxicology Program, Department of Health and Human Services. http://ntp-server.niehs.nih.gov/ntp/roc/toc11.html. January 11, 2008.

Olak, M., Gmurek, M., Miller, J.S., 2012. Phenolic compounds in the environment-occurrence and effect on living organisms. *Proceedings of ECOpole*, 6. OSHA. 2005. Air contaminants. Occupational safety and health standards for shipyard employment. https://www.osha.gov/laws-regs/regulations/standardnumber/1915/1915.1000.

OSHA. 2006. Code of Federal Regulations. 29CFR 1915.1000. Washington, DC: Occupational Safety and Health Administration. http://www.osha.gov/comp-links.html. March 08, 2006.

Petrasek, A. C., Kugelman, I. J., Austern, B. M., Pressley, T. A., Winslow, L. A., Wise, R. H., 1983. Fate of toxic organic compounds in wastewater treatment plants. *Journal (Water Pollution Control Federation)*, 1286–1296.

RAIS: Risk Assessment Information System. 2012. Retrieved on 9 October 2019 from http://www.rais.ornl.gov.in.

RAIS. 2016. Regional screening level-generic table. Retrieved on 27 March 2018 from https://www.epa.gov/risk/regional-screening-levels-rsls-generic-tables.

Reitzel, L. A., Ledin, A., 2002. Determination of phenols in landfill leachate-contaminated groundwaters by solid-phase extraction. *Journal of Chromatography A*, 972(2), 175–182.

Sahajanand, K., Meenal, M., Prasad, G., 2020. Flow and transport of Phenol in groundwater using PGWT equation. *Heliyon*, 6(2), e03413.

Sarafraz-Yazdi, A., Dizavandi, Z. R., Amiri, A., 2012. Determination of phenolic compounds in water and urine samples using solid-phase microextraction based on sol–gel technique prior to GC-FID. *Analytical Methods*, 4(12), 4316–4325.

Schweigert, N., Hunziker, R. W., Escher, B. I., Eggen, R. I., 2001. Acute toxicity of (chloro-) catechols and (chloro-) catechol-copper combinations in Escherichia coli corresponds to their membrane toxicity in vitro. *Environmental Toxicology and Chemistry: An International Journal*, 20(2), 239–247.

Sharma, S., Bhattacharya, A., 2017. Drinking water contamination and treatment techniques. *Applied Water Science*, 7(3), 1043–1067.

Sheets, F. 2007. Breast Cancer & the Environment Research Centers.

Shim, J., Mazumder, P., Kumar, M., 2018. Corn cob silica as an antibacterial support for silver nanoparticles: Efficacy on Escherichia coli and Listeria monocytogenes. *Environmental Monitoring and Assessment*, 190(10), 1–10.

Singh, A., Patel, A. K., Deka, J. P., Kumar, M., 2020. Natural recharge transcends anthropogenic forcing that influences arsenic vulnerability of the quaternary alluviums of the Mid-Gangetic Plain. *NPJ Clean Water*, 3(1), 1–12.

Stuermer, D.H., Ng, D.J. Morris, C.J., 1982. Organic contaminants in groundwater near an underground coal gasification site in northeastern Wyoming. *Environmental Science & Technology*, 16(9), 582–587.

Sun, R., Wang, Y., Ni, Y., Kokot, S., 2014. Spectrophotometric analysis of phenols, which involves a hemin–graphene hybrid nanoparticles with peroxidase-like activity. *Journal of Hazardous Materials*, 266, 60–67.

Thurman, C., 1982. Phenol. In: *Kirk-Othmerencyclopedia of Chemical Technology*, 3rd ed., Vol 17. New York: John Wiley and Sons, pp. 373–384.

USEPA. 1997. *Exposure Factors Handbook*. General factors, EPA/600/P-95/002Fa, 1, Washington, DC: USEPA.
USEPA. 1998. Hazardous waste management system; identification and listing of hazardous waste; solvents; final rule. Federal Register 63 FR 64371-402. (Retrieved on 29 December 2020)
USEPA. 1999. Integrated Risk Information System (IRIS) on Phenol. National Center for Environmental Assessment, Office of Research and Development, Washington, DC. (Retrieved on 29 December 2020).
USEPA. 2004. Risk assessment guidance for superfund volume I: Human health evaluation manual (Part E, Supplemental Guidance for Dermal Risk Assessment) Final. EPA/540/R/99/005. (Retrieved on 29 December 2020).
USEPA. 2005. Guidelines for Carcinogen Risk Assessment, EPA/630/P-03/001F. Risk Assessment Forum, Washington, DC. (Retrieved on 29 December 2020).
USEPA. 2008. Toxicological profile for Phenol. Agency for Toxic Substances and Disease Registry. Retrieved on 29 December 2020 from https://www.atsdr.cdc.gov/toxprofiles/tp115.pdf.
USEPA. 2015. Risk based screening table-generic table. Retrieved on 24 September 2015 from http://semspub.epa.gov/work/03/2218450.pdf#_ga=1.192660266.1609005446.1445688363.
Vázquez-Suñé, E., Carrera, J., Tubau, I., Sánchez-Vila, X., Soler, A., 2010. An approach to identify urban groundwater recharge.
Vom Saal, F. S., Hughes, C., 2005. An extensive new literature concerning low-dose effects of bisphenol A shows the need for a new risk assessment. *Environmental Health Perspectives*, 113(8), 926–933.
Wallis, P. M., Hynes, H. B. N., Telang, S. A., 1981. The importance of groundwater in the transportation of allochthonous dissolved organic matter to the streams draining a small mountain basin. *Hydrobiologia*, 79(1), 77–90.
Wolf, L., Held, I., Eiswirth, M., Hötzl, H., 2004. Impact of leaky sewers on groundwater quality. *Acta hydrochimica et hydrobiologica*, 32(4–5), 361–373.
York, R. G., 1997. Oral (gavage) developmental toxicity study of phenol in rats. Proctor &.
Zemann, A., Volgger, D., 1997. Separation of priority pollutant phenols with coelectroosmotic capillary electrophoresis. *Analytical Chemistry*, 69(16), 3243–3250.

Chapter 14

Sustainable governance of groundwater as drinking water in the post-COVID-19 world

An urban water security perspective

Subham Mukherjee
Freie Universität Berlin

Pradip Kumar Sikdar
Indian Institute of Social Welfare and Business Management

Sukdeb Pal
CSIR-National Environmental Engineering Research Institute (CSIR-NEERI)
Academy of Science and Innovative Research (AcSIR)

CONTENTS

DOI: 10.1201/9781003157465-17

14.1 INTRODUCTION: COVID-19 PANDEMIC AND URBAN WATER SECURITY

Novel coronavirus disease 2019 (COVID-19) is a respiratory infectious disease caused by a newly discovered or 'novel' coronavirus. There is no evidence as of now to prove COVID-19 is related to drinking water. Neither the novel coronavirus has been detected in drinking water. To keep the drinking water protected from any contamination and maintain its quality, guidelines for drinking water, in general, employ a multibarrier approach to ensure safe and quality drinking water for the citizens. They are designed to safeguard savouring drinking water from three different aspects:

1. **Safeguarding source water**: Water utilities get their drinking water from the best quality and most secured sources possible. This lessens or eliminates the danger of contaminating from entering the water system in any case.
2. **Treatment**: Water utilities use filtration and additional purification with chlorine to treat drinking water whenever necessary. Chlorine is exceptionally compelling in killing biological organisms including viruses, and new research suggests that this method will be compelling in eliminating COVID-19 too.
3. **Monitoring**: Water utilities assemble water test samples at any rate from hourly to monthly. In case any possible contamination is detected, the guidelines and further planning can be shaped to advise the beneficiaries and recommend steps necessary to guarantee their security. Unfortunately, numerous water systems utilizing groundwater wells do not have perpetual purification techniques inbuilt in India (Singh, 2017). This does not imply that these sources are in danger, yet they may show even more challenges to manage as the extra hindrance may be fatally disruptive for the recommended Water, Sanitation and Hygiene (WaSH) guidelines during and after COVID-19 pandemic (DOH, 2020).

14.1.1 Groundwater as drinking water

A recent study shows that globally around 2.5 billion individuals exclusively depend on groundwater to meet daily drinking water requirements (Grönwall and Danert, 2020; Singh et al., 2020a,b). The dependence on this subsurface resource and its centrality to appreciate the basic right to 'safe' drinking water has increased profoundly. Nevertheless, this appreciation has not been widely recognized at the national level by the governments and policymakers. Utilization of groundwater has been fundamental to human development, from both cultural and ecological perspectives, for their reliance on groundwater and its related ecosystem services since the beginning of the human history (Singh, 2017; Velis et al., 2017; Patel et al., 2020). For example, production of food and energy from groundwater-fed agriculture, groundwater as a main source of safe drinking water, sanitation and hygiene as well as recreational and cultural activities including some religious rituals associated with the oasis in the dry desert areas (Velis et al., 2017). Its vital significance for worldwide water and food security is most likely to get intensified under rapidly shifting environmental as well as sociocultural scenarios as more continuous and serious climate change-related fluctuation in precipitation, surface water availability and soil moisture will occur more frequently (Velis et al., 2017). The dependence on water from subsurface water sources, such as springs, has stretched out a more complex way over the past 50 years, with the advancement in technology to drill and pump (e.g. introduction and augmentation of better developed techniques such as jetting, percussion, auger and sludging) for instantaneous access to groundwater (van der Wal, 2010; World Bank, 2010; Kumar et al 2020a,b). The general increase in public health awareness and related societal values should have been more moved essentially towards the conservation of this vital resource. In many places on earth, the comprehension of the asset's vitality is exceptionally limited, especially the issues related to the availability, the maximum level of abstraction and the recharge to replenish even among the decision-makers. Lack of information implies that data on groundwater stay as obscure as the resource itself. These result in jeopardy as behaviour and extent of the contamination present in groundwater, their interrelationship and interaction with the present and shifting patterns of land management and prediction of the emerging contaminations in the drinking water are all not clear and confirmed yet. Till date, groundwater is being extensively used as the main source of drinking water worldwide, but a proper level of awareness and account for groundwater quantity and quality has not been achieved (Grönwall and Danert, 2020).

14.1.2 Crisis: pandemic-issues and water insecurity

A widely available and accessible WaSH provision is essential for the concealment and treatment of COVID-19 (Cooper, 2020b). Coronavirus is an infection with no demonstrated treatment or drug intervention and no demonstrated antibody detected so far. In such conditions, care and prevention are all we have. The part of handwashing in the avoidance of transferable diseases has been known for longer than a century, yet it remains seriously ignored as a general well-being mission, to be occasionally resurfaced during pandemic-scale diseases. Over 26% of the worldwide populace has no provision for a handwashing mechanism in their houses (Ray, 2020; Kumar et al., 2021); for some developing nations, this extent rises to even over a half of the total population (Ray, 2020). In some cases, the clean water is unreasonably expensive, or the compliantly has been stopped by virtue of unpaid bills (Mukherjee et al., 2020; Ray, 2020). In any case, when there is no water in the home or nearby premises, or no system for ensuring enough and safe water, widely advised handwashing is just incredibly difficult (Mukherjee et al., 2020; Ray, 2020). Even before COVID-19, various analyses worldwide already showed the fallacy of the absence of necessary funds, lack of investment in providing enough WaSH provisions and the low number of people under the available schemes for the benefits of having WaSH provisions at their acceptable range (UN, 2014; WWAP, 2018; Mukherjee et al., 2020), indicated the crisis from a different point of view which this current pandemic took advantage of and spread more widely. This aspect is not a local issue any further and proved to be the essential for international governance and investors to organize and substantially reserve reliable, affordable, acceptably safe and enough water to be supplied to the deprived areas of the world. COVID-19 pandemic is an unprecedented global phenomenon and needed to deal with enough dignity, respect and considerations (Cooper, 2020b).

This chapter analyses the present pandemic strategies and endeavours to identify the groundwater as drinking water within the framework of urban water security issues, focusing on India, which may face problems relating to groundwater and subsequently become a crucial factor for drinking water shortage in the vicinity. This study attempts to delve into governance aspects that mainly govern groundwater availability, accessibility, quality and related environmental risks in any urban India.

14.2 GROUNDWATER-RELATED ISSUES FOR URBAN AREAS

14.2.1 Availability

The availability and occurrence of sub-surface water in the Indian subcontinent is overly complex because of the existence of differentiated geological formations with significant variations in lithological and chronological characteristics, complex tectonic background, climatological variations and several hydrogeochemical conditions. Previous researches have showed that aquifer groups in alluvial or soft rocks even exceed the surface basin boundaries (CGWB, 2012). Some of the major basins (e.g. Brahmaputra, Ganga and Indus basins) in India (broadly, the Indian subcontinent) accommodate potential aquifers with high yielding capacity (Mukherjee et al., 2015; Kumar et al., 2017; Singh et al., 2020a,b). Nevertheless, the groundwater level that could be economically extracted varies considerably throughout the region and the continuous availability of pure water from many of these aquifers (e.g. part of Brahmaputra and Ganga basin in West Bengal, known as Bengal Basin) is hindered by the presence of geogenic contaminants (Sikdar, 2018). Moreover, the transboundary nature of some of these aquifers makes sub-surface water resources a potential politically sensitive issue. Being one of the heavily populated areas in the world, the use of groundwater is also extremely high in the Bengal Basin (Mukherjee et al., 2015). The major hydrogeological units in India are depicted in Figure 14.1. For details, please see Jain et al. (2007) and the Aquifer Atlas of India by Central Ground Water Board (CGWB) (2012).

Preventing and curing a pandemic requires enough safe water for all as hand hygiene is critical for lessening transmission of contagious diseases like COVID-19 and handwashing with polluted water may increase the risk of infection (Hannah et al., 2020). Therefore, water demand for all will increase during a pandemic for both the health issues as well as the general consumption. Demand for clean water increased during the previous global health crises such as Ebola and plague in West Africa during 2014 (Cooper, 2020a). This increased demand for water for domestic purposes and public health could pressure on both downstream for wastewater generation and upstream for negotiating trade-offs between agriculture and other industries. Redirecting water from agroindustry for use in urban communities could influence food creation, while keeping up water for farming could imply that urban areas do not have sufficient supplies to battle pandemics. The actual picture of the reliance on groundwater and its extent in the Indian urban areas is not still clear due to the lack of proper and valid data at the household level. In Indian urban households, groundwater is supplied through three main methods: (i) public supply by the local urban authority that is solely dependent on groundwater for the source of water, (ii) private access to borewells or shallow wells inside/close to premises and (iii) semiformal and casual trading of direct sub-surface water. While the official documents from the responsible authorities (including CGWB and the Ministry of Water Resources, River Development and

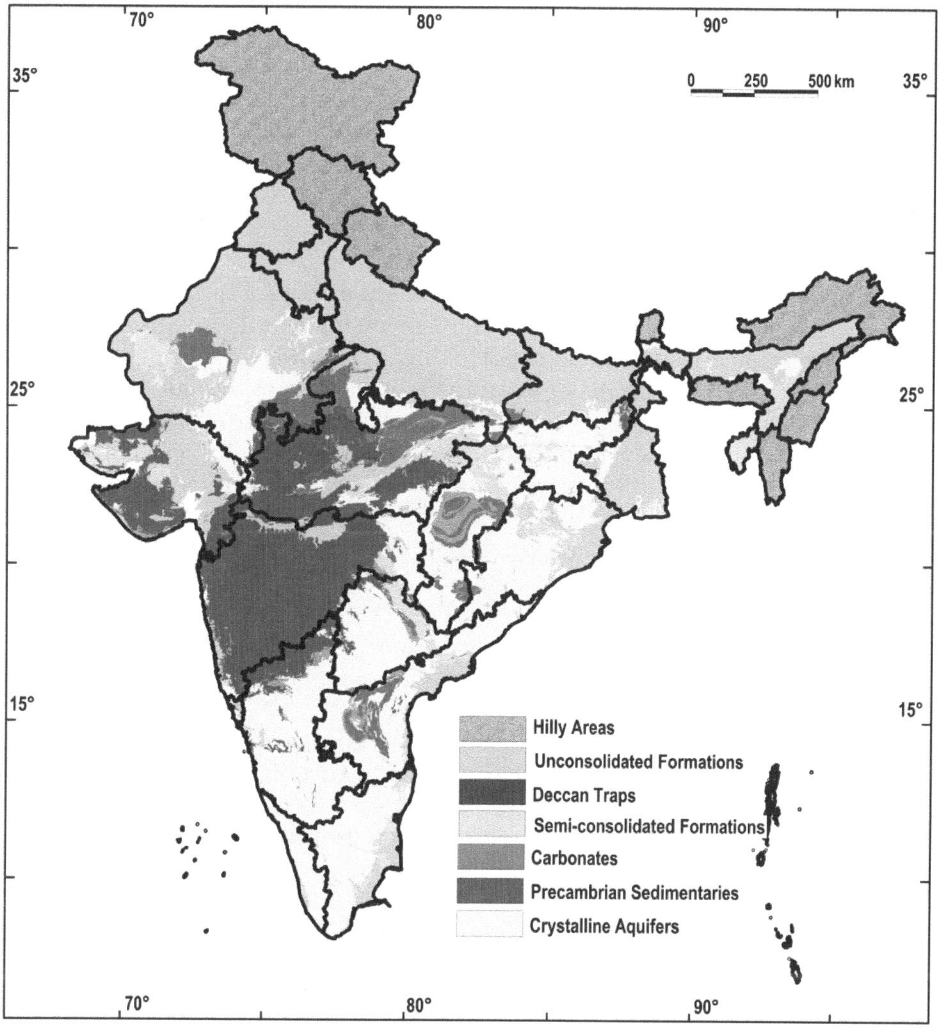

Figure 14.1 Hydrogeological map of India (Adapted from CGWB, 2012).

Ganga Rejuvenation) on urban water supply and its reliance on groundwater are having disagreements regarding actual and estimated demand–supply of water and generation of wastes, researchers found that more than half of the urban households in 71 studied urban areas in India during 2011–2012 were reliant on groundwater for various purposes including drinking. Furthermore, almost 33% of households across India are legitimately dependant on groundwater for drinking purposes, and the households in smaller urban areas are occupying maximum portion of it.

This sudden and huge increase in water demand may also increase more struggle among the already contesting drivers such as rapidly growing population, sporadic urbanization and shifting pattern of consumption (WWAP/UN-Water, 2018). Every year, water demand is raising at a rate of about 1% and the most of it goes for agriculture (WWAP/UN-Water, 2018). With the increasing pressure from the rapidly changing climate, continuous deforestation for both commercial and agriculture and encroachment of lands turning into paved areas is already put several burdens on the water availability and quality worldwide and the gravity of danger can already be felt during this pandemic (UNEP, 2016). Natural environmental change is reshaping the worldwide water cycle bringing about magnified unpredictability in freshwater availability, most of which is sourced from the groundwater. This incorporates shifting patterns of precipitation designs, melting glaciers, changing river courses and extended surges in both water scarcity and quality and inevitably, resulting stress (Smith et al., 2019; Cooper, 2020a).

The reliance on groundwater is essentially based on the subsurface geology of the urban areas. Urban areas situated on the alluvium plains and with canals have the best natural aquifers underneath and potential for high groundwater recharge capacity. On the contrary, urban areas on the unfractured crystalline rock are not capable of the natural development of sub-surface water tables (Banks et al., 2002). Therefore, these urban areas are bound to depend on surface waterbodies more than groundwater. Urban areas from the Deccan Peninsula are the distinct examples of this latter type.

The urban areas (cities/towns) in India situated on the rocky aquifers such as crystalline rocks, basalt and limestone, covering 30%, 15% and 3%, respectively, of a total geographical area of the country, lack enough groundwater to use, and hence, these urban areas widely depend upon surface water (Table 14.1).

Consequently, rapid and sporadic urbanization in these areas put direct stress on surface water that ultimately pressures heavily on the nearby irrigation reservoirs. These phenomena directly impact the water productivity through the inefficient use of water for the urban households and industries as well as indirectly impact the groundwater storage and quality because the growing urban area surpasses the supply capacity of the local aquifer, mainly for the drinking purposes; the quality of groundwater deteriorates with increased contamination through geogenic processes. Thus, inadequate (in both quantity and quality) local groundwater sources drive the urban areas to import water from outside the urban limits (World Bank, 2003).

Basin-wise groundwater supply of the urban areas in India (Table 14.2) reveals that Ganga, Indus and Brahmani-Baitarni basins have the highest groundwater supply (>60%). Among other basin-level urban areas in India, Tapti river basin has no

Table 14.1 Relationship between aquifer and reliance on groundwater

	Aquifers					
	Alluvium and sandstone discourse	*Aquifer in hilly areas*	*Basalt*	*Crystalline rocks*	*Extensive alluvium*	*Limestone*
No. of towns	78	19	43	70	84	2
Average of groundwater supply in overall water supply	44.1	47.34	8.23	21.1	74.84	42.12

Source: NIUA (2005).

Table 14.2 Basin wise groundwater supply in India

		Number of cities/towns	Average groundwater supply (%)
Basin	Barak	5	11
	Brahmani-Baitarni	3	66.7
	Brahmaputra	5	21.8
	Cauvery	17	7.4
	Ganga	109	67
	Godavari	18	5.4
	Indus	21	66.0
	Krishna	26	14.0
	Luni	16	36.0
	Mahanadi	5	27.6
	Mahi	4	51.0
	Narmada	5	28.2
	Pennar	8	48.0
	Sabarmati	3	40.9
	Tapti	5	0

Source: NIUA (2005).

groundwater supply. Nevertheless, Mahi, Sabarmati and Pennar have adequate groundwater supply (>40%). As groundwater supply heavily depends on the aquifer characteristics as mentioned earlier, Krishna, Cauvery and Godavari river basins that majorly pass through rocky aquifers in the Deccan Peninsula, due to the less retention capacity of subsurface, groundwater supply is poor.

14.2.2 Accessibility and human right to water

While groundwater is used practically to meet up 50% of all drinking water worldwide, around 2.5 billion individuals rely exclusively upon this asset to fulfil their daily water requirements (Kulkarni et al., 2015; Pallavi et al., 2015). Accordingly, it is essential to apprehend groundwater reliance, quantity and quality issues and estimate the requirement. However, such information is likely to be an underestimation as valid measurement often lacks data on all sources utilized at the household level (World Bank, 2006). While almost 85% of the Indian population depends on groundwater for drinking, hand pumps or borewells are the main sources of drinking water in rural areas of both India and Bangladesh (World Bank, 2006, 2010; WWAP/UN-Water, 2018). In developing sub-Saharan nations, including Somalia, Central African Republic, Burkina Faso, Nigeria, Chad, Ethiopia, South Sudan and Uganda, groundwater is the only source of drinking water for almost 80% of the total households (Grönwall and Danert, 2020). In wealthy countries, such as the USA, 38% of the population directly relies upon groundwater for drinking water purposes. The demand is met either from public wells or from privately installed wells. In Southeast Asian and Pacific countries, 79% of the population drink groundwater (WWAP/UN-Water, 2018; Grönwall and Danert, 2020).

Groundwater for domestic use can be drawn from various ways, particularly for deprived urban areas, commonly known as *slums*. It may be drawn or siphoned from a solitary well (i.e. scoop openings, burrowed wells or shallow or profound boreholes/handpumps) which is either private or shared with a network, a neighbour or a non-governmental organization (NGO). The well may be associated with a 'water ATM' (a computerized machine controlling drinking water abstraction) or groundwater may be conveyed through water booths, standpipes or casual delivery trucks and big haulers (Mukherjee et al., 2020). It can be devised to be disconnected from tube wells, borewells or springs and provided as a major part of a public/private water supply framework. In any case, individual groundwater-based water systems should not be encouraged and must be widely controlled by the governments or regulatory bodies in order to minimize the exploitation and increase the accessibility even to the poorest. Often, groundwater is utilized distinctly for other household purposes than drinking; consideration and control should be there on both quality and quantity which are mostly ignored or overlooked for the deprived areas, especially during the pandemic. Other than that, the human rights to water and related legal issues with groundwater should be confirmed and governed by the public authority(s) or responsible vendors acting on behalf of public–private partnership (PPP) arrangements, with the end goal that is 'self-provided' (Mukherjee et al., 2020).

Standard approaches of mainstream policy recommendations do not consider groundwater as the most critical source of drinking water for households. The significance of groundwater as a vital resource is not extremely obvious in the discourses on how the world is going to ensure enough, affordable and safe drinking water for all by 2030 or achieve the overall Sustainable Development Goals (SDG). Additionally, scarcity of good and reliable data on groundwater from the valid sources also play a crucial role in proper assessment and the absence of which can result in under-researched or incomplete policy documents. This crucial gap in groundwater data has jeopardized the research and policy documents and normative statements from the reputed international agencies such as the United Nations (UN) by prohibiting reports like *United Nations (UN) World Water Development Report for 2022 on groundwater*. The World Bank has made a thorough study on this data crisis in 2019 and reflected their concern over groundwater data availability and reliability issues (Mukherjee et al., 2020).

It is the human right that everyone should have physical access to sufficient, acceptable, affordable and safe water for individual and household use. Alternatively, human right to sanitation (a system for collection, transport, treatment, disposal or reuse of human excreta and associated hygiene, according to UN, 2014) empowers everyone to have physical access to safe, protected, hygienic, affordable, socially and culturally acceptable sanitation services which, most importantly, provide privacy and ensure dignity (UN, 2014; Mukherjee et al., 2020). Decades-long opinion about the content and implications of these rights has been formally recognized as the human right to water and sanitation (HRWS) in 2010 by UN member states (Mukherjee et al., 2020). According to the recent literature, water security can be referred to 'the secure, adequate, and sustainable access that people and ecosystems have to water, including the equitable distribution of advantages/disadvantages related to water use and development opportunities, the safeguarding against water-based threats, and the ways of sharing decision-making power in water governance' (Grey and Sadoff, 2007, p. 20; Boelens et al., 2013, p. 238; Thakur et al., 2020). Right to water security includes rudimentary rights and benefits, conceptualized in accordance with a given frame for interpreting them although lack of detailed data on groundwater fails the right that can respect, protect and fulfil groundwater use and management from the aspects of the specific content of the right or the responsibilities of governments (Grönwall and Danert, 2020).

14.2.3 Quality

The presence of contaminants in drinking water is one of the most concerning issues for every household. The sources and pathways of these contaminants as well as pathogens (including the novel coronavirus/SARS-CoV-2) are many (Kumar et al., 2020a; Kumar et al., 2020b; Mohapatra et al., 2020). It can come from the supplied water itself to the households due to the leakages in pipes or even old age of the entire pipe system which is a common phenomenon in the older urban areas across India. However, public water supply systems are accountable for and bound to check the filtration and distillation system thoroughly before the supply, whereas, for the private supply agencies, this accountability is not always guaranteed. For the households, where groundwater is directly used for drinking, the sources of contamination are different. The sources can be traced to the associated aquifers in these cases. Since traces of coronavirus have been found in wastewaters, it is crucial to notice that direct outlets of wastewater in the surface waterbodies, such as lakes, canals and rivers, can also pollute groundwater in the nearby unconfined aquifers. Possibility of this sort of pollution can be higher in the megacities or areas where excessive and illegal pumping groundwater occurs.

Besides many anthropogenic sources, such as contaminations through leakages in domestic sewage system underground or dumping of wastes in the surface waterbodies (e.g. lakes in megacities like Bengaluru) where artificial groundwater recharge system is installed (Gronwall et al., 2010; Kozhisseri, 2005; Singh et al., 2010), some natural traces of contaminants are also found in the aquifers. Geogenic substances like fluoride and arsenic in aquifers are a genuine reason to worry in parts of West Bengal, Andhra Pradesh, Rajasthan and Gujarat (Chowdhury et al., 2000; Sikdar, 2018). Additionally, effluents from the industries, such as textile industries in Tirupur (Senthilnathan and Azeez, 1999), small, medium and large industries in Ludhiana (Kakar and Bhatnagar, 2008) and Coimbatore (Mohanraj et al., 2000) pollute aquifers significantly. Other than these, anthropogenic pollutants like organochlorine pesticides including prohibited ones like DDT and HCH were detected in the groundwater of Delhi (Mukherjee and Gopal, 2001; Mutiyar et al., 2011), Hyderabad (Shukla et al., 2006) and Jaipur (Bakore et al., 2004).

Lack of infrastructural facilities can be linked with contamination in groundwater. Open defaecation or existence of pit latrines can be the source of organic contamination in the nearby aquifers particularly when the groundwater table is higher during the monsoon in India. These examples can be found in the added areas of a megacity or urban fringe areas. Moreover, there is the chance of contaminating drinking water inside the house even with improved public sources. Studies show that there is a probability of water pollution associated with storage systems at the household level (Jensen et al., 2002). Particularly in the deprived areas (generally designated as slums) in a city, this problem is more critical even if the source of the collection of the publicly supplied treated water is safe and properly monitored (Palit et al., 2012). According to the government and other international official reports, just half of the urban households have water filters or similar facilities (NSSO, 2013). Lack of data and information available on the public domain regarding the quality issues of supplied and used drinking water also aggravates the severity of the urban water security issues.

14.2.4 Environment and risk factors

At the household level, groundwater extraction from the private wells is the cheapest and the most reliable way for access to drinking water when there is either irregularity in public water supply services or affordability is an issue in the case of private water vendors. Most of these private wells are not even been known or notified by the outsiders than by the household(s) or the locality involved in the illegal existence and usage of the private wells. In Chennai, it was found that the price of the water from the privately owned tankers was even 50 times more than the municipal water supply or using a private well (Srinivasan et al., 2010). It is surely a question of affordability even for the owning of a private well for many households. It is also an issue of land, particularly for the deprived areas where land tenure is a big issue of insecurity. Despite having enough surface water for supply, to reduce the price for the treatment and related maintenance costs and heavier investment requirements, sometimes, public supply services also directly distribute groundwater to meet the increasing demand of growing urbanization (World Bank, 2010; Mukherjee et al., 2018; Mukherjee et al., 2020).

Thus, without sufficient and regular public water supply facilities, informal and illegal groundwater markets have been developed in certain pockets of the urban areas across India. These informal and basically unorganized business sectors have not been sufficiently studied and focused yet for further development at the policy level. Therefore, these unorganized and overlooked sectors continue to remain outside the domain of policy and planning guidelines. Nevertheless, departments of urban water in both Chennai and Kolkata municipal corporations have arranged private distributors to pay for groundwater for supplying in the added peri-urban areas in the vicinity of the main city through tankers (Srinivasan et al., 2010) to lessen the pressure on the existing and depleting water tables.

This widespread and almost abundant reliance on groundwater can be linked to some laws that authorize the landowners of households or private companies or even from the government body to extract limitless water from aquifers. In the Easement Act of 1882, water has been recognized as an easement of added benefit inseparably linked with land. This is further strengthened by the Transfer of Property Act of 1882 and the Land Acquisition Act of 1894 (Iyer, 2007; Saleth, 2009). It has been an uphill task for the Central Groundwater Authority to regulate such arbitrary withdrawal due to a large number of borewells and non-availability of alternate sources in lots of areas. Meanwhile, cities are increasingly going beyond their administrative limits in search of groundwater. For example, the water utility service in Chennai pays for groundwater to the farmers in peri-urban areas.

The increasing reliance on groundwater has numerous negative outcomes: the most obvious outcome is falling of urban water table. Water tables are typically depleting in newly added areas that are profoundly reliant on groundwater because of the lack of infrastructural facilities or services, while water levels can rise or stay unchanged in the other parts, predominantly the older parts where the water services are comparatively stable and often connected with piped water supply. This increase in groundwater reliance has a few outcomes: the most unmistakable one being exhaustion of metropolitan springs. With regard to metropolitan groundwater, the predominant story one finds in the mainstream media is one of quickly draining metropolitan springs, but considerably more nuanced. Inside a similar urban area, water tables are normally depleting in those areas that are intensely subject to groundwater, as water levels are ascending in the more seasoned areas that mostly connected with piped supply. This variety is fundamentally a result of high paces of spillage from supply pipes and wastewater pipelines (CGWB, 2011). Comparatively smaller urban areas (towns) than megacities, like Lucknow in Uttar Pradesh; Jodhpur (Paliwal and Baghela, 2006) and Ajmer in Rajasthan (Jat et al., 2009); Pune and Solapur (Naik et al., 2008) in Maharashtra; and Hubli-Dharwad (Hollingham, 2008), Bangalore (CGWB, 2008; CGWB, 2010) and Mulbagal in Karnataka appear to adjust to this characterization.

Excessive and unplanned groundwater extraction pressures on the aquifer continuously even beyond its limit. Mostly, coastal urban areas that are intensely reliant on groundwater face the danger of saltwater intrusion because of the sudden change in pressure brought about by continuous and increasing groundwater extraction. There are many pieces of evidence of the saltwater intrusion along the coastal urban areas in India for the last few decades, such as Tamil Nadu (Ramesh et al., 1995), Kerala and Gujarat (Moench, 1992; Shah, 2008). Extreme extraction of sub-surface water can equally cause interruption of polluting rivers, canals or lakes. Furthermore, the pressure on the groundwater increases, becoming the reason for land subsidence in the urban areas, consequently (Sikdar et al., 1996; Sahu and Sikdar, 2011).

14.3 PERSPECTIVES OF GROUNDWATER AS DRINKING WATER IN URBAN INDIA: SDGs, DRINKING WATER STANDARD AND GROUNDWATER LAWS

Previously, the development pathway of India has been directed by the key priorities of generating employment; enhancing economic growth; and meeting food, water and energy security, disaster resilience and poverty mitigation. Restoration of its natural capital and adopting transparent and robust governance following democratic lines have also been attempted (Mukherjee et al., 2018). Emerging challenges such as impacts due to climate change, increasing inequalities and lagging human development indices are also well recognized by both the citizens and the government. The post-2015 UN Sustainable Development Agenda framework offers an opportunity to regenerate and incorporate attempts to meet national and global objectives in a specified timeframe (Mukherjee et al., 2018; Mukherjee et al., 2020).

The SDG 6 demands for collaboration between the water and water users and shift beyond their conventional disintegrated 'silos' to integrate the ecosystems into consideration and manage water to achieve the already defined impact-oriented goals. For example, SDG 6.6 specifically requires nations to 'protect (by 2020) and restore water-related ecosystems, including mountains, forests, wetlands, rivers, aquifers and lakes' (Mukherjee et al., 2020).

While the overall proportion of households in India with access to improved sources of clean water was 68% in 1992–1993, after almost two decades the figure has increased to 90.6% in 2011–2012. On the contrary, still 8% of urban households did not have admittance to improved sanitation facilities in 2012. The number of open defaecations in India was almost 600 million—the highest in the world. Improving sanitation is a key priority of the government.

14.3.1 Drinking water quality

Drinking water includes treated or untreated water supplied or procured for human consumption. While specifications for drinking water quality in India (IS 10500:2012) are prescribed by the Bureau of Indian Standards, surface water classification for designated best use based on quality (Table 14.3) is provided by the Central Pollution Control Board (CPCB).

Indian subcontinent has diverse geological formations from oldest Achaeans to recent alluviums (CGWB, 2012). Various parts of the country are characterized by varying climatic conditions as well. Sub-surface geological formations and the depth of the sediments through which groundwater remains in contact influence the natural chemical content of groundwater. In the major part of the country, groundwater is of as such of good quality and suitable for drinking, agricultural or industrial purposes. Sub-surface water in shallow aquifers, predominantly of calcium bicarbonate and mixed type, is generally suitable for use for different purposes. Other types of water (such as sodium chloride-type water) are also available (CGWB, 2012). Although groundwater quality in deeper aquifers similarly fluctuates from place to place, it is usually found appropriate for public uses. Salinity problems in the coastal expanses and high incidences of geogenic contamination (fluoride and heavy metals, etc.) have also been reported. Table 14.4 provides a list of various geogenic contaminants found in groundwater in different parts of the country along with their health implications.

Table 14.3 Surface water classification based on designated best use set by the Central Pollution Control Board, India

Designated best use	*Class of treated water*	*Criteria*
Drinking water source without conventional treatment but after disinfection	**A**	• TC $\leq$ 50 MPN/100 mL • pH between 6.5 to 8.5 • DO $\geq$ 6mg/L • BOD (5 days 20°C) $\leq$ 2 mg/L
Organized outdoor bathing	**B**	• TC $\leq$ 500 MPN/100 mL • 6.5 $\leq$ pH $\leq$ 8.5 • DO $\geq$ 5 mg/L • BOD (5 days 20°C) $\leq$ 3mg/L
Drinking water source after conventional treatment and disinfection	**C**	• TC $\leq$ 5,000 MPN/100 mL • pH between 6.0 and 9.0 • DO $\geq$ 4 mg/L • BOD (5 days 20°C) $\leq$ 3mg/L
Propagation of wildlife and fisheries	**D**	• pH between 6.5 and 8.5 • DO $\geq$ 4 mg/L • NH_3 - N $\leq$ 1.2 mg/L
Irrigation, industrial cooling, controlled waste disposal	**E**	• pH between 6.0 and 8.5 • EC (at 25°C) $\leq$ 2,250 µS/cm • SAR $\leq$ 26 • B $\leq$ 2 mg/L
	Below-E	Not meeting any of the above criteria

TC, Total coliform; DO, Dissolved oxygen; BOD, Biochemical oxygen demand, EC, Electric conductivity; SAR, Sodium absorption ratio; B, Boron

Table 14.4 Various geogenic contaminants found in groundwater according to Central Ground Water Board (CGWB) (CGWB, 2012)

Contaminants			*Situation*
Salinity	inland salinity	• Prevalent mainly in arid and semi-arid areas of Rajasthan, Haryana, Punjab, Gujarat, Uttar Pradesh, Delhi, Andhra Pradesh, Maharashtra, Karnataka and Tamil Nadu • EC ≥ 10,000 mS/cm at 25°C in various places in Rajasthan and southern Haryana • Can also be caused due to surface water irrigation without considering the status of groundwater • Gradual rise in sub-surface water levels leads to waterlogging • Heavy evaporation in semi-arid regions leads to salinity problems	• Makes groundwater non-potable • Well water used directly for salt manufacturing by solar evaporation
	coastal salinity	• Dynamic coastline of India is ~ 7,500 km • Western coastal areas are characterized by wide continental shelf and marked with backwaters and mudflats • Eastern coastal areas have a narrow continental shelf and are characterized by deltaic and estuarine landforms • Groundwater in coastal areas occurs under unconfined or confined conditions at large scale of unconsolidated and consolidated formations • Origins of saline waterbodies: entrapped seawater (connate water), seawater ingress, leachates from navigation canals constructed along the coast, leachates from salt pans, etc.	• Salinity problems are observed at various locations of the coastal areas • Following situations are encountered in coastal areas: a. Saline water overlying freshwater aquifer b. Freshwater overlying saline water c. Alternating sequence of freshwater and saline water aquifers
Fluoride		• At places found in high concentration in groundwater beyond the permissible limit of 1.5 mg/L	• Consumption of fluoride-rich water poses health problem
Arsenic		• Occurs naturally as an element of the earth's crust • Reacts with oxygen, chlorine and sulphur to form inorganic arsenic compounds	• Transport and distribution of arsenic in environment is complex continuous

(*Continued*)

Table 14.4 (Continued) Various geogenic contaminants found in groundwater according to Central Ground Water Board (CGWB) (CGWB, 2012)

Contaminants		*Situation*
	• These compounds are widely used in medicine, agriculture, electronics, livestock feed, metallurgy, chemical warfare agents, etc. • During pedogenesis it gets released from parent rocks into the soil and further in groundwater due to weathering of rocks followed by leaching • May be released anthropogenically such as fertilizers, arsenical pesticides, irrigation, disposal of industrial and animal wastes and dust from burning of fossil fuel • Reports suggest that it is present in very high concentration in groundwater around Ganga–Brahmaputra plains • Found concentration exceed the permissible limit of 0.01 mg/L in groundwater in 18 states in India • Aquifers present below the alluvial plains entrenched within the Late Quaternary deposits are reportedly affected but a few exceptions prevail like in Chhattisgarh and Karnataka where hard rock aquifers are also affected • Concentration in groundwater is marked by wide spatial variability • The contaminated aquifer is generally found to be confined within the depth of 100 m in alluvial aquifers	cycling of different forms of arsenic through air, water and soil • In sub-surface water inorganic arsenic is present, commonly as arsenate (As V) and arsenite (As III)
Iron	• Observed in high concentration of (>1.0 mg/L) in groundwater in more than 0.11 million habitations in India, including states of Chhattisgarh, Assam, Goa, Bihar, Gujarat, Haryana, Karnataka, Jharkhand, Uttar Pradesh, Meghalaya, Tripura, Kerala, Madhya Pradesh, Maharashtra, Manipur, Tamil Nadu, Orissa, Andhra Pradesh, Punjab, Rajasthan, J and K, West Bengal and UT of Andaman and Nicobar	

(*Continued*)

Table 14.4 (Continued) Various geogenic contaminants found in groundwater according to Central Ground Water Board (CGWB) (CGWB, 2012)

Contaminants		*Situation*
Nitrate	• Quite common constituent especially in shallow aquifers groundwater • Main source is anthropogenic activities • Found in high concentration in groundwater in almost all hydrogeological formations in India	• Consumption of water having concentration above the permissible limit of 45 mg/L poses problem for human health

14.3.2 The National Water Policy (NWP)

India had revised the NWP in 2002 with the following significant elements relevant to urban water security:

- Formation of national- and state-level data banks to observe and scrutinize the demand and supply;
- Supporting the conversion of available water resources into utilizable water;
- Unconventional and improved methods for efficient water use;
- Supply of water from water excess areas to water-scarce areas;
- Judicious allocation of water for diverse uses and pricing of water to secure sustainable development;
- Regulation on groundwater exploitation and monitoring of water table using modern, improved and efficient scientific techniques;
- Sustainability of existing waterbodies, involving all the stakeholders especially local communities at all levels;
- The PPPs for water resource development and distribution;
- Master plan for flood control, by linking different rivers and promoting soil conservation measures;
- Development of drought-prone areas through watershed development, afforestation and sustainable farming practices; and
- Interstate water sharing policy and timely addressing of disputes.

14.3.3 State acts and regulations of groundwater

Most of the states and the union territories in India have either enforced laws regulating the use of groundwater or have Bills pending final approval. However, such legislations are inapplicable to some notified geographical areas. The system is generally based on the setting up of state-level authorities for implementation, and on user registration and prior approval of licenses.

In India, the groundwater right is not vested within the government and its authorities, neither considered as the Central Government's prerogative. It is a state subject and only individual states and union territories are responsible for legislation on water (both surface and groundwater) related issues (except the irrigation sector). The following list shows laws, regulations and bills (the legislature has been adopted but have not yet been passed or has not received the final assent) enacted by individual states, also including the union territories, in India:

- Andhra Pradesh Groundwater (Regulation for Drinking Water Purposes) Act, 1996, and the Andhra Pradesh Water, Land and Trees Act and Rules, 2002;
- Assam Groundwater Control and Regulation Act, 2012;
- Bihar Groundwater (Regulation and Control of Development and Management) Act, 2006;
- Chennai Metropolitan Area Groundwater (Regulation) Act, 1987;
- Delhi National Capital Territory (NCT) Groundwater Regulation Directions, 2010;
- Goa Groundwater Regulation Act, 2002;
- Himachal Pradesh Groundwater (Regulation and Control of Development and Management) Act, 2005;
- Jammu and Kashmir (J and K) Water Resources (Regulation and Management) Act, 2010, and J and K State Water Resources Regulatory Authority Regulations, 2013;
- Karnataka Groundwater (Regulation for Protection of Sources of Drinking Water) Act, 1999, and the Karnataka Groundwater (Regulation and Control of Development and Management) Act, 2011;
- Kerala Groundwater (Control and Regulation) Act, 2002;
- Lakshadweep Groundwater (Development and Control) Regulation, 2001;
- Maharashtra Groundwater (Regulation for Drinking Water Purposes) Act, 1993, and the Water Resources Regulatory Authority Act 2005, Maharashtra Groundwater (Development and Management) Bill (yet to be notified), 2009;
- Puducherry Groundwater (Control and Regulation) Act, 2002;
- West Bengal Ground Water Resources (Management, Control and Regulation) Act, 2005.
- Chhattisgarh Groundwater (Regulation and Control of Development and Management) Bill, 2012 (pending);
- Haryana Groundwater Management and Regulation Bill, 2011 (pending);
- Odisha Groundwater (Regulation, Development and Management) Bill, 2011 (pending);
- Uttar Pradesh Groundwater Conservation, Protection and Development (Management, Control and Regulation) Bill, 2010 (pending).

14.4 GROUNDWATER GOVERNANCE FRAMEWORK: TOWARDS AN INTEGRATED AND INCLUSIVE FRAMEWORK DURING PANDEMIC AND POST-COVID-19 RELATED TO GROUNDWATER CRISIS IN A CITY

The present scenario demands a differentiation between governance and management of groundwater resources. Management approach is based mainly on the daily activities necessary for intrinsic and spontaneous tasks to fulfil certain objectives to reach the organized achievement or goals. Consequently, the management supports the useful and compelling approaches to accomplish certain objectives or destinations but holistic. Conversely, groundwater governance facilitates a system that helps in viable administration of water assets through the advancement of capable aggregated activity to guarantee socially manageable usage and successful assurance of groundwater resource to support both human and environment. The fundamental principles of groundwater governance are transparency in legality and data, investment, participatory decision-making and the accountability for both the users and the controller of the resource.

Groundwater is a non-renewable resource, thus vulnerable against extreme use and abuse by stakeholders who may attempt to utilize it for their own momentary gains instead of contemplating long-haul sustainable necessities for others. Groundwater as a non-renewable resource along these lines requires the hands-on utilization of a strategic framework that incorporate contemplations such as transparent demarcation of limits through a clear set of resource assessment, management strategy and distribution policy, obligation, stakeholder interest, data accessibility, straightforwardness and legality. Groundwater governance accordingly incorporates components of socio-hydrogeological settings while making associations and coordinated efforts to protect and sustainable use of aquifers.

Groundwater governance requires considering the geographic scale at which decisive strategies should be prepared. Even though groundwater aquifers are generally widely distributed, there is a need to have a clear demarcation of the system boundary to implement the appropriate governance over the resources at different levels of geographical scales. As groundwater is not bound to a particular administrative boundary or, even, geographical area, the implementation of strategic governance arrangements can be, most likely, influenced by both physical environmental and sociopolitical factors to deal with the crisis of groundwater exploitation. In a country like India, it is exceptionally challenging to design at the state or provincial level as hydrogeological settings are varied even inside a single state to govern groundwater distinctively without ignoring the requirements at the other administrative levels. Additionally, because of complex hydrogeologic distribution and characters, in many cases, it is more important to comprehend the aquifer characteristics at the local level instead of the higher geographical levels to plan for the governance strategies.

The difficulties to overcome the overburdened, and ever-increasing, reliance on groundwater, mainly for drinking purposes, in India are numerous where groundwater overextraction is not just prompting fast exhaustion of the resource itself, yet additionally offering to ascend to water quality issues in a circumstance where the reaction at the degree of governance strategy keeps on being non-decisive. Current governance and management policy framework for groundwater in India for conjunctive use and management of groundwater is given in Figure 14.2.

Since aquifers are at the crucial level in which groundwater governance should be outlined, there is next to no reliable data and awareness about the study of aquifers and hydrogeology among the individuals, individuals who are the most reliant on groundwater for their daily needs, particularly for drinking purpose. Then again, the data on water table and hydrogeology has stayed limited to formal scholastic circles who have had remarkably less association with its application at the individual level. Comprehensive knowledge on aquifers and hydrogeology hence requires to shift from

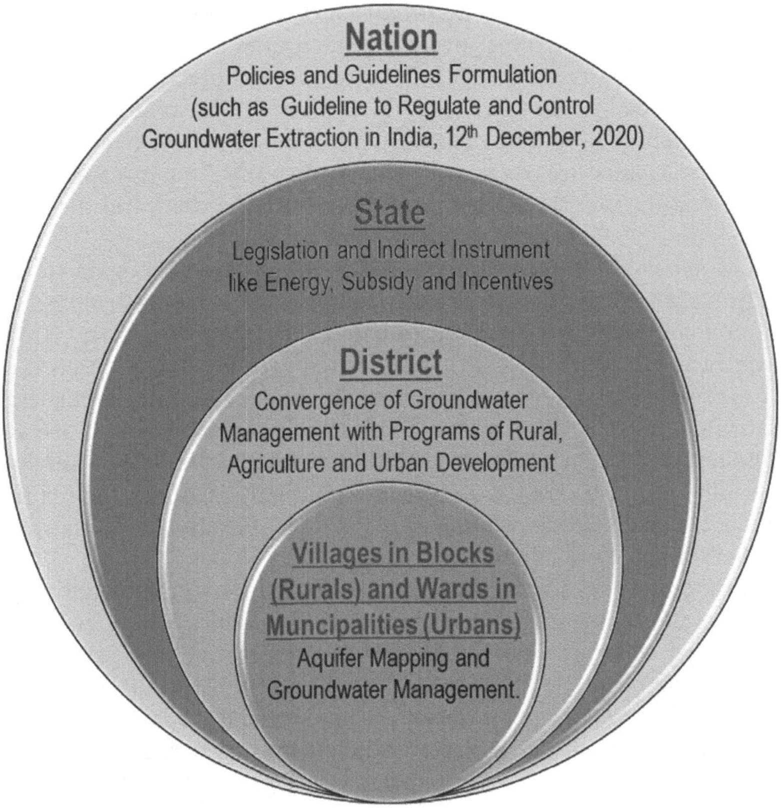

Figure 14.2 Current governance and management policy framework for groundwater in India (Source: Kulkarni et al., 2015).

areas of simple scholastic understanding to deciphering and interpreting it for individuals in a multidisciplinary and inclusive way for even the laypersons to be included in the governance framework for groundwater as an important stakeholder. Thus, a continuous effort needs to be implemented for capacity building through participatory processes.

The requirement for robustness at the aquifer-level data for comprehensive understanding of the ground-level scenarios is especially significant for structuring groundwater governance strategies. Additionally, the challenge of scaling up this data from the individual to higher geographical scale levels to design and draft suitable governance strategies for groundwater is a crucial task for a sustainable future. Data accessibility regarding the availability of groundwater (such as aquifer mapping) and potential discrepancies at the national level to be easily accessible at the individual level would be another challenge for the decision-makers. Transparency and accountability are the two important aspects in this regard that need to be properly addressed in the strategic framework of groundwater governance. This attempt to create an easily accessible data pool can also be advantageous for effective planning and management of the resource. Engaging local communities and increasing the ways of stakeholder participation in the entire planning process of groundwater governance would benefit the process of generating and validating local-level data which in turn ease the difficulties for various decision-making at the management level.

Coordinated efforts and bridging communication gaps within the partners and their allies are hence crucial for successful governance of groundwater at its springshed-level scale. Silo approach at the policy level to envision the challenges as compartments segregate groundwater from its vital components such as surface water. Considering them as a single entity will also be helpful for governing the entire water resources prioritizing their horizontal connectivity, such as connection between aquifers with watersheds and other essential components of surface waterbodies, ecosystem to manage the upstream and downstream water demand and usages, more holistic way.

There is a requirement to quit practising in storehouses and to stay away from the compartmentalized methodology at the arrangement level that manages sub-surface water and surface water as two separate assets. We additionally need to know that while we centre around vertical-level network of springs while getting them, the even availability of springs, for instance, their associations with watersheds, surface water, environments, just as downstream and upstream water should be perceived and considered. In this regard, quality issues, a critical source of risk to both public and environmental health, need to be properly referred since a continuous overexploitation of groundwater aggravates the contaminants to be present in the water. It possesses a great threat especially during this time of the pandemic. Therefore, even at the springshed-level scale data collection, management and analyses should be involved with reliable water quality data. Capacity building to regenerate awareness about water quality issues at the individual- and community-level scales is hence non-negotiable for good governance of groundwater resources.

Institutional mechanisms are necessary to meet the challenges regarding groundwater governance (Karar, 2017). The key role of stakeholder participation, their rights and reliable data are significant parts in the legitimate usage of the groundwater governance framework. Often stakeholders are neither involved in the selection of project ingredients nor encouraged to participate in various project activities (Mechlem, 2016). The current groundwater management framework in India (Figure 14.2) at both the administrative (governance) and watershed levels involves only Government departments and their favoured Contractors with a clear top-down approach neglecting the involvement of watershed communities. This lack of coordination and cooperation has resulted in overlooking local needs resulting inefficient implementation and inadequate sustainability.

The stakeholder participation should be incorporated directly at the necessary stages of the local, state and national levels of decision-making (Karar, 2017). It is a fact that the engaging local level of governance always emphasizes to enable the public participation more intrinsic from the technical/hydrogeological knowledge perspective. Capacity building for strong collective decision-making with different stakeholders with local, academic and technical knowledge should be encouraged and involved in the production of data for further analyses and management of groundwater resources for the urban areas.

Since groundwater is an exceptionally confined resource, suitable establishments may need to be formed, authorized and given rights to implement the strategic measures particularly at the ground level with a strong and integrated collaboration with individual/communities/city levels using government instruments. This local information can be used to generate reliable and evidence-based data that should be incorporated at every level of decision-making and steadily through the inclusive bottom-up approach. This way, various alternatives can be taken care of at the decision-making level to address the existing gaps that prevail between the local-level factors and alternatives or steps taken at the strategy level. Thus, administrative uphold for simple admittance to data is fundamental alongside consistent and decentralized information assortment and dynamic management of groundwater resources. Likewise, discovering methods to coordinate information from different NGOs, academia and research institutions and legislative councils and publish for public for further enhancing and refining are required.

Training and capacity building at each level of governance and over a scope of areas and partners is significant for guaranteeing solid cooperation in reconciliation of groundwater the board and administration techniques at different levels across different socio-hydrological typologies in India. These require to take care of into the process of ratification that holds the comprehension of groundwater as a vulnerable and essential resource as opposed to following just the control mechanism and order only. Strategy rules and regulations to govern groundwater resources, whether through social standards or through formal law-making, must be created with the point of securing the resource itself as well as building capacity among the end-users that incorporate various methods and instruments that advance impartial and productive utilization of groundwater resources. Control mechanism, wherever applicable, must be capable to compliment the participatory nature of groundwater governance.

Groundwater governance should perceive the connection between groundwater exploitation and land use change due to urbanization, impacts of climate change and other external drivers, like changing drinking water guidelines for urban water system and industrialization. It is imperative to recognize that the issue of groundwater exploitation is always associated with the shifting nature of the intensive socio-economic and political issues above ground. Here the role of science and technology-based and solution-oriented research can have an important contribution. Innovative scientific approach can explore the extent of groundwater beyond its visible geographic area and examine the position of socio-economic and political realm comprehensively. Therefore, the groundwater crisis can be related to a greater extent of political economy issue which governance should investigate and address in the policy to establish the interlinkages within the society and its continuous shifts. In this way, laws and regulations for protecting groundwater can be tailored to effective and representative groundwater governance strategies. Strategy guidelines, regardless of whether through accepted practices or through proper law-making, should be created with the point of securing the resource enabling best management practices that incorporate methods that advance fair and proficient utilization of groundwater resource. Guidelines should likewise have the option to approve participatory nature of groundwater governance.

Since groundwater is a vulnerable resource, planning and policy guidelines, systematic allocation and management strategies at each level of governance are significant. It is noteworthy to mention that the equity in participation and equal representation from every class, caste, gender, sexual orientation and economic group should be highly encouraged and heard while crating for a participatory capacity building for the good governance of groundwater in the city. Extending discoursed and guaranteeing everyone's portrayal at each level will guarantee majority rule interest in multilayered groundwater administration instruments even in the most deprived or underrepresented areas, which are essential for Indian cities keeping in mind their enormously crowded and diverse nature. Accordingly, strong coordination in participatory actions and decision-making for knowledge generation at every level of governance can be more effective in transparent, independent and accountable community-driven strategies to prepare for sustainable management of groundwater resources.

Hence, the current policy and governance framework for groundwater management can be modified to include the watershed-level management at each of the policy level. Watershed is defined as a region above a known drainage point beside a stream that provides water to the flow at that point (Ferguson and Maxwell, 2010). It can be therefore stated as a piece of land or a geohydrological unit that drains at a common point. Groundwater management at the watershed level necessitates the rational and optimal utilization of land, vegetation and water resources with minimum risk of hazard to natural and human resources. Thus, combining existing administrative governance framework with the watershed management can ensure inclusive guidance and management of sustainable use of groundwater resources to provide desired goods and services for all without adversely affecting the environment in an urban setting (the same can be argued for rural set-up as well). A proposed comprehensive and inclusive framework for management and governance of groundwater resources is shown in Figure 14.3. This framework can also be used for risk assessment at various spatial, administrative and policy levels, to investigate, analyse and determine the potential fallacy in the current management of groundwater and mitigate the problems, specifically for the emergency such as the COVID-19 pandemic.

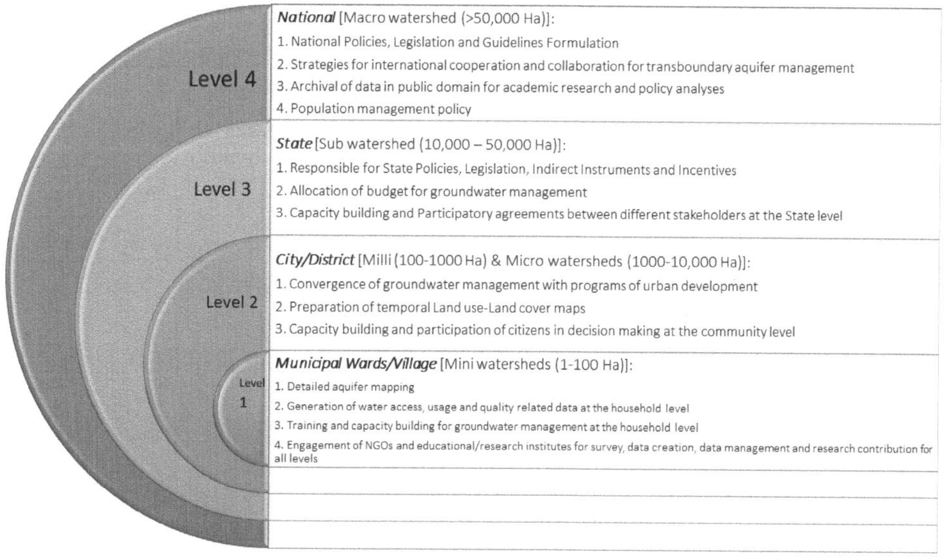

Figure 14.3 Proposed framework for comprehensive management and governance of groundwater resources.

14.5 BEST PRACTICES FOR GOVERNANCE AND MANAGEMENT OF GROUNDWATER IN A CITY DURING THE CORONAVIRUS PANDEMIC

Everyone is spending more time at home due to the COVID-19 global pandemic and being advised to keep washing hands and follow safe and legal practices for disposal of medicated stuff used in the household to avoid further contamination. There are important guidelines to follow to protect public health and contaminating water system. In addition to the potential human health impacts of leftover medications, these compounds can contaminate surface water and groundwater—when flushed, disposed in to drain or discarded in the trash. Most of the conventional water treatment facilities are not equipped to remove these compounds. Therefore, protective and conservative measures may be adopted:

1. Well casing should essentially be provided with a proper sanitary seal around to stop migrating contaminants from the land surface to enter the aquifer, bypassing the unsaturated zone that naturally helps in cleaning of groundwater.
2. Human enteric viruses (which could replicate in the human intestinal tract) are of greatest concern in well water microbial contamination as they shed in human excreta in enormous numbers and commonly tied with disease outbreaks. Usually, there are three processes that dictate the reduction of pathogens in the subsurface: adsorption, die-off/inactivation and filtration.

    a. Filtration is effective when pathogens size is large enough to get blocked by the soil or aquifer pores and cracks. Thus, due to extremely small sizes of viruses' physical removal by size exclusion is not highly effective for viruses. Efficiency of filtration is regulated by the property of soil and rocks through which groundwater flows. Likewise, silts are more effective than sands at trapping microorganisms (Velonakis et al., 2014).
    b. Adsorption takes place when microorganisms get attached to particles, and either remove or delay their transport in a water medium. In soils and groundwater, the adsorption of viruses on sediment grains is the primary removal mechanism. Such adsorption processes are governed by complex dependence on the chemistry of the sediment and water.
    c. Survival of the viruses or the extent of their infectivity can be reduced by increasing the residence time in the subsurface. pH, temperature and other factors also have profound influence on the survival of viruses in water media.

3. Soils have been found effective for virus removal. Removal efficiency depends on soil characteristics such as texture, composition and reactions that occur within soil layers. On the other hand, wells in certain types of aquifers, such as karst and fractured rock, are susceptible to enhanced virus transport.
4. Decentralized wastewater treatment (septic tanks) systems and open or leaked sewer lines are the potential sources of viruses for private wells. Hence, septic systems should be meticulously designed and operated to protect wells from pathogenic cross-contamination; otherwise, outbreaks associated with septic systems will continue to be reported especially in areas having high septic system densities.

5. A second line of defence is septic system and well maintenance. Cracks in the casing, riser of well and apron around the wellhead may allow floodwater to enter the well and the annular space around the casing below ground during flooding. This will make wells more vulnerable to microbial contamination after flooding as storm water can carry loads of microorganisms including viruses and other pathogens. The severity is more in the case of shallow wells that are dug or bored or submerged by floodwater for long periods of time. In such cases through disinfection of well water is essential to eliminate any microbial contamination and subsequently, the quality of the well water should be tested by certified/licensed laboratories.
6. Since 2000, despite the huge change in the socio-economic and political scenarios of India, the expected progress in the water sector has not yet been developed at the same pace. Therefore, it is proposed to bring changes in the National Water Policies to ensure Urban Water Security:

    a. Domestic Sector

        i. Introduction of a policy for mandatory rainwater harvesting in cities;
        ii. Propagation of efficient water usage;
        iii. Spreading of awareness about water conservation among the public.

    b. Industrial Sector

        i. Encourage reuse/recycle/reduce and treatment of effluent through regulations and subsidies;
        ii. Encourage the development of new technologies that require low water consumption

7. Finally, the proposed framework for comprehensive management and governance of groundwater resources for achieving Urban Water Security (Figure 14.3) ensures:

    water conservation by developing watersheds that result in an increment of the storage capacity (Koop et al., 2019),
    effective water uses through improved recycling of wastewater (Koop et al., 2019),
    deterring contamination of water by prohibiting the discharge of city's raw waste and sewage in river, the wise use of agrochemicals for urban gardens and supervision on over-exploitation of groundwater (Koop et al., 2019),
    water resources management and distribution through the proper and safe piped system, rainwater harvesting and artificial recharging of aquifers, metered water supply and pricing of water for sustainability (Koop et al., 2019; Mukherjee et al., 2020),
    awareness and orientation among the water users for change in lifestyle and conservation of water (Koop, 2019; Mukherjee et al., 2020),
    equitable access to safe and enough water for all (Mukherjee et al., 2020).

14.6 CONCLUSIONS

This chapter analyses the interface of urban water security, as related to the agreements, liabilities and tasks of different stakeholders, in relation to groundwater planning, management and concerns for security as drinking water during and after

COVID-19 pandemic. Usage of groundwater as drinking water from borewells or tube wells poses a minimal risk for coronavirus contamination, particularly contrasted with direct head-to-head transmission or by contacting an infected surface. Anxieties about the coronavirus in groundwater serve as a token of significance to ensure from contaminants/pathogens through legitimate consideration and preservation of wells and infrastructures of urban septic system. Reviewing the literature, this chapter discoursed the state's and individual's duties to respect, reserve and realize the right to water and sanitation especially in relation to the human right to use the available groundwater resources; the management and guideline of private service-providers and health and safety issues. Recognizing the public institutions' responsibility to deliver through direct drinking water service provision, this book chapter recommends that self-provision is the fundamental means to enjoy the basic human right to water and sanitation. This has substantial connotations for the role of the governance in generating awareness and building capacity for safeguarding point sources and allocating aquifer recharge. Disregarding self-provision, the governance will miss tremendous prospect of groundwater management by jeopardizing the water security of the urban areas for the future generations.

Last but not the least, it is difficult to ignore the necessity to control population growth in India. By the mid of this century, India will have 1.67 billion people. Currently, the available groundwater resources can sustain only a population of 1.50 billion people, if we immediately stop wasting and polluting the water with a condition that monsoon would never fail in the future, which is unlikely to meet (particularly in this rapidly changing climate era). Therefore, best practices for governance and management of groundwater in India must also include the population management strategy at the national level for the sustainable groundwater future for urban India.

ACKNOWLEDGMENTS

Subham Mukherjee acknowledges the German Academic Exchange Service (DAAD) for providing a fellowship to conduct this study. The authors also thank colleagues from Freie University Berlin who provided insight and expertise that greatly assisted the research. The Director of CSIR-NEERI, Nagpur, is thankfully acknowledged for giving the opportunity to pursue the work in CSIR-NEERI, Nagpur, India. The chapter is checked for plagiarism using the iThenticate software and recorded in the Knowledge Resource Center, CSIR-NEERI, Nagpur, for anti-plagiarism (KRC No.: CSIR-NEERI/KRC/2020/NOV/WWT/1).

REFERENCES

Bakore, N., John, P. J., Bhatnagar, P., 2004. Organochlorine pesticide residues in wheat and drinking water samples from Jaipur, Rajasthan, India. *Environmental Monitoring and Assessment*, 98, 381–389.

Banerjee, S., Sikdar, P. K., 2020. Land Subsidence due to leakage of aquitard-aquifer pore water in an under-construction tunnel of East-West Metro Railway Project, Kolkata. *Journal of Geological Society of India*, 96, 467−474.

Banks, D., Robins, N. S., Robins, N., 2002. *An Introduction to Groundwater in Crystalline Bedrock*. Trondheim: Norges geologiske undersøkelse, p. 64.

Boelens, R., Lankford, B., Bakker, K., Zeitoun, M., Conway, D., Eds., 2013. The Shotgun Marriage. Water Security, Cultural Politics, and Forced Engagements between Official and Local Rights Frameworks. In *Water Security: Principles, Perspectives and Practices.* New York, USA; London, UK: Routledge, pp. 239–255.

Bureau of Indian Standards (BIS), 2012. Drinking water- specification IS No. 10500;2004. 2005. 4., Bureau of Indian Standards, New Delhi, India.

CGWB, 2008. *Groundwater Information Booklet: Bangalore Urban District.* Bangalore: Ministry of Water Resources, Government of India, India.

CGWB, 2010. *Ground Water Scenario of India: 2009–10.* Faridabad: Central Groundwater Boar, Government of India, India.

CGWB, 2011. *Ground Water Scenario in Major Cities of India.* Faridabad: Ministry of Water, Resources, Government of India, India.

CGWB, 2012. *Ground Water Yearbook - India.* Faridabad: Central Ground Water Board, Government of India, India.

Chowdhury, U. K., Biswas, B. K., Chowdhury, T. R., Samanta, G., Mandal, B. K., Basu, G. C., Chakraborti, D., 2000. Groundwater arsenic contamination in Bangladesh and West Bengal, India. *Environmental Health Perspectives*, 108, 393–397

Cooper, R., 2020a. Water security beyond Covid-19.K4D Helpdesk Report 826. Brighton, UK: Institute of Development Studies. Available online: https://opendocs.ids.ac.uk/opendocs/handle/20.500.12413/15240 Accessed on 24.08.2020.

Cooper, R., 2020b. Water for the Urban Poor and Covid-19. K4D Helpdesk Report 826. Brighton, UK: Institute of Development Studies. Available online: https://reliefweb.int/sites/reliefweb.int/files/resources/826_Water_for_the_urban_poor_and_Covid19.pdf. Accessed on 24.08.2020.

Department of Health (DOH). 2020. Drinking Water and Novel Coronavirus Disease (COVID-19), Washington State, United States of America. Available online: https://www.doh.wa.gov/Portals/1/Documents/1600/coronavirus/DrinkingWater.pdf Accessed on 24.08.2020.

Ferguson, I. M., Maxwell, R. M., 2010. Role of groundwater in watershed response and land surface feedbacks under climate change. *Water Resources Research*, 46.

Grey, D., Sadoff, C.W., 2007. Sink or Swim? Water security for growth and development. *Water Policy*, 9, 545–571.

Grönwall, J., Danert, K., 2020. Regarding groundwater and drinking water access through a human rights lens: Self-supply as a norm. *Water*, 12, 419.

Hannah, D. M., Lynch, I., Mao, F., Miller, J. D., Young, S. L., Krause, S., 2020. Water and sanitation for all in a pandemic. *Nature Sustainability*, 3, 773–775.

Iyer, R. R., 2007. *Towards Water Wisdom: Limits, Justice, Harmony.* India: Sage Publications India Pvt Ltd.

Jain, S. K., Agarwal, P. K., Singh, V. P., 2007. *Hydrology and Water Resources of India.* Springer Science & Business Media.

Jat, M. K., Khare, D., Garg, P., 2009. Urbanization and its impact on groundwater: A remote sensing and GIS-based assessment approach. *The Environmentalist*, 29, 17–32.

Jensen, P. K., Ensink, J. H., Jayasinghe, G., Van Der Hoek, W., Cairncross, S., Dalsgaard, A., 2002. Domestic transmission routes of pathogens: The problem of in-house contamination of drinking water during storage in developing countries. *Tropical Medicine and International Health*, 7, 604–609.

Karar, E., 2017. *Freshwater Governance for the 21st Century.* Springer Nature.

Kakar, Y. P., Bhatnagar, N. C., 2008. Ground water pollution due to industrial effluents in Ludhiana, India, *Studies in Environmental Science*, 17, 265–272.

Koop, S. H. A., Van Dorssen, A. J., Brouwer, S., 2019. Enhancing domestic water conservation behaviour: A review of empirical studies on influencing tactics. *Journal of Environmental Management*, 247, 867–876.

Kozhisseri, D., 2005. Groundwater dip: Quantum, quality. Down to Earth. India.
Kulkarni, H., Shah, M., Shankar, P. V., 2015. Shaping the contours of groundwater governance in India. *Journal of Hydrology: Regional Studies*, 4, 172–192.
Kumar, M. (Ed.), 2012. *Problems, Perspectives and Challenges of Agricultural Water Management*. In-Tech International Publication, ISBN 979-953-307-706-1 https://www.intechopen.com/books/problems-perspectives-and-challenges-of-agricultural-water-management.
Kumar, M., Patel, A. K., Das, A., Kumar, P., Goswami, R., Deka, P., Das, N., 2017. Hydrogeochemical controls on mobilization of arsenic and associated health risk in Nagaon district of the central Brahmaputra Plain, India. *Environmental Geochemistry and Health*, 39(1), 161–178.
Kumar, M., Mohapatra, S., Mazumder, P., Singh, A., Honda, R., Lin, C., Kuroda, K., 2020a. Making waves perspectives of modelling and monitoring of SARS-CoV-2 in aquatic environment for COVID-19 pandemic. *Current Pollution Reports*, 1–12.
Kumar, M., Thakur, A. K., Mazumder, P., Kuroda, K., Mohapatra, S., Rinklebe, J., Gikas, P., 2020b. Frontier review on the propensity and repercussion of SARS-CoV-2 migration to aquatic environment. *Journal of Hazardous Materials Letters*, 1, 100001.
Kumar, M., Kuroda, K., A. K., Patel, N., Bhattacharya, P., Joshi, M., Joshi, C. G., 2021. Decay of SARS-CoV-2 RNA along the wastewater treatment outfitted with Upflow Anaerobic Sludge Blanket (UASB) system evaluated through two sample concentration techniques. *Science of the Total Environment*, 754, 142329.
Mechlem, K., 2016. Groundwater governance: The role of legal frameworks at the local and national level—Established practice and emerging trends. *Water*, 8, 347.
Moench, M. H., 1992. Chasing the watertable: Equity and sustainability in groundwater management. *Economic and Political Weekly*, A171–A177.
Mohanraj, R., Sathishkumar, M., Azeez, P. A., Sivakumar, R., 2000. Pollution status of wetlands in urban Coimbatore, Tamilnadu, India. *Bulletin of Environmental Contamination and Toxicology*, 64, 638–643.
Mohapatra, S., Menon, N. G., Mohapatra, G., Pisharody, L., Pattnaik, A., Menon, N. G., Gin, K. Y. H., 2020. The novel SARS-CoV-2 pandemic: Possible environmental transmission, detection, persistence and fate during wastewater and water treatment. *Science of The Total Environment*, 765, 142746.
Mukherjee, I., Gopal, M., 2001. Chromatographic techniques in the analysis of organochlorine pesticide residues. *Journal of Chromatography A*, 33–42.
Mukherjee, A., Saha, D., Harvey, C. F., Taylor, R. G., Ahmed, K. M., Bhanja, S. N., 2015. Groundwater systems of the Indian sub-continent. *Journal of Hydrology: Regional Studies*, 4, 1–14.
Mukherjee, S., Bebermeier, W., Schütt, B., 2018. An overview of the impacts of land use land cover changes (1980–2014) on urban water security of Kolkata. *Land*, 7, 91.
Mukherjee, S., Sundberg, T., Schütt, B., 2020. Assessment of water security in socially excluded areas in Kolkata, India: An approach focusing on water, sanitation and hygiene. *Water*, 12, 746.
Mutiyar, P. K., Mittal, A. K., Pekdeger, A., 2011. Status of organochlorine pesticides in the drinking water well-fields located in the Delhi region of the flood plains of the river Yamuna. *Drinking Water Engineering and Science Discussions*, 85–115.
Naik, P. K., Tambe, J. A., Dehury, B. N., Tiwari, A. N., 2008. Impact of urbanization on the groundwater regime in a fast-growing city in central India. *Environmental Monitoring and Assessment*, 146, 339–373.
National Institute of Urban Affairs (NIUA), 2005. *Status of Water Supply, Sanitation and Solid Waste Management in Urban Areas*. New Delhi: National Institute of Urban Affairs.
National Sample Survey Organisation (NSSO), 2013. Key Indicators of Drinking Water, Sanitation, Hygiene and Housing Condition in India - NSSO 69th Round. In NSSO (Ed.), *Key Indicators of Drinking Water, Sanitation, Hygiene and Housing Condition in India*. India: National Sample Survey Organisation.

Palit, A., Batabyal, P., Kanungo, S., Sur, D., 2012. In-house contamination of potable water in urban slum of Kolkata, India: A possible transmission route of diarrhea. *Water Science and Technology*, 66, 299–303.

Paliwal, B. S., Baghela, A., 2006. Alarming rise in groundwater levels beneath the city of Jodhpur: An example of ground and surface water interaction in the Thar Desert of western India. In K. F. W. Howard (Ed.), *Urban Groundwater, Meeting the Challenge: IAH selected papers on Hydrogeology*, pp. 239–258.

Patel, A. K., Bhagat, C., Taki, K., & Kumar, M. (2020). Microplastic vulnerability in the sediments of the Sabarmati river of India. In *Resilience, Response, and Risk in Water Systems*. (ISBN 978-981-15-4668-6) Springer, Singapore, pp. 127–138.

Pallavi, D., Kumar, M., Sarma, K. P., 2015. Speciation of heavy metals in surface sediment of the Brahmaputra river, Assam, India. *Journal of Environmental Research and Development*, 9, 944–952.

Ramesh, R., Kumar, K.S., Eswaramoorthi, S., Purvaja, G., 1995. Migration and contamination of major and trace elements in groundwater of Madras City, India. *Environmental Geology*, 25, 126–136.

Ray, I., 2020. Viewpoint| Handwashing and COVID-19: Simple, right there? *World Development*, 135: 105086. Available online: https://www.sciencedirect.com/science/article/pii/S0305750X20302126 Accessed on 24.08.2020.

Sahu, P., Sikdar, P.K., 2011. Threat of land subsidence in and around Kolkata City and East Kolkata Wetlands, West Bengal, India. *Journal of Earth System Science*, 120, 435–446.

Saleth, R.M., 2009. Land-dominated Approach to Water Policy. *India Infrastructure Report*, 29.

Senthilnathan, S., Azeez, P. A., 1999. Influence of dyeing and bleaching industries on ground water of Tirupur, Tamilnadu, India. *Bulletin of Environmental Contamination and Toxicology*, 62, 330–335.

Shah, T., 2008. *Taming the Anarchy: Groundwater Governance in South-Asia*. New Delhi: Routledge.

Shukla, G., Kumar, A., Bhanti, M., Joseph, P.E., Taneja, A., 2006. Organochlorine pesticide contamination of ground water in the city of Hyderabad. *Marine Pollution and Ecotoxicology*, 32, 244–247.

Sikdar, P. K. Ed., 2018. *Groundwater Development and Management: Issues and Challenges in South Asia*. Switzerland: Springer, p. 533.

Sikdar, P.K., Biswas, A.B., Saha, A.K., 1996. A study on the possible land subsidence in Calcutta and Howrah cities due to groundwater overdraft. *Indian Journal of Geology*, 68, 193–200.

Singh, A. K., 2017. *Water in India: Dimensions of Crisis-An Overview of the Water Related Problems in India*. New Delhi: Kepa-India Group and Centre for the study of Developing Societies.

Singh, M. J., Somashekar, R., Prakash, K., Shivanna, K., 2010. Investigation of heavy metals in crystalline aquifer groundwater from different valleys of Bangalore, Karnataka. *Journal of Geography and Regional Planning*, 3, 262–270.

Singh, A., Patel, A. K., Deka, J. P., Kumar, M. 2020a. Natural recharge transcends anthropogenic forcing that influences arsenic vulnerability of the quaternary alluviums of the Mid-Gangetic Plain. *NPJ Clean Water*, 3(1), 1–12.

Singh, A., Patel, A. K., Ramanathan, A., Kumar, M. 2020b. Climatic influences on arsenic health risk in the metamorphic Precambrian deposits of Sri Lanka: A re-analysis-based critical review. *Journal of Climate Change*, 6(1), 15–24.

Smith, L., 2001. The urban political ecology of water in Cape Town. *Urban Forum*, pp. 204–224.

Srinivasan, V., Gorelick, S. M., Goulder, L., 2010. Factors determining informal tanker water markets in Chennai, India. *Water International*, 35, 254–269.

Thakur, A. K., Vithanage, M., Das, D., Kumar, M., 2020. A review on design, material selection, mechanism, and modeling of permeable reactive materials for community-scale groundwater treatment. *Environmental Technology & Innovation*, 19, 100917.

UN Special Rapporteur's Realizing the Rights to Water and Sanitation: A Handbook, 2014.

UNEP, 2016. UNEP Frontiers 2016 Report: Emerging Issues of Environmental Concern. Nairobi, Kenya: United Nations Environment Programme. Available online: https://wedocs.unep.org/bitstream/handle/20.500.11822/7664/Frontiers_2016.pdf?sequence=1andisAllowed=y accessed on 30.10.2020.

van der Wal, A., 2010. Understanding groundwater and wells in manual drilling: Instruction handbook for manual drilling teams on hydrogeology for well drilling, well installation and well development. *PRACTICA Foundation*, 3, 1–41.

Velis, M., Conti, K. I., Biermann, F., 2017. Groundwater and human development: Synergies and trade-offs within the context of the sustainable development goals. *Sustainability Science*, 12, 1007–1017.

Velonakis, E., Dimitriadi, D., Papadogiannakis, E., Vatopoulos, A., 2014. Present status of effect of microorganisms from sand beach on public health. *Journal of Coastal Life Medicine*, 2, 746–756

World Bank, 2003. "Sustainable groundwater management: Concepts and tools", GW-MATE (Groundwater Management Advisory Team) Available online: http://www.worldbank.ogr/gwmate accessed on 30.10.2020.

World Bank, 2006. "India: Water supply and sanitation - bridging the gap between infrastructure and service" Background paper Urban water supply and sanitation, India Country Team, Energy and Infrastructure Department, South Asia Region, World Bank, Washington DC, USA

World Bank, 2010. *Deep Wells and Prudence: Towards Pragmatic Action for Addressing Groundwater overexploitation in India*. Washington, DC: International Bank for Reconstruction and Development/ The World Bank.

WWAP (United Nations World Water Assessment Programme)/UN-Water, 2018. *The United Nations World Water Development Report 2018: Nature-Based Solutions for Water*. Paris: UNESCO. Available online: https://unesdoc.unesco.org/ark:/48223/pf0000261424/PDF/261424eng.pdf.multi accessed on 30.10.2020.

Chapter 15

Comprehensive understanding of fate, transport and consequences of contaminants of emerging concerns (CECs) in urban environment

Shivaraju Harikaranahalli Puttaiah, Yashas Shivamurthy Ravindra, Prabagar Jijoe Samuel, Sneha Yadav, and Aiswarya Cholayampara Murukadasan
Academy of Higher Education and Research

CONTENTS

DOI: 10.1201/9781003157465-18

15.1 INTRODUCTION

From the past few decades due to an increase in industrialization and excess population, contaminants of emerging concerns (CECs) are discharged continuously into the environment. The introduction of these contaminants into environment especially the aquatic ecosystem is primarily due to discharge of wastes without proper treatment or entering directly without treatment in the environment. According to available reports, CECs were found in the environment from the early 1800s (Dey et al., 2019, Kumar et al., 2019). CECs can be defined as the substances that are discharged in the environment with no regulations for monitoring the environment (Kumar et al., 2020). It is termed as CECs because they were not recognized earlier and does not have standard guidelines for monitoring but now with the increase in accumulation in the various environmental system, but it causes adverse health effects. CECs can include pharmaceuticals, personal care products (PCPs), pesticides and any other compounds. Countries with less revenue have no proper water treatment infrastructure causing the contaminants entering the environment making it difficult to understand its source and transportation eventually affecting human health. CECs are found in low concentrations causing potential risk to aquatic organisms. India is considered to be a country with low revenue, where there are only a few studies related to CECs and the available reports show that harmful CECs, especially pharmaceuticals, PCPs, and many others, were found in major Indian water bodies (Philip et al., 2018). The sewage connectivity in urban regions is not efficient and treating huge loads of wastewater that is discharged from various sources seems to be an impossible task. The region with poor water availability and waste management is sensitive to CECs especially developing countries like India with an increase in population, no proper pollution control system and transitional economies. It is estimated that approximately 700 Indian cities produce 30,000 million litres of wastewater containing CECs in which two-thirds of wastes generated by Class-I cities, i.e. population greater than 100,000, and 90% generated by Class-II cities, i.e. population greater than 50,000 (Water and Sanitation Program (WSP) and Water Management Institute, 2016). The Indian pharmaceutical companies, for example, are increasing every day due to huge demand of various curative drugs as well as veterinary drugs to use for poultry and animal farming to produce processed meats or food causing a huge discharge of CECs into water bodies including the drinking water potentially disrupting the food chain. Protection of natural resources from being contaminated, especially water, is necessary for healthy living. Only a few studies regarding CECs are available and are hence necessary to conduct widespread research across the country to understand its source, transportation and fate.

15.2 OCCURRENCE (SOURCE) OF CECs

With the upsurge in manufacturing and improper disposal of chemicals and other compounds, they are easily found everywhere in our environment. Considering both natural and man-made compounds, millions of these compounds are present in the environment. Numerous CECs and their sources have been studied in the past years. The commonly found CECs in the environment include pharmaceuticals, pesticides, insecticides and other compounds (Menon et al., 2020; Mohapatra et al., 2018, 2021;

Priyanka and Mohapatra, 2020). Most of the CECs may not be harmful because they degrade quickly, but there are some CECs that are very harmful to humans and the environment. Understanding the source of CECs is important to reduce the health and ecological impacts. CECs enter into the environment via point or non-point sources through mining activities, pharmacy industries or irrigation. In an urban environment, CECs are found due to agricultural run-off, the excess flow of sewage water, household dumping, and the excretion of incompletely metabolized drugs by humans as well as animals.

These CEC occurrences may vary, for instance, pharmaceuticals ranging from one to many have been identified with various samples collected in a single location. In India alone, various CECs (such as pharmaceutical contaminants) are being released into the environment with high concentrations (Patel et al., 2019). Since India is one of the major pharmaceutical manufacturing countries, the manufacturers release major pharmaceutical contaminants into the environment without any treatment. A single water treatment plant in an Indian city receives water waste from approximately 90 different drug-producing firms (Larsson, 2014). The concentration of CECs is high in industrial wastes, hospital effluents and wastewater treatment plant effluents, slightly lower in surface water, groundwater and drinking water (Patel et al., 2019; Kumar et al., 2021). The concentration of CECs in the environment depends on the compound's sorption ability on solids, biological properties, and their interaction with microorganisms while undergoing treatment. According to a study, the occurrence of some CECs such as hybrid pesticides and antibiotics was not expected to be spread in the environment.

Seasons can also play an important role in the occurrence of these contaminants in various regions. Various studies have been conducted to examine the seasonality of these compounds. One such compound is 4-nonylphenol, a pesticide used in agriculture, was noticed to have higher concentrations during spring and late summer due to biological transformation of the compound that arises frequently in wastewater treatment plants in warmer periods. In urban cities, carbaryl concentration was found to be higher during summer and winter. The source for this compound was found to be excessive residential use, later reaching the wastewater treatment plants.

15.3 CLASSIFICATION OF CECs

Various chemical products such as pharmaceuticals, cosmetics, PCPs are being manufactured widely due to global demand. Insecticides, pesticides, nanoparticles, additives, etc., are some of the other widely manufactured chemicals around the world. Most of these chemical compounds are released into the environment untreated, creating health and ecological impacts. Studying the different categories of CECs is essential to understand their properties and their interactions with the environment. Figure 15.1 portrays the various classifications of CECs.

15.3.1 Pharmaceuticals

Pharmaceuticals are an artificial or natural compound that is manufactured to treat diseases and to improve the living standards for humans as well as animals. Pharmaceuticals are given to animals to increase their growth rate. Based on their

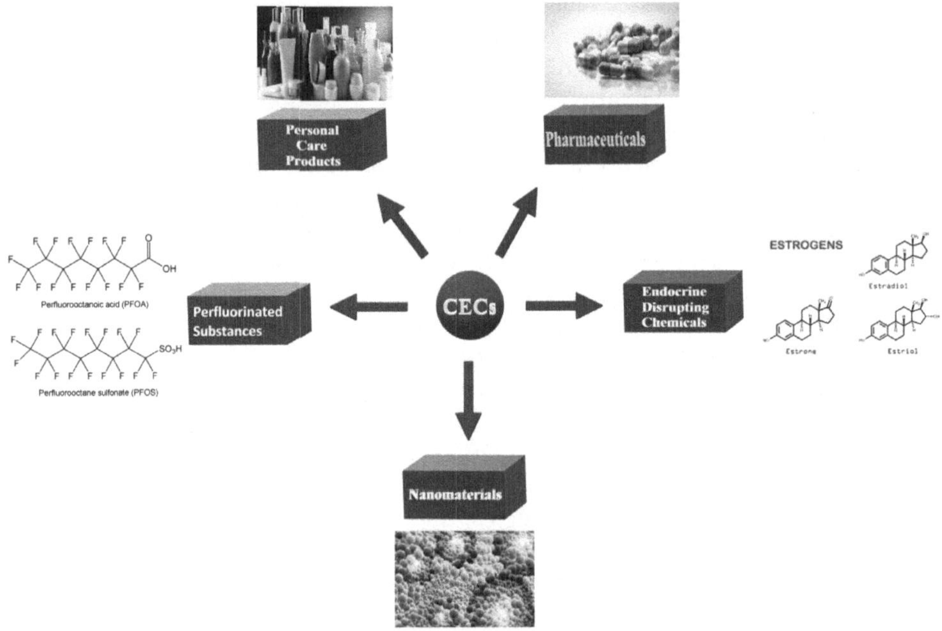

Figure 15.1 Classification of CECs.

usage, pharmaceuticals are categorized into antibiotics, antidiabetics, antiepileptics, anti-inflammatory, along with others (Tijani et al., 2016). Pharmaceuticals are labelled as emerging compounds due to their biological activities in the environment. More than 5,000 pharmaceuticals were identified in a particular report resulting in 100,000–200,000 tons of annual consumption of these drugs in countries like China and India (Van Doorslaer et al., 2014). When consumed, these drugs undergo partial metabolism inside the body and are excreted out through urine or faeces, ending up unaltered or as an active substance that may undergo biotransformation in wastewater treatment plants (Salimi et al., 2017). With the increase in the use of pharmaceuticals, various drugs are found in the environment. Pharmaceuticals are hard to degrade with the available technology to treat water due to their stability and complex structure making it unsafe for the environment. With the increase in production of pharmaceuticals day by day, less importance is given to study its adverse effects on the environment. They are referred to as pseudo-persistent pollutants, entering the environment endlessly with low concentration. The World Health Organization (WHO) has stated that further development of such drugs can be a threat to human health. The aquatic ecosystem is vulnerable against these pharmaceuticals due to the release of water wastes into the water bodies. Efficient sewage treatment plant/water treatment plant is necessary to monitor the contaminants as well as treat them. Some pharmaceuticals and their concentration are listed in Table 15.1.

Table 15.1 Concentration of pharmaceuticals and their contamination sites

Pharmaceuticals	*Concentration*	*Sites of contamination*	*References*
Amoxicillin	13.75 μg /L	River water (Europe)	Philip et al. (2018)
	0.31 ng/L	Tap water (Malaysia)	Praveena et al. (2021)
Atenolol	430 ng/g	Plants under greenhouse condition	Keerthanan et al. (2020)
Sulfamethoxazole	460 ng/g	Urban environment (Sri Lanka)	Guruge et al. (2019)
	934 ng/L		
Ciprofloxacin	31 mg/L	Surrounding environment of the pharmaceutical industry in Hyderabad (India)	Philip et al. (2018)
	28.23 ppb	Nag River (India)	Archana et al. (2016)
	5.08 ppb	Pili River (India)	
	0.32 ng/L	Tap water (Malaysia)	Praveena et al. (2021)
Triclosan	0.36 ng/L		
Ibuprofen	125 ng/L	Kaveri River (India)	Shanmugam et al. (2014)
	9.88 ng/L	Vellar River (India)	
	200 ng/L	Tamiraparani River (India)	
Diclofenac	103 ng/L	Kaveri River (India)	
	1.78 ng/L	Vellar River (India)	
	16.1 ng/L	Tamiraparani River (India)	
Parabens	124 ng/L	Pichavaram mangrove (India)	Philip et al. (2018)
Fluoroquinolones	100 ng/L	Groundwater in Hyderabad (India)	Rutgersson et al. (2014)
Acetaminophen	13,000 ng/L	Domestic and industrial area (Pakistan)	Ashfaq et al. (2019)
Clarithromycin	119 ng/L	Urban environment (Sri Lanka)	Guruge et al. (2019)
Indole	0.0284 μg/L	Laurentian Lake (North America)	Thomas et al. (2017)
Caffeine	1,313 ng/L	Water bodies (West Indies)	Edwards et al. (2018)
Trimethoprim	8.5 ng/L		
Paraxanthine	10.5 ng/L	Brisbane River (Australia)	Anim et al. (2020)
Paracetamol	6 ng/L		
Acesulfame	103 ng/L		

15.3.2 Personal care products (PCPs)

Personal care products are substances that include prescribed and non-prescribed pharmaceuticals used for personal maintenance purposes. Shampoos, soaps, fragrances, cosmetics, toiletries, etc., are some of the commonly used PCPs in our day-to-day life. Instead of ingesting into the body like pharmaceuticals, PCPs are applied to the body. PCPs are also referred to as pseudo-persistent due to their incessant presence in the water. Polycyclic musks and parabens are categories of PCPs that constrain the decay of bacteria (Fawell and Ong, 2012). Disinfectant compounds like triclosan and chlorophene are being used widely compared to pharmaceuticals. For many years, triclosan is used in various products such as toothpaste, soaps and toys. PCPs get into the aquatic ecosystem through showers, recreational activities and other processes via wastewater treatment plants. Various PCPs and their by-products have been identified in the wastewater treatment plants. When undergoing treatment in the wastewater treatment plants, most of them are converted into non-toxic inorganic compounds like carbon dioxide and water, some are adsorbed by sludges due to their lipophilic nature (Jiang et al., 2013). But some compounds do not undergo treatment that is carelessly discharged, ending up in an environment which is toxic (Dubey et al., 2021; Eregowda and Mohapatra, 2020). PCPs can enter into the food chain due to their lipophilic property that can be a threat to the environment.

15.3.3 Endocrine-disrupting chemicals (EDCs)

The human physiological activities like reproductive processes along with embryonic development, metabolic development and sex variation are regulated by the endocrine system that includes the endocrine glands, hormones and receptors (Flint et al., 2012). Hormones that are secreted by these endocrine glands are circulated inside the body through the bloodstream. A certain set of compounds are found to have been imitating or disrupting the endocrine glands causing health problems. These compounds are often referred to as EDCs. EDCs can occur naturally or can be made artificially, but when they interact with the oestrogenic receptors, it boosts or constrains the hormones from functioning efficiently (Jackson and Sutton, 2008). EDCs are classified into three classes, oestrogenic, i.e. the working of natural oestrogens is being changed or imitated by them; androgenic, i.e. testosterone is obstructed; thyroidal, i.e. the thyroid is affected (Snyder et al., 2003). These compounds are widely used in the manufacture of toys, plasticware, toothpaste, cosmetics and others. EDCs enter the environment through various human activities, animals and industrial discharge that undergo treatment eventually reaching the soil, surface water and groundwater (Singh et al., 2020). When exposed to EDCs, they cause adverse human health issues such as immune deficiency, neurological effects and psychological effects. The WHO has defined EDCs as "an exogenous substance or mixture that alters function(s) of the endocrine system and consequently causes adverse health effects in an intact organism, or its progeny" (Damstra et al., 2002). With no precautions and no regular monitoring, the concentration of ECDs is increasing in the environment especially in drinking water. To eliminate, these substances from the water system effective treatment strategies are required.

15.3.4 Perfluorinated substances

Compounds such as perfluorooctane sulfonate and perfluorooctanoic acid are referred to as perfluorinated substances that are widely used in non-stick cookware, fire extinguishing foams, dirt-repellent coating and more. Perfluorinated substances are considered to be harmful to the environment due to their toxic properties and can develop carcinogenicity (Authority, 2011). In major urban cities, these compounds are being identified in surface water and can be a threat to groundwater that can stay there for a long period though the source is immobilized.

15.3.5 Nanomaterials

Materials ranging from 1 to 100 nm are generally referred to as nanoparticles. Various nanomaterials are engineered for various reasons and numerous other novel nanomaterials may appear in forthcoming years. Nanoparticles can be of different shapes spherical, rod, and many others. Properties like high surface and small size make nanoparticles more feasible when compared to other materials for numerous applications. Because of the increase in the production of nanoparticles, they can easily accumulate in environmental systems. Production of nanomaterials can result in exposure through inhalation, ingestion or dermal routes. Nanomaterials are considered to be emerging contaminants due to their harmful effects on the environment. Frequent exposure to nanomaterials can be toxic to individuals and may even alter the physiological functions of the body (Armstead and Li, 2016). Titanium dioxide is a commonly used nanomaterial for different applications such as water treatment and sunscreens. But constant deposit of this material on water bodies can affect the aquatic ecosystem (Yan et al., 2010). Tungsten carbide cobalt is another such nanomaterial that is used in industries for spray coating, drilling equipment and sharp instruments. But the coating wears off over the period resulting in tungsten carbide cobalt dust accumulating in the surrounding work area. The higher concentration of this nanomaterial can be harmful to the people inhaling its difficulty in breathing and other respiratory problems (Armstead and Li, 2016). The behaviour of nanoparticles in the environment is complex and difficult to understand its effect on both environment and humans which makes it necessary to research extensively regarding the same.

15.4 TRANSPORTATION AND FATE OF CECs

Once discharged from diverse sources, contaminants travel via various environmental systems. Pharmaceuticals, PCPs, EDCs and other such emerging contaminants continue to accumulate in the environment at an alarming rate due to the increase in the human population and demand for more products. Concerns regarding transportation and the fate of these CECs are worrying the scientific community due to their constant presence in the environment, especially in water bodies. The fate of CECs and their concentration can occur by various mechanisms or routes depending on various aspects such as characteristics of a single compound, degradation of contaminants and period when the contaminants were introduced in the environment. It is, therefore, necessary to understand its transportation and fate after it being discharged into the environment. The fate and transportation of CECs are described in Figure 15.2.

15.4.1 Transportation in atmosphere and precipitation

Earth's atmosphere has been a place for both source and transportation of various harmful contaminants. Emerging contaminants that are reported most frequently in the atmosphere are fluorotelomer alcohols (FTOHs). Mean concentration of up to 241 pg/m^3 FTOHs was recorded in a study conducted in certain cities in Japan. FTOHs are mass-produced worldwide which are mostly used as water repellents (Oono et al., 2008). This compound has been identified in most urban cities during rainfall and eventually in surface water.

In a recent study, EDCs were found in the atmosphere that is responsible for the accumulation of particulate materials, eventually affecting human health. EDCs such as nonylphenol, di(2-ethylhexyl) phthalate and dibutyl phthalate were found at a higher concentration of 30, 45 and 55 ng/m^3, respectively, which are usually higher

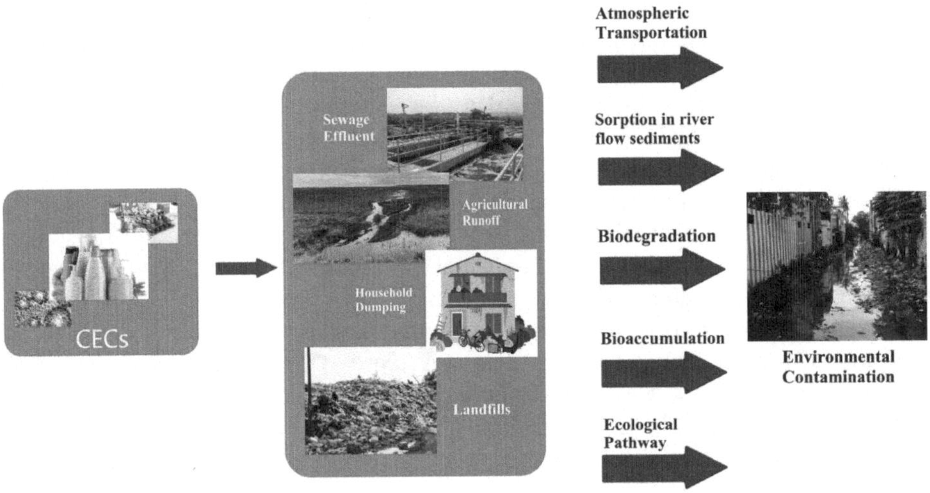

Figure 15.2 Fate and transportation of CECs.

near the industrial area (Salgueiro-González et al., 2015). Most of these contaminant materials are often found indoors rather than in outdoor air due to furnishing, use of perfumes and other cosmetics. At indoors, the concentration level of dibutyl phthalate and diethyl phthalate, used in PCPs, was found to be 2,300 and 4,300 ng/m^3, respectively. The higher concentration of these substances can be harmful to humans (Bergh et al., 2011).

As mentioned earlier, the use of PCPs and other related products has increased because of the increase in urbanization causing contaminants to enter the water cycle. This can be due to the direct disposal of products on surface water or from water treatment plants and can contaminate the environment in the form of precipitation (Pal et al., 2014). The concentration of these contaminants can vary geographically. It is still not clear how these contaminants enter the atmosphere before being accumulated on land through precipitation. Hence, it is necessary to do more research regarding this matter.

15.4.2 Sorption processes in river flow sediments

Pharmaceuticals and PCPs are utilized in excess every day by various communities and are discharged into the water bodies with or without treatment. It is observed that, due to the sorption ability of these compounds with the suspended sediments or solids, their concentration decreases (Osenbrück et al., 2007). This necessarily does not specify that the contaminants have been removed from the water bodies. Polycyclic aromatic hydrocarbons are used for various applications and due to its persistency, it is considered to be an emerging contaminant. These compounds are found in Willamette River (USA), Hai River Tianjin (China) and Seine Estuary (France) at different concentrations of 16,000, 5,540, 1,177 n ng/sediment, respectively. Transportation of such compounds is aided by such adsorption processes with suspended solids (Gregg et al., 2015).

If the suspended solids possess higher cation exchange capacity and organic matter, it can increase the CEC adsorption process that helps in decreasing the concentration of contaminants. But in the case of fast-flowing rivers, the adsorption process is decreased, resulting in an increasing concentration of CECs. Indian rivers like Ganges and Yamuna have been contaminated by various emerging contaminants, especially pharmaceuticals due to the discharge of effluents from industries and households. Since it is a flowing river, the adsorption process is weakened which in turn increases the concentration of the CECs. In the Yamuna, a high concentration of up to 172.6 ng/L of amoxicillin, an antibiotic, was reported.

The concentration reduced to 13.75 μg/L during the monsoon since the contaminated water is replaced by freshwater flowing from glaciers in the Yamuna River, mainly along the stretch of Delhi since there is less flow of freshwater (Philip et al., 2018).

15.4.3 Biodegradation

Once entering the environment, CECs will degrade into other simpler compounds. Most of the biological degradation is either by human or animal metabolism and by microorganisms that are present in sewage water treatment plants that can take place through hydrolysis, reduction and oxidation (Christensen and Li, 2014). Temperature, availability of nutrients, pH, the flow of river and salinity are such factors that play an important role in biodegradation which are usually found in swamps and wetlands. It is important to study the occurrence of the parent compound rather than studying its degradation procedure to understand its potential pathways. For example, pharmaceuticals, when consumed, undergo metabolism in the body that is released out through excretion, eventually reaching the sewage water treatment plants indicating the initial presence of degraded contaminants in sewage treatment inflowing.

The degradation of CECs is much higher when it interacts with microorganisms like bacteria, fungi and so on when compared to humans and animals (Kumar et al., 2012). Enzymes are known for their crucial role in organic contaminant degradation. The possible pathways of CECs when it is degraded through enzymes are as follows:

1. Bioavailability of microbes to CECs.
2. Uptake of CECs and diffusion through cell walls.
3. Complex of contaminant enzyme is formed.
4. Reaction of contaminant enzyme is facilitated by coenzymes and cofactors.
5. Discharge of converted products (Christensen and Li, 2014).

The reaction sites are, however, highly specific, the binding and route through the biological membrane are widely reliant on the polarity and chemical composition of a particular contaminant. For instance, binding of hydrophobic CECs to the membrane is easier and degrades swiftly when compared to hydrophilic CECs (Christensen and Li, 2014). The degradation process can occur during the activated sludge process in sewage treatment that is mediated by microorganisms. The toxicity of biodegraded CECs can be just as or highly toxic as the parent compound. The breakdown transitions of certain CECs and its pathway are yet to be studied in detail.

15.4.4 Bioaccumulation

Various CECs are produced for different intentions that are considered to be persistent and harmful to the ecosystem. Pharmaceuticals and perfluorinated substances are present in water sources that can cause potential environmental risks because they can be consumed by animals as well as humans and can be bioaccumulated in the food chain (Suja et al., 2009).

In a study, it was noticed that oestrogenic compounds have high bioaccumulation potential posing threat to aquatic species causing reproductive impairments and intensify unusual hormonal control (Topp et al., 2008). Bioaccumulation of CECs in aquatic species takes place predominantly at respiratory and gastrointestinal epithelia, which are biologically fabricated for efficient uptake (Maulvault et al., 2019).

Perfluorinated substances have been noticed in human blood and tissues due to the consumption of contaminated drinking water. Serum samples of humans collected from a different region in Japan contained perfluorooctanesulfonate and perfluorooctanoate at a concentration of 3.5–28 μg/L and 2.8–12.4 μg/L, respectively (Harada et al., 2004).

Bisphenol A (BPA) is an industrial chemical used for various applications such as food wrap-ups, pharma products, vehicles and coating of paper. BPA has accumulated in various environmental pathways due to improper discharge and has been identified in tissues of aquatic species in both freshwater and seas including various other amphibians, molluscs, algae and insects eventually entering the food chain. Data obtained from various studies showed that the concentration of BPA in fishes, amphibians and humans was found to be 13,000, 458 and 5.62 ng/g, respectively (Corrales et al., 2015).

The PCPs and pharmaceuticals are continuously discharged into the environment through various pathways. These compounds get accumulated in plant tissues through various environmental sources. Bioaccumulation of PCPs and pharmaceuticals is higher in greenhouse condition rather than natural conditions because microorganisms play a vital role in the persistence of these contaminants in the environment. Under greenhouse and natural conditions, the atenolol concentration in certain plants was reported to be 430 and 55 ng/g, respectively. Similarly, the concentration of sulfamethoxazole was 460 ng/g and 5.25 μg/kg, respectively (Keerthanan et al., 2020).

15.4.5 Ecological pathways

Due to the frequent discharge of CECs through various sources, they get transported through the soil, air, water, and other geological pathways in trace concentration which are persistent in the environment.

- **Soil:** CECs are exposed to soils due to various external activities such as agricultural activities. This can lead to groundwater pollution due to the transport of various contaminants through soil pores. For example, nanoparticles discharged on an open surface are taken up by the soil and may be degraded by the available microorganism. CECs such as di-n-butyl phthalate, di (2-ethylhexyl) phthalate and dimethyl phthalate were found in different concentrations of 1,100, 70 and

150 mg/kg in various soil samples (Yan et al., 2010). Fire retardants, which are commonly used as fire repellents, were also reported on soil reaching up to a concentration of 3.89 ng/g (Yan et al., 2010).

- **Air:** An important transport media for CECs in the air that can travel a far distance from its source of origin. For example, pesticides are found in the air due to due spraying of pesticides on plants (Mas et al., 2020). Since pesticide droplets are tiny, they can easily drift away to far of places. Nanoparticles are other such emerging contaminants that are bigger than air molecules. These particles can modify their physical properties such as shape, size and density when discharged into the air (Armstead and Li, 2016). Certain CECs released in the atmosphere can be harmful to humans since they can enter the bloodstreams of vital organs.
- **Water:** CECs can enter the water bodies via point sources such as landfills and non-point sources such as water run-off. Pesticides are commonly found CECs on water bodies since they are discharged with high concentrations affecting the aquatic habitat. Pesticides accumulated on the water surface can be transported to a greater distance (Mas et al., 2020). Transport of CECs occurs mainly through run-off and leaching processes. The presence of such contaminants in water can cause various adverse effects on humans and animals. For example, animals that consumed contaminated water containing DDTs or other substances have developed reproduction abnormalities (Pal et al., 2014). Similarly, humans consuming this water can cause cancer, decrease sperm count and various other complications (Fawell and Ong, 2012).

15.5 CECs IN SURFACE AND GROUNDWATER

15.5.1 Indian scenario

With the increase in the accumulation of CECs in the environment, it is necessary to consider it as an emergency issue that should be taken care of around the world. India particularly is responsible for the immense production and consumption of a wide variety of CECs. India is one of the top five countries that produce pharmaceuticals that are accountable for 50% of global exports for several vaccines and in terms of revenue the Indian pharmaceutical sector is responsible for 70% of market shares (IBEF, 2020). Many such CECs are widely produced in India. It was reported that out of 38,255 million litres per day of wastes generated from urban areas, only 11,787 million litres per day is being treated and the rest is released into the environment untreated causing a threat to the surface and groundwater system in India (CSE, 2016).

Trace levels of pharmaceuticals were found in rivers Ganges and Yamuna due to the discharge of non-treated sewage water (Mutiyar and Mittal, 2013). Ciprofloxacin, an antibiotic, has been reported in higher concentrations of 31 mg/L in surface waters, surrounding the pharmaceutical industry in Patancheru, Hyderabad (Philip et al., 2018). In a study conducted in Kaveri, Vellar, and Tamiraparani rivers, the concentration of ibuprofen was reported to be 125, 9.88 and 200 ng/L and the concentration of diclofenac was 103, 1.78 and 16.1ng/L, respectively (Khosravi et al., 2010; Shanmugam et al., 2014). Caffeine and ciprofloxacin were reported in rivers of Nag at a concentration of 46.97 and 28.23 ppb and Pili at a concentration of 36.67 and 5.08 ppb in Nagpur due to the discharge of wastes from various sources (Archana et al., 2016; Shivaraju, 2011). Perfluorinated substances were common in the Ganges due to heavy discharge of these substances, roughly estimated to be around hundreds of kilograms per year (Sharma et al., 2016). These substances were found in a high concentration ranging from 4 to 93 ng/L in the Cooum River in Chennai and were also reported in Kaveri and Noyyal rivers including lakes that are around Chennai (Sunantha and Vasudevan, 2016). Parabens were identified in water samples of Pichavaram mangrove at a concentration of 33.3–124 ng/L (Philip et al., 2018).

Drugs such as diclofenac and pitavastatin were reported in groundwater samples across few villages in Punjab with an average concentration of 100 µg/mL and 290 ng/mL, respectively (Jindal et al., 2015). Fluoroquinolones at a mean concentration of 100 ng/L were found in the groundwaters of Patancheru, Hyderabad, due to the discharge of waste from the industries surrounding the area (Rutgersson et al., 2014). Various types of perfluorinated substances were detected in groundwater near the Ganges River with high concentrations (Sharma et al., 2016). Pharmaceuticals and PCPs were found in groundwater around the Ganges river basin and the Himalayan regions with concentration ranging from 0.4 to 10.2 ng/L (Sharma et al., 2019).

CECs are being produced and distributed across the world from India. But on the negative side, the country's major ground and surface water resources are being exploited and contaminated. Hence, national-level monitoring of these CECs is necessary to prevent further environmental risks.

15.5.2 World scenario

a. Asia

Around half the world's population resides in Asia due to which there is a huge burden on environmental resources. In countries like Pakistan and Sri Lanka, pharmaceuticals and PCPs are being manufactured in abundance and are discharged without treatment into water bodies. In Lahore, one of the largest cities in Pakistan, acetaminophen was reported in the domestic and industrial area at a concentration ranging up to 13,000 ng/L are released into the aquatic ecosystem without any treatment causing risks both to the environment and humans (Ashfaq et al., 2019). A similar situation was reported in urban regions of Sri Lanka where sulfamethoxazole, *N, N*-diethyl-meta-toluamide and clarithromycin were reported at a concentration of 934, 202 and 119 ng/L (Guruge et al., 2019). In Indonesian seafood, various CECs are due to emissions from industries, which are found in low concentration but can cause health issues with long-term exposure (Dwiyitno, 2019).

b. Africa

Studies related to CECs are very limited in Africa covering only 17% of the continent. With the available studies, it is observed that pharmaceuticals and PCPs are the major CECs that are found in various water bodies. Neonicotinoid pesticides are currently being used widely in the different African regions and their environmental impacts are less understood. Various pharmaceuticals, PCPs and pesticides are identified in groundwaters in different regions at a concentration of 0.1 ng/L to

18.4 μg/L and 0.02ng/L to 34 μg/L in surface waters (K'oreje et al., 2020). CECs are found in groundwater due to no proper treatment plants in countries like Ghana, Ethiopia, Kenya, South Africa and Zambia but in low concentration when compared to surface water. Since there are no many studies related to emerging contaminants in the African context, interest should be shown in this area for a better understanding of CECs.

c. North America

Around 12% of the USA and Canada population depends on groundwater contaminated with CECs. Perfluorinated substances were found in groundwater due to the discharge from industries, army fire drill and various other anthropogenic activities with groundwater being the major source at an average concentration of 0.17–0.30 μg/L (Lee and Murphy, 2020). Laurentian Great Lake which is present in the upper-middle east region of North America has been contaminated with CECs such as indole and cholesterol at concentration level from 0.0284 to 72.2 μg/L due to human activities resulting in disruption of aquatic habitat. It was also reported that the presence of CECs in this lake may cause adverse effects on reproduction and sustainability in fishes (Thomas et al., 2017). In the Caribbean region, CECs, such as pharmaceuticals, and EDCs, such as caffeine, trimethoprim and oestrone, are released into water bodies at concentrations of 1313, 8.5 and 2.8 ng/L, respectively. With many tourists visiting the West Indies every year, the concentration of certain CECs such as pharmaceuticals, EDCs and caffeine was found to be higher (Edwards et al., 2018).

d. South America

In certain regions of Argentina, rainwater is the main source of water because of less surface water and low-quality groundwater due to an increase in land use for agriculture with no-till management and heavy use of pesticides making it non-consumable (Mas et al., 2020). Dams, cisterns and wells were found to be the major sources of CECs in this region. Perchlorate, a type of CECs, was reported in the Atacama Desert in Chile due to its use in various applications such as fireworks, rocket fuel and other industrial products. In drinking water, perchlorate concentration was 1,480 μg/L, and in breast milk, trace amount up to 100 μg/L of perchlorate was reported. Perchlorate was reported in a natural water spring that is used for recreation purposes, groundwater in a very low concentration and wells in the surrounding region of Atacama Desert causing potential risk to human health (Vega et al., 2018).

e. Europe

The PCPs and perfluorinated substances were reported to be the major contributors to water contamination in some European countries. A higher concentration of CECs such as musk with concentration up to 92 ng/L was reported in southern Europe due to the increase in tourism (Aminot et al., 2019). In separate studies conducted across Europe, CECs such as perfluorinated substances, musks and insect repellents were reported in Italy, Portugal, Spain and the United Kingdom (Munoz et al., 2017). Po river delta situated in Italy has exposed to industrial and domestic discharge especially pharmaceuticals, PCPs and pesticides that are considered to be a threat to the surrounding ecosystem, for instance, 80 ng/g N, N-diethyl-meta-toluamide was reported in the surrounding area of the river (Mascolo et al., 2019). Groundwaters in European countries were also reported to have been contaminated by CECs mostly consisting of pharmaceutical, perfluorinated, substances (Lapworth et al., 2019).

f. Australia

The Great Barrier Reef (GRB) is famous for its coral reefs and unique aquatic species. Commercial and non-commercial activities such as fishing, tourism, shipping and other activities are allowed here. But in the past few decades, the GRB ecosystem has been affected by various factors, CECs being one of the factors (Kroon et al., 2020). Perfluorinated compounds, nanomaterials, explosives, pharmaceuticals and PCPs were the major CECs that were reported with concentration ranging from 1.1 to 10,000 μg/L (Kroon et al., 2020). The source of these CECs was water discharge from the water treatment plants, landfills, tourism, aquaculture and agriculture. Pharmaceuticals (paraxanthine-10.5 ng/L, paracetamol-6 ng/L and acesulfame-103 ng/L), PCPs (triclosan-2.8 ng/L and salicylic acid-0.1 ng/L), and pesticides (triclopyr-226.44 ng/L and gabapentin-117.6 ng/L) were reported in Brisbane river estuary (Anim et al., 2020).

15.6 HEALTH IMPACTS

The occurrence of CECs in the environment is increasing at an alarming rate, posing a major risk to human health. The use of CECs especially pharmaceuticals and PCPs among humans have been drastically increased in past years. The exposure of these contaminated compounds to humans depends on their fate and behaviour in water treatment plants. Reports have shown abnormalities in newborns and children due to low semen quality and testosterone levels and an increase in thyroid cancer. Change in kidney structure and intestines including gene alterations affecting the metabolic activities of the body was also noticed (Stuart et al., 2012). The concentration of the CECs, exposure time, volume and metabolism rate possess possible risks when consuming water but due to the presence of previous natural health issues in the human body, it is hard to determine the effects of CECs on human health. Perfluorinated compounds exposed to humans have been widely studied and were observed that fertility may reduce. Among women, it was noticed that the concentration of perfluorinated reduced the capacity of a woman to lactate (Lei et al., 2015). Disinfection chemicals are a class of CECs that are used in swimming pools and water purification, creating unwanted by-products during the disinfection process. A study showed that these by-products may affect ovary function with reduced menstrual cycle length and follicular phase length (Windham et al., 2003).

Overuse of antibiotics for human and animal treatments has led to an increase in antibiotic-resistant genes and antibiotic-resistant bacteria in water bodies that are capable of altering the physiological properties of the body (Kim and Aga, 2007). These antibiotic-resistant substances are discharged into drinking water especially in developing countries causing major health issues or even death when consumed. Possible ways humans are affected by antibiotic resistance are: direct consumption of contaminated water or transmission from the person who is already affected to another non-affected person (Laxminarayan and Chaudhury, 2016). Klebsiella pneumoniae is an antibiotic-resistant substance that causes serious health issues such as urinary tract, respiratory tract and bloodstream infection. In countries like India where there are poor sanitation and unclean drinking water, Shigella species that are resistant to ciprofloxacin result in diarrhoea and other fatal infections, especially in infants (Sanganyado and Gwenzi, 2019).

Pharmaceuticals such as amoxicillin, ciprofloxacin and triclosan were reported to have been found in tap water in major cities. In Malaysia, the concentration of the mentioned pharmaceuticals in tap water was found to be 0.31, 0.32 and 0.36 ng/L, respectively (Praveena et al., 2021). When consumed, the remnants of pharmaceuticals present in tap water are absorbed in the biological system causing allergic reactions. Multiple remains of pharmaceuticals when in contact with a human can cause adverse reactions resulting in disruption of the metabolic system. Water system contaminated by pharmaceutical residues is reported to have contaminated the consumable plants. Traces of pharmaceutical were detected in potato, tomato, and carrot (Keerthanan et al., 2020). When regularly exposed to pharmaceuticals, particularly antibiotics will restrict the antimicrobial activity in the body eventually increasing the risk of death. Approximation of potential health risks by pharmaceuticals and other CECs is assessed by risk quotients (RQs) by obtaining the data from the survey. The sum of RQs from each pharmaceutical is referred to as total risk quotients (RQ_T). If the RQ_T value exceeds 1, it is indicated that human health is potentially at risk through various exposure routes and vice versa.

The major sources of drinking water in India and other developing countries are frequently discharged with various CECs without proper treatment causing a threat to human health. Some studies have shown that water-contaminated CECs may not pose a major threat to human health but can cause health complications to people who already have a history of having diseases.

15.7 ECOLOGICAL IMPACTS

Evaluating the level of exposure of various types of CECs on the ecosystem and their risks is crucial. From the existing records, the concentration of CECs that are detected is generally low, making it difficult to analyse its acute toxicity. The toxic effect of CECs especially pharmaceuticals and PCPs is much higher in aquatic species due to their constant exposure throughout their lifespan. Certain CECs does not metabolize causing them to enter the food chain (Fawell and Ong, 2012). The exposure of EDCs, especially organochlorines, has resulted in feminization in certain aquatic species, and sexual anomalies in alligators were also noticed. Various studies have shown that EDCs are responsible for low sperm count and malfunction in the reproductive system of aquatic species (Joffe, 2001). Prokaryotic cells are affected when interacted with antibiotics through various mechanisms. When exposed to non-therapeutically, antibiotics can cause severe impacts on aquatics species and environmental microbes. CECs can affect microorganisms as well as animals. Various issues, such as alterations in genes, unusual protein and enzyme activities, growth malfunction, were observed in certain vertebrates and amphibians due to exposure to antibiotics (Patel et al., 2019).

In a review, the toxicity of certain CECs was analysed for selected species. Predicted environmental concentration (PEC) and predicted no-effect concentration (PNEC) were calculated from the data obtained from the Korean Index of Medical Specialities (KIMS). In Korea, the amount of acetaminophen, carbamazepine, sulfamethoxazole and trimethoprim manufactured for the year 2003 was 1,068,921, 9,155,

61,272 and 13,553 kgs, respectively. PEC value was calculated from the mentioned values using the formula

$$PEC = A * \frac{(100 - R)}{365 * P * V * D * 100} * 10^9 \quad (15.1)$$

where

A manufactured substance per year
R sewage treatment plant removal rate in percentage
P population number of a particular region
V wastewater produced per capita
D dilution factor in environment (value = 10).

After calculation, the PEC values of acetaminophen, carbamazepine, sulfamethoxazole and trimethoprim were 16.5, 0.14, 0.95 and 0.21 μg/L, respectively. PNEC values for the same were found to be 9.2, 31.6, 0.146 and 120.7 μg/L, respectively (Kim et al., 2007). PEC is referred to as the concentration of CECs present in the environment and PNEC is the concentration of a particular CEC in an environment below which adverse long-term/short-term effects most likely will not occur. PNEC values are determined from lethal/mean effective concentration (L/EC_{50}). If the ratio of PEC and PNEC is less than 1, it signifies that there is no potential risk to the environment with the presence of the CECs. If the value is greater than 1, environmental risk is possible and further steps should be taken to the further contamination of the environment. Table 15.2 further explains the L/EC_{50} of *D. magna*, *Cyclotella meneghiniana* and *Selenastrum capricornutum*, PEC and PNEC values and the risk factor of a few CECs (de Oliveira et al., 2016; Vélez et al., 2019).

The decline in vulture population was reported in certain regions of India due to the feasting of carcasses of cattle that was medicated with diclofenac (Puri et al., 2018). A decline in the growth of tadpoles was also reported due to the high concentration of antibiotics in the water system. A study conducted in a South Indian fish farm showed that freshwater fishes were more vulnerable because of the toxic effect of pharmaceutical compounds (Philip et al., 2018). CECs enter the aquatic bodies beyond acceptable limits affecting the ecosystem, especially the aquatic environment. The impacts of

Table 15.2 List of model emerging contaminants with acute median effective concentration (L/EC_{50}), predicted environmental concentration (PEC) and predicted no-effect concentration (PNEC)

Emerging contaminants	*Description*	*LC_{50}/EC_{50} (mg/L)*	*PEC (μg/L)*	*PNEC (μg/L)*	*PEC/ PNEC*	*Risk factor*
Acetaminophen	Analgesic	30.1[a]	16.5	9.2	1.8	Likely
Carbamazepine	Anticonvulsant	31.6[b]	0.14	31.6	0.0044	Unlikely
Diclofenac	Anti-inflammatory	123.3[a]	0.14	10	0.014	Unlikely
Sulfamethoxazole	Antibiotics	0.146[c]	0.95	0.146	6.3	Likely
Trimethoprim	Antibiotic	167.4[a]	0.21	120.7	0.0017	Unlikely

[a] Daphnia magna (48 h)
[b] Cyclotella meneghiniana
[c] Selenastrum capricornutum

CECs on the ecosystem can depend on pH, soil/water type and various other factors. Hence, it is necessary to create a comprehensive framework to reduce the discharge of CECs into the environment.

15.8 RESILIENCE OF CECs REMEDIATION

As mentioned earlier, with an increase in demand for various products, CECs are discharged into the environment every day since they have no standard limit for discharge resulting in contamination of water resources and other environmental systems. Since CECs are present in the environment for decades, it is necessary to come up with various strategies to remove these contaminants completely. There is strong evidence that current conventional water treatment techniques do not effectively remove the CECs which eventually is released back to the environment (Salimi et al., 2017). Alternatively, various other techniques can minimize or completely remove CECs before entering the environment (Shivaraju et al., 2020, Shim et al., 2018).

Aerobic treatment such as the activated sludge process has shown promising results in the efficient removal of EDCs. A study conducted in Sweden revealed that a good amount of oestrone and 17β-oestradiol were removed using activated sludge process when compared to other EDC removal techniques (Svenson et al., 2003). Natural aerobic treatments such as waste stabilization ponds (WSPs) and constructed wetlands (CWs) are used to efficiently remove pharmaceuticals and PCPs. The average removal efficiency of pharmaceuticals and PCPs was observed to be 82% in WSPs (Li et al., 2013). CWs can be used as an option to remove pharmaceuticals in secondary wastewater treatment plants since they can remove various types of pharmaceuticals with high efficiency (Li et al., 2014).

Physical–chemical treatments such as activated carbon adsorption, coagulation, advanced oxidation process (AOP), ozonation, chlorination, ultraviolet irradiation and nanofiltration have shown promising results in the effective removal of CECs. Activated carbon adsorption can be used to treat water contaminated with pharmaceuticals including EDCs (Le-Minh et al., 2010). In the coagulation technique, colloidal particles are altered which causes aggregation by chemicals letting them flump gradually. Pharmaceuticals like betaxolol, warfarin and hydrochlorothiazide were reported to have a high removal efficiency in this technique (Huerta-Fontela et al., 2011). Notable removal rates were observed in the case of PCPs especially musks in the coagulation technique. AOP is another such method for effective removal of CECs which is achieved by the generation of hydroxyl radicals (Shivaraju et al., 2017), e.g. titanium dioxide-based photocatalysis (Shivaraju et al., n.d.). Various antibiotics were removed with ozonation treatment such as sulfonamides, fluoroquinolones and macrolides (Gogoi et al., 2018). The addition of chlorine can inactivate the functional group making the compound non-functional that is usually referred to as chlorination (Gogoi et al., 2018). Using this technique, antibiotics can be oxidized. Free chlorine and sufficient contact are necessary to remove CECs efficiently from water. It was reported that when exposed to chlorine for 10 days, antibiotics such as tetracyclines, trimethoprim, sulfonamides and fluoroquinolones were completely removed from water (Gibs et al., 2007). Another budding technique is nanofiltration which has shown almost 90% efficiency in the removal of pharmaceuticals. Various techniques and the removal efficiency are represented in Figure 15.3.

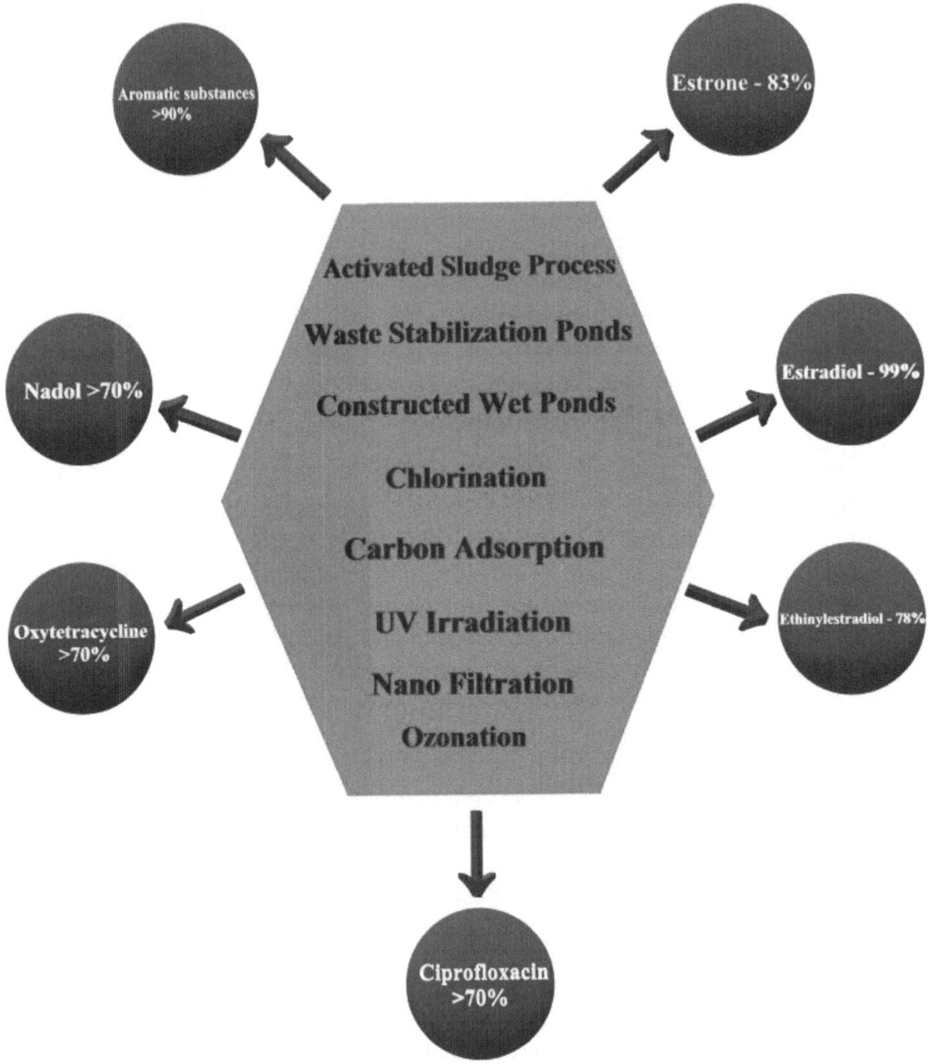

Figure 15.3 Removal techniques and efficiency on various CECs.

15.9 CONCLUSION

Contamination of CECs is inevitable as long as their uses remain a crucial factor of a modern and healthy society. But with an increase in the accumulation of various CECs every day, it is uncertain how it will affect humans and the environment. CECs have not shown significant health impacts, but the concentration of a single compound may cause significant risks to humans. Due to fewer studies regarding CECs and their impact on the aquatic ecosystem, their potential impact on aquatic organisms is still not understood clearly.

Other than the previously mentioned CECs, there are a new class of emerging contaminants that we are not aware of. The fate of CECs after their treatment and their toxicity is not known in detail. CECs exhibit a different ecotoxicity in the natural environment when compared to convention tests conducted in the laboratory. Hence, it is necessary to develop new procedures to understand the precise toxicity of CECs.

In India, the concept of CECs is not understood well due to limited research. The available studies are mostly reported by scientists from other countries. Hence, India needs high-end equipment and laboratories to study CECs without depending on other countries. Very less is known about the toxicity and risk assessment. Over 166 CECs were reported in India but only 15 were studied in detail regarding their toxicity (Philip et al., 2018). Many new contaminants may be identified in future and it is necessary to come up with different strategies to overcome this issue.

To prevent the CECs from entering the environment, an attempt should be made to prevent the discharge of CECs for the source. Steps such as enforcing strict laws, ideal operation of wastewater treatment and reducing the use of substances that are emerging contaminants should be followed. Efficient wastewater must be installed to ensure safe drinking water. It is also important to study the fate and treatment of man-made metabolites in the wastewater treatment plants. In future, to understand the transportation and fate of CECs, research should be focused on the development of an advanced framework to determine the risk and toxicity of CECs. Nevertheless, this topic is attracting many researchers nowadays, is considered to gain interest in further years and eventually helps us understand CECs better.

REFERENCES

Aminot, Y., Sayfritz, S.J., Thomas, K.V., Godinho, L., Botteon, E., Ferrari, F., Boti, V., Albanis, T., Köck-Schulmeyer, M., Diaz-Cruz, M.S., Farré, M., Barceló, D., Marques, A., Readman, J.W., 2019. Environmental risks associated with contaminants of legacy and emerging concern at European aquaculture areas. *Environmental Pollution* 252, 1301–1310. https://doi.org/10.1016/j.envpol.2019.05.133.

Anim, A.K., Thompson, K., Duodu, G.O., Tscharke, B., Birch, G., Goonetilleke, A., Ayoko, G.A., Mueller, J.F., 2020. Pharmaceuticals, personal care products, food additive and pesticides in surface waters from three Australian east coast estuaries (Sydney, Yarra and Brisbane). *Marine Pollution Bulletin* 153, 111014. https://doi.org/10.1016/j.marpolbul.2020.111014.

Archana, G., Dhodapkar, R., Kumar, A., 2016. Offline solid-phase extraction for preconcentration of pharmaceuticals and personal care products in environmental water and their simultaneous determination using the reversed phase high-performance liquid chromatography method. *Environmental Monitoring and Assessment* 188, 512.

Armstead, A.L., Li, B., 2016. Nanotoxicity: Emerging concerns regarding nanomaterial safety and occupational hard metal (WC-Co) nanoparticle exposure. *International Journal of Nanomedicine* 11, 6421–6433. https://doi.org/10.2147/IJN.S121238.

Ashfaq, M., Li, Y., Rehman, M.S.U., Zubair, M., Mustafa, G., Nazar, M.F., Yu, C.-P., Sun, Q., 2019. Occurrence, spatial variation and risk assessment of pharmaceuticals and personal care products in urban wastewater, canal surface water, and their sediments: A case study of Lahore, Pakistan. *Science of the Total Environment* 688, 653–663. https://doi.org/10.1016/j.scitotenv.2019.06.285.

Authority, E.F.S., 2011. Results of the monitoring of perfluoroalkylated substances in food in the period 2000-2009. *EFSA Journal* 9, 2016.

Bergh, C., Torgrip, R., Emenius, G., Östman, C., 2011. Organophosphate and phthalate esters in air and settled dust–a multi-location indoor study. *Indoor Air* 21, 67–76.

Christensen, E.R., Li, A., 2014. *Physical and Chemical Processes in the Aquatic Environment*. John Wiley & Sons, Hoboken, NJ.

Corrales, J., Kristofco, L.A., Steele, W.B., Yates, B.S., Breed, C.S., Williams, E.S., Brooks, B.W., 2015. Global assessment of bisphenol A in the environment: Review and analysis of its occurrence and bioaccumulation. *Dose-Response* 13. https://doi.org/10.1177/1559325815598308

CSE, 2016. Question raised in Lok Sabha on Management of Sewage [WWW Document]. India Environment Portal. URL http://www.indiaenvironmentportal.org.in/content/425690/question-raised-in-lok-sabha-on-management-of-sewage-24022016/ (Accessed 10.6.20).

Damstra, T., Barlow, S., Bergman, A., Kavlock, R., Van Der Kraak, G., 2002. *Global Assessment of the State-of-the-Science of Endocrine Disruptors*. Geneva: World Health Organization, 11–32.

de Oliveira, L.L.D., Antunes, S.C., Gonçalves, F., Rocha, O., Nunes, B., 2016. Acute and chronic ecotoxicological effects of four pharmaceuticals drugs on cladoceran Daphnia magna. *Drug and Chemical Toxicology* 39, 12–21. https://doi.org/10.3109/01480545.2015.1029048.

Dey, S., Bano, F., Malik, A., 2019. Pharmaceuticals and personal care product (PPCP) contamination—a global discharge inventory. In: *Pharmaceuticals and Personal Care Products: Waste Management and Treatment Technology*. Elsevier, pp. 1–26.

Dubey, M., Mohapatra, S., Tyagi, V.K., Suthar, S. and Kazmi, A.A., 2021. Occurrence, fate, and persistence of emerging micropollutants in sewage sludge treatment. *Environmental Pollution*, 273, 116515.

Dwiyitno, D., 2019. Emerging Pollutants from Industrial Emission and Their Health Hazards in Indonesian Seafood: A Review. *Squalen Bulletin of Marine and Fisheries Postharvest and Biotechnology* 14, 141–153. https://doi.org/10.15578/squalen.v14i3.399.

Edwards, Q.A., Sultana, T., Kulikov, S.M., Garner-O'Neale, L.D., Yargeau, V., Metcalfe, C.D., 2018. Contaminants of emerging concern in wastewaters in Barbados, West Indies. *Bulletin of Environmental Contamination and Toxicology* 101, 1–6. https://doi.org/10.1007/s00128-018-2346-0.

Eregowda T., Mohapatra S. 2020. Fate of micropollutants in engineered and natural environment. In: Kumar M., Munoz-Arriola F., Furumai H., Chaminda T. (eds.) *Resilience, Response, and Risk in Water Systems. Springer Transactions in Civil and Environmental Engineering*. Springer, Singapore. https://doi.org/10.1007/978-981-15-4668-6_15.

Fawell, J., Ong, C.N., 2012. Emerging contaminants and the implications for drinking water. *International Journal of Water Resources Development* 28, 247–263. https://doi.org/10.1080/07900627.2012.672394.

Flint, S., Markle, T., Thompson, S., Wallace, E., 2012. Bisphenol A exposure, effects, and policy: A wildlife perspective. *Journal of Environmental Management* 104, 19–34. https://doi.org/10.1016/j.jenvman.2012.03.021.

Gibs, J., Stackelberg, P.E., Furlong, E.T., Meyer, M., Zaugg, S.D., Lippincott, R.L., 2007. Persistence of pharmaceuticals and other organic compounds in chlorinated drinking water as a function of time. *Science of the Total Environment* 373, 240–249.

Gogoi, A., Mazumder, P., Tyagi, V.K., Tushara Chaminda, G.G., An, A.K., Kumar, M., 2018. Occurrence and fate of emerging contaminants in water environment: A review. *Groundwater for Sustainable Development* 6, 169–180. https://doi.org/10.1016/j.gsd.2017.12.009.

Gregg, T., Prahl, F.G., Simoneit, B.R., 2015. Suspended particulate matter transport of polycyclic aromatic hydrocarbons in the lower Columbia River and its estuary. *Limnology and Oceanography* 60, 1935–1949.

Guruge, K.S., Goswami, P., Tanoue, R., Nomiyama, K., Wijesekara, R.G.S., Dharmaratne, T.S., 2019. First nationwide investigation and environmental risk assessment of 72 pharmaceuticals and personal care products from Sri Lankan surface waterways. *Science of the Total Environment* 690, 683–695. https://doi.org/10.1016/j.scitotenv.2019.07.042.

Harada, K., Saito, N., Inoue, K., Yoshinaga, T., Watanabe, T., Sasaki, S., Kamiyama, S., Koizumi, A., 2004. The influence of time, sex and geographic factors on levels of perfluorooctane sulfonate and perfluorooctanoate in human serum over the last 25 years. *Journal of Occupational Health* 46, 141–147.

Huerta-Fontela, M., Galceran, M.T., Ventura, F., 2011. Occurrence and removal of pharmaceuticals and hormones through drinking water treatment. *Water Research* 45, 1432–1442.

IBEF, 2020. Indian Pharmaceutical Industry [WWW Document]. Indian Brand Equity Foundation. URL https://www.ibef.org/pages/pharmaceutical-india.aspx (Accessed 10.6.20).

Jackson, J., Sutton, R., 2008. Sources of endocrine-disrupting chemicals in urban wastewater, Oakland, CA. *Science of The Total Environment* 405, 153–160. https://doi.org/10.1016/j.scitotenv.2008.06.033.

Jiang, J.-Q., Zhou, Z., Sharma, V.K., 2013. Occurrence, transportation, monitoring and treatment of emerging micro-pollutants in waste water—A review from global views. *Microchemical Journal* 110, 292–300. https://doi.org/10.1016/j.microc.2013.04.014.

Jindal, K., Narayanam, M., Singh, S., 2015. A systematic strategy for the identification and determination of pharmaceuticals in environment using advanced LC–MS tools: Application to ground water samples. *Journal of Pharmaceutical and Biomedical Analysis* 108, 86–96.

Joffe, M., 2001. Are problems with male reproductive health caused by endocrine disruption? *Occupational and Environmental Medicine* 58, 281–281.

Keerthanan, S., Jayasinghe, C., Biswas, J.K., Vithanage, M., 2020. Pharmaceutical and Personal Care Products (PPCPs) in the environment: Plant uptake, translocation, bioaccumulation, and human health risks. *Critical Reviews in Environmental Science and Technology* 1–38. https://doi.org/10.1080/10643389.2020.1753634.

Khosravi, T., Shivaraju, H.P., Sajan, C.P., Byrappa, K., 2010. Impact assessment of effluent discharge on underground water qualities around Gemini Distillery, Nanjangud, Mysore District. *International Journal of Applied Environmental Sciences* 5, 617–627.

Kim, S., Aga, D.S., 2007. Potential ecological and human health impacts of antibiotics and antibiotic-resistant bacteria from wastewater treatment plants. *Journal of Toxicology and Environmental Health, Part B* 10, 559–573. https://doi.org/10.1080/15287390600975137.

Kim, Y., Choi, K., Jung, J., Park, S., Kim, P.-G., Park, J., 2007. Aquatic toxicity of acetaminophen, carbamazepine, cimetidine, diltiazem and six major sulfonamides, and their potential ecological risks in Korea. *Environment International* 33, 370–375. https://doi.org/10.1016/j.envint.2006.11.017.

K'oreje, K.O., Okoth, M., Van Langenhove, H., Demeestere, K., 2020. Occurrence and treatment of contaminants of emerging concern in the African aquatic environment: Literature review and a look ahead. *Journal of Environmental Management* 254, 109752. https://doi.org/10.1016/j.jenvman.2019.109752.

Kroon, F.J., Berry, K.L.E., Brinkman, D.L., Kookana, R., Leusch, F.D.L., Melvin, S.D., Neale, P.A., Negri, A.P., Puotinen, M., Tsang, J.J., van de Merwe, J.P., Williams, M., 2020. Sources, presence and potential effects of contaminants of emerging concern in the marine environments of the Great Barrier Reef and Torres Strait, Australia. *Science of the Total Environment* 719, 135140. https://doi.org/10.1016/j.scitotenv.2019.135140.

Kumar, M., Munoz-Arriola, F., Furumai, H., Chaminda, T. (Eds.), 2020. *Resilience, Response, and Risk in Water Systems: Shifting Management and Natural Forcings Paradigms, Springer Transactions in Civil and Environmental Engineering.* Springer, Singapore. https://doi.org/10.1007/978-981-15-4668-6.

Kumar, V., Johnson, A.C., Nakada, N., Yamashita, N., Tanaka, H., 2012. De-conjugation behavior of conjugated estrogens in the raw sewage, activated sludge and river water. *Journal of Hazardous Materials* 227–228, 49–54. https://doi.org/10.1016/j.jhazmat.2012.04.078.

Kumar, M., Snow, D. D., Li, Y., Shea, P. J. 2019. Perchlorate behavior in the context of black carbon and metal cogeneration following fireworks emission at Oak Lake, Lincoln, Nebraska, USA. *Environmental Pollution*, 253, 930–938.

Kumar, M., Kuroda, K., A. K., Patel, N., Bhattacharya, P., Joshi, M., Joshi, C. G. 2021. Decay of SARS-CoV-2 RNA along the wastewater treatment outfitted with Upflow Anaerobic Sludge Blanket (UASB) system evaluated through two sample concentration techniques. *Science of the Total Environment*, 754, 142329.

Lapworth, D.J., Lopez, B., Laabs, V., Kozel, R., Wolter, R., Ward, R., Amelin, E.V., Besien, T., Claessens, J., Delloye, F., 2019. Developing a groundwater watch list for substances of emerging concern: A European perspective. *Environmental Research Letters* 14, 035004.

Larsson, D.G.J., 2014. Pollution from drug manufacturing: Review and perspectives. *Philosophical Transactions of the Royal Society B: Biological Sciences* 369, 20130571. https://doi.org/10.1098/rstb.2013.0571.

Laxminarayan, R., Chaudhury, R.R., 2016. Antibiotic resistance in India: Drivers and opportunities for action. *PLoS Medicine* 13, e1001974.

Lee, D., Murphy, H.M., 2020. Private wells and rural health: Groundwater contaminants of emerging concern. *Current Environmental Health Reports* 7, 129–139. https://doi.org/10.1007/s40572-020-00267-4.

Lei, M., Zhang, L., Lei, J., Zong, L., Li, J., Wu, Z., Wang, Z., 2015. Overview of emerging contaminants and associated human health effects. *BioMed Research International* 2015, 1–12. https://doi.org/10.1155/2015/404796.

Le-Minh, N., Khan, S.J., Drewes, J.E., Stuetz, R.M., 2010. Fate of antibiotics during municipal water recycling treatment processes. *Water Research* 44, 4295–4323.

Li, X., Zheng, W., Kelly, W.R., 2013. Occurrence and removal of pharmaceutical and hormone contaminants in rural wastewater treatment lagoons. *Science of the Total Environment* 445, 22–28.

Li, Y., Zhu, G., Ng, W.J., Tan, S.K., 2014. A review on removing pharmaceutical contaminants from wastewater by constructed wetlands: Design, performance and mechanism. *Science of the Total Environment* 468, 908–932.

Mas, L.I., Aparicio, V.C., De Gerónimo, E., Costa, J.L., 2020. Pesticides in water sources used for human consumption in the semiarid region of Argentina. *SN Applied Sciences* 2, 691. https://doi.org/10.1007/s42452-020-2513-x.

Mascolo, G., Murgolo, S., Stefani, F., Viganò, L., 2019. Target and suspect contaminants of emerging concern in the Po River Delta lagoons. *Estuarine, Coastal and Shelf Science* 230, 106424. https://doi.org/10.1016/j.ecss.2019.106424.

Maulvault, A.L., Anacleto, P., Marques, A., Rosa, R., Diniz, M., 2019. Chemical contaminants in a changing ocean. *Ecotoxicology of Marine Organisms* 25.

Menon, N.G., Mohapatra, S., Padhye, L.P., Tatiparti, S.S.V., Mukherji, S. 2020. Review on occurrence and toxicity of pharmaceutical contamination in Southeast Asia. In: Kumar M., Snow D., Honda R. (eds.) *Emerging Issues in the Water Environment during Anthropocene*. Springer Transactions in Civil and Environmental Engineering. Springer, Singapore. https://doi.org/10.1007/978-981-32-9771-5_4.

Mohapatra, S., Padhye, L.P., Mukherji, S. 2018. Challenges in detection of antibiotics in wastewater matrix. In: Gupta T., Agarwal A., Agarwal R., Labhsetwar N. (eds.) *Environmental Contaminants. Energy, Environment, and Sustainability*. Springer, Singapore. https://doi.org/10.1007/978-981-10-7332-8_1.

Mohapatra, S., Menon, N., Padhye, L.P., Tatiparti, S.S.V., Mukherji, S. 2021. Natural attenuation of pharmaceuticals in the aquatic environment and role of phototransformation. In: Kumar M., Snow D., Honda R., Mukherjee S. (eds.) *Contaminants in Drinking and Wastewater Sources*. Springer Transactions in Civil and Environmental Engineering. Springer, Singapore. https://doi.org/10.1007/978-981-15-4599-3_3.

Munoz, G., Budzinski, H., Labadie, P., 2017. Influence of environmental factors on the fate of legacy and emerging per-and polyfluoroalkyl substances along the salinity/turbidity gradient of a macrotidal estuary. *Environmental Science & Technology* 51, 12347–12357.

Mutiyar, P.K., Mittal, A.K., 2013. Occurrences and fate of an antibiotic amoxicillin in extended aeration-based sewage treatment plant in Delhi, India: A case study of emerging pollutant. *Desalination and Water Treatment* 51, 6158–6164.

Oono, S., Harada, K.H., Mahmoud, M.A., Inoue, K., Koizumi, A., 2008. Current levels of airborne polyfluorinated telomers in Japan. *Chemosphere* 73, 932–937.

Osenbrück, K., Gläser, H.-R., Knöller, K., Weise, S.M., Möder, M., Wennrich, R., Schirmer, M., Reinstorf, F., Busch, W., Strauch, G., 2007. Sources and transport of selected organic micropollutants in urban groundwater underlying the city of Halle (Saale), Germany. *Water Research* 41, 3259–3270.

Pal, A., He, Y., Jekel, M., Reinhard, M., Gin, K.Y.-H., 2014. Emerging contaminants of public health significance as water quality indicator compounds in the urban water cycle. *Environment International* 71, 46–62. https://doi.org/10.1016/j.envint.2014.05.025.

Patel, M., Kumar, R., Kishor, K., Mlsna, T., Pittman, C.U., Mohan, D., 2019. Pharmaceuticals of emerging concern in aquatic systems: Chemistry, occurrence, effects, and removal methods. *Chemical Reviews* 119, 3510–3673. https://doi.org/10.1021/acs.chemrev.8b00299.

Philip, J.M., Aravind, U.K., Aravindakumar, C.T., 2018. Emerging contaminants in Indian environmental matrices – A review. *Chemosphere* 190, 307–326. https://doi.org/10.1016/j.chemosphere.2017.09.120.

Praveena, S.M., Mohd Rashid, M.Z., Mohd Nasir, F.A., Wee, S.Y., Aris, A.Z., 2021. Occurrence, human health risks, and public awareness level of pharmaceuticals in tap water from Putrajaya (Malaysia). *Expo Health* 13, 93–104. https://doi.org/10.1007/s12403-020-00364-7.

Priyanka M.S., Mohapatra S., 2020. Shifts and trends in analysis of contaminants of emerging concern: Sulfonamides. In: Kumar M., Munoz-Arriola F., Furumai H., Chaminda T. (eds.) *Resilience, Response, and Risk in Water Systems.* Springer Transactions in Civil and Environmental Engineering. Springer, Singapore. https://doi.org/10.1007/978-981-15-4668-6_14.

Puri, K., Yadav, V., Jha, V., Joshi, R., 2018. Abundance of vultures in shivalik foothills of North West India. *International Journal of Science, Environment and Technology* 7, 1750–1754.

Rutgersson, C., Fick, J., Marathe, N., Kristiansson, E., Janzon, A., Angelin, M., Johansson, A., Shouche, Y., Flach, C.-F., Larsson, D.J., 2014. Fluoroquinolones and QNR genes in sediment, water, soil, and human fecal flora in an environment polluted by manufacturing discharges. *Environmental Science & Technology* 48, 7825–7832.

Salgueiro-González, N., De Alda, M.L., Muniategui-Lorenzo, S., Prada-Rodríguez, D., Barceló, D., 2015. Analysis and occurrence of endocrine-disrupting chemicals in airborne particles. *TrAC Trends in Analytical Chemistry* 66, 45–52.

Salimi, M., Esrafili, A., Gholami, M., Jonidi Jafari, A., Rezaei Kalantary, R., Farzadkia, M., Kermani, M., Sobhi, H.R., 2017. Contaminants of emerging concern: A review of new approach in AOP technologies. *Environmental Monitoring and Assessment* 189, 414. https://doi.org/10.1007/s10661-017-6097-x.

Sanganyado, E., Gwenzi, W., 2019. Antibiotic resistance in drinking water systems: Occurrence, removal, and human health risks. *Science of the Total Environment* 669, 785–797. https://doi.org/10.1016/j.scitotenv.2019.03.162.

Shanmugam, G., Sampath, S., Selvaraj, K.K., Larsson, D.J., Ramaswamy, B.R., 2014. Nonsteroidal anti-inflammatory drugs in Indian rivers. *Environmental Science and Pollution Research* 21, 921–931.

Sharma, B.M., Bečanová, J., Scheringer, M., Sharma, A., Bharat, G.K., Whitehead, P.G., Klánová, J., Nizzetto, L., 2019. Health and ecological risk assessment of emerging contaminants (pharmaceuticals, personal care products, and artificial sweeteners) in surface and groundwater (drinking water) in the Ganges River Basin, India. *Science of the Total Environment* 646, 1459–1467. https://doi.org/10.1016/j.scitotenv.2018.07.235.

Sharma, B.M., Bharat, G.K., Tayal, S., Larssen, T., Bečanová, J., Karásková, P., Whitehead, P.G., Futter, M.N., Butterfield, D., Nizzetto, L., 2016. Perfluoroalkyl substances (PFAS) in river and ground/drinking water of the Ganges River basin: Emissions and implications for human exposure. *Environmental Pollution* 208, 704–713.

Shivaraju, H.P., 2011. Preparation and characterization of supported photocatalytic composite and its decomposition and disinfection effect on bacteria in municipal sewage water. *Research Journal of Chemical Sciences* 1, 56–63.

Shivaraju, H.P., Anilkumar, K.M., Yashas, S.R., Harini, R., Shahmoradi, B., Maleki, A., McKay, G., n.d. Facile synthesis of Mn-Ce/N-TiO_2 composite for CO_2 hydrogenation into methane and intensifying methane yield in biomethanation. *Biofuels, Bioproducts and Biorefining*. https://doi.org/10.1002/bbb.2150.

Shivaraju, H.P., Midhun, G., Kumar, K.A., Pallavi, S., Pallavi, N., Behzad, S., 2017. Degradation of selected industrial dyes using Mg-doped TiO_2 polyscales under natural sun light as an alternative driving energy. *Applied Water Science* 7, 3937–3948.

Shivaraju, H.P., Yashas, S.R., Harini, R., 2020. Application of Mg-doped TiO_2 coated buoyant clay hollow-spheres for photodegradation of organic pollutants in wastewater. *Materials Today: Proceedings*.

Snyder, S.A., Westerhoff, P., Yoon, Y., Sedlak, D.L., 2003. pharmaceuticals, personal care products, and endocrine disruptors in water: Implications for the water industry. *Environmental Engineering Science* 20, 449–469. https://doi.org/10.1089/109287503768335931.

Stuart, M., Lapworth, D., Crane, E., Hart, A., 2012. Review of risk from potential emerging contaminants in UK groundwater. *Science of the Total Environment* 416, 1–21.

Suja, F., Pramanik, B.K., Zain, S.Md., 2009. Contamination, bioaccumulation and toxic effects of perfluorinated chemicals (PFCs) in the water environment: A review paper. *Water Science and Technology* 60, 1533–1544. https://doi.org/10.2166/wst.2009.504.

Sunantha, G., Vasudevan, N., 2016. Assessment of perfluorooctanoic acid and perfluorooctane sulfonate in surface water-Tamil Nadu, India. *Marine Pollution Bulletin* 109, 612–618.

Svenson, A., Allard, A.-S., Ek, M., 2003. Removal of estrogenicity in Swedish municipal sewage treatment plants. *Water Research* 37, 4433–4443.

Singh, A., Patel, A. K., Ramanathan, A., Kumar, M. 2020. Climatic influences on arsenic health risk in the metamorphic Precambrian deposits of Sri Lanka: A re-analysis-based critical review. *Journal of Climate Change* 6(1), 15–24.

Shim, J., Mazumder, P., Kumar, M. 2018. Corn cob silica as an antibacterial support for silver nanoparticles: Efficacy on Escherichia coli and Listeria monocytogenes. *Environmental Monitoring and Assessment* 190(10), 1–10.

Thomas, L.M., Jorgenson, Z.G., Brigham, M.E., Choy, S.J., Moore, J.N., Banda, J.A., Gefell, D.J., Minarik, T.A., Schoenfuss, H.L., 2017. Contaminants of emerging concern in tributaries to the Laurentian Great Lakes: II. Biological consequences of exposure. *PLoS ONE* 12, e0184725. https://doi.org/10.1371/journal.pone.0184725.

Tijani, J.O., Fatoba, O.O., Babajide, O.O., Petrik, L.F., 2016. Pharmaceuticals, endocrine disruptors, personal care products, nanomaterials and perfluorinated pollutants: A review. *Environmental Chemistry Letters* 14, 27–49. https://doi.org/10.1007/s10311-015-0537-z.

Topp, E., Monteiro, S.C., Beck, A., Coelho, B.B., Boxall, A.B., Duenk, P.W., Kleywegt, S., Lapen, D.R., Payne, M., Sabourin, L., 2008. Runoff of pharmaceuticals and personal care products following application of biosolids to an agricultural field. *Science of the Total Environment* 396, 52–59.

Van Doorslaer, X., Dewulf, J., Van Langenhove, H., Demeestere, K., 2014. Fluoroquinolone antibiotics: An emerging class of environmental micropollutants. *Science of the Total Environment* 500–501, 250–269. https://doi.org/10.1016/j.scitotenv.2014.08.075.

Vega, M., Nerenberg, R., Vargas, I.T., 2018. Perchlorate contamination in Chile: Legacy, challenges, and potential solutions. *Environmental Research* 164, 316–326. https://doi.org/10.1016/j.envres.2018.02.034.

Vélez, V.P.P., Esquivel-Hernández, G., Cipriani-Avila, I., Mora-Abril, E., Cisneros, J.F., Alvarado, A., Abril-Ulloa, V., 2019. Emerging contaminants in trans-American waters. *Revista Ambiente e Água* 14, 1. https://doi.org/10.4136/ambi-agua.2436.

Water and Sanitation Program (WSP), W.B., Water Management Institute, I., 2016. Recycling and reuse of treated wastewater in urban India: A proposed advisory and guidance document. International Water Management Institute (IWMI). CGIAR Research Program on Water, Land and Ecosystems (WLE). https://doi.org/10.5337/2016.203.

Windham, G.C., Waller, K., Anderson, M., Fenster, L., Mendola, P., Swan, S., 2003. Chlorination by-products in drinking water and menstrual cycle function. *Environmental Health Perspectives* 111, 935–941.

Yan, S., Subramanian, S.B., Tyagi, R.D., Surampalli, R.Y., Zhang, T.C., 2010. Emerging contaminants of environmental concern: Source, transport, fate, and treatment. *Practice Periodical of Hazardous Toxic and Radioactive Waste Management* 14, 2–20. https://doi.org/10.1061/(ASCE)HZ.1944-8376.0000015.

Index

Note: **Bold** page numbers refer to tables and *italic* page numbers refer to figures.

Sustainable Water Developments

Book Series Editor: Jochen Bundschuh

ISSN: 2373-7506

Publisher: CRC Press/Balkema, Taylor & Francis Group

1. Membrane Technologies for Water Treatment: Removal of Toxic Trace Elements with Emphasis on Arsenic, Fluoride and Uranium
   Editors: Alberto Figoli, Jan Hoinkis & Jochen Bundschuh 2016
   ISBN: 978-1-138-02720-6 (Hbk)

2. Innovative Materials and Methods for Water Treatment: Solutions for Arsenic and Chromium Removal
   Editors: Marek Bryjak, Nalan Kabay, Bernabé L. Rivas & Jochen Bundschuh 2016
   ISBN: 978-1-138-02749-7 (Hbk)

3. Membrane Technology for Water and Wastewater Treatment, Energy and Environment
   Editors: Ahmad Fauzi Ismail & Takeshi Matsuura 2016
   ISBN: 978-1-138-02901-9 (Hbk)

4. Renewable Energy Technologies for Water Desalination
   Editors: Hacene Mahmoudi, Noreddine Ghaffour, Mattheus A. Goosen & Jochen Bundschuh 2017
   ISBN: 978-1-138-02917-0 (Hbk)

5. Application of Nanotechnology in Membranes for Water Treatment
   Editors: Alberto Figoli, Jan Hoinkis, Sacide Alsoy Altinkaya & Jochen Bundschuh 2017
   ISBN: 978-1-138-89658-1 (Hbk)

6. Geothermal Water Management
   Editors: Jochen Bundschuh & Barbara Tomaszewska 2017
   ISBN: 978-1-138-02721-3 (Hbk)
   ISBN: 978-1-138-74900-9 (Pbk)

7. Legacy, Pathogenic and Emerging Contaminants in the Environment
   Editors: Manish Kumar, Meththika Vithanage, Sanjeeb Mohapatra, Kishor Acharya 2021
   ISBN: 978-0-367-74363-5 (Hbk)
   ISBN: 978-0-367-74365-9 (Pbk)